Open Problems in Strongly Correlated Electron Systems

T0220830

NATO Science Series

A Series presenting the results of scientific meetings supported under the NATO Science Programme.

The Series is published by IOS Press, Amsterdam, and Kluwer Academic Publishers in conjunction with the NATO Scientific Affairs Division

Sub-Series

I. Life and Behavioural Sciences	IOS Press
II. Mathematics, Physics and Chemistry	Kluwer Academic Publishers
III. Computer and Systems Science	IOS Press
IV. Earth and Environmental Sciences	Kluwer Academic Publishers

The NATO Science Series continues the series of books published formerly as the NATO ASI Series.

The NATO Science Programme offers support for collaboration in civil science between scientists of countries of the Euro-Atlantic Partnership Council. The types of scientific meeting generally supported are "Advanced Study Institutes" and "Advanced Research Workshops", and the NATO Science Series collects together the results of these meetings. The meetings are co-organized bij scientists from NATO countries and scientists from NATO's Partner countries – countries of the CIS and Central and Eastern Europe.

Advanced Study Institutes are high-level tutorial courses offering in-depth study of latest advances in a field.
Advanced Research Workshops are expert meetings aimed at critical assessment of a field, and identification of directions for future action.

As a consequence of the restructuring of the NATO Science Programme in 1999, the NATO Science Series was re-organized to the four sub-series noted above. Please consult the following web sites for information on previous volumes published in the Series.

http://www.nato.int/science
http://www.wkap.nl
http://www.iospress.nl
http://www.wtv-books.de/nato-pco.htm

Series II: Mathematics, Physics and Chemistry – Vol. 15

Open Problems in Strongly Correlated Electron Systems

edited by

Janez Bonča

Peter Prelovšek

Anton Ramšak

Faculty of Mathematics and Physics,
Department of Physics,
Ljubljana, Slovenia,
Jožef Stefan Institute,
Ljubljana, Slovenia

Sarben Sarkar

Department of Physics,
King's College London,
London, United Kingdom

Kluwer Academic Publishers

Dordrecht / Boston / London

Published in cooperation with NATO Scientific Affairs Division

Proceedings of the NATO Advanced Research Workshop on
Open Problems in Strongly Correlated Electron Systems
Bled, Slovenia
26–30 April 2000

Library of Congress Cataloging-in-Publication Data

Open problems in strongly correlated electron systems / edited by Janez Bonca ... [et al.].
    p. cm. -- (NATO science series. Series II, Mathematics, physics, and chemistry ; vol. 15)
  Includes index.
   ISBN 0-7923-6895-9 (HB : acid-free paper)
    1. Electron configuration--Congresses. I. Bonca, Janez. II. Series.

QC176.8.E4 O64 2001
530.4'11--dc21
                                                                            2001023453
ISBN 0-7923-6895-9 (HB)
ISBN 0-7923-6896-7 (PB)

Published by Kluwer Academic Publishers,
P.O. Box 17, 3300 AA Dordrecht, The Netherlands.

Sold and distributed in North, Central and South America
by Kluwer Academic Publishers,
101 Philip Drive, Norwell, MA 02061, U.S.A.

In all other countries, sold and distributed
by Kluwer Academic Publishers,
P.O. Box 322, 3300 AH Dordrecht, The Netherlands.

Printed on acid-free paper

Printed in the Netherlands.

Part I CUPRATES: FERMI SURFACE AND SPECTRAL FUNCTIONS

Summary of ARPES Results on the Pseudogap in $Bi_2Sr_2CaCu_2O_{8+\delta}$ 3
J.C. Campuzano and M. Randeria

On the Breakdown of Landau-Fermi Liquid Theory in the Cuprates   13
C. Honerkamp, T.M. Rice, and M. Salmhofer
  1.   Introduction   13
  2.   Truncation of the Fermi Surface in Ladder Systems.   14
  3.   The Breakdown of Landau Theory in Two Dimensions   15
  4.   Conclusions   20

Metal-Insulator Transition and Many-body Band Structure of the Hubbard Model   23
R. Eder, C. Gröber, M. G. Zacher and W. Hanke
  1.   Introduction   23
  2.   Metal-Insulator Transition   25
  3.   Fermi Surface in the Doped Case   27
  4.   Conclusion   31

Spectral Properties of Underdoped Cuprates   33
A. Ramšak, P. Prelovšek, and I. Sega
  1.   Generalized t-J Model   33
  2.   Pseudo Gap in the Density of States   35
  3.   Large or Small Fermi Surface?   37
  4.   Summary and Analytical Results   40

High Resolution Fermi Surface Mapping of Pb-doped Bi-2212   43
S.V. Borisenko, M.S. Golden, T. Pichler, S. Legner,
C. Dürr, M. Knupfer, J. Fink, and H. Berger
    1.   Introduction   43
    2.   Fermi Surface Maps of Pb-doped Bi-2212   45
    3.   Conclusions   48

Single Particle Excitations in the *t-J* Model   51
M. Brunner, C. Lavalle, F.F. Assaad, and A. Muramatsu

Part II   CUPRATES: SPIN AND CHARGE FLUCTUATIONS

Magnetic Resonance Peak and Nonmagnetic Impurities   59
Y. Sidis, P. Bourges, B. Keimer, L. P. Regnault, J. Bossy,
A. Ivanov, B. Hennion, P. Gautier-Picard, and G. Collin
    1.   Introduction   59
    2.   INS Measurements   60
    3.   Discussion and Conclusion   64

Pseudogap and Kinetic Pairing Under Critical Differentiation   69
of Electrons in Cuprate Superconductors
M. Imada and S. Onoda
    1.   Introduction   69
    2.   Kinetic Pairing Derived from Electron Differentiation   72
    3.   Pseudogap Phenomena in Cuprates as Superconducting Fluctuations   75
    4.   Conclusion and Discussion   76

Density Response of Cuprates and Renormalization of   81
Breathing Phonons
P. Horsch and G. Khaliullin
    1.   Introduction   81
    2.   Slave Boson Theory of Density Response   82
    3.   Renormalization of Breathing Phonons   87
    4.   Summary   90

Phase Diagram of Spin Ladder Models and the Topology of   91
Short Range Valence Bonds
J. Sólyom
    1.   Introduction   91
    2.   Isotropic Spin Ladder Models   92
    3.   The Topology of Short Range Valence Bonds   94
    4.   Anisotropic Models   96
    5.   Summary   99

Diagrammatic Theory Of Anderson Impurity Models: Fermi   101
and Non-Fermi Liquid Behavior
J. Kroha and P. Wölfle
    1.   Introduction   101

2.   Auxiliary Particle Representation   102
3.   Conserving Slave Particle T-Matrix Approximation   106
4.   Conclusion   109

s+d Mixing in Cuprates: Strong electron correlations   111
and superconducting gap symmetry

N.M.Plakida and V.S. Oudovenko
1.   Introduction   111
2.   Model and Dyson Equation   112
3.   Weak Coupling Approximation   113
4.   Numerical Results and Discussion   115

Part III   CUPRATES: STRIPE AND CHARGE ORDERING

Fermi Surface, Pseudogaps and Dynamical Stripes in   119
$LA_{2-x}SR_xCUO_4$

A. Fujimori, A. Ino, T. Yoshida, T. Mizokawa, Z.-X. Shen,
C. Kim, T. Kakeshita, H. Eisaki, and S. Uchida
1.   Introduction   119
2.   Band Dispersion and Fermi Surface   120
3.   Large Pseudogap   124
4.   Small Pseudogap and Superconducting Gap   125
5.   Evolution of Electronic Structure with Hole Concentration   125

Stripes and Nodal Fermions as Two Sides of the Same Coin   129
J. Zaanen and Z. Nussinov
1.   The Paradox   129
2.   The Nodal Fermions as Dirac Spinons   131
3.   Stripe Duality   132
4.   The Faith of the Paradox   139

DMRG Studies of Stripes and Pairing in the *t-J* Model   141
S. R. White and D.J. Scalapino
1.   Introduction   141
2.   Convergence of DMRG   142
3.   Pairing   146
4.   Summary   149

Coexistence of Charge and Spin-Peierls Orders in the 1/4-filled   151
Ladder NaV_2O_5
D. Poilblanc and J. Riera
1.   Introduction   151
2.   Isolated Ladder: Charge Instability   153
3.   Isolated Ladder: Coexisting Charge and SP Orders   155
4.   Trellis Lattice: Charge Instability   157
5.   Conclusions   157

Part IV   CUPRATES: NUMERICAL METHODS AND QUANTUM HALL EFFECT

Normal State Properties of Cuprates: *t-J* Model vs. Experiment   163
P. Prelovšek   163
   1.   Introduction   164
   2.   Thermodynamics   167
   3.   Dynamics at Optimum Doping   171
   4.   Discussion

Stability of d-wave Superconductivity in the *t-J* Model   173
F. Becca, L. Capriotti, and S. Sorella   173
   1.   Introduction   175
   2.   Numerical Method   176
   3.   Numerical Tests   179
   4.   Larger Size Calculations   181
   5.   Conclusions   183
   Appendix: Variance estimate of the error on bulk correlation functions

A New Simulation Method for Infinite Size Lattices   187
H.G. Evertz and W. von der Linden   187
   1.   Introduction   187
   2.   Cluster Methods   188
   3.   New Method   190
   4.   Quantum Case   192
   5.   Discussion

Universality in 2-D Quantum Heisenberg Antiferromagnets   193
M. Troyer   193
   1.   Introduction   194
   2.   The Ordered Ground State   195
   3.   The Renormalized Classical Regime   197
   4.   The Quantum Phase Transition   198
   5.   The Quantum Disordered Regime   198
   6.   The Quantum Critical Regime   199
   7.   Crossovers
   8.   Bilayer Antiferromagnet in a Magnetic Field and Quantum Hall Bilayers   199
   9.   Conclusions   201

Stripes and Pairing in the $\nu = 5/2$ Quantum Hall Effect   203
F. D. M. Haldane   203
   1.   Introduction   203
   2.   Abelian and Non-Abelian Quantum Hall States   208
   3.   Paired Quantum Hall States   210
   4.   Transition to the Striped Phase   212
   5.   Summary

Part V MANGANITES, ORBITAL DEGENERACY

Theory of Manganites: The Key Role of Phase Segregation   217
E. Dagotto, A. Feiguin and A. Moreo

Magnetic and Orbital Ordering in Manganites   227
A.M. Oleś and L.F. Feiner
1.   Superexchange in $La_{1-x}Ca_xMnO_3$ 227
2.   Orbital Liquid with Complex Orbitals   231
3.   Magnons in Ferromagnetic Manganites   234
4.   Summary   235

Orbital Dynamics: The Origin of Anomalous Magnon Softening in Ferromagnetic Manganites   237
G. Khaliullin and R. Kilian
1.   Introduction   237
2.   The Model   238
3.   Double Exchange   239
4.   Superexchange   241
5.   Indirect Magnon-Phonon Coupling via the Orbital Sector   242
6.   Magnon Self Energies   242
7.   Summary   245

Field Induced Metal-Insulator Transition in $(Pr:Ca:Sr)MnO_3$ 247
J. Hemberger, M. Paraskevoupolos, J. Sichelschmidt, M. Brando, R. Wehn, F. Mayr, K. Pucher, P. Lunkenheimer, A. Loidl, A.A. Mukhin, and A.M. Balbashov
1.   Introduction   247
2.   Results and Discussion   248
3.   Summary   251

Triplet Pairing via Local Exchange in Correlated Systems   253
J. Spałek
1.   Introduction   253
2.   Nambu-De Gennes Method for Triplet Pairing in 2-band Case   254
3.   Final Remarks   258

Part VI LOW DIMENSIONAL SYSTEMS AND TRANSPORT

Dimensional Crossover, Electronic Confinement and Charge Localization in Organic Metals   263
G. Mihály, F. Zámborszky, I. Kézsmárki, and L. Forró
1.   Introduction   263
2.   Pressure Induced Delocalization   264
3.   Electron Correlations   268
4.   Summary   270

Drude Weight, Integrable Systems and the Reactive Hall Constant   273

X. Zotos, F. Naef, M. Long, and P. Prelovšek
   1.   The Drude Weight   273
   2.   Integrable Systems   275
   3.   Reactive Hall Response   279

Inhomogeneous Luttinger Liquids: Power-Laws and Energy Scales   283

V. Meden, W. Metzner, U. Schollwöck, and K. Schönhammer
   1.   Luttinger Liquids: Bulk Properties   283
   2.   Luttinger Liquids with Boundaries   285
   3.   The Hartree-Fock Approximation   287
   4.   Lattice Model of Spinless Fermions   288
   5.   Hubbard Model   289
   6.   Summary   291

Nodal Liquids and Duality   293

N. E. Mavromatos and S. Sarkar

Spin–Charge Separation in the Sr_2CuO_3 and $SrCuO_2$ Chain Materials   303

K. Penc and W. Stephan
   1.   Introduction   303
   2.   Photoemission Spectra   304
   3.   EELS Spectra   306

Frustrated Quantum Ising Model and Charged Kinks   311

M.V. Mostovoy, D.I. Khomskii, J. Knoester, and N.V. Prokof'ev
   1.   Introduction   311
   2.   Mean Field and Monte Carlo Results   312
   3.   Kink Crystal   313
   4.   Summary   316

Ergodic Properties of Quantum Spin Chains: Kicked Transverse Ising Model   317

T. Prosen

Part VII   MOTT-HUBBARD TRANSITION, INFINITE DIMENSION

Strongly Correlated Electrons: A Dynamical Mean Field Perspective   325

G. Kotliar
   1.   Strongly Correlated Materials   325
   2.   Dynamical Mean Field Theory   326

3. Insights from DMFT   327
4. Strongly Correlated Electron Systems, Renormalization Group Flows
    and Outlook   332

d-wave Pairing in the Strong-coupling $2D$ Hubbard Model   337
Th. Pruschke, Th. Maier, J. Keller, and M. Jarrell
    1. Introduction   337
    2. Method   338
    3. Results   340
    4. Summary   344

**Studies of the Mott–Hubbard Transition in One and Infinite
Dimensions**   347
R.M. Noack, C. Aebischer, D. Baeriswyl, and F. Gebhard
    1. Introduction   347
    2. The t–t' Hubbard Chain   348
    3. The Random Dispersion Approximation   353
    4. Summary   358

**Quantum Critical Behavior of Correlated Electrons: Resonant
States**   361
V. Janiš
    1. Introduction   361
    2. Classical vs. Quantum Criticality   363
    3. Parquet Approach   364
    4. Resonant States: Anomalous Two-particle Functions   368
    5. Summary   370

Theory of Valence Transitions in Ytterbium-based Compounds   371

V. Zlatić and J. K. Freericks
    1. Introduction   371
    2. Calculations   373
    3. Results and Discussion   375
    4. Summary   379

Strong Electronic Correlations and Low Energy Scales   381
R. Bulla and Th. Pruschke
    1. Introduction   381
    2. Results   382
    3. Conclusions   384

**Non-magnetic Mott Insulator Phase and Anomalous
Conducting States in Barium Vanadium Trisulphide**   387
*P. Fazekas, H. Berger, L. Forró, R. Gaál, I. Kézsmárki,
G. Mihály, M. Miljak, K. Penc, and F. Zámborszky*
    1. Introduction   387
    2. Experimental Results and Discussion   389

Part VIII   SHORT CONTRIBUTIONS

Low-Energy Excitations in Anisotropic Spin-Orbital Model   395
J. Bała, A. M. Oleś, and G. A. Sawatzky

Dynamical Mean-Field Selfconsistency Relation for Multiband Systems   399
G. Czycholl

Superfluidity in Fermi-Systems with Repulsion   403
M.S. Mar'enko, M.Yu. Kagan, D.V. Efremov, and M.A. Baranov
1.	Introduction	403
2.	Two-Dimensional Case	405
3.	Tc Enhancement in a Spin-Polarized Fermi-System	406
4.	Phase Diagram in Two Dimensional Case	407

Dynamic Jahn-Teller Effect and Distortional Disorder in Manganites   409
L.F. Feiner and A.M. Oleś

Diagrammatic Theory of the Anderson Impurity Model with Finite Coulomb Interaction   413
K. Haule, S. Kirchner, H. Kroha, and P. Wölfle
1.	Method	414
2.	Numerical Results	415
3.	Summary	415

Magnetotransport of the Cuprates in the Quantum Critical Point Scenario   417
R. Hlubina

Interplay Between Electron Correlations and Electron-Phonon Interaction in the Cuprates   421
J. Konior and P. Piekarz
1.	Introduction	421
2.	Effective Hole-Hole Interaction	422
3.	Summary	423

Superconductivity in Disordered Sr_2RuO_4   425
G. Litak, J.F. Annett, and B.L. Györffy
1.	Introduction	425
2.	The Model	425
3.	Density of States and Tc	426
4.	Summary	427

Fine Electronic Structure and Magnetism of LaMnO$_3$ and LaCoO$_3$ 429
R.J. Radwański and Z. Ropka

Anomalies in the Conduction Edge of Quantum Wires   433
T. Rejec, A. Ramšak, and J.H. Jefferson
1.   Introduction   433
2.   The Model   434
3.   Two-Electron Approximation   435
4.   Results   436
5.   Summary   438

Underdoped Region of the 2D t-J Model   441
A. Sherman and M. Schreiber

Lattice Fermions with Optimized Wave Functions: Exact Results   443
J. Spałek, A. Rycerz, W. Wójcik, and R. Podsiadły

Entropy Saturation and the Brinkmann-Rice Transition in a Random-Tiling Model   447
D. K. Sunko

Diamagnetic Properties of Doped Antiferromagnets   451
D. Veberič, P. Prelovšek, and H.G. Evertz

Index   455

Group Photo   461

Preface

In the last decade it has become increasingly evident that strong correlations between electrons are an essential and unifying factor in such diverse phenomena within solid state physics as high-temperature superconductivity, colossal magnetoresistance, the quantum Hall effect, heavy-fermion metals and Coulomb blockade in single-electron transistors. A new paradigm of non-Fermi Liquid behaviour is also emerging and, in a number of systems, replacing the Fermi liquid, which has been the cornerstone of the physics of metals and superconductors for the past decades.

In spite of major achievements, the theoretical studies and understanding of strongly-correlated electrons seems to be still in its infancy. Anomalous electron properties have been studied in some generic models of correlated electrons, such as the Hubbard and t-J models, the Anderson and Kondo impurity models, and their lattice equivalents. New insights into the behaviour of these, and related models is emerging from the introduction of powerful numerical methods to study such many-body models, including approximate techniques of many-body theory and exact results in low- and high-dimensional systems. These all show convincing evidence for breakdown of the Fermi liquid concept.

The Bled workshop focused on several major open questions in the theory of anomalous metals with correlated electrons. These theoretical advances were complemented by the latest experimental results in related materials, presented by leading experimentalists in the field. The main emphasis was on the following topics:

- physics of cuprates and high-temperature superconductors,

- charge- and spin-ordering and fluctuations,

- manganites and colossal magnetoresistance,

- low-dimensional systems and transport,

- Mott-Hubbard transition and infinite dimensional systems,

- quantum Hall effect.

The major funding of the Workshop was from the NATO grant, with further substantial Slovenian support from the Ministries of Defense, Science and Technology, and Foreign Affairs, the Institute Jožef Stefan, and the Faculty of Mathematics and Physics, University of Ljubljana. The organizers particularly acknowledge the personal support of the Slovenian Minister of Defense Dr. Franci Demšar, who gave the Workshop opening address.

The organizers also wish to acknowledge T.M. Rice for his encouragement and valuable advice concerning the program. Finally, we express our gratitude to colleagues, in particular to I. Sega for financial management, T. Rejec, D. Veberič and K. Haule for help and technical assistance before and during the Workshop, and M. Budiša for the work done on the final version of the book.

J. Bonča
P. Prelovšek
A. Ramšak
S. Sarkar

Ljubljana, November 2000

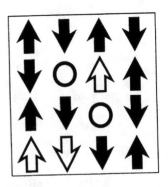

I
CUPRATES:
FERMI SURFACE
AND SPECTRAL FUNCTIONS

SUMMARY OF ARPES RESULTS
ON THE PSEUDOGAP IN Bi$_2$Sr$_2$CaCu$_2$O$_{8+\delta}$

J.C. Campuzano[1,2]

[1]*University of Illinois at Chicago, 845 W. Taylor St., Chicago, IL, 60607*
[2]*Argonne National Laboratory, 9400 S. Cass Ave. Argonne, IL 60439*

M. Randeria

Tata Institute for Fundamental Research, Mumbai 400005, India

Abstract   We summarize our angle-resolved photoemission (ARPES) results on the pseudogap observed in the high temperature superconductor Bi$_2$Sr$_2$CaCu$_2$O$_{8+\delta}$, such as temperature, momentum and doping dependencies.

There have been many theoretical proposals (some presented at this conference) attempting to explain the pseudogap observed in the underdoped high T_c superconductors. Here we summarize our extensive ARPES experimental results on the pseudogap in Bi$_2$Sr$_2$CaCu$_2$O$_{8+\delta}$ (Bi2212) in order to aid in this theoretical development. Due to space limitations, we leave the details to the original articles referenced here.

The leading-edge pseudogap effect [1, 2, 3] is shown in Fig. 1 for an underdoped 83 K sample at the edge of the Brillouin zone. One can see that as the temperature increases, the sharp peak signalling the coherent superconducting state becomes smaller, until it disappears exactly at T_c for underdoped samples. But as the temperature increases above T_c, the gap in the excitation spectrum persists. Finally, the gap fills in at some higher temperature T^*. A detailed examination of this phenomena leads us to the following conclussions:

1) **The pseudogap is tied to the Fermi surface**. One of the fundamental questions is wether the pseudogap is tied to the Fermi momentum k_F, or some other unrelated momentum $\mathbf{Q}$. We determine the detailed momentum dependence of the pseudogap by plotting what we call the minimum gap locus [4]. As one follows the dispersion perpendicular to the Fermi surface in the pseudogap state, one finds that the dispersion approaches the chemical potential, and then recedes, as shown in Fig. 2a. The minimum occurs at k_F, as shown in Fig. 2b, unlike say charge density waves, which are tied to other characteristic $\mathbf{Q}$

3

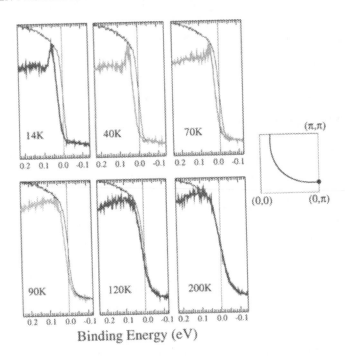

Figure 1   ARPES spectra along the $(\pi, 0) \rightarrow (\pi, \pi)$ direction for an 85 K underdoped sample at various temperatures (solid curves). The dotted curves are reference spectra from polycrystalline Pt, used to accurately determine the chemical potential.

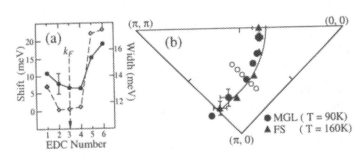

Figure 2   (a) Midpoint shifts (dots) and widths (diamonds) for an 83K sample at a photon energy of 22 eV at 90K for a cut shown by the open dots in (b). (b) Fermi surface at 160K (solid triangles) and minimum gap locus at 90K (solid dots). Notice that the two surfaces coincide within error bars. The error bars represent uncertainties of Fermi crossings as well as possible sample misalignment. The solid curve is a rigid band estimate of the Fermi surface.

vectors. But here, just as in the case of the superconducting gap, the pseudogap is tied to the Fermi surface [4], even in underdoped samples with a T_c of only 15K (a doping of only ≈ 0.05 electrons/Cu).

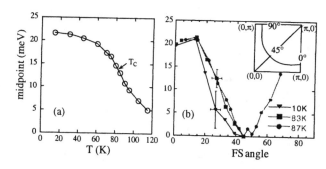

Figure 3 (a) Leading edge shift of an underdoped 83 K sample vs. T. T_c is indicated by an arrow, where no discontinuity in the shift is visible. (b) Momentum dependence of the gap estimated from the leading-edge shift in samples with T_c's of 87K, 83K and 10K, measured at 14K. The inset shows the Brillouin zone with a large Fermi surface closing the (π, π) point with the occupied region on the (0,0) side.

2) **There is no discontinuity as the pseudogap evolves into the superconducting gap.** Fig. 3a shows a plot of the leading edge shift of the spectra in Fig. 1 as a function of temperature, a measure of the gap size. The pseudogap appears to continuously evolve into the superconducting gap, with no evidence of a discontinuity at T_c (This can also be seen in more detail in Fig. 4a).

3) **The magnitude of the pseudogap varies as the magnitude of a d-wave superconducting gap just above** T_c. Fig. 3b shows the momentum dependence of the mid-point of the leading edge, in both the superconducting state for samples with T_c's of 87K and 83K, and in the pseudogap state, for a highly underdoped 10K sample [7], all measured at 14K. Remarkably, the k-dependence of the superconducting and pseudogap is similar for the three samples, even though T_c and T^* change appreciably. We emphasize that, as far as ARPES is concerned, there is no difference between the gap seen above and below T_c, except for the fact that the spectral function below T_c is sharp.

4) **The pseudogap does not close, but gradually fills in.** The detailed temperature behavior of the pseudogap can be more clearly seen by eliminating the Fermi function from the spectra. Given ARPES data described by [5] $I(\omega) = \sum_k I_0 f(\omega) A(\mathbf{k}, \omega)$ (with the sum over a small momentum window about the Fermi momentum $\mathbf{k}_F$) [6], we can generate the symmetrized spectrum $I(\omega) + I(-\omega)$. Making the reasonable assumption of particle-hole symmetry for a small range of ω and $\epsilon_\mathbf{k}$, we have $A(\epsilon_\mathbf{k}, \omega) = A(-\epsilon_\mathbf{k}, -\omega)$ for $|\omega|, |\epsilon|$ less than few tens of meV. It then follows, using the identity $f(-\omega) = 1 - f(\omega)$, that $I(\omega) + I(-\omega) = \sum_k I_0 A(\mathbf{k}, \omega)$ which is true even in the presence of a (symmetric) energy resolution function. This symmetrized spectrum coincides with the raw data for $|\omega| \leq -2.2T_{eff}$, where $4.4T_{eff}$ is the 10%-90% width of the Pt leading edge, which includes the effects of both temperature and

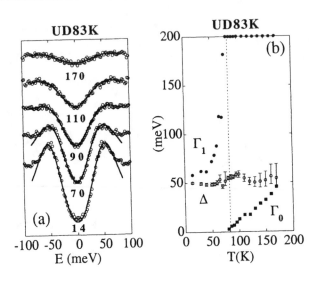

Figure 4   a) Symmetrized data for a $T_c = 83$ K underdoped sample at the $(\pi, 0) \rightarrow (\pi, \pi)$ Fermi crossing at five temperatures, compared to the model fits described in the text. b) Δ (open circles), $\Gamma 1$ (solid circles), and $\Gamma 0$ (solid squares) versus T for the data in (a). The dashed line marks T_c. The error bars for Δ are based on a 10% increase in the rms error of the fits.

resolution. Non-trivial information is obtained for the range $|\omega| \leq 2.2 T_{eff}$, which is then the scale on which p-h symmetry has to be valid [7].

Fig. 4a shows symmetryzed data for an underdoped 83K sample as a function of temperature [8]. The pseudogap can be seen filling in, while the maximum in the spectral function does not change with temperature, i.e. the gap does not close, it just fills in. From fit to the data, using a phenomenological self-energy of the form $\Sigma(k, \omega) = -i\Gamma_1 + \Delta^2[\omega + \varepsilon(k) + i\Gamma_0]$, [8] we can obtain some indication of how the gap, the scattering rate, and the pair fluctuations vary with temperature, shown in Fig 4b.

It is important to note that the gap does not behave as shown in Fig. 4b everywhere around the Fermi surface [8]. As can be seen in Fig. 5b, moving away from the zone edge, the gap closes with temperature, more akin to what one expects from a BCS-like gap. However, the gap is still closing at a temperature higher than T_c. This remarkable difference can be seen directly from the data in Fig. 5a, as indicated by the lines joining the maxima in the spectral functions obtained at points 1 and 2, drawn to guide the eye.

5) **Below** T^*, **the Fermi surface breaks up into arcs** [7]. In Fig. 6 we show ARPES spectra for an underdoped 85K sample at three **k** points on the Fermi surface for various temperatures. The superconducting gap, as estimated by the position of the sample leading edge midpoint, is seen to decrease as one moves from point a near $(\pi, 0)$ to b to c, closer to the diagonal $(0, 0) \rightarrow (\pi, 0)$ direction,

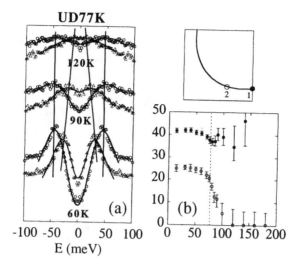

Figure 5 (a) Symmetrized data for a T_c =77 K underdoped sample for three temperatures at (open circles) k_F point 1 in the zone inset, and at (open triangles) k_F point 2, compared to the model fits. (b) $\Delta(T)$ for these two **k** points (filled and open circles), with T_c marked by the dashed line.

consistent with a $d_{x^2-y^2}$ order parameter. At each **k** point the quasiparticle peak disappears above T_c as T increases, with the pseudogap persisting well above T_c, as noted earlier.

The striking feature which is apparent from Fig. 6 is that the pseudogap at different **k** points closes at different temperatures, with larger gaps persisting to higher T's. At point a, near $(\pi, 0)$, there is a pseudogap at all T's below 180K, at which the Bi2212 leading edge matches that of Pt. We take this as the definition of T^* [2] above which the the largest pseudogap has vanished within the resolution of our experiment, and a closed contour of gapless excitations – a Fermi surface – is obtained [4]. The surprise is that if we move along this Fermi surface to point b the sample leading edge matches Pt at 120K, which is smaller than T^*. Continuing to point c, about halfway to the diagonal direction, we find that the Bi2212 and Pt leading edges match at an even lower temperature of 95K. In addition, spectra measured on the same sample along the Fermi contour near the $(0,0) \rightarrow (\pi, \pi)$ line shows no gap at any T, even below T_c, consistent with $d_{x^2-y^2}$ anisotropy [2].

One simple way to quantify the behavior of the gap is to plot the midpoint of the leading edge of the spectrum (Fig. 6e). We will say that the pseudogap has closed at a **k** point when the midpoint equals zero energy, in accordance with the discussion above. From this plot, we find that the pseudogap closes at point a at a T above 180K, at point b at 120 K, and at point c just below 95 K.

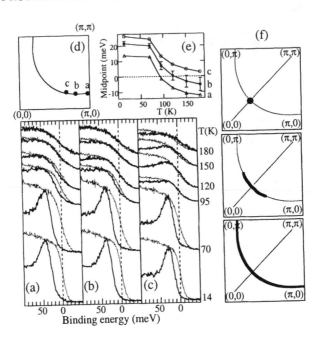

Figure 6   Shown are spectra taken at three **k** points in the Y quadrant (d) of the Brillouin zone for an 85K underdoped Bi2212 sample at various temperatures (solid curves). The dotted curves are reference spectra from polycrystalline Pt (in electrical contact with the sample) used to determine the chemical potential (zero binding energy). Note the closing of the spectral gap at different T for different **k**. This feature is also apparent in the plot (e) of the midpoint of the leading edge of the spectra as a function of T. Panels (f) show a schematic illustration of the temperature evolution of the Fermi surface in underdoped cuprates. The d-wave node below T_c (top panel) becomes a gapless arc above T_c (middle panel) which expands with increasing T to form the full Fermi surface at T^* (bottom panel).

If we now view these data as a function of decreasing T, the picture of Fig. 6f clearly emerges. The pseudogap suppression first opens up near $(\pi, 0)$ and progressively gaps out larger portions of the Fermi contour, leading to gapless arcs which shrink with decreasing T. It is worth noting that midpoints with negative binding energy, particularly for **k** point c, indicate the formation of a peak in the spectral function at $\omega = 0$ as T increases.

6) **There is a large energy pseudogap** [9]. Fig. 8b shows the T evolution of the spectra of an 83K sample near the $(\pi, 0)$ point of the Brillouin zone. For temperatures above the pseudogap temperature scale T^* [2] we see a broad peak which is chopped off by the Fermi function. While there is only weak dispersion $(\pi, 0) \rightarrow (\pi, \pi)$ for $T > T^*$, there is definite loss of integrated spectral weight [5], and one can identify the $(\pi, 0) \rightarrow (\pi, \pi)$ "Fermi surface" crossing [4]. As the temperature is reduced below T^*, but still above T_c, we see that the spectral function remains completely incoherent, as shown in

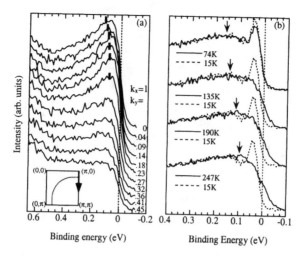

Figure 7 Dispersion and temperature dependence of spectra near $(\pi, 0)$: (a) Spectra along $(\pi, 0) \rightarrow (\pi, \pi)$ for an under-doped 83 K sample at 200 K (above T^*), with the thick vertical bar indicating the peak position. The curves are labeled in units of π/a. The Brillouin zone is shown as an inset, with the Fermi surface as a dotted line. (b) Temperature evolution of the spectra at the $(\pi, 0)$ point for an underdoped 89 K sample, with the positions of the high energy feature marked by arrows.

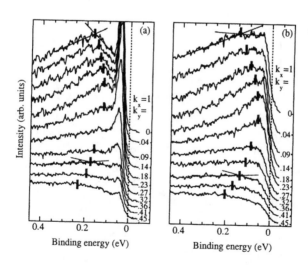

Figure 8 Spectra along $(\pi, 0) \rightarrow (\pi, \pi)$ in (a) the superconducting state ($T = 60$ K), and (b) the pseudogap state ($T = 100$ K) for an underdoped 75 K sample (curves are labeled in units of π/a)). The thick vertical bar indicates the position of the higher energy feature, at which the spectrum changes slope as highlighted by the intersecting straight lines.

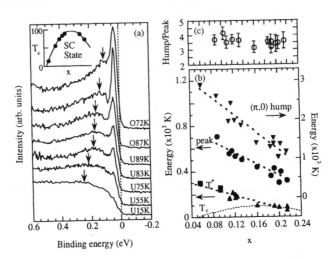

Figure 9   Dependence of energy scale on carrier density: (a) Doping dependence of the spectra (T_c =15 K) at the $(\pi, 0)$ point (U55K is a film). The inset shows T_c vs doping. (b) Doping dependence of T^*, and the peak and hump binding energies in the superconducting state along with their ratio (c), as a function of doping, x. The empirical relation between T_c and x is given by $T_c/T_c^{max} = 1 - 82.6(x - 0.16)^2$ [10] with T_c^{max} =95 K. For T^*, solid squares represent lower bounds.

Fig. 7b for an underdoped 89 K sample. The leading edge pseudogap which develops below T^* is difficult to see on the energy scale of Fig. 1b (the midpoint shift at 135 K is 3 meV). However, a higher energy feature, the high energy pseudogap, can easily be identified by a change in slope of the spectra as function of energy (see Fig. 8b). On further reduction of the temperature below T_c, a coherent quasiparticle peak begins to grow at the position of the leading edge gap, accompanied by a redistribution of the incoherent spectral weight leading to a dip and hump structure [5]. The high energy pseudogap feature is closely related to the hump below T_c, as seen from a comparison of their dispersions. We show data along $(\pi, 0) \rightarrow (\pi, \pi)$ for an underdoped 75 K sample in the superconducting state (Fig. 8a) and in the pseudogap regime (Fig. 8b). Below T_c, the sharp peak at low energy is essentially dispersionless, while the higher energy hump rapidly disperses from the $(\pi, 0)$ point towards the $(\pi, 0) \rightarrow (\pi, \pi)$ Fermi crossing [4] seen above T^*. Beyond this, the intensity drops dramatically, but there is clear evidence that the hump disperses back to higher energy. In the pseudogap state, the high energy feature also shows strong dispersion, much like the hump below T_c, even though the leading edge is nondispersive like the sharp peak in the superconducting state.

7) **The pseudogap temperature T^* and the large pseudogap scaless with the superconducting gapΔ** [9]. In Fig. 9b we show the dependence of T^* and the large energy pseudogap (labelled hump) with doping. It is observed

that they scale with the gap energy Δ, which is obtained from the maximum in the spectral function, the quasiparticle peak energy, shown in Fig. 9a, again pointing to the intimate relation of the pseudogap to the superconducting gap.

Acknowledgments

This work was supported by the National Science Foundation DMR 9974401, and the U. S. Dept. of Energy, Basic Energy Sciences, under contract W-31-109-ENG-38. The Synchrotron Radiation Center is supported by NSF DMR 9212658. MR is supported in part by the Indian DST through the Swarnajayanti scheme.

References

[1] D.S. Marshall, *et al.*, Phys. Rev. Lett. **76**, 4841 (1996).

[2] H. Ding, *et al.*, Nature **382**, 51 (1996).

[3] A.G. Loeser, *et al.*, Science **273**, 325 (1996).

[4] H. Ding *et al.*, Phys. Rev. Lett. **78**, 2628 (1997).

[5] M. Randeria *et al.*, Phys. Rev. Lett. **74**, 4951 (1995).

[6] The experimental signal is a convolution of $I(\mathbf{k}, \omega)$ with the energy resolution and a sum over the momentum window, plus an additive (extrinsic) background.

[7] M.R. Norman, *et al.*, Nature **392**, 157 (1998).

[8] M.R. Norman, *et al.*, Phys. Rev. B **57**, R11093 (1998).

[9] J.C. Campuzano, *et al.*, Phys. Rev. Lett. **83**, 3709 (1999).

[10] M.R. Presland, *et al.*, Physica C **176**, 95 (1991).

ON THE BREAKDOWN OF LANDAU-FERMI LIQUID THEORY IN THE CUPRATES

C. Honerkamp and T.M. Rice
Theoretische Physik, ETH–Hönggerberg, CH-8093 Zürich, Switzerland

M. Salmhofer
Mathematik, ETH–Zürich, CH-8092 Zürich, Switzerland

1.   INTRODUCTION

One of the key features of the high temperature cuprate superconductors that distinguishes them from other metals is the clear breakdown of Landau-Fermi liquid behavior in the normal state [1]. Although Landau theory is based on a perturbative treatment of the interactions, in practice it is found to be very robust and applicable even in cases where naive estimates would have it fail completely e.g. the normal phase of ^{3}He and of the related transition metal oxide, Sr_2RuO_4. The breakdown of Landau behavior in the cuprates is already clearly demonstrated at optimal doping by the linear temperature dependence of the resistivity and it becomes even more pronounced in underdoped cuprates with the appearance of the spin gap at the temperature scale T^* which exceeds the critical temperature for superconductivity, T_c. There has been much debate over the meaning of this temperature, T^* [2]. One school argues for an actual phase transition to a state with a hidden order parameter which accounts for the failure to observe a symmetry breaking at T^*. However there are no reports of singular behavior in thermodynamic properties at T^*, even in well characterized underdoped cuprates such as $YBa_2Cu_4O_8$. This raises the question whether there can be a breakdown of Landau behavior without symmetry breaking and if so, what would be the driving force that is so special to the cuprates. In this lecture, this possibility will be examined with emphasis on the role of Umklapp electron–electron scattering. First, the simpler and well understood case of ladder systems will be reviewed. Then the case of two dimensions will be examined and the results of a recent series of numerical calculations based on a 1-loop renormalization group treatment will be discussed. The last section is devoted to the conclusions.

J. Bonča et al. (eds.), Open Problems in Strongly Correlated Electron Systems, 13–22.

2.   TRUNCATION OF THE FERMI SURFACE IN LADDER SYSTEMS

2.1   SINGLE CHAINS

The study of interacting fermions confined to ladders, i.e. systems of coupled chains, has been an active topic in recent years [3]. While Landau theory breaks down to give a Luttinger liquid in a single chain, it is still possible to define a Fermi wave vector with the standard form, $k_{F,\sigma} = \pi \nu_\sigma$, where ν_σ is the density of electrons with spin, σ. A Luttinger liquid is a conductor except for the case of exactly 1 electron/site or a half-filled band [4]. Repulsive interactions open up a gap in the spectrum of charge excitations at this density creating an insulator. This behavior is independent of the interaction strength and even for arbitrarily weak coupling, the interactions flow to strong coupling. This insulator is a form of Mott insulator which in this 1-dimensional system occurs already at weak coupling. An examination of the RG (renormalization group) flow shows immediately that the origin of the Mott gap is the presence of elastic Umklapp scattering processes at the Fermi energy. These are responsible for the opening of the charge gap and the power law spin–spin correlations do not play a direct role.

2.2   2–LEG LADDERS

In the case of the 2–leg ladder at half-filling the spin–spin correlations have a very different form and decay exponentially with separation, characteristic of a spin liquid. However again a Mott charge gap opens up for arbitrarily small coupling at half-filling and again this is driven by the presence of elastic Umklapp scattering at the Fermi energy [5, 6]. The spin–spin correlations are strictly short range and there is no breaking of translational symmetry along the ladder. Yet the Fermi surface is now completely truncated and again the origin is the flow to strong coupling of the Umklapp scattering terms across the Fermi surface. The resulting phase can be regarded as a form of the RVB state (Resonant Valence Bond) first introduced by Anderson [7] which causes an ISL (Insulating Spin Liquid) groundstate.

Bosonization gives a clear description of the low energy behavior at and near half-filling [5, 6]. Umklapp scattering generates a pinning term for the translation degree of freedom, θ_+. Near to half-filling doped charges appear as charge $+2|e|$ solitons in the θ_+-variable and, as shown by Schulz [8], universally map on to a dilute gas of hard core bosons with predominantly superconducting correlations. This description of the system in term of hole pairs applies at low energies, i.e. below the energy scale at which the Umklapp terms have scaled to strong coupling. The latter is clearly a prerequisite for a description as a doped Mott insulator.

2.3   3–LEG AND WIDER LADDERS

As the width of a ladder grows, the number of transverse channels increases. The simplest case is the 3–leg ladder [9, 10, 11]. The one electron bands split into two classes, with even and odd parity with respect to reflection about the central leg. The two outermost bands have even parity and pair to form an ISL at half-filling, while the central odd parity band forms a single chain gapless Luttinger liquid. Again this low energy behavior does not depend on whether one starts with weak or strong coupling and corresponds to the effective single chain behavior of the 3–leg $S = 1/2$ AF Heisenberg ladder at low energies.

When we examine the doping behavior one must distinguish in weak coupling between the order of the limits of small doping and small interaction. Taking the limit of small doping away from half-filling and finite interaction strength, then one finds a finite difference in the hole chemical potential in the odd and even parity channels [11]. The former case has gapless spin excitations and doped holes enter first only in this channel up to finite critical hole concentration. In this region the low energy properties are those of a single channel Luttinger liquid. The Fermi surface is truncated in the even parity channels altho' again there is no translational symmetry breaking. This example demonstrates that a partial truncation of the Fermi surface is possible without symmetry breaking. This behavior continues up to a critical concentration when the holes enter in the even parity channels.

The analysis has been continued for wider ladders for weak coupling by Ledermann et al. [11]. Using a renormalization group (RG) approach they could show that a similar decoupling into band pairs with differing chemical potentials occurs also in wider ladders. Again the outermost bands pair on the highest energy scale since they have the largest Fermi velocity along the ladder. Therefore there is a progressive truncation of the Fermi surface due to the hierarchy of energy gaps associated with the band pairs. Again at half-filling a complete Mott charge gap opens but its magnitude differs between the different pair channels. As a result holes will not enter all the transverse channel immediately upon doping.

3.   THE BREAKDOWN OF LANDAU THEORY IN TWO DIMENSIONS

3.1   INTRODUCTION

The case of 2-dimensions has been extensively studied in recent years, stimulated by the high-T_c cuprates [2]. Since the hole doping is relatively small, most attempts seek to construct a non-Landau theory for Fermi liquids starting from a Mott insulating state. Relatively little attention has been devoted to the approach from the overdoped side of the phase diagram. Most of these efforts have focussed on the growing antiferromagnetic spin fluctuations. However there are reasons to believe that there must be additional effects than just the antiferromagnetic instability alone. One reason is the clear separation in doping between the breakdown of Landau behavior and the onset of AF order. Another is the fact that the breakdown of Landau theory more or less coincides with the transition into a region where the transport behavior is that of a hole-doped insulator rather than that of an electron metal. This crossover appears without a symmetry breaking. We have seen such behavior in the ladder system where the key effect was the flow to strong coupling of the Umklapp scattering as a precursor of the Mott charge gap at half-filling. Since the stoichiometric cuprates are Mott insulators it is clearly necessary to explore the effects of Umklapp scattering. Of course in a 2-dimensional system with a general Fermi surface, one cannot expect a complete Mott charge gap to open for arbitrarily weak repulsion as it does in the 1-dimensional systems. This is an argument in favor of using strong coupling but then analytic approaches are quite limited and numerical approaches are restricted by the notorious fermion sign problem. One compromise is to use a weak coupling RG approach but keep the value of the interaction moderately strong and to look for the appearance of precursors to the Mott charge gap away from half-filling.

In a 2-dimensional square lattice, elastic Umklapp scattering with momentum change $(2\pi, 2\pi)$, occurs across a square connecting the X-points $(\pm\pi, 0)$ and $(0, \pm\pi)$. Since this U-surface approximately coincides with the Fermi surface at or near half-filling for a single tight binding band, it will be clearly relevant even for moderately strong coupling in this density region. The exact form of the Fermi surface will also enter. In the cuprates the energy band can be approximately represented in a tight-binding form with both n.n. and n.n.n. hopping matrix elements t and t'.

$$\varepsilon(\mathbf{k}) = -2t \left(\cos k_x + \cos k_y \right) + 4t' \cos k_x \cos k_y \tag{1}$$

with ratio $t'/t = 0.3$. In this case the Fermi surface with increasing electron density first touches the U-surface at the density at which the saddle points in the band structure at $(\pm\pi, 0)$ and $(0, \pm\pi)$ are at the Fermi energy. Increasing the density further leads to a crossing of the two surfaces at 8 points (see Fig. 1).

3.2   RG TREATMENT

The RG method has been widely used for 1-dimensional interacting fermions where the Fermi surface consists of a small number of Fermi points. In 2-dimensions the application of the method is greatly complicated by the continuous Fermi surface leading to wave vector dependent coupling constants. This difficulty can be overcome by a discretization of the Fermi curves into a finite set of patches or line segments (see Fig. 1). For the cuprate type of Fermi surface with the Fermi energy at or near to the saddle point energy, the

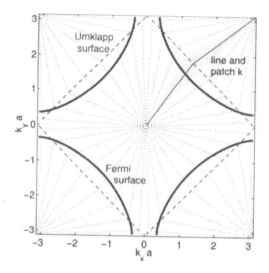

Figure 1 The Brillouin zone, Fermi and Umklapp surface. Also shown are 32 lines in the centers of the patches used in the numerical implementation of the RG equations.

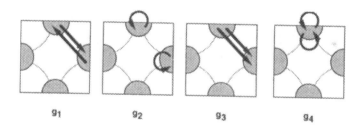

Figure 2 The relevant scattering processes in the two-patch model. The gray semi-circles denote the phase space patches around the saddle points. The interactions are assumed to be spin-independent and constant over the patches. In this notation the spin of the initial and final particle connected by an arrow has to be the same.

patches that cover these saddle points will be the most important, since these contain the extra logarithmic singularities in the density of states. A simple approximation then is to limit the integrations to just these two patches at the saddle points.

This 2-patch model was considered by a number of authors (Dzyaloshinskii [12], Schulz [13] and Lederer et al. [14]) already shortly after the discovery of the high-T_c cuprates. The relevant coupling constants $g_1 \ldots g_4$ are illustrated in Fig. 2. The g_3-scattering process for incoming and outgoing particles right at the saddle points correspond both to Cooper and Umklapp processes. Away

a) *Overdoped regions*

This is the density region of relatively low electron density when the Fermi surface lies inside the Umklapp surface. In this case the Umklapp scattering processes play no special part in the RG flow. Instead the only divergence in the flow appear in the Cooper channel, i.e. when the two incoming wave vectors add up to zero. The dominant intermediate processes are particle-hole scattering with momentum transfer near (π, π). This type of flow to strong coupling can be viewed as a Kohn-Luttinger type of Cooper instability driven in the d-wave channel by particle-hole processes which in turn saturate at low energies as they run out of phase space. A particular example of this flow for a chemical potential $\mu = -1.2t$ can be found in Ref. 18. In this region the d-wave pairing instability clearly diverges faster than the AF spin susceptibility at (π, π).

b) *Strongly underdoped region*

Next they examined the flows when close to half-filling. In this case there is an approximate nesting of the Fermi surface and the Umklapp surface. Here the Umklapp processes grow all around the Fermi surface and are not especially strong near the saddle points. As a consequence the coupling of the Umklapp processes into the pairing channel is not substantial. Instead they feed strongly into the AF spin susceptibility at (π, π) which now becomes the dominantly divergent susceptibility. The critical energy scale Λ_c at which the first divergence occurs, is much larger than in the overdoped regime. Detailed examples of the flow in this region can be found in Honerkamp et al. [18] when the chemical potential μ is set to $\mu = -0.8\,t$. Extrapolation of the RG flow points to an AF ordered groundstate.

c) *The saddle point region*

The most interesting region lies in between these two extremes. In particular when a substantial value is chosen for the n.n.n. hopping parameter t', a distinct region of RG flows emerges at intermediate densities. The RG flows analyzed by Honerkamp et al. for the choice $\mu = -t$ illustrate this. At this density the van Hove singularity lies slightly below the Fermi energy and there is a large phase space region for quasi-elastic Umklapp scattering near the saddle points. This causes a strong coupling between these Umklapp processes and the Cooper pairing processes. For example, in addition to strongly repulsive g_3-type processes, strongly attractive g_4-type processes appear. There is a clear mutual reinforcement of the Umklapp and d-wave Cooper processes. The critical energy scale lies between the two extremes but is evidently enhanced by the Umklapp processes relative to the overdoped region.

Another feature for the RG flows in this density region is the behavior of the forward scattering of g_2-type by which particles are scattered in the vicinity of saddle points. Again Umklapp scattering enters and these g_2-type processes are driven to strong repulsive values. These processes strongly modify the local charge compressibility. Honerkamp et al. study the local compressibility evaluated in a RPA scheme at low energy scales using the renormalized coupling constants. This RPA scheme was tested on the 2-leg ladder at half-filling where a charge gap is known to open, which is signaled in the RG flow by a suppression of the charge compressibility. Note the divergent density of states at the saddle point favors an enhanced compressibility but this tendency is overwhelmed by the strong repulsion. Honerkamp et al. [18] interpret these flows as signs of an ISL appearing at the saddle points. This in turn can be viewed as a precursor to the Mott charge gap which opens up completely at half-filling. They use the analogy between the RG-flows near the saddle points in this density region and the RG-flows in the 2-leg ladder problem to bolster their case. The appearance of an ISL fixed point would arise because the strong d-wave pairing instability is being driven in part by Umklapp scattering and this causes the Fermi surface to be truncated by both a charge and spin gap opening up along the Umklapp surface. This change in the topology of the Fermi surface is consistent with the change in transport properties from electron-like to hole-like in this density region.

In Fig. 3 an approximate phase diagram shows the approximate regions which delineate the three types of behavior described above.

4.   CONCLUSIONS

The chief limitation of the 1-loop RG scheme is that it cannot be extended below the critical energy scale. In this lower energy regime a true strong coupling theory is required. Usually an appropriate mean field approximation can be introduced which gives at least a good qualitative description of the strong coupling phase. However in the case of an ISL of the type found in the 2-leg ladder and postulated in the saddle point region, there is no simple broken symmetry order parameter. Translational symmetry is not broken nor is there a pairing order parameter since the insulating character and Mott charge gap make the system incompressible. Therefore it is not straightforward to construct a qualitatively correct strong coupling theory which could be applied on the low energy scale.

Nonetheless some qualitative remarks can be made on the consequences that follow when an ISL forms over part of the Fermi surface. Since holes enter an ISL preferentially in pairs, a strong coupling will emerge in the Cooper channel between the ISL and the remnant open Fermi surface. This situation is similar to a model put forward earlier by Geshkenbein et al. [19] in which preformed pairs, but with infinite mass, are formed on a higher energy and temperature scale near the saddle points. Again the exchange of hole pairs gives a new mechanism for pairing and can be the origin of superconductivity with a superconducting density determined by the hole doping. The difference between the models is that the preformed pairs here become insulating through Umklapp scattering rather than through an infinite mass. Alternatively one can view the task of constructing a strong coupling theory which applies at low energies as that of constructing a theory of a doped d-wave RVB state.

Acknowledgments

The work described above in large part is the result of a collaboration with N. Furukawa, and of discussions with U. Ledermann, K. Le Hur, S. Haas, M. Sigrist and F.C. Zhang. The support of Schweiz. Nationalfonds is gratefully acknowledged.

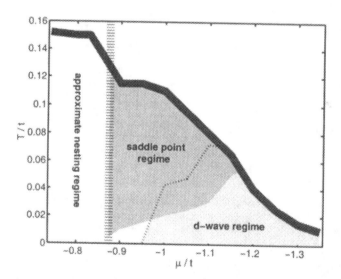

Figure 3 Dependence of the flow to strong coupling on the chemical potential μ and temperature T for $t' = 0.3t$ and initial interaction $U = 3t$. Above the thick line we can integrate the flow down to zero scale without reaching an instability. Below the thin broken line the d-wave pairing susceptibility χ_{dw} exceeds the AF susceptibility $\chi_s(\pi, \pi)$ when the largest couplings have reached the order of the band width. Above this line, $\chi_s(\pi, \pi)$ is larger than χ_{dw}. The darker gray region denotes the *saddle point regime* where the charge coupling of the saddle point regions goes to zero and the total charge compressibility is suppressed. The lightly shaded region represents the *d-wave dominated regime*. Left to the thick vertical line the instability is increasingly dominated by couplings away from the saddle points, we refer to this region as the *approximate nesting regime*.

References

[1] Anderson, P.W. (1997) The Theory of High-T_c Superconductivity in the Cuprates, *Princeton University Press.*

[2] Proc. of M^2S Conference Houston TX (2000), *Physica C* (in press).

[3] Dagotto, E. and Rice, T.M. (1996) *Science* **271**, 618.

[4] Schulz, H.J. (1996) in Proc. Les Houches Summer School LXI, Mesoscopic Quantum Physics (ed. E. Akkermans et al., *Elsevier Amsterdam*).

[5] Balents, L. and Fisher, M.P.A. (1996) *Phys. Rev. B* **53**, 12133.

[6] Fisher, M.P.A. (1998) *cond-mat/9806164* and references therein.

[7] Anderson, P.W. (1987) *Science* **235**, 1196.

[8] Schulz, H.J. (1999) *Phys. Rev. B* **59**, R2471.

[9] Rice, T.M., Haas, S., Sigrist, M. and Zhang, F.C. (1997) *Phys. Rev. B* **56**, 14655.

[10] White, S.R. and Scalapino, D.J., (1998) *Phys. Rev. B* **57**, 3031.

[11] Ledermann, U., Le Hur, K. and Rice, T.M., (2000) *cond-mat 0002445.*

[12] Dzyaloshinskii, I. (1987) *Sov. Phys. JETP* **66**, 848 and *J. Phys. (France)* **6**, 1190996.

[13] Schulz, H.J. (1987) *Phys. Rev. B* **4**, 609.

[14] Lederer, P., Montabaux, G. and Poilblanc, D. (1987) *J. Phys.* **48**, 1613.

[15] Furukawa, N., Rice, T.M. and Salmhofer, M. (1998) *Phys. Rev. Lett.* **81**, 3195.

[16] Zanchi, D. and Schultz, H.J. (1997) *Europhys. Lett.* **44**, 235 and *cond-mat/9812303.*

[17] Halboth, C.J. and Metzner, W. (1999) *cond-mat/9908471.*

[18] Honerkamp, C., Salmhofer, M., Furukawa, N. and Rice, T.M. (1999) *cond-mat/9912358* .

[19] Geshkenbein, V.B., Ioffe, L.B. and Larkin, A.I. (1997) *Phys. Rev. B* **55**, 3173.

METAL-INSULATOR TRANSITION AND MANY-BODY BAND STRUCTURE OF THE HUBBARD MODEL

R. Eder, C. Gröber, M. G. Zacher and W. Hanke

Institut für Theoretische Physik, Universität Würzburg

Am Hubland, 97074 Würzburg

Germany

Abstract   We study the excitation spectra of the Hubbard model by means of Quantum-Monte-Carlo simulations of finite clusters. We focus on two issues: first the metal-insulator transition in the half-filled 2D Hubbard model as a function of t/U This is done at elevated temperature $T = t/3$, where no more long range antiferromagnetic spin correlations remain, and on lattices up to 24×24. We discuss the evolution of the single-particle spectrum and the self-energy as the system goes through the transition. Second, we study the single-particle band structure at $U/t = 8$ and $T = t/3$ focussing on the interpretation of the 'band structure' and the evolution of the Fermi surface with doping. A simple generalization of the Hubbard-I approximation reproduces the band structure at half-filling quite well, and the Fermi surface volume evloves with doping in a very similar way as predicted by the Hubbard-I approximation.

1. INTRODUCTION

Since the pioneering works of Hubbard [1], the metal-insulator transition in a paramagnetic metal and the physics of a doped insulator has been a subject of intense study. The simplest model which incorporates the key physics is the Hubbard model, which describes electrons on a lattice:

$$H = \sum_{\mathbf{k},\sigma} \epsilon_{\mathbf{k}} c_{\mathbf{k},\sigma}^{\dagger} c_{\mathbf{k},\sigma} + U \sum_{i} n_{i,\uparrow} n_{i,\downarrow}, \tag{1}$$

whereby usually the single-electron dispersion $\epsilon_{\mathbf{k}}$ is taken to correspond to simple nearest neighbor hopping with hybridization integral $-t$. The main challenge lies in the fact that perturbation expansions in U may not be expected to describe either the phenomenon of the metal-insulator transition at half-filling (i.e. at an electron density of $n = 1$/lattice site), nor the physics of the doped system at large U/t.

J. Bonča et al. (eds.), Open Problems in Strongly Correlated Electron Systems. 23–32.

A possible way towards a solution is Hubbard's idea of splitting the electron creation operator into two 'particles' which are exact eigenstates of the interaction term $H_U = U \sum_i n_{i,\uparrow} n_{i,\downarrow}$:

$$
\begin{aligned}
c_{i,\sigma}^\dagger &= c_{i,\sigma}^\dagger n_{i,\bar\sigma} + c_{i,\sigma}^\dagger (1 - n_{i,\bar\sigma}) \\
&= \hat{d}_{i,\sigma}^\dagger + \hat{c}_{i,\sigma}^\dagger,
\end{aligned}
\tag{2}
$$

with $[H_U, \hat{d}_{i,\sigma}^\dagger] = U \hat{d}_{i,\sigma}^\dagger$, and $[H_U, \hat{c}_{i,\sigma}^\dagger] = 0$. The interaction term is therefore treated exactly, approximations are made to the kinetic energy. This is precisely the opposite situation as compared to the perturbation expansion in U, which leads for example to the Luttinger theorem. At half-filling the two 'particles' $\hat{d}_{i,\sigma}^\dagger$ and $\hat{c}_{i,\sigma}^\dagger$, whose energy of formation differs by U, then form the two separate Hubbard bands in the insulating phase.

One way to view the metal-insulator transition then is like this: one can also introduce the antisymmetric combination of the Hubbard particles: $\tilde{c}_{i,\sigma}^\dagger = \hat{d}_{i,\sigma}^\dagger - \hat{c}_{i,\sigma}^\dagger$. For $U = 0$, the spectral function of this operator turns out to be a connected three-particle bubble, that means it consists of a completely incoherent continuum. On the other hand, in the insulating phase $U \to \infty$ this operator, being a linear superposition of the two Hubbard-particles should have well defined poles in its spectral function. The metal-insulator transition thus may be viewed as a change of the operator $\tilde{c}_{i,\sigma}^\dagger$ from a completely incoherent excitation into a spectrum with quasiparticle-like poles.

Another question is the effect of doping the system away from half-filling. The effect of doping in the Hubbard I approximation consists in the chemical potential cutting gradually into the top of the lower Hubbard band, in much the same fashion as in a doped band insulator. On the other hand the spectral weight along the lower Hubbard band deviates from the free-particle value of 1 per momentum and spin so that the Fermi surface volume (obtained from the requirement that the integrated spectral weight up to the Fermi energy be equal to the total number of electrons) is not in any 'simple' relationship to the number of electrons - the Luttinger theorem must be violated.

In this work we wish to address the question as to what really happens at the metal-insulator in the paramagnetic phase and when a paramagnetic insulator is doped away from half-filling, by a Quantum Monte Carlo (QMC) study of the 2D Hubbard model. We use the value $U/t = 8$ and work throughout at the moderately high temperature $T = 0.33t$. This temperature is small compared to both the bandwidth, $W = 8t$, and the gap in the single particle spectrum (see Figure 1). The main effect of T is the destruction of antiferromagnetic order - in fact the magnetic correlation length is < 1.5 lattice spacings for all U under consideration. We therefore believe that our study realizes to good approximation the situation for which Hubbard's solutions were originally designed: a paramagnetic system in the limit of large U, at a temperature which is small on the relevant energy scales.

2.   METAL-INSULATOR TRANSITION

We first turn to a scan of U/t at half-filling. Figure 1 shows the single-particle spectral density $A(\mathbf{k}, \omega)$ for a sequence of U/t values where the transition occurs. While for $U=4t$ the most intense 'peaks' still can be fitted roughly by a standard cosine band with undiminished bandwidth $\geq 8t$, there now appears some additional fine structure in the spectra. First, faint bands with weak

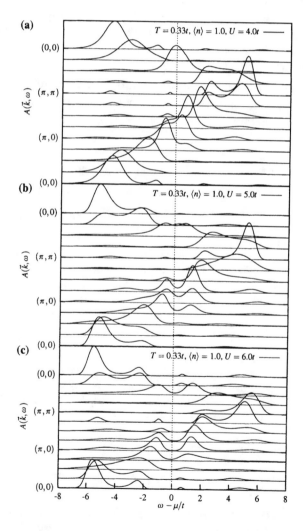

Figure 1   Single particle spectral function $A(\mathbf{k}, \omega)$ Along high-symmetry directions of the Brillouin zone, for the 8×8 clusters and different values of U/t.

intensity form close to the Fermi energy (they can be seen most clearly at energies $\omega = \pm 2t$ for $(0,0)$ and (π, π)), and also a flat low-intensity band at $\omega = \pm 4t$ appears. The topmost and lowermost parts of the original cosine band now appear to be split, which increases the total bandwidth to $\approx 10t$. Secondly, and most spectacularly, the large low-energy peaks around $(\pi, 0)$ now show an unambiguous splitting. The peak at $(\pi/2, \pi/2)$ on the other hand is rather broad, which may indicate a tendency towards splitting, but it still has its maximum right at E_F.

Inspection of the further development of $A(\mathbf{k}, \omega)$ for larger U (Figures 3b,c) shows that what actually happens at $U=4t$ is the formation of a total of 4 'bands'. There are two weak bands with a width of $\approx 2t$ which form the first ionization/affinity states, and two essentially dispersion-less bands at higher energy. Increasing U makes this 4-band structure more and more obvious and increases the Mott-Hubbard gap. Apart from a more or less rigid shift the dispersion of each of the 4 bands remains unaffected by the increase of U.

For very large U/t the 4-band structure becomes quite pronounced, see figure 2, which was computed on a 20×20 cluster. For comparison the two bands

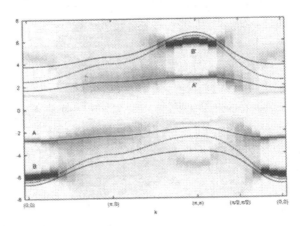

Figure 2   Gray-scale plot of the single particle spectral density for the 20×20 cluster at half-filling. The gray scale gives the intensity of spectral weight at the respective $\mathbf{k} - \omega$ point. Also shown are the Hubbard I bands (dashed lines) and the 4-bands obtained by solving (4) (full lines).

predicted by the Hubbard I approximation,

$$E_{\pm}(\mathbf{k}) = \frac{1}{2} \left[(\epsilon_{\mathbf{k}} + U) \pm \sqrt{\epsilon_{\mathbf{k}}^2 + U^2} \right] \tag{3}$$

are also shown as the dashed dispersive lines ($\epsilon_{\mathbf{k}} = -2t(\cos(k_x) + \cos(k_y))$ is the noninteracting dispersion). These provide at best a rough fit to those parts of the spectral function which have high spectral weight. Inspection of the numerical spectra shows quite a substantial difference between the numerical and the Hubbard-type band structures: the latter always give two bands, whereas in the numerical spectra one can rather unambiguously identify 4 of them, denoted as as B, A, A' and B' (see the Figure). None of these bands shows any indication of antiferromagnetic symmetry; together with the short spin correlation length this shows that we are really in the paramagnetic phase. We found that to model this 4-band structure one can introduce two additional dispersionless bands at energies of $\bar{E}_\pm = \frac{U}{2} \pm \epsilon$. We now allow mixing between each of these dispersionless bands and the respective Hubbard band, as would be described by the Hamilton matrix

$$H_\pm = \begin{pmatrix} E_\pm(\mathbf{k}) & V \\ V & \bar{E}_\pm(\mathbf{k}) \end{pmatrix}. \tag{4}$$

Using the values $\epsilon = 3t$ and $V = t$, the resulting 4-band structure provides at least a qualitatively correct fit to the numerical data. An explanation for these additional bands may be given in terms of additional quasiparticles which correspond to 'dressed holes'. Commuting the Hubbard particles with the kinetic energy yields operator products [3], which may be viewed as a Hubbard-particles dressed by a spin- density- or pairing fluctuation. These additional quasiparticles then are quite 'heavy' and a simple extension of the Hubbard-I approximation [3] reproduces the 4-band structure quite well. We also note that the a 4-band structure which has quite some similarity with our results has recently been obtained by Pairault *et al.* using a strong coupling expansion [2]. Moreover, a numerical study [4] of the spectral functions of these 'dressed hole operators' provides further evidence for the correctness of this picture.

3.   FERMI SURFACE IN THE DOPED CASE

We do not, however, pursue this issue further but turn to our main subject, the effect of hole doping. Figure 3 shows the development of $A(\mathbf{k}, \omega)$ with doping. Thereby the $A(\mathbf{k}, \omega)$ for different hole concentrations have been overlaid so as to match dominant features, and the chemical potentials for the different hole concentrations are marked by lines. It is quite obvious from this Figure that the 2 bands seen at half-filling in the photoemission spectrum persist with an essentially unchanged dispersion. The chemical potential gradually cuts deeper and deeper into the A band, forming a hole-like Fermi surface centered on (π, π), the top of the lower Hubbard band. The only deviation from a rather simple rigid-band behavior is an additional transfer of spectral weight: the part of the A-band near (π, π) gains in spectral weight, whereas the B-band looses weight. The loss of the B band cannot make up for the increase of the A band, but rather there is an additional transfer of weight from the upper Hubbard bands, predominantly the A' band. This effect is quite well understood [5]. The A' band seems to be affected strongest by the hole doping and in fact the rather clear two-band structure visible near (π, π) at half-filling rapidly gives

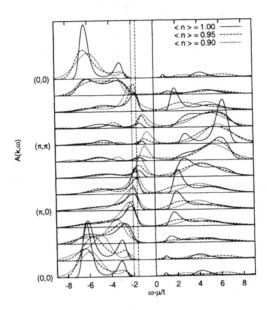

Figure 3 Single particle spectral function for the 8 × 8 cluster and different electron density ⟨n⟩. The chemical potential at half filling is the zero of energy, the spectra for different ⟨n⟩ are rigidly shifted relative to one another so as to match dominant features The chemical potentials for the different ⟨n⟩ are given by vertical lines.

way to one broad 'hump' of weight. Apart from the spectral weight transfer, however, the band structure on the photoemission side is almost unaffected by the hole doping - the *dispersion* of the A-band becomes somewhat wider but does not change appreciably. In that sense we see at least qualitatively the behavior predicted by the Hubbard I approximation. Next, we focus on the Fermi surface volume. Some care is necessary here: first, we cannot actually be sure that at the high temperature we are using there is still a well-defined Fermi surface. Second, the criterion we will be using is the crossing of the A band through the chemical potential. It has to be kept in mind that this may be quite misleading, because band portions with tiny spectral weight are ignored in this approach (see for example Ref. [9] for a discussion). When thinking of a Fermi surface as the constant energy contour of the chemical potential, we have to keep in mind that portions with low spectral weight may be overlooked. On the other hand the fact that a peak with appreciable weight crosses from photoemission to inverse photoemission at a certain momentum is independent of whether we call this a 'Fermi surface' in the usual sense, and should be reproduced by any theory which claims to describe the system. It therefore has to be kept in mind that in the following we are basically studying a 'spectral weight Fermi surface', i.e. the locus in **k** space where an apparent quasiparticle band with high spectral weight crosses the chemical potential.

With these *caveats* in mind, Figure 4 shows the low-energy peak structure of

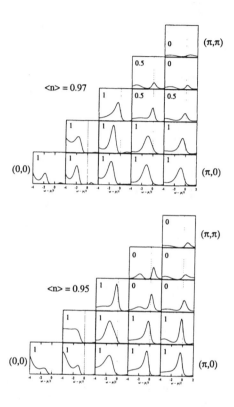

Figure 4   Single particle spectral function for all **k**-points of the 8 × 8 cluster in the irreducible wedge of the Brillouin zone. For each **k** the weight $w_{\mathbf{k}}$ is given.

$A(\mathbf{k}, \omega)$ for all allowed momenta of the 8 × 8 cluster in the irreducible wedge of the Brillouin zone, and for different hole concentrations. In all of these spectra there is a pronounced peak, whose position shows a smooth dispersion with momentum. Around (π, π) the peak is clearly above μ, whereas in the center of the Brillouin zone it is below. The locus in **k**-space where the peak crosses μ forms a closed curve around (π, π) and it is obvious from the Figure that the 'hole pocket' around (π, π) increases very rapidly with δ. To estimate the Fermi surface volume V_F we assign a weight $w_{\mathbf{k}}$ of 1 to momenta **k** where the peak is below μ, 0.5 if the peak is right at μ and 0 if the peak is above μ. Our assignments of these weights are given in Figure 4. The fractional Fermi surface volume then is $V_F = \frac{1}{N} \sum_{\mathbf{k}} w_{\mathbf{k}}$, where $N = 64$ is the number of momenta in the 8 × 8 cluster. Of course, the assignment of the $w_{\mathbf{k}}$ involves a certain degree of arbitrariness. It can be seen from Figures 4, however, that our $w_{\mathbf{k}}$ would in any way tend to underestimate the Fermi surface volume, so that the obtained V_F data points rather have the character of a lower bound to the true V_F. Even if we take into account some small variations of V_F due to different assignments of the weight factors, however, the resulting V_F versus δ curve never can be made consistent with the Luttinger volume, see Figure 5. The deviation from the Luttinger volume is quite pronounced at low doping.

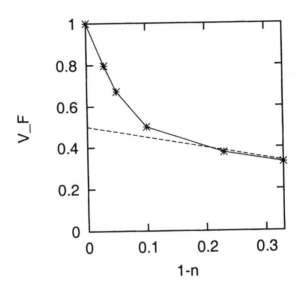

Figure 5   Fermi surface volume as estimated from the single particle spectral function, plotted versus the concentration of holes in the half-filled band. The dashed line gives the value predicted by the Luttinger theorem, $V_F = \frac{n}{2}$.

V_F approaches the Luttinger volume for dopings $\approx 20\%$, but due to our somewhat crude way of determining V_F we cannot really decide when precisely the Luttinger theorem is obeyed. The Hubbard I approximation approaches the Luttinger volume for hole concentrations of $\approx 50\%$, i.e. the steepness of the drop of V_F is not reproduced quantitatively. The latter is somewhat improved in the so-called 2-pole approximation [6, 7]. For example the Fermi surface given by Beenen and Edwards [7] for $\langle n \rangle = 0.94$ obviously is very consistent with the spectrum in Figure 3 for $\langle n \rangle = 0.95$.

A recent study of the momentum distribution in the t-J model by high-temperature series expansion [8] has also provided some evidence for a 'Fermi surface' which encloses a larger value than predicted by the Luttinger theorem. The criterion used there was a maximum of $|\nabla_k n_k|$, i.e. the locus of the steepest drop of n_k. This would in fact be quite consistent with the present results. However, the same *caveat* as in the present case applies, i.e. this criterion will overlook Fermi level crossings of bands with low spectral weight [9].

4.   CONCLUSION

In summary, we have studied the evolution of the single particle spectral function for the paramagnetic phase of the 2D Hubbard model, with both U and with increasing hole doping from the insulator obtained for large U/t. With increasing U/t at half-filling the Mott-Hubbard gap opens continuously, and in fact nonuniformly on the Fermi surface. In the very first stage of gap formation the gap seems to open at $(\pi, 0)$, which may suggest a connection with the 'pseudogap' in cuprate superconductors, as described by the 'patch model' of Furukawa and Rice [10]. There is no preformed gap or any discontinuity in the Hubbard gap. This is consistent with some recent results in the framework of the $d \to \infty$ approach: a numerical study by Noack and Gebhard [11] where the infinite dimension limit was simulated by a random dispersion has also shown a continuous opening of the gap, and a recent re-investigation in the framework of the infinite-d-limit [12] has also shown that the transition is continuous down to rather low temperatures. In our data the gap in fact opens in the 'most continous way possible', in that the gap is already open at $(\pi, 0)$ while being closed at $(\pi/2, \pi/2)$.

As concerns the behaviour upon doping we found as a surprising result that in this situation the Hubbard I and related approximations give a qualitatively quite correct picture. The main discrepancy between the Hubbard I and the so-called 2-pole approximation and our numerical spectra is the number of 'bands' of high spectral weight, which is 4 in the numerical data. This is no reason for concern, because we have seen that adding two more bands allows for an quite reasonable fit to the numerical band structure and one might expect that finding a somewhat more intricate decoupling scheme for the Hubbard I approximation or a suitable 4-pole approximation should not pose a major problem [3, 2]. The greatest success of the Hubbard-type approximations, however, is a qualitatively quite correct description of the evolution of the 'Fermi surface'. The effect of doping consists of the progressive shift of the chemical potential into the topmost band observed at half-filling, accompanied by some transfer of spectral weight. The Fermi surface volume, determined in an 'operational way' from the band crossings, violates the Luttinger theorem for low hole concentrations and does not appear to be in any simple relationship to the electron density. The Luttinger sum rule is recovered only for hole concentrations around 20%.

In our opinion the strange dependence of V_f on electron density makes it questionable whether the 'spectral weight Fermi surface' in our data is a true constant energy contour for a system of 'quasiparticles'. It may be possible that at the temperature we are studying a Fermi surface in the usual sense no longer exists, and that the Hubbard I approximation merely reproduces the *spectral weight distribution* in this case. As our data show, however, for that purpose the approximation is considerably better than commonly believed.

Zero temperature studies for the doped t-J and Hubbard model are only possible by using exact diagonalization [13], in which case the shell-effects due to the small system size require special care [14, 15]. One crucial point is the very different shape of the quasiparticle dispersion at zero temperature. Whereas the A band is at least topologically equivalent to a *nearest neighbor* hopping dispersion, with minimum at $(0,0)$ and maxiumum at (π, π), the zero temperature data [13] show a *second-nearest neighbor* dispersion with a nearly degenerate band maximum along the antiferromagnetic zone boundary, and a shallow absolute maximum at $(\pi/2, \pi/2)$. The effect of hole doping at zero temperature, however, has a qualitatively very similar effect as in the present case [14]: the chemical potential simply cuts into the quasiparticle band for the insulator, which thus is populated by hole-like quasiparticles [15]. Again, these 'hole pockets' violate the Luttinger theorem, indicating again the breakdown of adiabatic continuity in the low doping regime persists also at low temperatures.

References

[1] J. Hubbard, Proc. Roy. Soc. A **276**, 238 (1963); J. Hubbard, Proc. Roy. Soc. A **277**, 237 (1964); J. Hubbard, Proc. Roy. Soc. A **281**, 401 (1964).

[2] S. Pairault, D. Senechal, and A.-M. S. Tremblay, Phys. Rev. Lett. **80**, 5389 (1998).

[3] A. Dorneich *et al.* Phys. Rev. B **61**, 12816 (2000).

[4] C. Gröber, R. Eder, W. Hanke, cond-mat/0001366.

[5] H. Eskes and A. M. Oles, Phys. Rev. Lett. **73** 1279 (1994).

[6] B. Mehlig, *et al.* Phys. Rev. B **52**, 2463 (1995).

[7] J. Beenen and D. M. Edwards, Phys. Rev. B **52**, 13636 (1995).

[8] W. O. Putikka, M. U. Luchini, and R. R. P. Singh, Phys. Rev. Lett. **81**, 2966 (1998).

[9] R. Eder and Y. Ohta, Phys. Rev. Lett. **72**, 2816 (1994).

[10] N. Furukawa and T. M. Rice, J. Phys. C, **L381**, 1998.

[11] R. M. Noack and F. Gebhard, Phys. Rev. Lett **82**, 1915 (1999).

[12] J. Schlipf, *et al.* Phys. Rev. Lett **82**, 4890 (1999).

[13] E. Dagotto, Rev. Mod. Phys. **66**, 763 (1994).

[14] R. Eder, Y. Ohta, and T. Shimozato, Phys. Rev. B **50**, 3350 (1994); R. Eder and Y. Ohta, Phys. Rev. B **51**, 6041 (1994).

[15] S. Nishimoto, Y. Ohta, and R. Eder, Phs. Rev. B **57**, R5590 (1998).

SPECTRAL PROPERTIES OF UNDERDOPED CUPRATES

A. Ramšak[1,2], P. Prelovšek[1,2], and I. Sega[1]

[1] *J. Stefan Institute, SI-1000 Ljubljana, Slovenia*

[2] *Faculty of Mathematics and Physics, University of Ljubljana, SI-1000 Ljubljana, Slovenia*

Abstract   In the framework of the planar t-J model for cuprates we analyze the development of a pseudo gap in the density of states, which at low doping starts to emerge for temperatures $T < J$ and persists up to the optimum doping. The analysis is based on numerical results for spectral functions obtained with the finite-temperature Lanczos method for finite two-dimensional clusters. Numerical results are additionally compared with the self consistent Born approximation (SCBA) results for hole-like (photoemission) and electron-like (inverse photoemission) spectra at $T = 0$. The analysis is suggesting that the origin of the pseudo gap is in short-range antiferromagnetic (AFM) spin correlations and strong asymmetry between the hole and electron spectra in the underdoped regime.

We analyze also the electron momentum distribution function (EMD). Our analytical results for a single hole in an AFM based on the SCBA indicate an anomalous momentum dependence of EMD showing "hole pockets" coexisting with a signature of an emerging large Fermi surface (FS). The position of the incipient FS and the structure of the EMD is determined by the momentum of the ground state. The main observation is the coexistence of two apparently contradicting FS scenarios. On the one hand, the δ-function like contributions at $(\pi/2, \pi/2)$ indicate, that for finite doping a pocket-like small FS evolves from these points, provided provided that AFM long range order persists. On the other hand, the discontinuity which appears at the same momentum is more consistent with with infinitesimally short arc (point) of an emerging large FS.

1.   GENERALIZED t-J MODEL

Spectral properties of underdoped cuprates are of current interest, in particular results for the electron spectral functions as obtained with the angle resolved photoemission (ARPES) [1, 2, 3]. A remarkable feature of the ARPES data is the appearance of a pseudogap already at temperature T^* well above the superconducting T_c. Other quantities, e.g., the uniform susceptibility, the Hall constant and the specific heat also show a pseudogap consistent with energy scale T^* [4]. At very low temperature $T \ll T^*$ the Fermi surface and the

33

J. Bonča et al. (eds.), Open Problems in Strongly Correlated Electron Systems, 33–42.
© 2001 *Kluwer Academic Publishers. Printed in the Netherlands.*

corresponding electron spectral functions change dramatically with doping of planar cuprate systems with holes, where a transition from "small" to "large" FS seems to be consistent with ARPES, but is not adequately understood from the theoretical point of view.

There have been also several theoretical investigations of this problem, using the exact diagonalization (ED) of small clusters [5, 6, 7, 8], string calculations [9], slave-boson theory [10] and the high temperature expansion [11]. While a consensus has been reached about the existence of a large Fermi surface in the optimum-doped and overdoped materials, in the interpretation of ARPES experiments on *underdoped* cuprates the issue of the debate is (i) why are experiments more consistent with the existence of parts of a large FS – Fermi arcs or Fermi patches [3, 12] – rather than with a hole pocket type small FS, predicted by several theoretical methods based on the existence of AFM long range order in cuprates, (ii) how does a partial FS eventually evolve with doping into a large closed one.

The main emphasis of the present study is on the pseudo gap found in ARPES and also in some exact diagonalization studies [13, 14, 15]. We employ the standard t-J model to which we add a nearest neighbor repulsion term V,

$$H = -t \sum_{<ij>,\sigma} (c_{i,\sigma}^\dagger c_{j,\sigma} + \text{H.c.}) +$$
$$+ \sum_{<ij>} [JS_i^z S_j^z + \frac{\gamma}{2} J(S_i^+ S_j^- + S_i^- S_j^+) + (V - \frac{J}{4})n_i n_j]. \quad (1)$$

Here i, j refer to planar sites on a square lattice and $c_{is}^\dagger$ represent projected fermion operators forbidding double occupation of sites. S_i^α are spin operators. For convenience we treat the anisotropy γ as a free parameter, with $\gamma = 0$ in the Ising case, and $\gamma \to 1$ in the Heisenberg model.

Numerical results presented here were obtained with Lanczos exact diagonalization (ED) technique on small clusters with $N = 16 \sim 32$ sites, for temperature in the range $T < J$. The method is simple: we take into account only the lowest ≈ 100 Lanczos states and evaluate the corresponding thermal averages. The method is compared with a more elaborate finite-temperature Lanczos method (FTLM) [13] and the agreement in here studied $T < J$ regime is excellent. It should be noted that for low temperatures the results of the diagonalization of small clusters always have to be examined with caution, because low energy scale exhibits very strong finite size effects.

Our analytical approach is based on a spinless fermion – Schwinger boson representation of the t-J Hamiltonian [16] and on the SCBA for calculating both the Green's function [16, 17, 18] and the corresponding wave function [19, 20]. The method is known to be successful in determining spectral and other properties of the quasi particles (QP). In contrast to other methods the

SCBA is expected to correctly describe the *long-wavelength* physics, the latter being determined by the linear dispersion of spin waves, whereas the *short-wavelength* properties can be studied with various other methods. Here we compare the SCBA results with the corresponding ED, as shown further-on.

In the SCBA fermion operators are decoupled into hole and pseudo spin – local boson operators: $c_{i,\uparrow} = h_i^\dagger$, $c_{i,\downarrow} = h_i^\dagger S_i^+ \sim h_i^\dagger a_i$ and $c_{i,\downarrow} = h_i^\dagger$, $c_{i,\uparrow} = h_i^\dagger S_i^- \sim h_i^\dagger a_i$ for i belonging to A- and B-sublattice, respectively. The effective Hamiltonian emerges

$$\tilde{H} = N^{-1/2} \sum_{\mathbf{kq}} (M_{\mathbf{kq}} h_{\mathbf{k-q}}^\dagger h_{\mathbf{k}} \alpha_{\mathbf{q}}^\dagger + \text{H.c.}) + \sum_{\mathbf{q}} \omega_{\mathbf{q}} \alpha_{\mathbf{q}}^\dagger \alpha_{\mathbf{q}}, \qquad (2)$$

where $h_{\mathbf{k}}^\dagger$ is the creation operator for a (spinless) hole in a Bloch state. The AFM boson operator $\alpha_{\mathbf{q}}^\dagger$ creates an AFM magnon with the energy $\omega_{\mathbf{q}}$, and $M_{\mathbf{kq}}$ is the fermion-magnon coupling.

We calculate the Green's function for a hole $G_{\mathbf{k}}(\omega)$ within the SCBA [16, 17, 18]. This approximation amounts to the summation of non-crossing diagrams to all orders and the corresponding ground state wave function with momentum $\mathbf{k}$ and energy $\epsilon_{\mathbf{k}}$ [19, 20, 21] is represented as

$$|\Psi_{\mathbf{k}}\rangle = Z_{\mathbf{k}}^{1/2} \Big[h_{\mathbf{k}}^\dagger + \dots + N^{-n/2} \sum_{\mathbf{q}_1,\dots,\mathbf{q}_n} M_{\mathbf{kq}_1} G_{\bar{\mathbf{k}}_1}(\bar{\omega}_1) \dots M_{\bar{\mathbf{k}}_{n-1}\mathbf{q}_n} \times$$

$$\times G_{\bar{\mathbf{k}}_n}(\bar{\omega}_n) h_{\bar{\mathbf{k}}_n}^\dagger \alpha_{\mathbf{q}_1}^\dagger \dots \alpha_{\mathbf{q}_n}^\dagger + \dots \Big] |0\rangle. \qquad (3)$$

Here $\bar{\mathbf{k}}_m = \mathbf{k} - \mathbf{q}_1 - \dots - \mathbf{q}_m$, $\bar{\omega}_m = \epsilon_{\mathbf{k}} - \omega_{\mathbf{q}_1} - \dots - \omega_{\mathbf{q}_m}$ and $Z_{\mathbf{k}}$ is the QP spectral weight.

2.   PSEUDO GAP IN THE DENSITY OF STATES

We study here the planar density of states (DOS), defined as $\mathcal{N}(\omega) = 2/N \sum_{\mathbf{k}} A_{\mathbf{k}}(\omega - \mu)$, where $A_{\mathbf{k}}(\omega)$ is the electron spectral function [14, 13], and μ denotes the chemical potential. First we calculate the DOS with the finite-temperature Lanczos method for clusters of $N = 18, 20$ sites doped with one hole, $N_h = 1$. Here we denote with $\mathcal{N}^-(\omega)$ the density of states corresponding to adding a hole into the system and thus to the photoemission experiments, while $\mathcal{N}^+(\omega)$ represents the inverse photoemission (IPES) spectra.

In Fig. 1(a) we present $\mathcal{N}(\omega)$ for two different $J/t = 0.3$ and $J/t = 0.6$ on a $N = 18$ sites cluster for $V = 0$. We observe that the pseudo gap scales approximately as $2J$. The analysis at elevated temperatures shows that the gap slowly fills up and disappears at $T \sim J$. In Fig. 1(b) results for different values V are presented. The gap remains robust also in the presence of the V term, which enhances ($V < 0$) or suppresses ($V > 0$) the binding of hole pairs. It is thus evident that the effect of V is only of qualitative character and

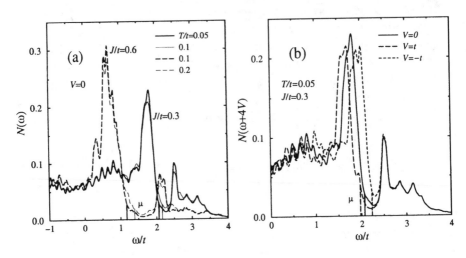

Figure 1 $\mathcal{N}(\omega)$ for one hole on $N = 18$ sites, presented for different J/t, V/t and T/t. (a) $V = 0$. (b) $J = 0.3t$. Broadening of peaks is taken $\delta/t = 0.04$.

not relevant for the existence of the pseudo gap. We therefore believe that this analysis suggests that the origin of the pseudo gap is in short-range AFM spin correlations rather than in the binding tendency of doped holes.

AFM spin correlations are correctly taken into account in the SCBA. Therefore for the limiting case of low doping, $c_h \to 0$, and $T \to 0$ we approximate $\mathcal{N}^-(\omega)$ with

$$\mathcal{N}^-(\omega) \propto \sum_k \mathrm{Im} G_k(-\omega), \tag{4}$$

where $G_k(\omega)$ is the SCBA Green's function for adding one hole (ARPES) to an AFM reference system (instead of adding one hole to the state with one hole). The corresponding DOS for *removing* a hole (IPES) from the state with one hole, $\mathcal{N}^+(\omega) = \frac{2}{N} \sum_k A_k^+(\omega)$, can be calculated accurately in the SCBA as follows. First the spin averaged hole-like spectral function,

$$A_k^+(\omega) = -\frac{1}{2\pi} \mathrm{Im} \sum_\sigma \langle \Psi_{k_0} | c_{k,\sigma} \frac{1}{\omega - \tilde{H}} c_{k,\sigma}^\dagger | \Psi_{k_0} \rangle, \tag{5}$$

is expressed in terms of holon and magnon operators and the result for $\mathcal{N}^+(\omega)$ emerges,

$$\mathcal{N}^+(\omega) = \frac{1}{N} \sum_i \langle \Psi_{k_0} | h_i^\dagger [\delta(\omega - \tilde{H}) + a_i \delta(\omega - \tilde{H}) a_i^\dagger] h_i | \Psi_{k_0} \rangle. \tag{6}$$

Here $| \Psi_{k_0} \rangle$ represents a weakly doped AFM, i.e., it is the ground state (GS) wave function of a planar AFM with one hole and the GS wave vector k_0.

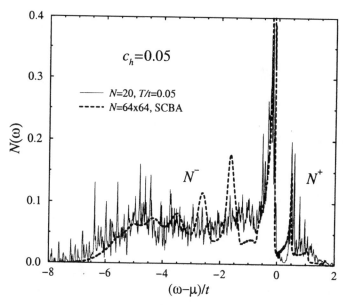

Figure 2 $\mathcal{N}(\omega)$ for $N_h = 1$ on $N = 20$ sites, with $J/t = 0.3$, $V = 0$, $T/t = 0.05$ (full line). Dashed heavy line represents the SCBA result on large lattice obtained as a sum of $\mathcal{N}^-(\omega)$ and $\mathcal{N}^+(\omega)$. The SCBA result is obtained on a $N = 64 \times 64$ cluster and for undoped reference system. Note the "string states" resonances, absent in the finite doping Green's function. $\mathcal{N}^+(\omega)$ corresponds to IPES. Reference hole concentration is $c_h = 1/N$. The SCBA result is normalized to $c_h = 1/20$. Broadening of peaks is taken $\delta/t = 0.01$.

The normalization (sum rule) of $A_{\mathbf{k}}^{\pm}(\omega)$ and $\mathcal{N}^{\pm}(\omega)$ is discussed in detail in Ref. [13]. It should be noted that the normalization of $A_{\mathbf{k}}^{+}(\omega)$ is not trivial and is related to the EMD presented in the next section.

In Fig. 2 are shown spectra $\mathcal{N}(\omega)$ obtained with the ED on a $N = 20$ sites cluster. We compare these spectra with the DOS within the SCBA. The peaks in $\mathcal{N}^+(\omega)$ can well be explained with magnon structure of single hole ground state, while peaks in $\mathcal{N}^-(\omega)$ are string states known in the single hole case and are for the present study of the pseudo gap and FS not relevant. As seen in Fig. 2 the total DOS obtained with the SCBA and in particular $\mathcal{N}^+(\omega)$ remarkably accurately resemble the ED DOS.

3.   LARGE OR SMALL FERMI SURFACE?

The electron momentum distribution function $n_{\mathbf{k}} = \langle \Psi_{\mathbf{k}_0} | \sum_\sigma c_{\mathbf{k},\sigma}^{\dagger} c_{\mathbf{k},\sigma} | \Psi_{\mathbf{k}_0} \rangle$ is the key quantity for resolving the problem of the Fermi surface. Numerically $n_{\mathbf{k}}$ can be determined by exact diagonalization of the model in small clusters, where only a restricted number of momenta $\mathbf{k}$ is allowed. The GS wave vector due to finite size effects varies with N. Therefore we present here

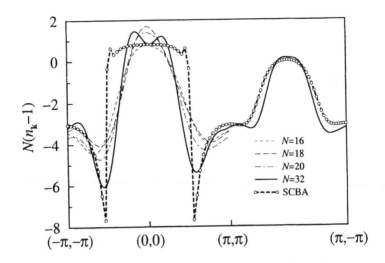

Figure 3 $N(n_\mathbf{k} - 1)$ obtained from ED for various systems $N = 16, 18, 20, 32$ and $J/t = 0.3$. For the SCBA $N = 64 \times 64$, $\gamma = 0.999$ and note that delta-function contributions at $\mathbf{k} = \pm\mathbf{k}_0$ are not shown. In plotting the curve for $N = 32$ data from Ref. [8] were used.

results obtained with the method of twisted boundary conditions [22], where $t_{jj'} \rightarrow t_{jj'} \exp i\theta_{jj'}$. Since $n_\mathbf{k} \equiv n_\mathbf{k}(\mathbf{k}_0, \theta)$ depends both on $\mathbf{k}_0$ and θ it follows from Peierls construction that $n_\mathbf{k}(\mathbf{k}_0, 0) = n_{\mathbf{k}+\mathbf{k}_0}(0, \mathbf{k}_0)$ for $\theta = \mathbf{k}_0$. This allows us to study $n_\mathbf{k}$ for arbitrary $\mathbf{k}$ and $\mathbf{k}_0$. Furthermore, the finite size effects of the results are suppressed if we fix $\mathbf{k}_0$ for *all clusters* here studied to the symmetry point $\mathbf{k}_0 = (\frac{\pi}{2}, \frac{\pi}{2})$.

In Fig. 3 we present for $J/t = 0.3$ ED results for clusters with different N and $\gamma = 1$. The EMD obeys the sum rule $\sum_\mathbf{k} n_\mathbf{k} = N - 1$ and, for the allowed momenta, the constraint $N(n_\mathbf{k} - 1) \leq 1$. We show here the quantity $N(n_\mathbf{k} - 1)$, which for different N scales towards the same curve. Results are presented for some selected directions in the Brillouin zone (BZ) and should be averaged over all four possible ground state momenta when discussed, e.g., in connection with ARPES data.

Analytically we study the EMD again in the spinless fermion – Schwinger boson representation. The wave function Eq. (3) corresponds to the projected space of the model Eq. (1) and $n_\mathbf{k} = \langle \Psi_{\mathbf{k}_0} | \hat{n}_\mathbf{k} | \Psi_{\mathbf{k}_0} \rangle$ with the projected *electron* number operator $\hat{n}_\mathbf{k} = \sum_\sigma c^\dagger_{\mathbf{k},\sigma} c_{\mathbf{k},\sigma}$. Consistent with the SCBA approach, we decouple $\hat{n}_\mathbf{k}$ into hole and magnon operators,

$$\hat{n}_\mathbf{k} = \frac{1}{N} \sum_{ij} h_i h_j^\dagger \left(\eta_{ij}^+ [1 + a_i^\dagger a_j (1 - \delta_{ij})] + \eta_{ij}^-(a_i^\dagger + a_j) \right), \qquad (7)$$

where $\eta_{ij}^\pm = e^{-i\mathbf{k}\cdot(\mathbf{R}_i - \mathbf{R}_j)}(1 \pm e^{-i\mathbf{Q}\cdot(\mathbf{R}_i - \mathbf{R}_j)})/2$ with $\mathbf{Q} = (\pi, \pi)$. Local $a_i^\dagger$ are further expressed with proper magnon operators $\alpha_\mathbf{q}^\dagger$. In general the expectation

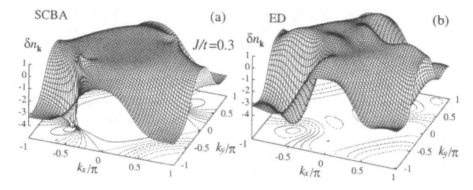

Figure 4   (a) SCBA result for $N = 64 \times 64$ and $J/t = 0.3, \gamma = 0.99$. (b) Exact diagonalization result for $N = 32$ sites as in Fig. 3.

value $n_{\mathbf{k}}$ for a single hole has the following structure [23]

$$n_{\mathbf{k}} = 1 - \frac{1}{2} Z_{\mathbf{k}_0} (\delta_{\mathbf{k}\mathbf{k}_0} + \delta_{\mathbf{k}\mathbf{k}_0 + \mathbf{Q}}) + \frac{1}{N} \delta n_{\mathbf{k}}. \tag{8}$$

Here the second term proportional to δ-functions corresponds to hole pockets. Note that $\delta n_{\mathbf{k}}$, for the case of a single hole fulfills the sum rule $\frac{1}{N} \sum_{\mathbf{k}} \delta n_{\mathbf{k}} = Z_{\mathbf{k}_0} - 1$ and $\delta n_{\mathbf{k}} \leq 1$. The introduction of $\delta n_{\mathbf{k}}$ is convenient as it allows the comparison of results obtained with different methods and on clusters of different size N.

In Fig. 3 we also present the SCBA result. We have also checked the convergence of $\delta n_{\mathbf{k}}$ with the number of magnon lines, n. For $J/t > 0.3$ we find for all $\mathbf{k}$ that the contribution of terms $n > 3$ amounts to less than few percent. This is in agreement with the convergence of the norm of the wave function, which is even faster [20].

In Fig. 4(a) we present this $\delta n_{\mathbf{k}}$ for the whole BZ. The important ingredient of the SCBA is the gapless magnons with linear dispersion and a more complex ground state of the planar AFM. $G_{\mathbf{k}}(\omega)$ and $\epsilon_{\mathbf{k}}$ are strongly $\mathbf{k}$-dependent. As a consequence $n_{\mathbf{k}}$ in general depends both on $\mathbf{k}$ and $\mathbf{k}_0$. The ground state is for the t-J model fourfold degenerate and we again choose $\mathbf{k}_0 = (\frac{\pi}{2}, \frac{\pi}{2})$. In Fig. 4(b) is presented also the result of exact diagonalization of a $N = 32$ sites cluster [8], but generalized to the whole BZ.

The main conclusion regarding the EMD is the coexistence of two apparently contradicting Fermi-surface scenarios in EMD of a single hole in an AFM. (i) On one hand, the δ-function contributions in Eq. (8) seem to indicate that at finite doping a delta-function might develop into small Fermi surface, i.e., a hole pocket, provided that AFM long range order persists. (ii) A novel feature is that also $\delta n_{\mathbf{k}}$ is singular in a particular way, i.e., it shows a discontinuity at $\mathbf{k} = \mathbf{k}_0$ with a strong asymmetry with respect to $\mathbf{k}_0$. It is therefore more consistent with infinitesimally short arc (point) of an emerging large FS. For finite doping the discontinuity could possibly extend into such a finite arc (not closed) FS. Note that as long-range AFM order is destroyed by doping, hole pocket contributions should disappear while the singularity in $\delta n_{\mathbf{k}}$ could persist. The results of the two methods, the SCBA and the ED agree quantitatively at all points in the BZ. However, the SCBA result is *symmetric* around Γ point in the direction $\mathbf{k} \parallel \mathbf{k}_0$, while small system results show a weak asymmetry for $\mathbf{k} = \pm\mathbf{k}_0$, respectively. From our analysis of the SCBA results for $N \to \infty$ and long range AFM spin background it follows that in the thermodynamic limit $c_h \to 0$ $n_{\mathbf{k}}$ is symmetric. The asymmetry is in Ref. [8] attributed to the opening of the gap in the magnon spectrum at $\mathbf{q} \sim \mathbf{Q}$ in finite systems. Within the SCBA the asymmetry also appears if the EMD is evaluated with $\mathbf{k}_0$ *displaced* from $(\frac{\pi}{2}, \frac{\pi}{2})$ by a small amount $\delta\mathbf{k}_0$ (not shown here).

4.   SUMMARY AND ANALYTICAL RESULTS

Full numerical results are captured with a simple analytical expansion which gives more insight into the structure of $A_{\mathbf{k}}^+(\omega)$ and $\delta n_{\mathbf{k}}$. We simplify the wave function, Eq. (3), by keeping only the one-magnon contributions and take $J/t \gg 1$ and the leading order contributions are then

$$A_{\mathbf{k}}^+(\omega) \sim A_{\mathbf{k}}^{(1)}(\omega) = [\tfrac{1}{2}Z_{\mathbf{k}_0}(\delta_{\mathbf{kk}_0}+\delta_{\mathbf{kk}_0+\mathbf{Q}}) + \tfrac{1}{N}(1 - \delta n_{\mathbf{k}}^{(1)})]\delta(\omega - \omega_{\mathbf{q}}),$$

$$\delta n_{\mathbf{k}}^{(1)} = -Z_{\mathbf{k}_0}M_{\mathbf{k}_0\mathbf{q}}G_{\mathbf{k}_0}(\epsilon_{\mathbf{k}_0}-\omega_{\mathbf{q}})[2u_{\mathbf{q}}+M_{\mathbf{k}_0\mathbf{q}}G_{\mathbf{k}_0}(\epsilon_{\mathbf{k}_0}-\omega_{\mathbf{q}})]$$

$$\sim -8Z_{\mathbf{k}_0}^2 J\frac{\mathbf{q}\cdot\mathbf{v}}{\omega_{\mathbf{q}}^2}(1 + Z_{\mathbf{k}_0}\frac{\mathbf{q}\cdot\mathbf{v}}{\omega_{\mathbf{q}}}), \qquad q \to 0, \tag{9}$$

with $\mathbf{q} = \mathbf{k} - \mathbf{k}_0$ (or $\mathbf{k} - \mathbf{k}_0 - \mathbf{Q}$) and $\mathbf{v} = t(\sin k_{0x}, \sin k_{0y})$.

A surprising observation is that the EMD exhibits for momenta $\mathbf{k} \sim \mathbf{k}_0, \mathbf{k}_0 + \mathbf{Q}$ a discontinuity $\sim Z_{\mathbf{k}_0}N^{1/2}$ and $\delta n_{\mathbf{k}}^{(1)} \propto -(1 + \text{sign}\, q_x)/q_x$. These discontinuities are consistent with ED results. One can interpret this result as an indication of an emerging *large* Fermi surface at $\mathbf{k} \sim \pm\mathbf{k}_0$. The discontinuity appears only as *points* $\pm\mathbf{k}_0$, not *lines* in the BZ. Note, however, that this result is obtained in the extreme low doping limit, i.e., $c_h = 1/N$ and it is not straightforward to generalize it to the finite doping regime.

A direct reflection of the anomaly in $\delta n_{\mathbf{k}}$ is the structure of $A_{\mathbf{k}}^+(\omega)$ as presented in Fig. 5(a), where $A_{\mathbf{k}}^+(\omega)$ exhibits a typical asymmetry in momentum dependence. This shows the instability towards the large FS. The electron part is in this figure approximated with $A_{\mathbf{k}}^-(\omega) \sim Z_{\mathbf{k}_0}\delta(\epsilon_{\mathbf{k}} - \epsilon_{\mathbf{k}_0} - \omega)$. In Fig. 5(b) $\delta n_{\mathbf{k}}^{(1)}$ is presented and the anomaly discussed above is clearly seen.

We conclude by stressing that the origin of the pseudo gap found in cuprates seems to be in the short range spin correlations of the reference AFM system, as well as in the strong asymmetry between the hole-like and electron-like spectra

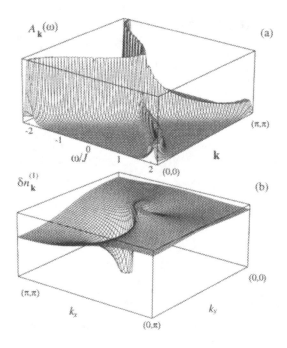

Figure 5   The tendency of formation of the large FS as seen in (a) $A_{\mathbf{k}}(\omega)$ and (b) $\delta n_{\mathbf{k}}^{(1)}$ for $Z_{\mathbf{k}_0} t/J \sim 1$. The figures are in arbitrary units and clipped for convenience.

in underdoped systems. From the present SCBA analysis it is clear that the gap size is a natural consequence of magnons with dispersion $\sim 2J$, and the first peak in $\mathcal{N}^+(\omega)$ thus corresponds to the "van Hove" high density of magnon states. In addition in making contact with ARPES experiments we should note that ARPES measures the imaginary part of the electron Green's function. We must note that using these experiments in underdoped cuprates $n_{\mathbf{k}}$ can be only qualitatively discussed since the latter is extracted only from rather restricted frequency window below the chemical potential. Nevertheless our results are not consistent with a small hole pocket FS (at least only a part of presumable closed FS is visible), but rather with partially developed arcs resulting in FS which is just a set of disconnected segments at low temperature collapsing to the point [1, 3]. The SCBA results for singular $\delta n_{\mathbf{k}}$ seem to allow for such a scenario. It should also be stressed that the SCBA approach is based on the AFM long-range order, still we do not expect that finite but longer-range AFM correlations would entirely change our conclusions.

References

[1] J.C. Campuzano and M. Randeria, this volume, p. 3.

[2] Z.-X. Shen and D.S. Dessau, Phys. Rep. **253**, 1 (1995); B.O. Wells *et al.*, Phys. Rev. Lett. **74**, 964 (1995); D.S. Marshall *et al.*, Phys. Rev. Lett. **76**, 4841 (1996).

[3] M.R. Norman *et al.*, Nature **392**, 157 (1998).

[4] For a review see, e.g., M. Imada, A. Fujimori, and Y. Tokura, Rev. Mod. Phys. **70**, 1039 (1998).

[5] For a review see, e.g., E. Dagotto, Rev. Mod. Phys. **66**, 763 (1994).

[6] W. Stephan and P. Horsch, Phys. Rev. Lett. **66**, 2258 (1991).

[7] R. Eder and Y. Ohta, Phys. Rev. B **57**, R5590 (1998).

[8] A.L. Chernyshev, P.W. Leung and R.J. Gooding, Phys. Rev. B **58**, 13594 (1998).

[9] R. Eder, Phys. Rev. B, **44**, R12609 (1991).

[10] X.-G. Wen and P.A. Lee, Phys. Rev. Lett. **76**, 503 (1996).

[11] W.O. Putikka, M.U. Luchini, and R.R.P. Singh, J. Phys. Chem. Solids **59**, 1858 (1998); Phys. Rev. Lett. **81**, 2966 (1998).

[12] N. Furukawa, T.M. Rice, and M. Salmhofer, Phys. Rev. Lett., **81**, 3195 (1998).

[13] For a review see J. Jaklič and P. Prelovšek, Adv. Phys. **49**, 1 (2000).

[14] P. Prelovšek, J. Jaklič, and K. Bedell, Phys. Rev. B **60**, 40 (1999).

[15] P. Prelovšek, A. Ramšak, and I. Sega, Phys. Rev. Lett. **81**, 3745 (1998).

[16] S. Schmitt-Rink, C.M. Varma, and A.E. Ruckenstein, Phys. Rev. Lett. **60**, 2793 (1988).

[17] A. Ramšak and P. Prelovšek, Phys. Rev. B **42**, 10415 (1990).

[18] G. Martínez and P. Horsch, Phys. Rev. B **44**, 317 (1991).

[19] G.F. Reiter, Phys. Rev. B **49**, 1536 (1994).

[20] A. Ramšak and P. Horsch, Phys. Rev. B **48**, 10559 (1993); *ibid.* **57**, 4308 (1998).

[21] J. Bała, A. M. Oleś, and J. Zaanen, Phys. Rev. B **52**, 4597 (1995).

[22] X. Zotos, P. Prelovšek and I. Sega, Phys. Rev. B **42**, 8445 (1990).

[23] A. Ramšak, P. Prelovšek, and I. Sega, Phys. Rev. B **61**, 4389 (2000).

HIGH RESOLUTION FERMI SURFACE MAPPING OF Pb-DOPED Bi-2212

Sergey V. Borisenko, Mark S. Golden, Thomas Pichler*, Sibylle Legner, Christian Dürr, Martin Knupfer, and Jörg Fink
Institute for Solid State Research, IFW Dresden, Postfach 270016, D-01171 Dresden, Germany
*and Institut für Materialphysik, Universität Wien, Strudlhofgasse 4, A-01090 Wien, Austria

Helmut Berger
Institut de Physique Appliquée, Ecole Politechnique Féderale de Lausanne, CH-1015 Lausanne, Switzerland

Abstract   We illustrate the effectiveness of high resolution Fermi surface (FS) mapping of the Bi-2212 based materials in the investigation of the electronic structure of the the high temperature superconductors. In particular, the advantages of the simultaneous combination of the momentum and energy distribution modes of angle-resolved photoemission will be illustrated in the elucidation of the normal state FS topology of these key systems. We show that the use of unpolarised radiation and modulation-free Pb-doped Bi-2212 samples enables the exclusion of a Γ-centred Fermi surface in these systems. The true Fermi surface topology is, as physical intuition insists, independent of photon energy, hole-like and takes the form of barrels centred at the (π,π) and $(-\pi,\pi)$ points of the two dimensional Brillouin zone.

1.   INTRODUCTION

The development of a widely accepted microscopic theory underlying both the normal state and superconducting properties of the high temperature cuprate superconductors (HTSC) remains, despite more than thirteen years' research by a large community world-wide, one of the outstanding problems in solid state physics. The question of the experimental determination of the electronic structure of the HTSC has been at the forefront of this research activity, as this is one of the main interfaces at which experiment and theory meet most directly, thus providing a test-track for aspiring theories of superconductivity in these systems.

J. Bonča et al. (eds.), Open Problems in Strongly Correlated Electron Systems, 43–50.
© 2001 *Kluwer Academic Publishers. Printed in the Netherlands.*

In the last few years, a number of surprising results have been reported for ARPES experiments carried out at photon energies in the range of 30-35 eV. These include the existence of a FS with missing segments, and additional set of one-dimensional states and, most controversially, the existence of a Γ-centred, electron-like Fermi surface in the Bi-2212 system [1, 2, 3, 4]. If this last point were true, it would constitute a revolution in our thinking about the normal state FS of the HTSC, and we would, in fact, be forced not only to question the authority of ARPES data from the HTSC, but also to throw out all of the theoretical models developed to date and search for new paradigms for the electronic structure of the HTSC materials. This prompted us to re-visit this issue and to experimentally determine the Fermi surface topology in Bi-2212 based HTSC using high resolution Fermi surface mapping.

FS mapping has been shown to be a powerful method of providing a more-or-less direct image of the FS of the HTSC [9]. However, until recently, the information attainable was limited the photoemission intensity in a narrow energy window centred at $E=E_F$. Recently, we extended the FS mapping method by adding to it, apart from k_x and k_y, a third dimension - namely binding energy [5]. In this way, one is able to obtain a detailed dataset from a *single* cleave of a HTSC crystal in which the photoemission intensity is measured as a function of three variables (k_x, k_y and binding energy). In order to be able to display this 4-dimensional dataset on a piece of paper, the following procedure can be followed. Firstly, the ARPES intensity is represented using a grey (or color) scale. We then have a 3D block of data, whose horizontal cross-section gives a constant energy surface - a momentum distribution map (MDM) - which describes the FS if we choose $E=E_F$. An intensity linescan across this surface is called a momentum distribution curve (MDC) [6]. The local maxima of these curves indicate points which belong to the Fermi surface. The right panel of Fig. 1 shows such a data-block from Pb-doped Bi-2212, and how it relates to the bandstructure of the system, which is schematically depicted in the left part of Fig. 1.

The remaining degree of freedom is that of binding energy, and thus a slice down through the block will give an energy distribution map (EDM) which is the (E,k) plot illustrating the band structure. For fixed momentum co-ordinates, a line-scan along the vertical (energy) axis gives a photoemission energy distribution curve (EDC). In this way the concept of using ARPES EDCs to experimentally determine the band structure is clearly exemplified.

The right part of Fig. 1 also shows two instructive and important types of scenario. The sides of the wedge show the dispersion relation of the CuO_2-plane-related band. In one case (ΓX) this clearly crosses E_F, thus identifying the location of a k_F-point. In the other case ($\Gamma \overline{M}$), the band approaches to within about 50 meV of E_F but then does not cross E_F and remains at the

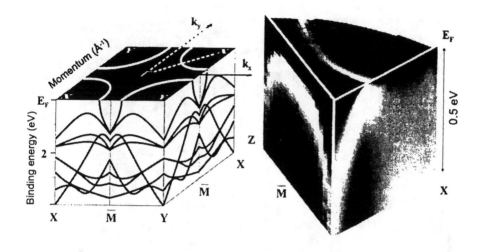

Figure 1 Left: schematic of the band structure of Bi-2212. The black surface plane represents the MDM for E=E_F. The right panel illustrates a three dimensional ARPES dataset in which the intensity is grey-scaled. The position of this data-wedge in the band structure is illustrated in the left panel. The experimentally determined MDM for E=E_F (Fermi surface map) is clearly visible on the upper face of the wedge. For details see text.

same energy over a wide range of momenta - i.e. indicating the presence of the well-known extended saddle point singularity.

2.   FERMI SURFACE MAPS OF Pb-DOPED Bi-2212

In Fig. 2 we show Fermi level MDMs (i.e. constant energy surfaces, in this case with E = E_F) of Pb-doped Bi2212 recorded (left panel) at room temperature and (right panel) at 120 K. Each of these maps is based upon the equivalent of more than 10^3 EDC's per Brillouin zone quadrant, making subsequent massage of the data (by interpolation, smoothing or differentiation, for example) unnecessary. We stress that *every* pixel in such a map represents a particular binding energy in a real EDC underlying the map, whose energy range extends to some 400 meV into the occupied electronic states. A further vital point is the choice of Pb-doped Bi-2212 as this material does not possess the strong incommensurate structural modulation along the b-axis which is characteristic of pristine Bi2212, thus enabling a degree of certainty in the interpretation of the data [5, 7] which is impossible in pristine Bi-2212.

The room temperature FS map clearly illustrates the nature and topology of the main Fermi surface in the Bi-2212 family of materials to be hole-like, having the form of tubes centered around the X,Y points [5], with no indications

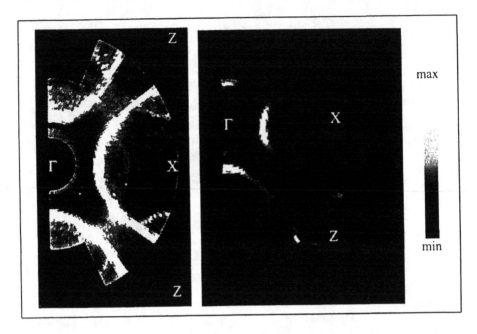

Figure 2   FS maps of Pb-doped Bi2212 recorded at room temperature (left panel) and at 120 K (right panel). The grey scale indicates the photoemission intensity at E_F. The main FS clearly has the topology of rounded tube-sections, and is hole-like, centered upon the X,Y points. The shadow FS is also clearly visible. For details see text.

of a Γ-centered, electron-like FS. These data enable a thorough discussion of the fermiology of the HTSC - a few points of which we are able to mention in the following.

Firstly, it is interesting to note that the Fermi surface does not exhibit strong nesting [the barrel shape is well rounded, without long *parallel* straight sections near $(\pi,0)$].

Secondly, the true FS topology for this system (seen in Fig. 2) is also quite unlike that seen for $La_{1.28}Nd_{0.6}Sr_{0.12}CuO_4$, in which static stripes are known to exist. In this case, long , parallel, straight sections straddling the $(0,0)$-$(\pi,0)$ lines are observed, with reduced intensity along the Brillouin zone diagonal $(0,0)$-(π,π) [8]. We emphasize here that the FS topology is robust with respect to using different criteria for determining k_F or to using different normalization procedures for the EDC's which form the foundation of the FS maps.

A further major point is the existence of a further features, which resemble the shadow Fermi surface (SFS) first seen in the pioneering FS mapping experiments [9]. Previously, these features were explained in terms of copies of the main FS barrels shifted by a vector of (π,π). Our data suggest that the shadow FS are not full barrels, but form lenses enclosing the $(\pi/2,\pi/2)$ point

[5]. The observation of these structures, which go beyond the predictions of band structure calculations, for both Pb-doped and pristine Bi-2212 argues in favor of a non-structural origin. The most compelling explanation is that originally proposed in 1990 [10], and taken up in Ref. [9] that the SFS is due to the the presence of short-range antiferromagnetic correlations, even though these samples are well into the metallic and superconducting regime of the canonical HTSC phase diagramme.

Figure 2 also indicates the use of temperature as a further parameter. It is clear comparing the two FS maps that something dramatic has happened. Although at 120K we are still almost 50K above the superconducting transition the Luttinger theorem FS seen at room temperature has vanished. We now no longer have a continuous FS contour centred around X,Y, but merely FS arcs centred around the nodal $(0,0)$-(π,π) direction [11]. Thus Fig. 2 is a graphical illustration of the opening of the pseudo-gap at certain locations on the FS of the HTSC above T_c, which is felt to be one of the keys to the understanding of the high Tc mechanism and has been suggested to be a consequence of the electrons pairing-up above T_c.

The data of Fig. 2 have illustrated the power of high resolution ARPES operated in the combined FS mapping / EDC mode. We now turn to the controversy mentioned in the introduction regarding the true Fermi surface topology of the Bi-2212 family, which was sparked by measurements carried out with photon energies around 30-35 eV. As already mentioned, these measurements have been intepreted in terms of an electron-like FS centred around the Γ point [1, 2, 3]. Taking a stock-take of what *is* agreed upon, we see firstly that there is a consensus that the 'traditional' FS picture is correct for ARPES data recorded with low photon energies ($h\nu \leq 22eV$), as can be seen clearly in Fig. 2 above [2, 5]. Secondly, the same is true for the use of much higher photon energies (e.g. 50 eV) [2, 7], proving that there is no general problem that lower energy data are somehow singular and thus unrepresentative. Thirdly, with respect to the high symmetry directions in **k**-space, the main FS crossing along the ΓX direction is also generally accepted to be valid for all photon energies used to date.

In an earlier paper, we showed that the ARPES data from pristine Bi-2212 recorded along $(\overline{\Gamma M})$ showed quite different behaviour as a function of photon energy [7]. For $h\nu$=50eV, the characteristic saddle-point behaviour (as can be seen in the right part of Fig. 1 for excitation with 21.2 eV photons) is observed. For $h\nu$=32 and to a slightly lesser extent for $h\nu$=40 eV, the photoemission peak intensity if significantly reduced shortly before the $(\pi,0)$ point, with an intensity increase again after $(\pi,0)$ [7]. It is this kind of behaviour that has been taken as evidence for a main FS crossing along Γ-$(\pi,0)$ in Bi-2212. Thus, 40eV photons appear to be within the 'critical' regime where the data appear to deviate from the saddle-point scenario.

In Refs. [5, 7] we argued that the observed anomalous behaviour along Γ-$(\pi,0)$ is a consequence of the conjunction of two factors. Firstly, sophisticated calculations have shown that the photoemission intensity for the states near $(\pi,0)$ is suppressed for photon energies around 30-35 eV [14]. Secondly, the incommensurate modulation of the Bi-O planes of pristine Bi-2212 results in a highly complex picture around $(\pi,0)$ with the main FS, the SFS and a host of diffraction replicas (DR) of these intrinsic features all overlapping in a small region of k-space. The DR features are caused by the refraction of the outgoing CuO_2 plane photoelectrons as they pass the modulated Bi-O cleavage surface [12].

Thus, combining the use of the FS mapping method presented here (high E, k resolution, unpolarised radiation) with Pb-doped Bi-2212 (no incommensurate Bi-O modulation) we can design a simple yet powerful test of the Γ-centred FS hypothesis. The result is shown in Fig. 3. After first characterising the FS using 21.2 eV photons in the left panel of the figure (hole-like barrels centred at X,Y), we have recorded smaller MDMs with He II radiation ($h\nu =40.8$ eV) centred at the two regions in k-space marked by the dotted lines. As can be seen from Fig. 3, the situation in the FS region centred on the $(0,0)$-(π,π) line is, as ever, uncontroversial - with the main FS being clearly visible.

For the k-space region near to the $(\pi,0)$ point the picture is equally clear - even for a photon energy sensitive to the proposed Γ-centred FS - a DR-free Fermi surface map contains no hint of a FS crossing on or close to the $(0,0)$-$(\pi,0)$ line. Therefore, the FS maps presented in Fig. 3, taken together with the other data presented here offer overwhelming support to arguments that the alleged FS crossings along the $\Gamma \overline{M} Z$ direction in pristine Bi2212 are, in fact, due to extrinsic DR features [5, 13]. These dominate the ARPES spectra as a result of the matrix element-related supression of the saddle-point emission near $\overline{M}$ for photon energies around 30eV [14].

3.   CONCLUSIONS

Here we've given a brief overview of the kind of detailed, high resolution ARPES datasets that can be recorded from a single cleave of a HTSC single crystal. We have stressed the need for a high density mesh in k-space and the use of real experimental data as well as pointing out the distinct advantages of the use of radiation from laboratory sources characterised by a low degree of polarisation.

Taking Pb-doped Bi-2212 as an ideal, simple model system, we have discussed the normal state Fermi surface topology of these systems, including its temperature dependence and photon energy independence. The true Fermi surface topology of the Bi-2212 family is hole-like, taking the form of rounded barrels centred at the corners of the 2D Brillouin zone. Such detailed FS

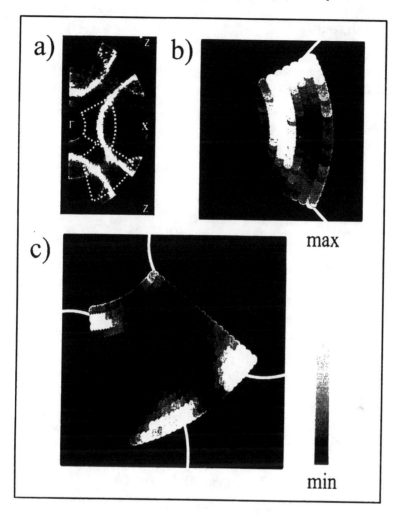

Figure 3 FS maps of Bi-O modulation-free Pb-doped Bi-2212 recorded using unpolarised radiation at room temperature. The grey scale indicates the photoemission intensity in a 20meV window centred at E_F. (a) hν=21.22 eV (He I): The upper [lower] dotted areas indicate the portions of k-space examined with hν=40.8 eV (He II) radiation in panels (b)[(c)]. The He I and He II FS maps were from consecutive cleavages of the same single crystal. Note the complete absence of any sign of a FS crossing along, or near to the $\overline{\Gamma M}$ line in (c).

maps also form the basis for discussion of further aspects of HTSC fermiology. As examples we presented data showing the shadow Fermi surfaces and the destruction of the Luttinger theorem FS for lower temperatures with unprecedented clarity.

Acknowledgments

We are grateful to the the BMBF (05 SB8BDA 6), the DFG (Graduiertenkolleg 'Struktur- und Korrelationseffekte in Festkörpern' der TU-Dresden) and the SMWK (4-7531.50-040-823-99/6) for financial support, to U. Jännicke-Rössler and K. Nenkov for characterisation of the crystals.

References

[1] Y.-D. Chuang *et al.*, Phys. Rev. Lett. **83**, 3717 (1999).

[2] D. L. Feng *et al.*, preprint, cond-mat / 9908056.

[3] A. A. Zakharov *et al.*, Phys. Rev. B **61**, 115 (2000).

[4] N.L. Saini *et al.*, Phys. Rev. B **57**, R11101 (1998).

[5] S. V. Borisenko *et al.*, Phys. Rev. Lett., **84** 4453 (2000).

[6] T. Valla *et al.*, Science **285**, 2110 (1999).

[7] S. Legner *et al.*, Phys. Rev. B, 1 July 2000, in press, cond-mat / 0002302. at $2\pi/\mathbf{a(b)} \approx 1.16\text{Å}^{-1}$. $\overline{M}$ is the midpoint of XY in the 2D BZ. We use the generally accepted shorthand: $\Gamma \equiv (0,0)$, X $\equiv (\pi,-\pi)$, Y $\equiv (\pi,\pi)$, $\overline{M}$ $\equiv (\pi,0)$ and Z $\equiv (2\pi,0)$. Lett. **75**, 1955 (1999). Berlin 1995). **152**, 251 (1988). Relat. Phenom, **76** 127 (1995).

[8] X. J. Zhou et al., Science **286**, 268 (1999).

[9] P. Aebi *et al.*, Phys. Rev. Lett. **72** 2757 (1994).

[10] A. P. Kampf and J. R. Schrieffer, Phys. Rev. B **42**, 7967 (1990).

[11] M. R. Norman *et al.*, Nature **392**, 157 (1998).

[12] H. Ding *et al.*, Phys. Rev. Lett. **76** 1533 (1996).

[13] H.Fretwell *et al.*, Phys. Rev. Lett., **84** 4449 (2000).

[14] A. Bansil and M. Lindroos, Phys. Rev. Lett. **83** 5154 (1999).

SINGLE PARTICLE EXCITATIONS IN THE t-J MODEL

M. Brunner, C. Lavalle, F.F. Assaad, and A. Muramatsu

Institut für Theoretische Physik III, Universität Stuttgart, D-70550 Stuttgart, Germany

Abstract   We present a new quantum Monte Carlo method for the determination of the one-particle propagator in the t-J model. The method can be used both at zero doping, where it is free from the notorious sign problem, and at finite dopings. Here, we present results in two dimensions. As obtained by other numerical methods, we observe a dispersion as predicted by self-consistent Born approximation. We observe an extremely flat band at $\vec{k} = (\pi, 0)$, and a minimum of the dispersion at $\vec{k} = (\pi/2, \pi/2)$. We further show the existence of string excitations by considering the excitations above the quasiparticle peak at $\vec{k} \approx (\pi/2, \pi/2)$. As opposed to the one-dimensional case, the quasiparticle weight is finite in the thermodynamic limit in two dimensions.

The t-J model is one of the paradigmatic models for strongly correlated electrons. On one side, the t-J model can be obtained from the Hubbard model in the large coupling limit. On the other side, it is the relevant one to simulate the cuprates, as shown by Zhang and Rice [1]. Its Hamiltonian is given by

$$H_{t-J} = -t \sum_{<i,j>,\sigma} \tilde{c}_{i,\sigma}^{\dagger} \tilde{c}_{j,\sigma} + J \sum_{<i,j>} \left(\vec{S}_i \cdot \vec{S}_j - \frac{1}{4} \tilde{n}_i \tilde{n}_j \right) , \qquad (1)$$

where $\tilde{c}_{i,\sigma}^{\dagger}$ are projected fermion operators $\tilde{c}_{i,\sigma}^{\dagger} = (1 - c_{i,-\sigma}^{\dagger} c_{i,-\sigma}) c_{i,\sigma}^{\dagger}$, $\tilde{n}_i = \sum_{\alpha} \tilde{c}_{i,\alpha}^{\dagger} \tilde{c}_{i,\alpha}$, $\vec{S}_i = (1/2) \sum_{\alpha,\beta} c_{i,\alpha}^{\dagger} \vec{\sigma}_{\alpha,\beta} c_{i,\beta}$, and the sum runs over nearest neighbors only.

A very successful QMC approach for Hubbard models are the so called projector and grand canonical Monte Carlo algorithms [2]. The basic idea in both cases is the introduction of auxiliary bosonic fields to transform the Hamilton operator to a form, where the fermions only appear in bilinear form, which is equivalent to noninteracting fermions coupling to the bosonic field.

In the case of the t-J model we follow a different *Ansatz*, leading to the hybrid loop algorithm. Instead of introducing auxiliary bosonic fields, we perform a canonical transformation [3] given by $c_{i\uparrow}^{\dagger} = \gamma_{i,+} f_i - \gamma_{i,-} f_i^{\dagger}$, $c_{i\downarrow}^{\dagger} = \sigma_{i,-}(f_i + f_i^{\dagger})$, where $\gamma_{i,\pm} = (1 \pm \sigma_{i,z})/2$ and $\sigma_{i,\pm} = (\sigma_{i,x} \pm i\sigma_{i,y})/2$. The spinless fermion operators fulfill the canonical anticommutation relations $\{f_i^{\dagger}, f_j\} =$

J. Bonča et al. (eds.), Open Problems in Strongly Correlated Electron Systems, 51–56.
© 2001 *Kluwer Academic Publishers. Printed in the Netherlands.*

$\delta_{i,j}$, and $\sigma_{i,a}$, $a = x, y$, or z, are the Pauli matrices. The Hamiltonian becomes

$$\tilde{H}_{t-J} = +t \sum_{\langle i,j \rangle} P_{ij} f_i^\dagger f_j + \frac{J}{2} \sum_{\langle i,j \rangle} \Delta_{ij}(P_{ij} - 1), \qquad (2)$$

where $P_{ij} = (1 + \vec{\sigma}_i \cdot \vec{\sigma}_j)/2$, $\Delta_{ij} = (1 - n_i - n_j)$, and $n_i = f_i^\dagger f_i$.

In order to study the dynamics of the holes we calculate the time displaced one-particle Green's function for spin up,

$$G(i - j, \tau) = -\langle T \tilde{c}_{i,\uparrow}(\tau) \tilde{c}_{j,\uparrow}^\dagger \rangle, = -\langle T f_i^\dagger(\tau) f_j \rangle \qquad (3)$$

where T corresponds to the time ordering operator.

It is now possible to address two different problems: the dynamics of a single hole in an antiferromagnet and the physics of the model at finite doping (both thermodynamic and dynamic quantities). The first step in both cases is a so called checkerboard decomposition, where the Hamiltonian is split up in commuting terms $H_1, \ldots H_p$ and a Trotter decomposition of the partition sum. Inserting complete sets of spin states in the case of zero doping, Eq. (3) transforms as

$$
\begin{aligned}
G(i - j, -\tau) &= \frac{\sum_{\sigma_1} \langle v| \otimes \langle \sigma_1 | e^{-(\beta - \tau)\tilde{H}_{t-J}} f_j e^{-\tau \tilde{H}_{t-J}} f_i^\dagger | \sigma_1 \rangle \otimes |v\rangle}{\sum_{\sigma_1} \langle \sigma_1 | e^{-\beta \tilde{H}_{t-J}} | \sigma_1 \rangle} \\
&= \sum_{\vec{\sigma}} P(\vec{\sigma}) G(i, j, \tau, \vec{\sigma}) + \mathcal{O}(\Delta \tau^2). \qquad (4)
\end{aligned}
$$

Here $m\Delta\tau = \beta$, $n\Delta\tau = \tau$, $\Delta\tau t \ll 1$, and $\exp(-\Delta\tau \tilde{H}(\sigma_1, \sigma_2))$ is the evolution operator for the holes, given the spin configuration (σ_1, σ_2). $|v\rangle$ is the vacuum state for holes, and $P(\vec{\sigma})$ is the probability distribution of a Heisenberg antiferromagnet for the configuration $\vec{\sigma}$, where $\vec{\sigma}$ is a vector containing all intermediate states $(\sigma_1, \ldots \sigma_n, \ldots, \sigma_1)$. The sum over spins is performed in a very efficient way by using a world-line loop-algorithm [4] for a Heisenberg antiferromagnet with discretized imaginary time. As the evolution operator for the holes is a bilinear form in the fermion operators, $G(i, j, \tau, \vec{\sigma})$ can be calculated exactly, in contrast to a direct implementation in the loop algorithm [5] where fermion paths are sampled stochastically. $G(i, j, \tau, \vec{\sigma})$ contains a sum over all possible fermion paths between $(i, 0)$ and (j, τ). Since at half filling $P(\vec{\sigma})$ is the probability distribution for the quantum antiferromagnet, the algorithm does not suffer from sign problems on bipartite lattices and non-frustrated magnetic interactions in any dimension. For more details on the above algorithm see Refs. [6, 7].

In the case of finite doping the trial wave function, that in the previous case was $|\Psi_T\rangle = \sum_{\sigma_1} |\sigma_1\rangle |v\rangle$, is substituted by $|\Psi_T\rangle = \sum_{\sigma_1} |\sigma_1\rangle |\Psi_{Tf}\rangle$

where $| \Psi_{Tf} \rangle$ is a product of one-particle states. On one side the fermionic dynamic is still fixed by the spin background which is simulated via the loop algorithm. On the other side we now have a non trivial trial wave function for the fermions, for which the ground state -given the spin background- has to be filtered out.

Let $| \Psi_{Tf} \rangle = \prod_{l=1}^{N_p} [\sum_{j=1}^{L} P_{jl} f_j^\dagger] | v \rangle$ where N_p is the number of holes and L the number of sites in the system. We then obtain for every realization of the spin field σ_n:

$$
\langle \Psi\{\sigma_n\} \mid \Psi\{\sigma_n\} \rangle = \langle v | [\prod_{k,l=1}^{N_p} \sum_{i,j=1}^{L} P_{ik}^T f_i P_{jl}^e f_j^\dagger] | v \rangle
$$

$$
= \det[P^T B_M^\sigma .... B_{M/2}^\sigma .... B_1^\sigma P] \tag{5}
$$

where the notation stands for $P_{jl}^e \equiv \prod_{n=1}^{M/2} e^{-\Delta_T H(\sigma_n)} P_{jl} \equiv \prod_{n=1}^{M/2} B_n^\sigma P_{jl}$.

From this follows that the weight of each configuration is given by

$$
\langle \Psi \mid \Psi \rangle = \sum_{\{\vec{\sigma}\}} P(\vec{\sigma}) \det[P^T B_M^\sigma .... B_{M/2}^\sigma .... B_1^\sigma P]. \tag{6}
$$

So we have finally a two level simulation: First the loop algorithm determines the spin dynamics. Once that the spin configuration is proposed the integration of the fermions is done via determinantal algorithms [2]. In one dimension, the algorithm reproduces known results obtained by exact diagonalizations [8].

The quasiparticle dispersions and the quasiparticle weight can directly be extracted from the Green's function; for the full spectral function the *Maximum Entropy* method is used. Details about *Maximum Entropy* can be found in the comprehensive review article by J.E. Gubernatis and M. Jarrell [9].

We now address the dynamics of a single hole in two dimensions. Quite recently, the dynamics of a single hole in an antiferromagnetic two-dimensional background became experimentally accessible by angle resolved photoemission spectroscopy (ARPES) experiments in undoped materials like $Sr_2CuO_2Cl_2$ [10, 11] and $Ca_2CuO_2Cl_2$ [12]. The main features observed there are a minimum of the dispersion at $\vec{k} = (\pi/2, \pi/2)$ together with a vanishing of spectral weight beyond this point along the (1,1) direction. The obtained spectra show that the very flat portion around $(\pi, 0)$, that in optimally doped materials is almost degenerate with the bottom of the spectrum at $(\pi/2, \pi/2)$ [13], is shifted upwards (in a hole representation) by approximately 300 meV. This contradicts the single hole spectra found theoretically so far, where essentially the lower edge of the spectrum at $\vec{k} = (\pi/2, \pi/2)$ and $(\pi, 0)$ are almost degenerate, such that additional second and third nearest neighbor hopping terms were suggested [11, 14].

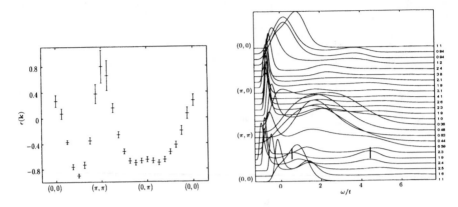

Figure 1   Quasiparticle dispersion and full spectral function in a 16×16 lattice for $J/t = 0.6$.

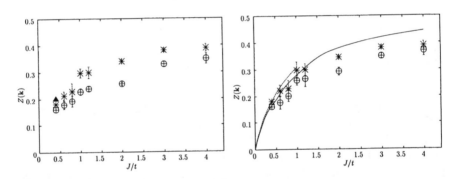

Figure 2   Quasiparticle weight in the thermodynamic limit, and for a 16×16 lattice. The comparison of the results in the finite lattice with self consistent Born approximation show a good qualitative agreement. The upper curves correspond to $\vec{k} = (\pi, 0)$, the lower curve represent $\vec{k} = (\pi/2, \pi/2)$.

In our approach, the lower edge of the spectrum is obtained directly from the asymptotic form of the imaginary time Green's function. The resulting dispersion agrees with previous results obtained within SCBA [16] and series expansions [15] for $J/t < 1$, whereas for $J/t > 1$ only agreement with series expansions is found (see Fig. 1). In particular, a flat dispersion is obtained around $\vec{k} = (\pi, 0)$ very close in value to the bottom of the band at $\vec{k} = (\pi/2, \pi/2)$, in contrast to the experiments [10, 11, 12, 17].

The asymptotics of the imaginary time Green's function delivers also the quasiparticle weight for that band. Results of finite size scaling is presented in Fig. 2 showing that $Z(\vec{k})$ is finite for the parameter range considered, such that the lower edge of the spectrum corresponds to a coherent quasiparticle. This is in agreement with SCBA [16, 18], but clearly disproves other analytical

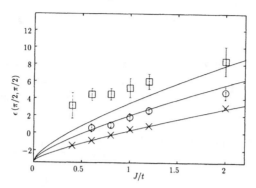

Figure 3 The first three excitations at the minimum of the dispersion $\vec{k} = (\pi/2, \pi/2)$. The lines represent the solutions for the linear string potential for one hole in the t-J_z model [7].

approaches and *Ansätze* [19], which predicted a vanishing quasiparticle weight in the thermodynamic limit.

The spectral function $A(\vec{k}, \omega)$ is calculated by analytic continuation with *Maximum Entropy* [9]. Overall agreement of our results is found with exact diagonalizations. At the supersymmetric point $J/t = 2$, the delta function predicted by Sorella [20, 6, 7] for the wave vector $\vec{k} = (0, 0)$ is exactly reproduced. By extracting the contribution of the quasiparticle from the imaginary time Green's function, a resonance above the quasiparticle band is made evident, that together with the lower edge of the spectrum scales as $a_n J^{2/3}$, in agreement with the string picture [21] used to described the excitations for a hole in an antiferromagnetic Ising-background. Remarkably, also the factors $a_n = 2.33, 4.08$ of the corresponding Airy functions are needed in order to properly describe the distance between the resonance and the quasiparticle band, as shown in Fig. 3. For $J/t \to 0$ our scaling is consistent with the result from SCBA, which gives $E_0(J \to 0) = -3.28t$. The third peak which can be resolved cannot be explained by the string picture, as its distance to the lower edge is independent of J and has a value of about $4t$.

Acknowledgments

This work was supported by Sonderforschungsbereich 382 in Tübingen-Stuttgart. The numerical calculations were performed at HLRS Stuttgart.

References

[1] F. Zhang and T. M. Rice, Phys. Rev. B **37**, 3759 (1988).

[2] A. Muramatsu, in *Quantum Monte Carlo Methods in Physics and Chemistry*, edited by M. P. Nightingale and C. J. Umrigar (Kluwer Academic Press, Dordrecht, 1999).

[3] G. Khaliullin, JETP Lett. **52**, 389 (1990); A. Angelucci, Phys. Rev. B **51**, 11580 (1995).

[4] H. G. Evertz, M. Marcu, and G. Lana, Phys. Rev. Lett **70**, 875 (1993).

[5] N. V. Prokof'ev, B. V. Svistunov, and I. S. Tupitsyn, JETP **87**, 310 (1998); R. Brower, S. Chandrasekharan, and U.-J. Wiese, Physica A **3-4**, 520 (1998).

[6] M. Brunner, F. F. Assaad, and A. Muramatsu, to be published in Eur. Phys. J. B.

[7] M. Brunner, F. F. Assaad, and A. Muramatsu, cond-mat/0002321; M. Brunner, Ph.D. thesis, Universität Stuttgart, 2000.

[8] K. von Szczepanski, P. Horsch, W. Stephan, and M.Ziegler, Phys. Rev. B **41**, 2017 (1990).

[9] M. Jarrell and J. Gubernatis, Phys. Rep. **269**, 133 (1996).

[10] B. O. Wells *et al*, Phys. Rev. Lett. **74**, 964 (1995).

[11] C. Kim *et al*, Phys. Rev. Lett. **80**, 4245 (1998).

[12] F. Ronning *et al* Science **282**, 2067 (1998).

[13] D. Marshall *et al*, Phys. Rev. Lett. **76**, 4841 (1996).

[14] T. Tohyama *et al*, J. Phys. Soc. Jpn. **69**, 9 (2000).

[15] C. J. Hamer, Z. Weihong, and J. Oitmaa, Phys. Rev. B **58**, 15508 (1998).

[16] G. Martinez and P. Horsch, Phys. Rev. B **44**, 317 (1991); Z. Liu and E. Manousakis, Phys. Rev. B **45**, 2425 (1992).

[17] S. LaRosa *et al*, Phys. Rev. B **56**, R525 (1997).

[18] A. Ramšak and P. Horsch, Phy. Rev. B **57**, 4308 (1998).

[19] Z. Y. Weng *et al*, Phys. Rev B **55**, 3894 (1997); P. Anderson, Phys. Rev. Lett **64**, 1839 (1990).

[20] S. Sorella, Phys. Rev. B **53**, 15119 (1996).

[21] L. N. Bulaevskii, E. L. Nagaev, and D. I. Khomskii, JETP **27**, 836 (1968); C. L. Kane, P. A. Lee, and N. Read, Phys. Rev. B **39**, 6880 (1989); B. Shraiman and E. Siggia, Phys. Rev. Lett. **60**, 740 (1988).

II
CUPRATES:
SPIN AND CHARGE
FLUCTUATIONS

MAGNETIC RESONANCE PEAK AND NONMAGNETIC IMPURITIES

Y. Sidis[1], P. Bourges[1], B. Keimer[2], L. P. Regnault[3], J. Bossy[4], A. Ivanov[5], B. Hennion[1], P. Gautier-Picard[1], and G. Collin[1]

[1] *Laboratoire Léon Brillouin, CEA-CNRS, CE-Saclay, 91191 Gif sur Yvette, France.*

[2] *Max-Planck-Institut für Festkörperforschung, 70569 Stuttgart, Germany.*

[3] *CEA Grenoble, Département de Recherche Fondamentale sur la Matière Condensée, 38054 Grenoble cedex 9, France.*

[4] *CNRS-CRTBT, BP 156, 38042 Grenoble cedex 9, France.*

[5] *Institut Laue Langevin, 156X, 38042 Grenoble cedex 9, France.*

Abstract   Nonmagnetic Zn impurities are known to strongly suppress superconductivity. We review their effects on the spin excitation spectrum in $YBa_2Cu_3O_7$, as investigated by inelastic neutron scattering measurements.

1.   INTRODUCTION

In optimally doped $YBa_2Cu_3O_{6+x}$, the spin excitation spectrum is dominated by a sharp magnetic excitation at an energy of ~ 40 meV and at the planar antiferromagnetic (AF) wave vector $(\pi/a, \pi/a)$, the so-called magnetic resonance peak [1, 2, 3, 4, 5, 6]. Its intensity decreases with increasing temperature and vanishes at T_c, without any significant shift of its characteristic energy E_r. In the underdoped regime, E_r monotonically decreases with decreasing hole concentration [7, 8, 9, 10], so that $E_r \simeq 5\,k_B T_c$. Besides, it is possible to vary T_c without changing the carrier concentration through impurity substitutions of Cu in the CuO_2 planes. In $YBa_2(Cu_{0.97}Ni_{0.03})_3O_7$ (T_c=80 K), the magnetic resonance peak shifts to lower energy with a preserved E_r/T_c ratio [11].

In optimally doped $Bi_2Sr_2CaCu_2O_{8+\delta}$ (T_c=91 K), a similar magnetic resonance peak has been recently observed at 43 meV [12]. Furthermore, E_r shifts down to 38 meV in the overdoped regime (T_c=80 K) [13], preserving a constant ratio with T_c: $E_r \simeq 5.4\,k_B T_c$. Thus, whatever the hole doping, the energy position of the magnetic resonance peak always scales with T_c.

In underdoped $YBa_2Cu_3O_{6+x}$ (x=0.6,T_c=63 K, E_r=34 meV), recent INS measurements provide evidence for incommensurate-like spin fluctuations at 24

J. Bonča et al. (eds.), Open Problems in Strongly Correlated Electron Systems, 59–68.
© *2001 Kluwer Academic Publishers. Printed in the Netherlands.*

meV and low temperature (seemingly similar to those observed $La_{2-x}Sr_xCuO_4$) [14, 15]. These incommensurate-like spin fluctuations are also observed at higher oxygen concentrations: x=0.7 [16, 17], x=0.85 [18]. As a function of temperature [16, 18] and energy [18], the incommensurability increases below T_c with decreasing temperature and decreases upon approaching E_r in the superconducting state. The results point towards an unified description of both incommensurate spin excitations and magnetic resonance peak in terms of an unique (dispersive) collective spin excitation mode, as predicted in Ref. [19].

In this paper, we review effects of nonmagnetic Zn impurities on the magnetic resonance peak in $YBa_2Cu_3O_7$. Among all candidates for substitution to Cu in the CuO_2 planes of $YBa_2Cu_3O_7$, nonmagnetic Zn^{2+} ions ($3d^{10}$, S=0) induce the strongest T_c reduction ($\sim$ -12 K/% Zn) [20, 21]. Furthermore, low Zn substitution preserves the doping level and introduces only minimal structural disorder. We compare the spin excitation spectra reported from inelastic neutron scattering (INS) measurements performed in $YBa_2(Cu_{1-y}Zn_y)_3O_7$ for various Zn/Cu substitution rates [6, 11, 22, 23, 24] (Characteristics of single crystals used for INS measurements are listed in Table 1). Through Zn substitution, the magnetic resonance peak magnitude strongly decreases and its energy position slightly shifts to lower energy, so that the ratio E_r/T_c increases. In contrast to the Zn free system, where the normal state magnetic response is not experimentally discernible, nonmagnetic impurities restore or enhance AF spin fluctuations above T_c and up to $\sim$250 K.

2.   INS MEASUREMENTS

Throughout this review, the wave vector **Q** is indexed in units of the reciprocal tetragonal lattice vectors $2\pi/a = 2\pi/b = 1.63$ Å^{-1} and $2\pi/c = 0.53$ Å^{-1}. In this notation the $(\pi/a, \pi/a)$ wave vector parallel to the CuO_2 planes corresponds to points of the form (h/2,k/2) with h and k odd integers. Because of the well known intensity modulation of the low energy spin excitations due to interlayer interactions [1, 2, 3, 4, 5, 6], data were taken close to L=1.7 l where l is an odd integer.

In pure $YBa_2Cu_3O_7$, the magnetic resonance peak appears at $E_r \sim$40 meV [4]. Figure 1.a shows the difference between two energy scans, performed with a momentum transfer fixed at Q_{AF}=(1.5,0.5,1.7). The former is measured deep in the superconducting state and the latter, in the normal state, close to T_c. The magnetic resonance peak gives rise to a positive contribution to the difference spectrum, with a maximum at E_r (Fig. 1.a). Besides, a negative difference at low energy stems from phonon scattering (determined independently through constant-energy scans). The magnetic intensity has been converted to the imaginary part of the dynamical magnetic susceptibility, χ", after correction by the detailed balance and magnetic form factors and calibrated against optical

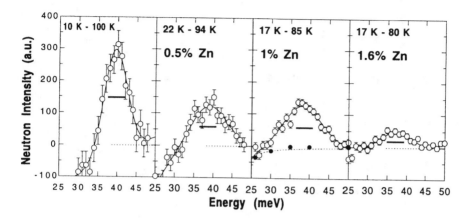

Figure 1   Difference spectrum of the neutron intensities at low temperature and T≥ T$_c$, in YBa$_2$(Cu$_{1-y}$Zn$_y$)$_3$O$_7$ (open symbols): a) y=0 [4] , b) y=0.005 [22], c) y=0.01 [11], d) y=0.016. Measurements have been normalized against phonons. Full symbols correspond to the reference level of magnetic scattering and is determined from the difference of constant-energy scans at both temperatures. This level becomes slightly negative with decreasing energy owing to the thermal enhancement of the nuclear background. Dotted lines are guides to the eye.

phonons according to a standard procedure [22]. The maximum at E$_r$ in Fig. 1.a then corresponds to an enhancement of χ" at the AF wave vector (hereafter $\Delta\chi$"($\mathbf{Q}_{AF}$, E$_r$)) of $\sim$ 300 μ_B^2.eV^{-1}. A fit to a Gaussian profile of the positive part of the difference spectrum (Fig 1) provides an estimate of the energy distribution of $\Delta\chi$"($\mathbf{Q}_{AF}$, E) around E$_r$. The full width at half maximum of the difference spectrum is ΔE $\sim$6 meV, of the same order of magnitude as the instrumental resolution ($\sim$5 meV), yielding an intrinsic energy width of at most $\sim$3 meV. In YBa$_2$Cu$_3$O$_7$, the magnetic resonance peak is thus almost resolution limited in energy [3, 4, 6]. The temperature dependence of χ"($\mathbf{Q}_{AF}$,40 meV), shown in Fig. 2.a [5], exhibits a marked change at T$_c$, and an order-parameter-like curve in the superconducting state: the telltale signature of the resonance peak. Above T$_c$, magnetic fluctuations are not sizeable anymore. According to Ref. [4, 16], the magnitude of spin fluctuations left above T$_c$ cannot exceed $\sim$70 μ_B^2.eV^{-1}. At 40 meV, the ratio, R, between the intensities of AF spin fluctuations above T$_c$ and at low temperature ranges from 0 to $\sim$ 20% (see Table. 1).

The same kind of INS measurements have been performed on Zn substituted YBa$_2$Cu$_3$O$_7$ samples. The difference spectrum of neutron intensities for each Zn content is determined from energy scans performed at wave vector $\mathbf{Q}_{AF}$=(-1.5,-0.5,1.7), following the procedure described above for the Zn free sample. For each Zn content, the difference spectra still exhibit a positive maximum

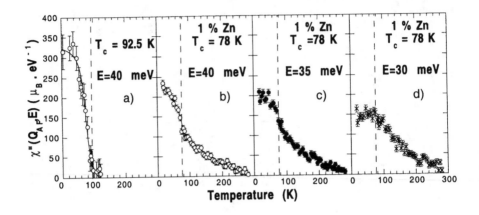

Figure 2 Temperature dependence of $\chi''(\mathbf{Q}_{AF}, E)$ in $YBa_2(Cu_{1-y}Zn_y)_3O_7$: a) y=0 [5], E=40meV, b) y=0.01 [11], E=40 meV, E, c)y=0.01, E=35 meV, d) y=0.01, E=30 meV. Data are given in absolute units. A ~30% overall in absolute unit calibration is not included in the error bars. Solid lines are guides to the eye.

around ~40 meV (Fig. 1), that accounts for an enhancement of the magnetic response at low temperature. The intensity at the maximum drops down with increasing Zn substitution (Table 1). Nonmagnetic impurities, that strongly reduce T_c, therefore significantly weaken the enhancement of the magnetic response in the superconducting state, ascribed to the magnetic resonance peak. The energy position of the maximum slightly moves to lower energy, giving rise to a progressive increase of the ratio E_r/T_c from ~ 5 to ~ 6. In addition, the energy distribution ΔE (~8 meV) broadens with Zn substitution, as compared to the difference spectrum of the Zn free sample (Fig. 1.a). The magnetic resonance peak thus exhibits an intrinsic energy width that accounts for a disorder induced broadening [11, 22].

In $YBa_2(Cu_{0.99}Zn_{0.01})_3O_7$, the temperature dependence of $\chi''(\mathbf{Q}_{AF}, 40meV)$ shows an upturn at T_c and displays remnants of an order parameter lineshape in the superconducting state, that characterize the magnetic resonance peak (Fig. 2.b). The change of slope at T_c is hardly visible in the 0.5% and 2% Zn substituted samples [6, 22] (where data quality is not as high). Indeed, the hallmark of the magnetic resonance peak in the temperature dependence of χ'' is partially scrambled by AF spin fluctuations in the normal state which are enhanced or restored by Zn. These fluctuations persist up to ~250 K (Fig. 2). Close to 40 meV, their relative weight with respect to the magnetic intensity at low temperature, R, increases with increasing Zn substitution (Table 1). Notice that the slight anomaly at T_c in the 1% Zn substituted sample is visible only because of the high quality of the data and that the improvement of data quality in other Zn substituted samples could reveal the same feature.

The temperature dependence of $\chi''(\mathbf{Q}_{AF}, E)$ has been measured at 30 meV and 35 meV in $YBa_2(Cu_{0.99}Zn_{0.01})_3O_7$ (Fig. 2.c-d). In this system, the magnetic resonance peak appears precisely at 38 meV ($\Delta E \simeq 8$ meV) and the enhancement of the magnetic response around T_c can be observed in the temperature dependences of χ'' at 35 meV and 40 meV (Fig. 2.b-d). On the contrary, $\chi''(\mathbf{Q}_{AF}, 30\text{meV})$ saturates or even slightly decreases below T_c. A detailed analysis of Fig. 2.b-d reveals that the intensity of the magnetic response left in the normal state is actually larger at 30 and 35 meV than at 40 meV. This implies a possible redistribution of the magnetic spectral weight in the normal state.

Fig. 3.a-b show constant-energy scans at 40 meV in the (H,H/3,1.7) zone at 17 K and 275 K. At low temperature, the magnetic response displays a Gaussian momentum distribution centered at the AF wave vector, on top of a background that is slightly curved due to a contribution from phonons. At 275 K, the magnetic response is not sizeable anymore (Fig. 3.a) and an energy scan performed at $\mathbf{Q}_{AF}=(-1.5,-0.5,1.7)$ characterizes the energy dependence of the background at high temperature (Fig. 3.c). In the energy range E=30-50 meV, its lineshape is well approximated by a third order polynomial fit of data at {30, 35, 40, 50} meV. At lower temperature, the same fit of background intensities determined from a set of constant-energy scans at {30, 35, 40, 50}meV defines an effective background (Fig. 3.d). Its subtraction from the raw intensity leads to the magnetic excitation spectrum (open symbols in Fig. 3.e). Figure 3.e shows the magnetic excitation spectrum at $\mathbf{Q}_{AF}$ from 30 to 50 meV in $YBa_2(Cu_{0.99}Zn_{0.01})_3O_7$: in the superconducting state (17 K), close to T_c (85 K) and well above T_c (200 K). In the normal state, the maximum of $\chi''(\mathbf{Q}_{AF}, E)$

Table 1 $YBa_2(Cu_{1-y}Zn_y)_3O_7$ single crystals used in INS measurements: (a) Zn content,y, (b) volume,V, (c) superconducting critical temperature, T_c. The samples were heat-treated to achieve full oxygenation and the Zn/Cu substitution rate was deduced from the reduction of T_c as compared to the pure system. In each sample, the magnetic resonance peak can be characterized by the following parameters: (d) the energy position, E_r, (e) the enhancement, between T→ 0 and T_c, of dynamical spin susceptibility at E_r and $\mathbf{Q}_{AF}$, $\Delta\chi''(\mathbf{Q}_{AF}, E_r)$. R corresponds to the ratio between AF intensities left just above T_c and at low temperature: $R=\chi''(\mathbf{Q}_{AF}, E_r)_{T_c} / \chi''(\mathbf{Q}_{AF}, E_r)_{T\to 0}$. R is given at 39 or 40 meV. (for further details, see text)

y (%)	V (cm^3)	T_c (K)	E_r (meV)	$\Delta\chi''$ ($\mu_B^2.eV^{-1}$)	R (%)	Ref.
0	10	93	40	~300	≤20	[4]
0.5	1.7	87	39	~130	~50	[22]
1	~2	78	38	~130	~50	[11]
1.6	~2	73	37	~50	-	-
2	0.2	69	-	-	≥70	[23, 6]

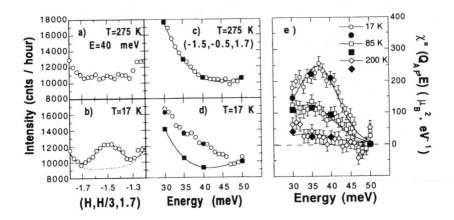

Figure 3 $YBa_2(Cu_{0.99}Zn_{0.01})_3O_{7-\delta}$. Constant energy scans at 40 meV in the (H,H/3,1.7) zone: a) 275 K, b) 17 K). Energy scan at the wave vector (-1.5,-0.5,1.7): c) 275 K, d) 17 K. Full circles and squares account for the magnetic intensity and background intensities determined from constant energy scans at different energies. The lineshape of the background is fitted to a third order polynomial function. e) $\chi''(Q_{AF}, E)$ in the energy range 30-50 meV, at different temperatures. The shaded area corresponds to $\Delta\chi''(Q_{AF}, E)$ reported in Fig.1.c. Solid lines are guides to the eye.

moves inside the energy range 30-35 meV, whereas the maximum intensity is still peaked around ~38 meV in the superconducting state.

We can summarize the experimental observations in $YBa_2(Cu_{1-y}Zn_y)_3O_7$ as follows. In the superconducting state, the magnetic resonance peak broadens in energy and slightly moves to lower energy, but remains located close to ~40 meV, the energy position of the resonance peak in pure $YBa_2Cu_3O_7$. A broad peak with a characteristic energy comparable to (but somewhat lower than) the energy of the resonance peak appears in the normal-state response of Zn-substituted systems. While the normal state AF spin fluctuations develop with increasing Zn substitution, the enhancement of the magnetic response, associated with magnetic resonance peak in the superconducting state, fades away.

3.   DISCUSSION AND CONCLUSION

A comparison of our measurements in the pure YBCO system with current models of the spin dynamics in cuprates is given in [10]. However, most of these models do not incorporate disorder. Here we restrict the discussion to theoretical works where the interplay between (collective) spin excitations and quantum impurities in high temperature superconductors is expressly considered.

In BCS d-wave superconductors, nonmagnetic impurities cause the decay of the quasi-particle states due to a strong scattering rate (close to the unitary limit), and then, give rise to a pair breaking that reduces strongly the superconducting order parameter [25, 26, 27, 28, 29, 30, 31, 32, 33, 34]. The resonant scattering by non magnetic impurities qualitatively account for most of the Zn substitutions effects in High-T_c superconductors: i) the strong alteration the bulk superconducting properties, such as the critical temperature [20, 21] and the superfluid density [35, 36], ii) the increase of the in-plane residual resistivity, iii) the reduction of the microwave surface resistance [37] and iv) the appearance of a finite density of state at the Fermi level below T_c [38]. Therefore, as a consequence of the reduction of the superconducting order parameter, the threshold of the electron hole spin flip continuum at the AF wave vector moves in principle to low energy as, in the superconducting state, that threshold energy is basically proportional to twice the maximum of the d-wave gap. Thus, in any models where the resonance appears at or below the continuum threshold, nonmagnetic impurities should lead to a shift of the resonance to lower energy and the occurrence of damping (so that no clear resonance is observed at large impurity concentrations) [39, 40]. These results provide an explanation for the broadening of the magnetic resonance peak. Similar broadening can also occur due to disorder in the paramagnetic state of quantum antiferromagnet [41]. However, the magnitude of the E_r-renormalization strongly depends on the model used to account for the magnetic resonance peak in the Zn-free system, and then, would be important to discriminate between the different models for the magnetic resonance peak. Furthermore, the strong scattering by non magnetic impurities, so crucial in the superconducting state, also modifies the normal state properties. In the normal state, the effect of nonmagnetic impurities on the the spin fluctuation spectral weight at the AF wave vector has been studied in the framework of the the 2D Hubbard model using the random phase approximation [42]. The main effect of dilute impurities on the noninteracting dynamical spin susceptibility is a weak smearing. On the contrary, for an interacting system, the scattering of spin fluctuations by the (static and extended [43]) impurity potential with a finite momentum transfer ("umklapp" processes) becomes essential. Indeed, the $(\pi/a, \pi/a)$ spin fluctuations become mixed with other wave vector components, and a new peak in $\chi"(\mathbf{Q}_{AF}, \omega)$ can appear.

Scanning tunneling microscopy (STM) in Zn substituted $Bi_2Sr_2CaCu_2O_{8+\delta}$ [44] confirms the existence of a strong quasi-particle scattering rate by impurities. Indeed, STM shows intense quasi-particle scattering resonances at Zn sites, coincident with strong suppression of the superconducting coherence peaks at the Zn site. Furthermore, the superconducting peaks then are progressively restored over a distance, ξ, of about 15 Å. STM supports the proposal that the superfluid density reduction can be explained by non-superconducting regions of area $\sim \pi \xi^2$ around each impurity atoms, the so-called "Swiss cheese" model, introduced to account for the decrease of the superconducting condensate density from μSR measurements [35]. On a phenomenological level, our data are actually consistent with this scenario in which Zn impurities are surrounded by extended regions in which superconductivity never develops [35, 44]. Within this picture, superconductivity is then confined to (perhaps only rather narrow) regions far from the Zn impurities. This would explain why Zn impurities all but eradicate the effect of superconductivity on the spin excitations which is so readily apparent in the pure system. Since INS is a bulk measurement and the magnetic resonance peak is an intrinsic feature of the superconducting state, one may speculate that its intensity may be suppressed as the fraction of the system that becomes superconducting and, thus, may scale with the superfluid condensate density. In addition, in non-superconducting regions around Zn impurities, magnetic properties are strongly modified already far above T_c. According to nuclear magnetic resonance measurements, local magnetic moments develop on Cu sites surrounding a Zn impurity (up to the third nearest neighbors) [45].

The modification of the local magnetic properties and the resonant scattering from Zn impurities arise in a natural way from the strong correlation of the host as shown in exact diagonalizations of small clusters performed on the framework of the t-J model [46, 47, 43, 48, 49]. According to these calculations, electrons form a bound state around the impurity site and the magnitude of the local moment is enhanced, as observed experimentally [45, 21]. When J/t becomes larger than ~ 0.3, a mobile hole is trapped by the impurity potential induced by the local distortion of the AF background. Below T_c, a pair breaking effect occurs due to the binding of holes to the impurity. Likewise, a new magnetic excitation corresponding to the singlet-triplet excitation of the singlet impurity-hole bound state is predicted [49]. The nucleation of staggered magnetic moment where superconductivity is suppressed and/or the singlet-triplet excitation of the singlet impurity-hole bound state [49] may contribute to the a broad peak observed in the normal state up to ~ 250 K, with a characteristic energy smaller than E_r (but of the same order of magnitude). In this picture, the magnetic resonance peak and spin fluctuations intrinsic to Cu spins surrounding Zn impurities coexist in the superconducting state.

In conclusion, our INS data show that the interplay between non magnetic quantum impurities and spin dynamics in the cuprates is a surprisingly rich field of investigation, that emphasized the importance of strong correlation and the competition between the superconducting ground state and antiferromagnetism. We hope that this review will stimulate further theoretical and experimental work.

Acknowledgments

The authors wish to acknowledge P. Hirschfeld, J. Bobroff, P. Pfeuty and F. Onufrieva for helpful discussions.

References

[1] J. Rossat-Mignod *et al.*, Physica C **185-189**, 86 (1991).

[2] H.A. Mook *et al.*, Phys. Rev. Lett. **70**, 3490 (1993).

[3] L. P. Regnault *et al.*, Physica C **235-240**, 59 (1994); Physica B **213-214**, 48 (1994).

[4] H.F. Fong *et al.*, Phys. Rev. Lett. **75**, 316 (1995); Phys. Rev. B **54**, 6708 (1996).

[5] P. Bourges *et al.*, Phys. Rev. B **53**, 876 (1996).

[6] L. P. Regnault *et al.*, in *Neutron Scattering in Layered Coper-Oxide Superconductors*, A. Furrer (Kluwer, Amsterdam, 1998), p 85.

[7] P. Dai *et al.*, Phys. Rev. Lett. **77**, 5425 (1996).

[8] H.F. Fong *et al.*, Phys. Rev. Lett. **78**, 713 (1997).

[9] P. Bourges *et al.*, Europhys. Lett. **38**, 313 (1997).

[10] H.F. Fong *et al.*, Phys. Rev. B **61**, 14774 (2000).

[11] Y. Sidis *et al.*, Phys. Rev. Lett. **84**, 5900 (2000) (cond-mat/9912214).

[12] H.F. Fong *et al.*, Nature **398**, 588 (1999).

[13] H. He *et al.*, cond-mat/0002013.

[14] P. Dai *et al.*, Phys. Rev. Lett. **80**, 1738 (1998).

[15] H.A. Mook *et al.*, Nature **395** , 580 (1998).

[16] P. Bourges *et al.*, in *High Temperature Superconductivity* S.E. Barnes *et al* Eds. (CP483 AIP, Amsterdam, 1999) pp 207-212 (cond-mat/9902067).

[17] M. Arai *et al.*, Phys. Rev. Lett. **83**, 608 (1999).

[18] P. Bourges *et al.*, Science **288**, 1234 (2000); cond-mat/0006085.

[19] F. Onufrieva and P. Pfeuty, cond-mat/9903097.

[20] M. Tarascon *et al.*, Phys. Rev. B **37**, 7458 (1988).

[21] P. Mendels *et al.*, Europhysics Lett. **46**, 678 (1999).

[22] H.F. Fong *et al.*, Phys. Rev. Lett. **82**, 1939 (1999).

[23] Y. Sidis *et al.*, Phys. Rev. B **53**, 6811 (1996).

[24] Y. Sidis *et al.*, Int. J. Mod. Phys. B **12**, 3330 (1998).

[25] D. Poilblanc *et al.*, Solid. St. Com **59**, 111 (1986).

[26] L.S. Borkowski *et al.*, Phys. Rev. B **49**, 15404 (1994).

[27] T. Hotta *et al.*, Phys. Rev. B **62**, 274 (1993).

[28] Y. Sun *et al.*, Phys. Rev. B **51**, 6059 (1994).

[29] A.V. Balatsky *et al.*, Phys. Rev. B **51**, 15547 (1995).

[30] H. Kim *et al.*, Phys. Rev. B **52**, 13576 (1995).

[31] R. Fehrenbacher, Phys. Rev. Lett. **77**, 1849 (1996).

[32] A.V. Balatsky and M.I. Salkola, Phys. Rev. Lett. **76**, 2386 (1996).

[33] M. Franz *et al.*, Phys. Rev. B **54**, R6887 (1996).

[34] G. Harań *et al.*, condmat/9904263.

[35] B. Nachumi *et al.*, Phys. Rev. Lett. **77**, 5421 (1996).

[36] C. Bernhard *et al.*, Phys. Rev. Lett. **77**, 2303 (1996).

[37] D.A. Bernhard *et al.*, Phys. Rev. B **50**, 4051 (1994).

[38] K. Ishida *et al.*, Physica C **179**. 29 (1991); J. Phys. Soc. Jpn **62**, 2803 (1993).

[39] J.X. Li *et al.*, Phys. Rev. B **58**, 2895 (1998).

[40] D. Morr and D. Pines, Phys. Rev. Lett. **81**, 1086 (1998).

[41] M. Vojta *et al.*, Phys. Rev. B **61**, 15152 (2000) (cond-mat/9912020).

[42] N. Bulut, cond-mat/9909437, cond-mat/9908266.

[43] W. Ziegler *et al.*, Phys. Rev. B **53**, 8704 (1996).

[44] S.H. Pan *et al.*, Nature **403**, 748 (2000).

[45] V. A. Mahajan *et al.*, Phys. Rev. Lett. **72**, 3100 (1994).

[46] D. Poilblanc *et al.*, Phys. Rev. Lett **72**, 884 (1994); Phys. Rev. B **50**, 13020 (1994).

[47] Y. Ohta *et al.*, Physica C **263**, 94 (1996).

[48] S. Odashima *et al.*, Phys. Rev. B **56**, 126 (1997).

[49] J. Riera *et al.*, Phys. Rev. B **54**, 7441 (1996).

PSEUDOGAP AND KINETIC PAIRING UNDER CRITICAL DIFFERENTIATION OF ELECTRONS IN CUPRATE SUPERCONDUCTORS

Masatoshi Imada and Shigeki Onoda

Institute for Solid State Physics, University of Tokyo, 5-1-5, Kashiwanoha, Kashiwa, Chiba, 277-8581, Japan

Abstract   Superconducting mechanism of cuprates is discussed in the light of the proximity of the Mott insulator. The proximity accompanied by suppression of coherence takes place in an inhomogeneous way in the momentum space in finite-dimensional systems. Studies on instabilities of metals consisted of such differentiated electrons in the momentum space are reviewed from a general point of view. A typical example of the differentiation is found in the flattening of the quasiparticle dispersion discovered around momenta $(\pi, 0)$ and $(0, \pi)$ on 2D square lattices. This flattening even controls the criticality of the metal-insulator transition. Such differentiation and suppressed coherence subsequently cause an instability to the superconducting state in the second order of the strong coupling expansion. The d-wave pairing interaction is generated from such local but kinetic processes in the absence of disturbance from the coherent single-particle excitations. The superconducting mechanism emerges from a direct kinetic origin which is conceptually different from the pairing mechanism mediated by bosonic excitations as in magnetic, excitonic, and BCS mechanisms. Pseudogap phenomena widely observed in the underdoped cuprates are then naturally understood from the mode-mode coupling of d-wave superconducting (dSC) fluctuations repulsively coupled with antiferromagnetic (AFM) ones. When we assume the existence of a strong d-wave channel repulsively competing with AFM fluctuations under the formation of flat and damped single-particle dispersion, we reproduce basic properties of the pseudogap seen in the magnetic resonance, neutron scattering, angle resolved photoemission and tunneling measurements in the cuprates.

1.   INTRODUCTION

Magnetism in strongly correlated electron systems has been a subject of extensive studies for a long time. In many examples, the energy gain from electron kinetic energy term is a crucial driving force for stabilizing magneti-

J. Bonča et al. (eds.), Open Problems in Strongly Correlated Electron Systems, 69–80.
© 2001 *Kluwer Academic Publishers. Printed in the Netherlands.*

cally symmetry broken states. Even in the Mott insulating state, this is equally true. Although the itinerancy of electrons is lost by the Coulomb interaction in the Mott insulator, electrons still find an optimized way to gain the energy from the kinetic energy term through a virtual process, where the strong coupling expansion becomes an adequate description under totally suppressed coherent motion of electrons. Anderson's mechanism of the superexchange interaction thus proposed [1] has long been one of the most fruitful concepts in the condensed matter physics. The double exchange mechanism proposed by de Gennes [2], Zenner [3] and Anderson and Hasegawa [4] is another example, where a ferromagnetic (FM) metal is stabilized by the kinetic energy gain under a strong Hund's rule coupling between degenerate orbitals.

Superconductivity, on the other hand, has been analyzed mainly from a different view since the dramatic success of BCS theory [5]. Various types of pairing interactions mediated by bosonic excitations have been considered to be responsible for the mechanisms. To say nothing of the electron-phonon interaction, roles of AFM and FM paramagnons and excitons have been studied in various contexts. In these proposals, the superconducting ground state is the consequence of pair formation by the mediated attraction in the presence of coherent and metallic quasiparticles.

In this report, we discuss in more general perspective the importance of the kinetic origin for various symmetry breakings. We particularly discuss the kinetic mechanism of the superconductivity under the suppressed coherence of quasiparticles near the Mott insulator. The mechanism of superconductivity we discuss in this paper is conceptually different from the conventional magnetic mechanism discussed in the spin fluctuation theories and in the t-J model, because the origin is not in the mediated bosons but in the direct kinetic process as in the superexchange for the mechanism of antiferromagnetism. Although the superexchange interaction is the only process for the energy lowering by the kinetic energy term in the Mott insulating phase, this constraint is lost in the presence of doped carriers and various other processes may equally or even more importantly contribute to the kinetic energy gain. The existence of various other processes is in fact easily confirmed by the strong coupling expansion of the Hubbard-type models, where in the second order, namely in the order of t^2/U, it not only generates the superexchange interaction but also other two-particle terms including the pair hopping processes. Here in the strong coupling expansion, the onsite Coulomb repulsion U is assumed to be larger than the kinetic transfer term t. These second-order terms usually have secondary importance in the presence of the kinetic energy term in the first order of t in the normal metal. However, as similarly to the superexchange interaction in the Mott insulating phase, they become the primary origin of the kinetic energy gain even in metals when the single-particle processes are suppressed by some reason but the amplitude of the second-order terms are retained. In the study on the strong-coupling expansion of the d-p model, it was shown that the d-wave pair-hopping process appears in the third order, although the superexchange interaction appears only in the fourth order [6]. In the next section we first discuss the origin of the unusual suppression of coherent single-particle motion in the proximity of the Mott insulator before considering the mechanism of superconductivity.

High-temperature cuprate superconductors show a variety of unusual properties in the normal state [7]. Among others, we concentrate on two remarkable properties widely observed in the cuprates for the purpose of examining the relevance of the kinetic mechanism as the driving mechanism of the high-Tc superconductivity. One of the remarkable properties in the cuprates is the flat dispersion in single-particle excitations. Angle resolved photoemission spectra (ARPES) in Y and Bi based high-Tc cuprates show an unusual dispersion which is far from weak correlation picture [8]. The dispersion around $(\pi, 0)$ and $(0, \pi)$ is extremely flat beyond the expectations from usual van Hove singularities. The flat dispersion also shows rather strong damping.

The other remarkable property we discuss is the pseudogap phenomenon observed in the underdoped region [7, 9]. It is observed both in spin and charge excitations in which gap structure emerges from a temperature T_{PG} well above the superconducting transition point T_c. The gap structure is observed in various different probes such as NMR relaxation time, the Knight shift, neutron scattering, tunneling, ARPES, specific heat, optical conductivity, and DC resistivity. The ARPES [10, 11] data have revealed that the pseudogap starts growing first in the region around $(\pi, 0)$ and $(0, \pi)$ from $T = T_{PG}$ much higher than T_c. Therefore, the pseudogap appears from the momentum region of the flattened dispersion and it is likely that the mechanism of the pseudogap formation is deeply influenced from the underlying flatness. The superconducting state itself also shows a dominant gap structure in this flat spots, $(\pi, 0)$ and $(0, \pi)$, due to the $d_{x^2-y^2}$ symmetry. In fact, the pseudogap structure above T_c appears continuously to merge into the $d_{x^2-y^2}$ gap below T_c. To understand the superconducting mechanism and the origin of the high transition temperatures, a detailed understanding of the physics taking place in the flat dispersion region is required.

The emergence of the flat dispersion around $(\pi, 0)$ and $(0, \pi)$ has also been reported in numerical simulation results rather universally in models for strongly correlated electrons such as the Hubbard and t-J models on square lattices [12, 13, 14]. As we see in the next section, this criticality is interpreted from a strong proximity of the Mott insulator where strong electron correlation generates suppressed dynamics and coherence.

2.   KINETIC PAIRING DERIVED FROM ELECTRON DIFFERENTIATION

In a simple picture, the correlation effects emerge as the isotropic mass renormalization, where the Coulomb repulsion from other electrons makes the effective mass heavier. This effect was first demonstrated by Brinkman and Rice [19] in the Gutzwiller approximation and refined in the dynamical mean field theory [20].

In the numerical results on a square lattice, as discussed above, the correlation effects appear in more subtle way where the electrons at different momenta show different renormalizations. When the Mott insulator is approached and the doping concentration becomes small, the mass renormalization generally becomes stronger. However, once the renormalization effect gets relatively stronger in a part of the Fermi surface, it is further enhanced at that part in a selfconsistent fashion because the slower electrons become more and more sensitive to the correlation effect. This generates critical differentiation of the carriers depending on the portion of the Fermi surface.

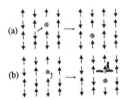

Figure 1   Intuitive picture to understand anisotropic renormalization effects. An electron moving in diagonal directions under the AFM correlations are not severely renormalized as in (a) while they are for horizontal or vertical directions as in (b). In (b), frustrations are generated after the hole motion (denoted by circles) as shown by wavy bonds. These differences induce the differentiation between electrons around $(\pi, 0)$ and $(\pi/2, \pi/2)$.

On square lattices, the stronger renormalization happens around $(\pi, 0)$ and $(0, \pi)$. A part of this anisotropic correlation effect concentrating near $(\pi, 0)$ and $(0, \pi)$ is intuitively understood from the carrier motion under the background of AFM correlations. As we see a real space picture in Fig.1a, the carrier motion in the diagonal directions does not disturb the correlations due to the parallel spin alignment, while the motion in horizontal and vertical directions strongly disturbs the AFM backgrounds as expressed as the wavy bonds and the motion itself is also disturbed as a feedback. Such strong coupling of charge dynamics to spin correlations causes flattening and damping of electrons around $(\pi, 0)$ and $(0, \pi)$, but not around the diagonal direction $(\pm\pi/2, \pm\pi/2)$. The anisotropic renormalization effect eventually may generate a singularly flat dispersion on particular region of the Fermi surface, which accepts more and more doped holes in that region due to the enhanced density of states.

The transition to the Mott insulator is then governed by that flattened part, since the carriers reside predominantly in the flat region. The criticality of the metal-insulator transition on the square lattice is thus determined from the doped carriers around the flat spots, $(\pi, 0)$ and $(0, \pi)$. The hyperscaling relation becomes naturally satisfied because singular points on the momentum space govern the transition. In fact, the hyperscaling relations are numerically supported in various quantities and shows agreements with experimental indications. For example, the electronic compressibility critically diverges as $\kappa \propto 1/\delta$ with decreasing doping concentration δ while the Drude weight is unusually suppressed as $D \propto \delta^2$ [15, 16]. The coherence temperature (the effective Fermi temperature) is also scaled as $T_F \propto \delta^2$ and indicates unusual suppression. In more comprehensive understanding, all the numerical data are consistent with the hyperscaling relations with a large dynamical exponent $z = 4$ for the metal-insulator transition [7, 18]. Such large exponent opposed to the usual value $z = 2$ for the transition to the band insulator is derived from the slower electron dynamics even at $T = 0$ generated by the flat dispersion.

We, however, should keep in mind that the relaxation time of quasiparticles and the damping constant of magnetic excitations do not have criticality at the transition point to the Mott insulator. A general remark is that the relaxation time is critical only in the case of the Anderson localization transition and not in the case of the transition to the Mott insulator. The DC transport properties and magnetic relaxation phenomena are contaminated by such noncritical relaxation times τ and are influenced by the carriers in the other portion than the flat part because the flat part has stronger damping and less contributes to the DC properties. Large anisotropy of τ masks the real criticality and makes it difficult to see the real critical exponents in the τ-dependent properties. Relevant quantities to easily estimate the criticality is the τ independent quantities such as the Drude weight and the compressibility.

Near the metal-insulator transition, the critical electron differentiation and selective renormalization may lead to experimental observations as if internal degrees of freedom of the carriers such as spin and charge were separated because each degrees of freedom can predominantly be conveyed by carriers in different part of the Fermi surface. Another possible effect of the electron differentiation is the appearance of several different relaxation times which are all originally given by a single quasiparticle relaxation in the isotropic Fermi liquids, but now depend on momenta of the quasiparticles.

Another aspect one might ask in connection to the relevance of the flat part to the metal-insulator transition is the observed level difference between $(\pi, 0)$ and $(\pm\pi/2, \pm\pi/2)$ in the undoped and underdoped cuprates [17]. The level at $(\pi, 0)$ is substantially far from the Fermi level, and at a first glance, it does not have a chance to contribute to the metal-insulator transition. However, it has been clarified [6] that this level difference is likely to be absent before the d-wave interaction channel starts growing where the flat part indeed governs the criticality, while it is developed by renormalization of the single-particle level accompanied by the d-wave interaction in the lower energy scale. It is remarkable that this renormalization exists even in the insulator.

If the mass renormalization would happen in an isotropic way as in the picture of Brinkman and Rice, the renormalization can become stronger without disturbance when the insulator is approached. However, if the singularly renormalized flat dispersion emerges critically only in a part near the Fermi surface but the whole band width is retained, that flattened part has stronger instability due to the coupling to larger energy scale retained in other part of the momentum space. The instability can be mediated by local and incoherent carrier motion generated from two-particle processes derived in the strong coupling expansion [6]. The local two-particle motion is given in the order of t^2/U with the bare t while the single-particle term is renormalized to t^* which can be smaller than t^2/U at the flattened portion. The instability of the flat dispersion was studied by taking account such local and incoherent terms in the Hubbard and t-J models [21, 22, 16]. The inclusion of the two-particle terms drives the instability of the flat part to the superconducting pairing and the formation of the d-wave gap structure. In fact, even at half filling, the two-particle process stabilizes the d-wave superconducting state and reproduces the basic feature of the pseudogap formation [21, 22] observed in the BEDT-TTF compounds [23].

The paired bound particles formed from two quasiparticles at the flat spots have different dynamics from the original quasiparticle. In fact, when the paired singlet becomes the dominant carrier, the criticality changes from $z = 4$ to $z = 2$, resulting in the recovery of coherence and kinetic energy gain [16]. It generates a strong pairing interaction from the kinetic origin. This pairing mechanism is a consequence of suppressed single-particle coherence and electron differentiation due to strong correlations.

The instability of the flat dispersion coexisting with relatively large incoherent process was further studied [6, 24, 25, 26]. It has turned out that promotion of the above scaling behavior and the flat dispersion offers a way to control potential instabilities. Even when a flat *band* dispersion is designed near the Fermi level by controlling lattice geometry and parameters, it enlarges the critical region under the suppression of single-particle coherence in the proximity of the Mott insulator mentioned above. In designed lattices and lattices with tuned lattice parameters, it was reported that the superconducting instability and the formation of the spin gap have been dramatically enhanced [24].

3.   PSEUDOGAP PHENOMENA IN CUPRATES AS SUPERCONDUCTING FLUCTUATIONS

As is mentioned in §1, the pseudogap in the high-Tc cuprates starts growing from the region of the flat dispersion. When the single-particle coherence is suppressed, the system is subject to two particle instabilities. As clarified in §2, the superconducting instability in fact grows. However, the AFM and charge order correlations are in principle also expected to grow from other two-particle (particle-hole) processes and may compete each other. In particular, the AFM long-range order is realized in the Mott insulator and its short-range correlation is well retained in the underdoped region. Therefore, to understand how the superconducting phase appears in the underdoped region, at least competition of dSC and AFM correlations has to be treated with underlying suppressed coherence in the region of $(\pi, 0)$ and $(0, \pi)$. The authors have developed a framework to treat the competition by employing the mode-mode coupling theory of dSC and AFM fluctuations where these two fluctuations are treated on an equal footing [27, 28].

It should be noted that the strong dSC pairing interaction is resulted from a highly correlated effect with electron differentiation while the critical differentiation has not been successfully reproduced from the diagrammatic approach so far. Then, within the framework of the mode-mode coupling theory, at the starting point, we have assumed the existence of correlation effects leading to the flattened dispersion and the d-wave pair hopping process. The AFM and dSC fluctuations are predominantly generated by the contributions from the quasiparticle excitations in the flattened regions $(\pi, 0)$ and $(0, \pi)$. These fluctuations are treated in a set of selfconsistent equations with mode couplings of dSC and AFM. From the selfconsistent solution, the pseudogap formation is well reproduced in a region of the parameter space. The pseudogap emerges when the mode coupling between dSC and AFM is repulsive with a severe competition and dSC eventually dominates at low temperatures. Such competition suppresses T_c, while above T_c it produces a region where pairing fluctuations are large. This region at $T_{PG} > T > T_c$ shows suppression of $1/T_1 T$ and the pseudogap formation around $(\pi, 0)$ and $(0, \pi)$ in $A(k, \omega)$. These reproduce the basic feature of the pseudogap phenomena experimentally observed in the underdoped cuprates. The pseudogap formation is identified as coming from the superconducting fluctuations. The momentum dependence shows that the pseudogap formation starts around $(\pi, 0)$ from higher temperatures and the formation temperature becomes lower with increasing distance from $(\pi, 0)$. All of the above reproduce the experimental observations.

We, however, note a richer structure of the gap formation observed in the transversal NMR relaxation time T_{2G} and the neutron resonance peak. One puzzling experimental observation is that the pseudogap structure appears in $1/T_1 T$ [9, 29, 30, 31, 32], while in many cases $1/T_{2G}$, which measures $\mathrm{Re}\chi(Q, \omega = 0)$ at $Q = (\pi, \pi)$, continuously increases with the decrease in temperature with no indication of the pseudogap. In addition, the so called resonance peak appears in the neutron scattering experiments [33]. A resonance peak sharply grows at a finite frequency below T_c with some indications even at $T_c < T < T_{PG}$. This peak frequency ω^* decreases with lowering doping concentration implying a direct and continuous evolution into the AFM Bragg peak in the undoped compounds. The neutron and T_{2G} data support the idea that the AFM fluctuations are suppressed around $\omega = 0$ but transferred to a nonzero frequency below T_{PG}.

To understand these features, a detailed consideration on damping of the magnetic excitations is required. With the increase in the pairing correlation length ξ_d, the pseudogap in $A(k, \omega)$ is developed. Since the damping is mainly from the overdamped Stoner excitations, the gap formation in $A(k, \omega)$ contributes not only to suppress growth of AFM correlation length ξ_σ but also to reduce the magnetic damping because, inside the domain of the d-wave order, the AFM excitations are less scattered due to the absence of low-energy quasiparticle around $(\pi, 0)$. If the quasiparticle damping is originally large around $(\pi, 0)$, the damping γ can be reduced dramatically upon the pseudogap formation. Under this circumstance, our calculated result reproduces the resonance peak and the increase in $1/T_{2G}$ with lowering temperature at $T > T_c$ in agreement with the experimental observations in YBa$_2$Cu$_3$O$_{6.63}$, YBa$_2$Cu$_4$O$_8$ and some other underdoped compounds [27, 28].

A subtlety arises when the damping around $(\pi/2, \pi/2)$ starts contributing. This is particularly true under the pseudogap formation. If contributions from the $(\pi/2, \pi/2)$ region would be absent, the damping of the magnetic excitation would be strongly reduced when the pseudogap is formed around $(\pi, 0)$ as we mentioned above. However, under the pseudogap formation, the damping can be determined by the Stoner continuum generated from the $(\pi/2, \pi/2)$ region and can remain overdamped. This process is in fact important if the quasiparticle damping around the $(\pi/2, \pi/2)$ region is large as in the case of La 214 compounds [34]. The formation of the pseudogap itself is a rather universal consequence of the strong coupling superconductors. However, the actual behavior may depend on this damping. If the damping generated by the $(\pi/2, \pi/2)$ region is large, it sensitively destroy the resonance peak structure observed in the neutron scattering experimental results.

4.   CONCLUSION AND DISCUSSION

Electron critical differentiation is a typical property of the proximity of the Mott insulator. The flattening of the quasiparticle dispersion appears around momenta $(\pi, 0)$ and $(0, \pi)$ on square lattices and determines the criticality of the metal-insulator transition with the suppressed coherence in that momentum region of quasiparticles. Such coherence suppression subsequently causes an instability to the superconducting state when a proper incoherent kinetic process is retained. The d-wave superconducting state is stabilized from such retained microscopic process derived from the strong correlation expansion. The origin of the superconductivity is ascribed to the kinetic energy gain.

By assuming the d-wave channel and the presence of strongly renormalized flat quasi-particle dispersion around the $(\pi, 0)$ region, we have constructed the mode-mode coupling theory for the AFM and dSC fluctuations. The pseudogap in the high-T_c cuprates is reproduced as the region with enhanced dSC correlations and is consistently explained from precursor effects for the superconductivity. The existence of the flat region plays a role to suppress the effective Fermi temperature E_F. This suppressed E_F and relatively large local pair hopping process both drive the system to the strong coupling superconductor thereby leading to the pseudogap formation. The pseudogap formation is also enhanced by the AFM fluctuations repulsively coupled with dSC fluctuations.

Several similar attempts have also been made to reproduce the pseudogap phenomena observed in the cuprates. A common conclusion inferred from these calculations including those by the present authors is that the pseudogap is reproduced when the d-wave channel is explicitly assumed [27, 28, 35] while crucial features such as the pseudogap formation around $(\pi, 0)$ cannot be well reproduced if one tries to derive the superconducting channel itself from the AFM spin fluctuations [36]. This difficulty is summarized: if one desires to stabilize a strongly fluctuating pseudogap region well above T_c generated originally by the magnetic interaction, one has to treat the strong-coupling superconductivity and hence even more strong-coupling magnetic interaction. In such a circumstance, it is dificult to escape from the magnetic instability before the pseudogap formation. The difficulty is naturally interpreted from our analysis: The kinetic d-wave channel we derived in the strong coupling expansion is not ascribed to the magnetic origin and is not contained neither in the spin-fluctuation mechanism nor in the t-J model. This superconducting channel has a comparable amplitude to the magnetic one, namel y t^2/U. Furthermore the channel is enhanced by the flattened dispersion and the electron critical differentiation. Such effects are far beyond the one-loop level of the weak coupling approach. The change in the quasiparticle dispersion in the portion of the Fermi surface is not well reproduced for the moment and the momentum dependent selfenergy appears to be significantly underestimated in the existing diagrammatic evaluations.

The strong coupling Hamiltonian was considered before [37] and also in terms of the pairing arising from the attractive superexchange interaction [38]. The superconductivity from direct kinetic origin was also discussed before in a specific model [39] as well as in the interlayer tunneling mechanism [40], where interlayer kinetic energy gain is required to stabilize the superconductivity. If we identify the intralayer charge incoherence observed in the normal state as the crossover phenomena above the unusually suppressed Fermi temperature in the proximity of the Mott insulator, and notice that no quantum coherence is reached even within the layer, the kinetic origin of the superconductivity is found solely in a two-dimensional plane.

Under the suppression of coherence in the proximity of the Mott insulator, the second-order process in the strong-coupling expansion may also drive other type of ordered state in addition to the superconductivity. In particular, because the charge compressibility is enhanced due to the flattened dispersion, the charge fluctuation may be strongly enhanced and the charge ordering and stripes can be triggered both by the kinetic energy gain and the intersite Coulomb interaction. When the competition between the superconductivity and the charge ordering becomes serious, presumable repulsive mode coupling of them may become another origin of the pseudogap formation, although such evidence is not clear in the present experimental results of the cuprates.

The present theory of the kinetic superconductor predicts a specific form of the kinetic energy gain to be seen in the optical conductivity [41] and the single particle spectra [42]. Qualitatively, it is expected that the in-plane kinetic energy starts gained even in the pseudogap region. The energy gain dominantly coming from the $(\pi, 0)$ and $(0, \pi)$ region of the single particles must have a significant doping dependence associated to the change in the dynamical exponent z from 4 to 2.

References

[1] P.W. Anderson: Phys. Rev. **115** (1959) 2.

[2] P.G. deGennes: Phys. Rev. **118** (1960) 141.

[3] C. Zener: Phys. Rev. **82** (1951) 403.

[4] P.W. Anderson and H. Hasegawa: Phys. Rev. **100** (1955) 675.

[5] J. Bardeen, L. N. Cooper and J.R. Schrieffer: Phys. Rev. **108** (1957) 1175

[6] H. Tsunetsugu and M. Imada: J. Phys. Soc. Jpn. **68** (1999) 3162.

[7] For a recent review see M. Imada, A. Fujimori and Y. Tokura: Rev. Mod. Phys. **70** (1998) 1039, Sec. IV.C.

[8] K. Gofron, J. C. Campuzano, A. A. Abrikosov, M. Lindroos, A. Bansil, H. Ding, D. Koelling and B. Dabrowski: Phys. Rev. Lett. **73** (1994) 3302. D. S. Marshall, D. S. Dessau, A. G. Loeser, C-H. Park, A. Y. Matsuura, J. N. Eckstein, I. Bozovic, P. Fournier, A. Kapitulnik, W. E. Spicer and Z.-X. Shen: Phys. Rev. Lett. **76** (1996) 4841.

[9] H. Yasuoka, T. Imai and T. Shimizu: "Strong Correlation and Super-conductivity" ed. by H. Fukuyama, S. Maekawa and A. P. Malozemoff (Springer Verlag, Berlin, 1989), p.254.

[10] Z.-X. Shen and D. S. Dessau: Physics Reports **253** (1995) 1; A. G. Loeser, Z.-X. Shen, D. S. Dessau, D. S. Marshall, C. H. Park, P. Fournier and A. Kapitulnik: Science **273** (1996) 325.

[11] H. Ding, T. Yokoya, J. C. Campuzano, T. Takahashi, M. Randeria, M. R. Norman, T. Mochiku, K. Kadowaki and J. Giapintzakis: Nature **382** (1996) 51.

[12] E. Dagotto, A. Nazarenko and M. Boninsegni: Phys. Rev. Lett. **73** (1994) 728.

[13] N. Bulut, D.J. Scalapino and S.W. White: Phys. Rev. B **50** (1994) 7215.

[14] F. F. Assaad and M. Imada: Eur. Phys. J. B **10** (1999) 595.

[15] N. Furukawa and M. Imada: J. Phys. Soc. Jpn. **61** (1992) 3331; *ibid.* **62** (1993) 2557.

[16] H. Tsunetsugu and M. Imada: J. Phys. Soc. Jpn. **67**, (1998)1864.

[17] C. Kim et al.:Phys. Rev. B **56** (1998) 15589.

[18] M. Imada, J. Phys. Soc. Jpn.**64**(1995)2954.

[19] W.F.Brinkman, and T.M. Rice, Phys. Rev. B**2**,(1970) 4302.

[20] A.Georges, G. Kotliar, W. Krauth, and M. J. Rozenberg, Rev. Mod. Phys. **68**(1996) 13.

[21] F.F. Assaad, M. Imada and D.J. Scalapino:Phys. Rev. B **56** (1998) 15001.

[22] F.F. Assaad and M, Imada: Phys. Rev. B. **58**, (1998) 1845.

[23] K. Kanoda: Hyperfine Interact. **104** (1997) 235.

[24] M. Imada and M. Kohno:Phys. Rev. Lett. **84** (2000) 143.

[25] M. Imada, M. Kohno and H. Tsunetsugu: Physica B **280**(2000) 303.

[26] M. Kohno and M. Imada: J. Phys. Soc. Jpn. **69**(2000) 25.

[27] S. Onoda and M. Imada: J. Phys. Soc. Jpn. **68** (1999) 2762.

[28] S. Onoda and M. Imada: J. Phys. Soc. Jpn. **69** (2000) 312.

[29] H. Zimmermann, M. Mali, D.Brinkmann, J. Karpinski, E. Kaldis and S. Rusiecki: Physica C **159** (1989) 681; T. Machi, I. Tomeno, T. Miyataka, N. Koshizuka, S. Tanaka, T. Imai and H. Yasuoka: Physica C **173** (1991) 32.

[30] K. Ishida, Y. Kitaoka, K. Asayama, K. Kadowaki and T. Mochiku: Physica C **263** (1996) 371.

[31] Y. Itoh, T. Machi, A. Fukuoka, K. Tanabe, and H. Yasuoka: J. Phys. Soc. Jpn. **65** (1996) 3751.

[32] M.-H. Julien, P. Carretta, M. Horvatić, C. Berthier, Y. Berthier, P. Ségransan, A. Carrington and D. Colson: Phys. Rev. Lett. **76** (1996) 4238.

[33] H. F. Fong, B. Keimer, D. L. Milius and I. A. Aksay: Phys. Rev. Lett. **78** (1997) 713.

[34] A. Ino, C. Kim, T. Mizokawa, Z.-X. Shen, A. Fujimori, M. Takabe, K. Tamasaku, H. Eisaki and S. Uchida: J. Phys. Soc. Jpn. 68 (1999) 1496; A. Ino, T. Mizokawa, K. Kobayashi, A. Fujimori, T. Sasagawa, T. Kimura, K. Kishio, K. Tamasaku, H. Eisaki, and S. Uchida: Phys. Rev. Lett. **78** (1998) 2124.

[35] Y. Yanase and K. Yamada: J. Phys. Soc. Jpn. **68** (1999) 2999.

[36] A. Kobayashi, A. Tsuruta, T. Matsuura and Y. Kuroda: J. Phys. Soc. Jpn. **68** (1999) 2506.

[37] K.A.Chao, J. Spalek and A.M. Oles: J. Phys. C **10** (1977) L271.

[38] J. E. Hirsch: Phys. Rev. Lett. **54** (1985) 1317.

[39] J. E. Hirsch: Physica C **201** (1992) 347.

[40] J.M. Wheatley, T.C. Hsu, and P.W. Anderson: Phys. Rev. Lett. **37** (1988) 5897.

[41] D.N. Basov et al.: Science **283** (1999) 49.

[42] N.R. Norman, M. Randeria, B. Janko and J. C. Campuzano: Phys. Rev. B **61** (2000) 14742.

DENSITY RESPONSE OF CUPRATES AND RENORMALIZATION OF BREATHING PHONONS

P. Horsch and G. Khaliullin

Max-Planck-Institut für Festkörperforschung, D-70569 Stuttgart, Germany

Abstract

We analyse the dynamical density fluctuation spectra for cuprates starting from the $t - J$ model in a slave-boson $1/N$ representation. The results obtained are consistent with diagonalization studies and show novel low-energy structure on the energy scale $J + \delta t$ due to the correlated motion of holes in a RVB spin liquid. The low-energy response implies an anomalous renormalization of several phonon modes. Here we discuss the renormalization of the highest breathing phonons in $La_{2-x}Sr_xCuO_4$.

1. INTRODUCTION

High-temperature superconductors are doped Mott-Hubbard insulators, therefore the low-energy density response is proportional to the doping, due to transitions in the lower Hubbard band. In one dimension the Hubbard physics is characterized by charge and spin separation, which implies that the density response is spinless fermion like, i.e., showing vanishing excitation energy at $4k_F$. Exact diagonalization studies [1, 2] for the t-J model have revealed that the dynamical density response $N(\mathbf{q}, \omega)$ for the 2D model relevant for the cuprates is very different from the 1D case. On the other hand these calculations show several features also unexpected from the point of view of weakly correlated fermion systems: (i) a strong suppression of low energy $2k_F$ scattering in the density response, (ii) a broad incoherent peak whose shape is rather insensitive to hole concentration and exchange interaction J, (iii) a very different form of $N(\mathbf{q}, \omega)$ compared to the spin response function $S(\mathbf{q}, \omega)$, which share common features in usual fermionic systems. Finite temperature diagonalization studies [3] show only weak temperature dependence for $T < 0.3t$ even at low energy.

While considerable analytical work has been done to explain the spin response of the t-J model only few authors analysed $N(\mathbf{q}, \omega)$. Wang *et al.* [4] studied collective excitations in the density channel and found sharp peaks at large momenta corresponding to free bosons. Similar results were obtained

J. Bonča et al. (eds.), Open Problems in Strongly Correlated Electron Systems, 81–90.
© 2001 *Kluwer Academic Publishers. Printed in the Netherlands.*

by Gehlhoff and Zeyher [5] using the X-operator formalism. Lee et al [6] considered a model of bosons in a fluctuating gauge field and found a broad incoherent density fluctuation spectrum at finite temperature, due to the coupling of bosons to a quasistatic disordered gauge field.

Starting from a slave-boson representation we show that the essential features observed in the numerical studies can be obtained in the framework of the Fermi-liquid phase of the t-J model at zero temperature [7]. Our main findings are: (i) at low momenta the main effect of strong correlations is to transfer spectral weight from particle-hole excitations into a pronounced collective mode. Because of the strong damping of this mode (linear in q) due to the coupling to the spinon particle-hole continuum, this collective excitation is qualitatively different from a sound mode. (ii) At large momenta we find a strict similarity of $N(\mathbf{q}, \omega)$ with the spectral function of a single hole moving in a uniform RVB spinon background. In this regime $N(\mathbf{q}, \omega)$ consists of a broad peak at high energy whose origin is the fast, incoherent motion of bare holes. (iii) The polaronic nature of dressed holes leads to the formation of a second peak at lower energy, which is more pronounced in $(\pi, 0)$ direction in agreement with diagonalization studies [1, 2].

The anomalous renormalization of certain phonon modes as observed in inelastic neutron scattering provides a sensitive test of the pecularities of the low-energy density response. In this contribution we shall analyse the strong, doping-dependent renormalization of the highest breathing phonon modes which is a generic feature in the high-T_c compounds.

2.   SLAVE BOSON THEORY OF DENSITY RESPONSE

Following Kotliar and Liu [8] and Wang et al [4] we start from the N-component generalization of the slave-boson t-J Hamiltonian, $H_{tJ} = H_t + H_J$, which is obtained by replacing the constrained electron creation operators $\tilde{c}_{i,\sigma}^{+} = c_{i,\sigma}^{+}(1 - n_{i,-\sigma}) \rightarrow f_{i,\sigma}^{+} b_i$:

$$H_t = -\frac{2t}{N} \sum_{<i,j>\sigma} (f_{i\sigma}^{+} h_j^{+} h_i f_{j\sigma} + h.c.), \tag{1}$$

$$H_J = \frac{J}{N} \sum_{<i,j>\sigma\sigma'} f_{i\sigma}^{+} f_{i\sigma'} f_{j\sigma'}^{+} f_{j\sigma}(1 - h_i^{+} h_i)(1 - h_j^{+} h_j), \tag{2}$$

where $f_{i,\sigma}^{+}$ is a fermionic (spinon) operator, $\sigma = 1, \cdots, N$ is the fermionic flavor index, and h_i denotes the bosonic holes. These operators obey standard commutation rules, yet the number of these auxiliary particles must obey the constraint $\sum_{\sigma} f_{i\sigma}^{+} f_{i\sigma} + h_i^{+} h_i = N/2$. The original t-J model is recovered for $N = 2$.

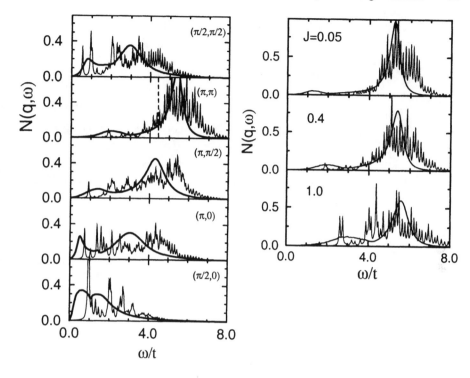

Figure 1   Comparison of $N(\mathbf{q}, \omega)$ obtained by the slave-boson theory (solid lines) with diagonalization data for a periodic 4×4 cluster with $J/t = 0.4$ and doping $\delta = 0.25$. The dashed line
in the (π, π) spectrum indicates the δ-function collective peak obtained when polaron effects
are neglected. Right figure shows the weak J-dependence for the (π, π) spectrum.

The slave boson parametrization provides a straightforward description of
the strong suppression of density fluctuations of constrained electrons through
the representation of the density response in terms of a dilute gas of bosons.
A common treatment of model (1) is the density-phase representation ("radial" gauge [9]) of the bosonic operator $h_i = r_i \exp(i\theta_i)$ with the subsequent
$1/N$-expansion around the Fermi-liquid saddle point. While this gauge is particularly useful to study the low energy and momentum properties, it is not
very convenient for the study of the density response in the full ω and $\mathbf{q}$ space.
Formally the latter follows in the radial gauge from the fluctuations of r_i^2.
If one considers for example convolution type bubble diagrams, one realizes
that their contribution to the static structure factor is correctly of order $1/N$,
but is not proportional to the density of holes δ as it should be. According
to Arrigoni et al [10] such unphysical results originate from a large negative
pole contribution in the $\langle r_{-\mathbf{q}} r_{\mathbf{q}} \rangle_\omega$ Green's function of the real field r, which
is hard to control by a perturbative treatment of phase fluctuations. We follow
therefore Popov [11] using the density-phase treatment only for small momenta

$q < q_0$, while keeping the original particle-hole representation of the density operator, b^+b, at large momenta. More precisely $h_i = r_i \exp(i\theta_i) + b_i$, where $b_i = \sum_{|\mathbf{q}|>q_0} h_{\mathbf{q}} \exp(i\mathbf{q}\mathbf{R}_i)$. The cutoff q_0 is introduced dividing "slow" (collective) variables represented by r and θ from "fast" (single-particle) degrees of freedom. As explained by Popov [11] this "mixed" gauge is particularly useful for finite temperature studies to control infrared divergences. We start formally with "mixed" gauge and keep only terms of order δ and $1/N$ in the bosonic self energies. In this approximation our zero temperature calculations become quite straightforward: The cutoff $q_0 < \delta$ actually does not enter in the results and we arrive finally at the Bogoliubov theory for a dilute gas of bosons moving in a fluctuating spinon background.

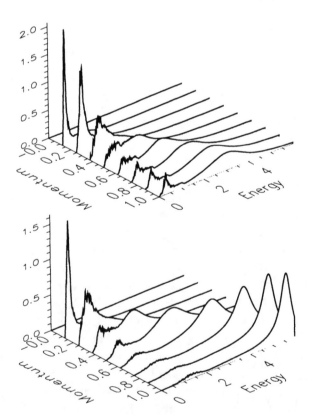

Figure 2 Density fluctuation spectra $N(\mathbf{q},\omega)$ for $J/t = 0.4$ and $\delta = 0.15$ along $(\pi,0)$ (top) and (π,π) directions (bottom). Energy in units of t.

The Lagrangian corresponding to the model (1) is then given by (the summation over σ is implied)

$$
L = \sum_i \left(f_{i\sigma}^+ \left(\frac{\partial}{\partial \tau} - \mu_f \right) f_{i\sigma} + b_i^+ \left(\frac{\partial}{\partial \tau} - \mu_b \right) b_i \right) + H_t + H_J
$$

$$
+ \frac{i}{\sqrt{N}} \sum_i \lambda_i \left(f_{i\sigma}^+ f_{i\sigma} + (r_i + b_i^+)(r_i + b_i) - \frac{N}{2} \right),
\tag{3}
$$

$$
H_t = -\frac{2t}{N} \sum_{\langle ij \rangle} f_{i\sigma}^+ f_{j\sigma} \left(b_j^+ b_i + r_i r_j + r_j b_i + b_j^+ r_i \right) + h.c.
\tag{4}
$$

Here the λ field is introduced to enforce the constraint, and μ_f, μ_b are fixed by the particle number equations $\langle n_f \rangle = \frac{N}{2}(1 - \delta)$ and $\langle r_i^2 + b_i^+ b_i \rangle = \frac{N}{2} \delta$, respectively. The uniform mean field solution $r_i = r_0 \sqrt{N/2}$ leads in the large N limit to the renormalized narrow fermionic spectrum $\xi_{\mathbf{k}} = -z \tilde{t} \gamma_{\mathbf{k}} - \mu_f$, with $\tilde{t} = J\chi + t\delta$, $\gamma_{\mathbf{k}} = \frac{1}{2}(\cos k_x + \cos k_y)$, $\chi = \sum_\sigma \langle f_{i\sigma}^+ f_{j\sigma} \rangle / N$, and $z = 4$ the number of nearest neighbors. In the $N = \infty$ limit $\chi_\infty \simeq 2/\pi^2$ is given by that of free fermions, while for the original t-J model its value should be larger [12] due to Gutzwiller projection. In the following $\chi = \frac{3}{2}\chi_\infty$ will be used. Distinct from the finite-temperature gauge-field theory of Nagaosa and Lee [13] the bond-order phase fluctuations acquire a characteristic energy scale in this approach [4], and the fermionic ("spinon") excitations can be identified with Fermi-liquid quasiparticles. The mean field spectrum of bosons is $\omega_{\mathbf{q}} = 2z\chi t(1 - \gamma_{\mathbf{q}})$. Thus the effective mass of holes $m_h^0 \propto 1/t$ is much smaller than that of the spinons.

Due to the diluteness of the bosonic subsystem, $\delta \ll 1$, the density correlation function $\chi_{\mathbf{q},\omega} = \langle \delta n^h \delta n^h \rangle_{\mathbf{q}\omega}$ is mainly given by the condensate induced part which is represented by the Green's function $\langle (b_{\mathbf{q}}^+ + b_{-\mathbf{q}})(b_{\mathbf{q}} + b_{-\mathbf{q}}^+) \rangle_\omega$ for $q > q_0$, and $2\langle r_{-\mathbf{q}} r_{\mathbf{q}} \rangle_\omega$ for $q < q_0$, respectively:

$$
\chi_{\mathbf{q}\omega} = \frac{N}{2} r_0^2 \left(\langle (b_{\mathbf{q}}^+ + b_{-\mathbf{q}})(b_{\mathbf{q}} + b_{-\mathbf{q}}^+) \rangle_{q>q_0} + 2\langle r_{-\mathbf{q}} r_{\mathbf{q}} \rangle_{q<q_0} \right).
\tag{5}
$$

The $1/N$ self-energy corrections to these functions are calculated in a conventional way [9, 8] expanding $r_i = (r_0 \sqrt{N} + (\delta r)_i)/\sqrt{2}$ and considering Gaussian fluctuations around the mean field solution. Neglecting all terms of order δ/N and q_0^2/N, only one relevant $1/N$ contribution remains which corresponds to the dressing of the slave-boson Green's function by spinon particle-hole excitations. Within this approximation and at zero temperature no divergences occur at low momenta, thus one can take the limit $q_0 \to 0$. The final result for the dynamic structure factor (normalized by the hole density) is:

$$
N_{\mathbf{q},\omega} = \frac{2}{\pi} Im \left(\left(\omega_{\mathbf{q}} a + S_{\mathbf{q},\omega}^{(1/N)} - \mu_b \right) / D_{\mathbf{q},\omega} \right),
\tag{6}
$$

$$D_{\mathbf{q},\omega} = \left(\omega_{\mathbf{q}}a + S_{\mathbf{q},\omega}^{(1/N)} - \mu_b\right)\left(\omega_{\mathbf{q}} + S_{\mathbf{q},\omega}^{(1)} + S_{\mathbf{q},\omega}^{(1/N)} - \mu_b\right) - \left(\omega a - A_{\mathbf{q},\omega}^{(1/N)}\right)^2. \tag{7}$$

The origin of the contribution

$$S_{\mathbf{q},\omega}^{(1)} = zt r_0^2 \left(\frac{(1+\Pi_2)^2}{\Pi_1} - \Pi_3\right)_{\mathbf{q},\omega}, \tag{8}$$

$$\Pi_m = zt \sum_{\mathbf{k}} \frac{n(\xi_{\mathbf{k}}) - n(\xi_{\mathbf{k+q}})}{\xi_{\mathbf{k+q}} - \xi_{\mathbf{k}} - \omega - i0^+} (\gamma_{\mathbf{k}} + \gamma_{\mathbf{k+q}})^{m-1},$$

is the indirect interaction of bosons via the spinon band due to the hopping term (which gives Π_3 in (4)) and due to the coupling to spinons via the constraint field λ. The latter channel provides a repulsion between bosons, making $S^{(1)}(\omega = 0)$ positive and therefore ensuring the stability of the uniform mean-field solution. The $1/N$ self energies $S^{(1/N)}$ and $A^{(1/N)}$ are essentially a single boson property. They are given by the symmetric and antisymmetric combinations (with respect to $\omega + i0^+ \to -\omega - i0^+$) of the self energy

$$\Sigma_{\mathbf{q},\omega}^{(1/N)} = \frac{4}{N} \sum_{|\mathbf{k}|<k_F<|\mathbf{k'}|} (zt\gamma_{\mathbf{k'}-\mathbf{q}})^2 G_{\mathbf{q+k-k'}}^0(\omega + \xi_{\mathbf{k}} - \xi_{\mathbf{k'}}). \tag{9}$$

Here $G_{\mathbf{q}}^0(\omega) = (\omega - \omega_{\mathbf{q}} - \Sigma_{\mathbf{q},\omega}^{(1/N)} + \mu_b)^{-1}$ is the Green's function for a single slave boson moving in a uniform RVB background. Although in the context of $1/N$ theory the G^0 function in (5) should be considered as a free propagator, we shall use here the selfconsistent polaron picture for a single hole [14]. This is crucial when comparing the theory for $N = 2$ with diagonalization studies. Finally, the constants a and μ_b in (3) are given by $(1 - tr_0^2/\tilde{t})$ and $S^{(1/N)}(\omega = \mathbf{q} = 0)$, respectively. The parameter r_0^2 in Eq.8, which formally corresponds to the condensate fraction in our theory, is determined selfconsistently from $r_0^2 = \delta - \sum_{\mathbf{q}\neq 0} \tilde{n}_{\mathbf{q}}$. The momentum distribution $\tilde{n}_{\mathbf{q}} = \langle b_{\mathbf{q}}^+ b_{\mathbf{q}} \rangle$ is calculated from the corresponding bosonic Green's function for finite hole-density.

In the small $\omega, \mathbf{q}$ limit $N(\mathbf{q},\omega)$ (3) is mainly controlled by the interaction of bosons represented by the $S^{(1)}$ term ($\propto r_0^2$), while the internal polaron structure of the boson determined by $S^{(1/N)}$ is less important. $N(\mathbf{q},\omega)$ consists of a weak spinon particle-hole continuum with cutoff $\propto v_F q$, and a very pronounced linear collective mode which nearly exhausts the sum rule. The velocity of this mode is always somewhat smaller than the spinon Fermi velocity, $v_s \leq v_F \simeq z\tilde{t}$, which implies a strong damping $\propto \omega$ (or q) of this mode (Fig.2).

The density response $N(\mathbf{q},\omega)$ at large momenta, $q > \delta$, which we can compare with diagonalization results, is dominated by the properties of a single boson selfenergy $S^{(1/N)}$. The calculated density response of the t-J model has three characteristic features on different energy scales: (i) The main spectral weight of the excitations at large momenta is located in an energy region of order of several t. This high energy peak is very broad and incoherent as a result of the strong coupling of bosons to low-energy spin excitations. The position of this peak and its shape are rather insensitive to the ratio $J/t \leq 1$ in agreement with conclusions of [1, 2]. This is simply due to the fact that the high-energy properties of the t-J model are controlled by t. (ii) The theory predicts also a second peak at lower energy (Fig. 2) which is more pronounced in the direction $(\pi, 0)$, while its weight is strongly suppressed for $\mathbf{q}$ near (π, π). The origin of this excitation is due to the formation of a polaron-like band of dressed bosons. The relative weight of this contribution increases with J as a result of the increasing spinon bandwidth. (iii) In addition there is the spinon particle-hole continuum which is generated by $S^{(1)}$ with relatively small weight ($\propto \delta^2$). At $(\pi, 0)$ the high energy cutoff of the spinon continuum is at $z(\chi J + \delta t)$ (Fig. 2), while the polaron peak is at about $(\chi J + \delta t)$.

We note that the polaron peak in $N(\mathbf{q}, \omega)$ is in the same energy range as the high energy phonons in cuprates.

3.   RENORMALIZATION OF BREATHING PHONONS

Inelastic neutron scattering experiments on high-T_c superconductors have shown that in particular the highest energy longitudinal optic phonons near $(\pi, 0)$ soften and broaden strongly as holes are doped into the insulating parent compound. Whereas the corresponding breathing mode at (π, π) shows much smaller softening and no anomalous broadening. This effect seems to be generic for cuprates and detailed neutron scattering studies have been reported for $La_{2-x}Sr_xCuO_4$ [16, 17, 18] and $YBa_2Cu_3O_{6+x}$ [18].

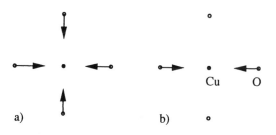

Figure 3   Atomic displacements of oxygen ions (a) for the $\mathbf{q} = (\pi, \pi)$ breathing phonon and (b) for the $(\pi, 0)$ half-breathing mode.

The renormalization of these phonons can be calculated in the framework of the t-J model, since these modes modulate the energy of a hole in a Zhang-Rice singlet state, and therefore couple directly to the density of doped holes. Expanding the Zhang-Rice energy $E_{ZR} = 8\frac{t_{pd}^2}{\Delta\epsilon}$ with respect to the oxygen displacements u_x^i, v_y^i (see Fig.3) of the four O-neighbors of the Cu-hole yields the linear electron-phonon coupling

$$H_{e-ph} = g \sum_i (u_x^i - u_{-x}^i + v_y^i - v_{-y}^i) h_i^+ h_i. \tag{10}$$

We assume that the resonance integral obeys the Harrison relation $t_{pd} \propto r_0^{-7/2}$, where r_0 is the Cu-O distance, and obtain $g = 7E_{ZR}/4r_0$, i.e., $g \approx 4\text{eV/Å}$. The lattice part of the Hamiltonian is determined by the force constant $K \approx 25\text{eV/Å}^2$ for the longitudinal O-motion. Due to the structure of H_{e-ph} the breathing modes couple directly to $\chi_{\mathbf{q},\omega}$.

We have studied the renormalization of the phonon Green's functions along $(\pi, 0)$ and (π, π) directions

$$D_{\mathbf{q},\omega}^{ph} = \frac{\omega_{q,0}}{\omega^2 - \omega_{q,0}^2(1 - \alpha_{\mathbf{q}}\chi_{\mathbf{q},\omega})}, \tag{11}$$

where $\omega_{q,0}$ is the bare phonon frequency, i.e. measured in the undoped parent compound, and $\alpha_{\mathbf{q}} = \frac{4g^2}{K}(\sin^2 q_x/2 + \sin^2 q_y/2)$. Based on the parameters of the pd-model we estimate for the dimensionless coupling constant $\xi = g^2/ztK \sim 0.3 - 0.5$.

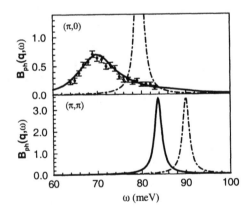

Figure 4 Calculated phonon spectral function B_{ph} for $(\pi, 0)$ and (π, π) breathing phonons for $\delta = 0.15$, $t = 0.4$ eV, $J = 0.12$ eV and $\xi = 0.25$ (solid lines) compared to undoped system (dash-dotted lines) and the inelastic neutron scattering data for $La_{1.85}Sr_{0.15}CuO_4$.

Figure 4 shows the strong renormalization of the $(\pi, 0)$ half-breathing mode for $La_{1.85}Sr_{0.15}CuO_4$ with a twice as large shift as for the (π, π) breathing phonon. The large damping of the $(\pi, 0)$ phonon results from the hybridization with the large polaron peak in $N(\mathbf{q}, \omega)$ at this momentum [19] and is consistent with the experimental data [17]. The phonon energies of the undoped parent compound $\omega_{q,0} = 80(90)$ meV for $(\pi, 0)$ and (π, π), respectively, are taken from Ref. [16]. The strong doping dependence of this effect is shown in Fig.5.

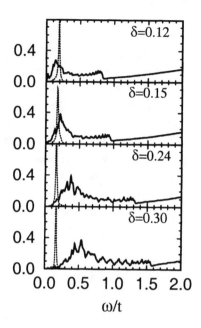

Figure 5 Doping dependence of low-energy density response at $(\pi, 0)$ (solid lines). As a consequence of the scaling of the polaron structure $\propto (\chi J + \delta t)$ there is a strong change in the renormalization and damping of the $(\pi, 0)$ half-breathing phonon, which is at $\omega_0 = 0.2t$ in the undoped system.

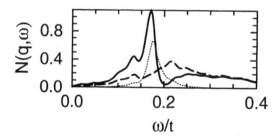

Figure 6 Fano structure in fully renormalized $N(\mathbf{q}, \omega)$ (solid line) due to the coupling to the $(\pi, 0)$ phonon. The bare $N(\mathbf{q}, \omega)$ and the phonon spectral function are indicated by dashed and dotted lines, respectively (parameters as in previous figure).

Finally we have studied the changes of the density response $\chi_{\mathbf{q},\omega}$ due to the additional coupling to the breathing phonon modes. This effect is displayed in Fig.6, which shows a rather strong Fano structure in $N(\mathbf{q}, \omega)$ for $\mathbf{q} = (\pi, 0)$.

4.   SUMMARY

We have outlined a 1/N slave-boson theory for the density response of the *t-J* model, which explains the data obtained by exact diagonalization. We demonstrated that the predicted low energy polaron structure in the density response, which is particularly pronounced along $(\pi, 0)$, explains the anomalous doping induced line width and shift of the longitudinal planar $(\pi, 0)$ phonon. The energy of the polaron peak is determined by the spinon energy scale, therefore we predict a nontrivial doping dependence for the phonon renormalization. In that respect further neutron scattering studies of the doping dependence of phonons would provide a sensitive test for the low energy density response as well as for the spin structure in the different doping regimes.

References

[1] T. Tohyama, P. Horsch, and S. Maekawa, Phys. Rev. Lett. **74** 980 (1995).

[2] R. Eder, Y.Ohta, and S. Maekawa, Phys. Rev. Lett. **74** 5124 (1995).

[3] J. Jaklic and P. Prelovsek, Adv. Phys. **49**, 1 (2000).

[4] Z. Wang, Y. Bang, and G. Kotliar, Phys. Rev. Lett. **67**, 2733 (1991).

[5] L. Gehlhoff and R. Zeyher, Phys. Rev. **52**, 4635 (1995).

[6] D.K.K. Lee, D.H. Kim and P.A. Lee, Phys. Rev. Lett. **76**, 4801 (1996).

[7] G. Khaliullin and P. Horsch, Phys. Rev. B **54**, R9600 (1996).

[8] G. Kotliar and J. Liu, Phys. Rev. B **38**, 5142 (1988).

[9] N. Read and D.M. Newns, J. Phys. C **16**, 3273 (1983).

[10] E. Arrigoni *et al.*, Physics Reports **241**, 291 (1994).

[11] V.N. Popov, Functional Integrals in Quantum Field Theory and Statistical Physics, (D. Reidel, Dordrecht, 1983).

[12] F.C. Zhang, C. Gros, T.M. Rice, and H. Shiba, Supercond. Sci. Technol. **1**, 36 (1988); R.B. Laughlin, J. Low Temp. Phys. **90**, 443 (1995).

[13] N. Nagaosa and P. Lee, Phys. Rev. Lett. **64**, 2450 (1990).

[14] C.L. Kane, P.A. Lee and N. Read, Phys. Rev. B **39**, 6880 (1989).

[15] N. Nücker *et al.*, Phys. Rev. B **39**, 12379 (1989).

[16] L. Pintschovius and W. Reichardt, *Physical Properties of High Temperature Superconductors IV*, edited by D. Ginsberg (World Scientific, Singapore,1994), p. 295.

[17] R. J. McQueeney, T. Egami, G. Sirane, and Y. Endoh, Phys. Rev. B **54**, R9689 (1996); R. J. McQueeney *et al.*, Phys. Rev. Lett. **82**, 628 (1999).

[18] L. Pintschovius and M. Braden, Phys. Rev. B **60**, R15039 (1999).

[19] G. Khaliullin and P. Horsch, Physica C **282-287**, 1751 (1997).

PHASE DIAGRAM OF SPIN LADDER MODELS AND THE TOPOLOGY OF SHORT RANGE VALENCE BONDS

J. Sólyom

Research Institute for Solid State Physics, H-1525 Budapest, P.O.Box 49, Hungary

Abstract   We study the ground state phase diagram of undoped spin ladders. It is shown that in two leg spin ladders depending on the sign of the interchain coupling there exist two types of the gapfull Haldane phase differing in the topology of the short range valence bonds. These phases can be distinguished by the absence or presence of appropriately defined string order parameters. It is also shown that the mechanism of gap generation is different in two-leg and four-leg ladders. This explaines the narrow Haldane regime in $S > 1$ spin systems.

1.   INTRODUCTION

Recently several classes of new materials have been discovered in which the ions having localized magnetic moment are arranged in ladder like configurations. Although the magnetic ions are part of a real three dimensional structure, the interaction between them is such that due to frustration effects the system could be considered as consisting of decoupled spin ladders with two, three or four legs. The schematic arrangement of the spins sitting in a block of atoms and the dominant interactions between them are shown in Fig. (1).

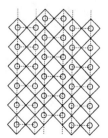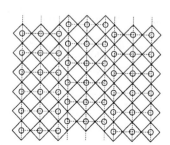

Figure 1   Schematic arrangement of the localized spins in two- and three-leg ladders.

J. Bonča et al. (eds.), Open Problems in Strongly Correlated Electron Systems, 91–100.

This special geometry of the spin arrangement may lead to unusual magnetic properties [1]. If the exchange interaction along the chain is antiferromagnetic and the weak couplings between the ladders can be neglected, the disordered ground state of the spin system can be visualized as a superposition of short range singlet valence bonds. Due to the finite binding energy of the valence bonds, the spectrum of magnetic excitations has a finite gap both for ferromagnetic [2] and antiferromagnetic interchain couplings [3].

In the extreme case when the interchain coupling is ferromagnetic and stronger than the intrachain coupling, the interchain coupling binds the two $s = 1/2$ spins on the same rung into a spin triplet. In this limit the two-leg ladder is equivalent to an antiferromagnetic spin-1 chain, for which a finite gap was conjectured by Haldane [4]. Therefore the gapped phase of the spin ladder, when the interchain interaction is ferromagnetic, is called a Haldane phase.

In the other extreme case, when the interchain coupling is strongly antiferromagnetic, spin singlets are formed on the rungs. Although the ground state of the spin-1 chain has a hidden order in the Haldane phase, and a string order parameter [5] can be defined, while the rung singlet state does not show this string order, nevertheless the gapped rung singlet state was also considered to be a Haldane state.

It is shown here, based on the results obtained in Ref. [6] that in fact in even-leg ladders two types of the gapped Haldane phase can be distinguished, of which the above mentioned two extreme cases are typical examples. These two types differ in the topology of the valence bonds. Isotropic interactions conserve the topological quantum number, therefore, by varying the parameters of the model one cannot go continuously, without a phase transition, from one type to the other.

At the end of the paper we will study the generation of the Haldane gap in four-leg ladders. It is shown that the mechanism of gap generation is different from that in two-leg ladders.

2.   ISOTROPIC SPIN LADDER MODELS

We will consider a system of spin-1/2 ions arranged along two chains, and interacting along the chain as well as between the chains. The intrachain coupling is assumed to be an isotropic antiferromagnetic Heisenberg exchange,

$$\mathcal{H}_0 = J \sum_{i=1}^{N} \left(\vec{S}_{i,1} \cdot \vec{S}_{i+1,1} + \vec{S}_{i,2} \cdot \vec{S}_{i+1,2} \right), \tag{1}$$

where $\vec{S}_{i,1}$ ($\vec{S}_{i,2}$) is the spin operator at site i on chain 1 (chain 2). In what follows we will assume $J > 0$, and the energy scale will be fixed by taking $J = 1$.

Assuming that the dominant interaction between the two legs is between spins on the same rung, the interchain coupling is described by

$$\mathcal{H}_1 = J_1 \sum_{i=1}^{N} \vec{S}_{i,1} \cdot \vec{S}_{i,2}. \tag{2}$$

This type of ladder is shown in Fig. (2).

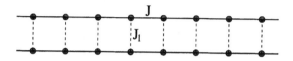

Figure 2   Schematic representation of a two-leg spin ladder with the intra- and interchain exchange couplings.

Alternatively we could consider the so-called diagonal ladder, where the spins on neighbouring rungs are coupled as shown in Fig. (3). The corresponding interchain Hamiltonian has the form

$$\mathcal{H}_2 = J_2 \sum_{i=1}^{N} \left[\vec{S}_{i,1} \cdot \vec{S}_{i+1,2} + \vec{S}_{i,2} \cdot \vec{S}_{i+1,1} \right]. \tag{3}$$

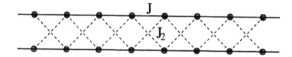

Figure 3   Schematic representation of the two-leg diagonal ladder with the intra- and interchain exchange couplings.

The ground state of the two-leg spin ladder with ferromagnetic interchain coupling and that of the diagonal ladder with antiferromagnetic interchain coupling are similar. They are closely related to the AKLT state [7] of the spin-1 chain. In that state short range valence bonds are formed predominantly between spins on nearest neighbouring rungs, as shown in Fig. (4).

Figure 4   The valence bond structure of the AKLT state.

On the other hand, if the interchain rung coupling is antiferromagnetic, the singlet valence bonds are formed predominantly on the rungs [8], as shown in Fig. (5). Similar situation occurs in diagonal ladders for ferromagnetic interchain coupling.

Figure 5   Typical valence bond structure of the rung singlet phase.

3.   THE TOPOLOGY OF SHORT RANGE VALENCE BONDS

White [9] has pointed out, that the two above mentioned states, the AKLT-like state and the rung singlet state belong to the same universality class when periodic boundary condition is used. Shifting one leg by one lattice spacing, one of the diagonal couplings becomes rung coupling, while the other can be shown to be irrelevant.

For ladders with open ends there is, however, a marked difference between the two cases. The AKLT-like phase has a four-fold degenerate ground state, while the rung singlet state has a unique ground state. This is already an indication that these phases are different. When the coupling constants are changed and seemingly irrelevant couplings are introduced, the rung singlet phase cannot evolve continuously from the AKLT-like phase. The two phases have to be separated by a phase transition line.

The difference between the two phases can be best seen if we count how many valence bonds are cut if the system is cut into two by vertical lines running between neighbouring rungs. Clearly in the AKLT state this number is exactly 1, wherever the cut is made, while in the state where all singlet bonds are on the rungs, this number is always exactly zero. Except for special situations these states are, however, not eigenstates of the Hamiltonian. The interaction between the spins in neighbouring bonds will mix in other configurations. It can be easily seen, that the Hamiltonian has finite matrix elements only between such configurations, in which the number of valence bonds between any neighbouring rungs has the same parity. Thus starting from the AKLT state, in all new configurations that have to be considered to build up a complete basis, the number of valence bonds cut in this way is always an odd number. Similarly, starting from the strict rung single state, in all new configurations this number is even, as seen in Fig. (6).

It can also be easily seen, that this topological distinction using the parity of the number of valence bonds does not hold for odd-leg ladders. As shown in Fig. (7), the parity of the number of valence bonds alternates. This indicates that the ground state of odd-leg ladders is generically two-fold degenerate, dimerized. Although this should be the typical situation, it may happen that long range valence bonds are also mixed in due to the interaction, in which case the gap may vanish, as is the case when only isotropic nearest neighbour exchange is present. The isotropic three-leg ladder has a gapless spectrum.

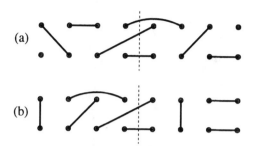

(a)

(b)

Figure 6   Counting the valence bonds between neighbouring rungs (a) in the AKLT-like state and (b) in the rung singlet state.

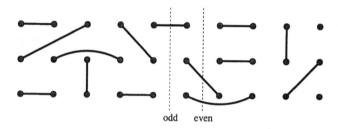

odd   even

Figure 7   The alternation of even and odd number of valence bond in three-leg ladders.

Since in even-leg ladders the parity of the number of valence bonds is a topological invariant, provided the interaction between the spins is isotropic, one cannot go continuously from an even parity state to an odd parity state by changing the coupling constants. There has to be a phase transition between them. This happens at $J_1 = 0$ for the usual ladder, or at $J_2 = 0$ for the diagonal ladder. Allowing for rung, as well as diagonal couplings, the transition occurs at $J_2 = J_1/2$. In the first two cases the transition takes place at the point when the two chains become decoupled, and infinitely long ranged valence bonds appear. The gap vanishes at that point, and the transition is of second order. When the rung and diagonal couplings compete, the transition could still be of second order, but first order transition may also occur, when the energies of an even and an odd states become equal.

Since the topological quantum number is not directly accessible for measurements, it is useful to have another signature of the type of the state. As mentioned before, the ground state of the spin-1 chain has a hidden order in the Haldane phase, which can be detected by the string order parameter [5]. Since there is a one to one correspondence between the low lying part of the energy spectrum of the spin-1 chain and that of the antiferromagnetic diagonal ladder [10], it is natural to expect that a similar string order exists in the AKLT-like Haldane phase of two-leg spin ladders. As a natural generalization of the string order parameter for the spin-1 chain, we write the order parameter for the case when the number of valence bonds is odd, in the form

$$\mathcal{O}^\alpha_{\text{odd}} = - \lim_{|i-j|\to\infty} \left\langle S^\alpha_{i,\text{odd}} \exp\left(i\pi \sum_{l=i+1}^{j-1} S^\alpha_{l,\text{odd}} \right) S^\alpha_{j,\text{odd}} \right\rangle, \qquad (4)$$

where $S^\alpha_{i,\text{odd}}$ is the α component of the total spin on rung i,

$$S^\alpha_{i,\text{odd}} = S^\alpha_{i,1} + S^\alpha_{i,2}. \qquad (5)$$

If the rung singlet (even number of valence bonds) state could be considered as a Haldane phase, a hidden order should exist in this phase as well. Instead of looking at the sequence of the total spin components on the rungs, it should be defined as

$$\mathcal{O}^\alpha_{\text{even}} = - \lim_{|i-j|\to\infty} \left\langle S^\alpha_{i,\text{even}} \exp\left(i\pi \sum_{l=i+1}^{j-1} S^\alpha_{l,\text{even}} \right) S^\alpha_{j,\text{even}} \right\rangle, \qquad (6)$$

where $S^\alpha_{i,\text{even}}$ is the sum of two diagonally lying spins,

$$S^\alpha_{i,\text{even}} = S^\alpha_{i+1,1} + S^\alpha_{i,2}. \qquad (7)$$

Numerical calculations were performed [11] to check if in fact the even or odd order parameters have finite value in the two kinds of Haldane phase, respectively. It was found that the two string order parameters are mutually exclusive, a phase transition line separates the region in the parameter space, where $\mathcal{O}_{\text{odd}}$ is finite, from the region where $\mathcal{O}_{\text{even}}$ takes on finite values. It should be mentioned, however, that although these string order parameters distinguish the even and odd parity states, their finite value does not necessarily indicate a Haldane like phase. Assuming four-spin interactions on plaquettes in the form

$$\begin{aligned}
\mathcal{H}_3 &= J_3 \sum_{i=1}^{N} \left(\vec{S}_{i,1} \cdot \vec{S}_{i+1,1} \right) \left(\vec{S}_{i,2} \cdot \vec{S}_{i+1,2} \right) \\
&+ J_4 \sum_{i=1}^{N} \left(\vec{S}_{i,1} \cdot \vec{S}_{i+1,2} \right) \left(\vec{S}_{i,2} \cdot \vec{S}_{i+1,1} \right),
\end{aligned} \qquad (8)$$

new phases may appear [12]. One of them is a dimerized state where the rungs are bound predominantly pairwise by two valence bounds into almost independent plaquettes. This gapped state is different from the Haldane phase, since it is dimerized. Nevertheless it has a finite even string order.

4.   ANISOTROPIC MODELS

Until now we have considered isotropic models only, where in even-leg ladders the Haldane gap is generated for arbitrary weak ferromagnetic or antiferromagnetic interchain coupling. The question arises if this is true for anisotropic spin ladders, and particularly if all even-leg ladders behave in the same way in this respect.

The Hamiltonian of the ladder is again written as the sum of intrachain and interchain terms. Assuming an XXZ-like anisotropy, the Hamiltonian of the intrachain interaction between spins in the lth chain is

$$H_l = \sum_i \left[\left(S^x_{i,l} S^x_{i+1,l} + S^y_{i,l} S^y_{i+1,l} \right) + \Delta S^z_{i,l} S^z_{i+1,l} \right]. \qquad (9)$$

For simplicity we will assume diagonal interchain coupling, and the same kind of anisotropy in the interchain coupling. The Hamiltonian describing the coupling between spins in chains l and l' is

$$
\begin{aligned}
H_{ll'} &= J_\perp \sum_i \left[\left(S^x_{i,l} S^x_{i+1,l'} + S^x_{i,l'} S^x_{i+1,l} + S^y_{i,l} S^y_{i+1,l'} + S^y_{i,l'} S^y_{i+1,l} \right) \right. \\
&\quad \left. + \Delta \left(S^z_{i,l} S^z_{i+1,l'} + S^z_{i,l'} S^z_{i+1,l} \right) \right].
\end{aligned}
\qquad (10)
$$

When $J_\perp = 1$, this ladder model with p legs becomes equivalent, at least in the low energy sector, to a spin chain of spin $p/2$. That means that at $J_\perp = 1$ the Haldane phase exists for a relatively large range of anisotropy in two-leg ladders, while in four-leg ladders the Haldane phase is restricted to a narrow range of anisotropy around $\Delta = 1$. Depending on whether the interchain couplings are relevant or irrelevant, two possible scenarios for the phase diagram of the two- and four-leg ladders are shown in Figs. (8) and (9), respectively.

In order to study analytically the effect of the interchain coupling, whether it is relevant or irrelevant, in other words whether it can generate a gap or not, it is useful to express the spin variables in terms of bosonic fields [13]. The intrachain terms give harmonic bosonic Hamiltonians,

$$\mathcal{H}_l = \frac{u}{2} \int dx \left[K \Pi_l^2 + \frac{1}{K} (\partial_x \Phi_l)^2 \right]. \qquad (11)$$

where

$$u = \frac{\pi \sqrt{1 - \Delta^2}}{2 \arccos \Delta}, \qquad K = \frac{\pi}{2(\pi - \arccos \Delta)}, \qquad (12)$$

while the interchain coupling becomes

$$
\begin{aligned}
\mathcal{H}_{ll'} &= \int \frac{dx}{(2\pi a)^2} \left[g_1 \cos \left(\sqrt{4\pi}(\Phi_l + \Phi_{l'}) \right) + g_2 \cos \left(\sqrt{4\pi}(\Phi_l - \Phi_{l'}) \right) \right. \\
&\quad \left. + g_3 \cos \left(\sqrt{\pi}(\Theta_l - \Theta_{l'}) \right) \right] + \frac{2 J_\perp \Delta}{\pi} \int dx \partial_x \Phi_l \partial_x \Phi_{l'} \\
&\quad + \int \frac{dx}{(2\pi a)^2} \left[g_4 \cos \left(\sqrt{\pi}(\Theta_l - \Theta_{l'}) \right) \cos \left(\sqrt{4\pi}(\Phi_l + \Phi_{l'}) \right) \right. \\
&\quad \left. + g_5 \cos \left(\sqrt{\pi}(\Theta_l - \Theta_{l'}) \right) \cos \left(\sqrt{4\pi}(\Phi_l - \Phi_{l'}) \right) \right],
\end{aligned}
\qquad (13)
$$

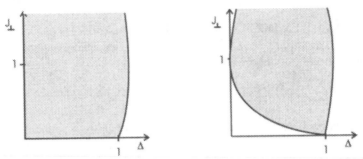

Figure 8   Possible phase diagram of the anisotropic two-leg ladder. The shaded region is the Haldane phase.

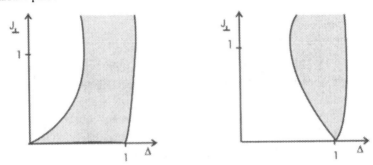

Figure 9   Possible phase diagram of the anisotropic four-leg ladder. The shaded region is the Haldane phase.

where

$$g_1 = 4J_\perp\Delta, \quad g_2 = -4J_\perp\Delta, \quad g_3 = -4\pi J_\perp, \quad g_4 = g_5 = 2\pi J_\perp. \quad (14)$$

The interaction term containing $\partial_x\Phi_l\partial_x\Phi_{l'}$ can be eliminated by taking symmetric and antisymmetric combinations of the field, but the remaining interaction will still be highly complicated, and a simple dimensional analysis is not sufficient to decide, which of the various terms are relevant. The scaling equations have been derived and solved in Ref. [14]. It was shown that in two-leg ladders in the whole interesting range of anisotropy ($0 < \Delta \leq 1$) the interchain coupling is relevant irrespective of its sign. Both the symmetric and antisymmetric combinations of the phase fields become pinned and both modes are gapped.

In four-leg ladders the situation is different. Out of the four phase fields belonging to the four chains one symmetric and three antisymmetric combinations can be formed. Although the antisymmetric modes are gapped, for weak $J_\perp$ the symmetric mode does not develop a gap in the whole $0 \leq \Delta < 0$ range. The interchain coupling is relevant at the isotropic point $\Delta = 1$ only. Otherwise in a very narrow range of anisotropy gap can be generated in the symmetric mode above a certain critical value of the interchain coupling. Looking back to Figs. (8) and (9), we see that for two-leg ladders the scenario shown on the left panel is realized, while for four-leg ladders the one on the right panel.

5. SUMMARY

In summary, we have shown that in even-leg ladders, if the ground state can be represented as a superposition of short range valence bonds, a topological quantum number can be defined. It is obtained by counting the parity of the number of valence bonds between any two neighbouring rungs. Using this quantum number two different Haldane phases are found, which have to be separated by phase transition lines in the phase diagram of spin ladder models. Two string order parameters can be defined to measure which of the two phases are obtained for a given set of parameters.

We have also shown using analytic results obtained for the bosonized version of the spin ladder models, that the mechanism of gap generation is different in two-leg and four-leg ladders. In anisotropic two-leg ladders any coupling between the two legs will generate a gap, since both the symmetric and anti-symmetric spin excitations become massive, while in four-leg ladders where there are four possible excitations (one symmetric and three antisymmetric), one of them (the symmetric mode) remains gapless. This explains the narrow Haldane regime in $S > 1$ spin systems.

Acknowledgments

The author wish to acknowledge G. Fáth, Ö. Legeza, E.H. Kim and D.J. Scalapino for their contribution in the work on which this report is based. This work was partly funded by the Hungarian Science Foundation (OTKA) under Grant No. 30173.

References

[1] For reviews see T.M. Rice, Z. Phys. B **103**, 165 (1997); E. Dagotto, cond-mat/9908250.

[2] K. Hida, J. Phys. Soc. Jpn. **60**, 1347 (1991); S. Takada and H. Watanabe, J. Phys. Soc. Jpn. **61**, 39 (1992); T. Barnes and J. Riera, Phys. Rev. B **50**, 6817 (1994).

[3] E. Dagotto, J. Riera, and D.J. Scalapino, Phys. Rev. B **45**, 5644 (1992); T. Barnes, E. Dagotto, J. Riera, and E.S. Swanson, Phys. Rev. B **47**, 3196 (1993); H. Watanabe, Phys. Rev. B **50**, 13 442 (1994).

[4] F.D.M. Haldane, Phys. Rev. Lett. **50**, 1153 (1983); Phys. Lett. **93A**, 464 (1983).

[5] M.P.M. den Nijs and K. Rommelse, Phys. Rev. B **40**, 4709 (1989).

[6] E.H. Kim, G. Fáth, J. Sólyom, and D.J. Scalapino, cond-mat/9910023.

[7] I. Affleck, T. Kennedy, E.H. Lieb, and H. Tasaki, Phys. Rev. Lett. **59**, 799 (1987).

[8] S.R. White, R.M. Noack, and D.J. Scalapino, Phys. Rev. Lett. **73**, 886 (1994).

[9] S.R. White, Phys. Rev. B **53**, 52 (1996).

[10] J. Sólyom and J. Timonen, Phys. Rev. B **34**, 487 (1986).

[11] Ö. Legeza, G. Fáth, and J. Sólyom, to be published.

[12] Ö. Legeza, G. Fáth, and J. Sólyom, Phys. Rev. B **55**, 291 (1997).

[13] A.M. Tsvelik, Quantum Field Theory in Condensed Matter Physics, Cambridge University Press, Cambridge 1995.

[14] E.H. Kim and J. Sólyom, Phys. Rev. **60**, 15 230 (1999).

DIAGRAMMATIC THEORY OF ANDERSON IMPURITY MODELS:

FERMI AND NON-FERMI LIQUID BEHAVIOR

Johann Kroha and Peter Wölfle

Institut für Theorie der Kondensierten Materie
Universität Karlsruhe, P.O. Box 680, 76128 Karlsruhe, Germany

Abstract   We review a recently developed method, based on a pseudoparticle representation of correlated electrons, to describe both Fermi liquid and non-Fermi liquid behavior in quantum impurity systems. The role of the projection onto the physical Hilbert space and the impossibility of slave boson condensation are discussed. By summing the leading coherent spin and charge fluctuation processes in a fully self-consistent and gauge invariant way one obtains the correct infrared behavior of the pseudoparticles. The temperature dependence of the spin susceptibility for the single channel and two-channel Anderson models is calculated and found to agree well with exact results.

1.   INTRODUCTION

Highly correlated electron systems are characterized by a strong repulsion between electrons on the same lattice site, effectively restricting the dynamics to the Fock subspace of states without double occupancy of sites. The prototype model for such systems is the Anderson impurity model, which consists of an electron in a localized level $\varepsilon_d < 0$ (called d-level in the following) with on-site repulsion U, hybridizing via a transition matrix element V with one or several degenerate conduction electron bands or channels [1]. Depending on the number of channels M, the model exhibits the single- or the multi-channel Kondo effect, where at temperatures T below the Kondo temperature T_K the local electron spin is screened ($M = 1$) or overscreened ($M \geq 2$) by the conduction electrons, leading to Fermi liquid (FL) or to non-Fermi liquid (NFL) behavior [2] with characteristic low-temperature singularities, respectively.

As perhaps the simplest model to investigate the salient features of correlations induced by short-range repulsion, the Anderson model plays a central role for the description of strongly correlated electron systems: In the limit of large spatial dimensions [3] strongly correlated lattice systems reduce in general to

J. Bonča et al. (eds.), Open Problems in Strongly Correlated Electron Systems, 101–110.
© 2001 *Kluwer Academic Publishers. Printed in the Netherlands.*

a single Anderson impurity hybridizing with a continuum of conduction electron states whose properties are determined from a self-consistency condition imposed by the translational invariance of the system [4]. Quantum impurity models have received further interest due to their relevance for mesoscopic systems like single electron transistors or defects in quantum point contacts.

The above-mentioned systems call for the development of accurate and flexible theoretical methods, applicable to situations where exact solution methods are not available. We here present a general, well-controlled auxiliary boson technique which correctly describes the FL as well as the NFL case of the generalized SU(N)×SU(M) Anderson impurity model. As a standard diagram technique it has the potential to be generalized for correlated lattice problems as well as for non-equilibrium situations in mesoscopic systems.

2.   AUXILIARY PARTICLE REPRESENTATION

2.1   SU(N)×SU(M) ANDERSON IMPURITY MODEL

The auxiliary particle method [5] is a powerful tool to implement the effective restriction to the sector of Fock space with no double occupancy imposed by a large on-site repulsion U. The creation operator for an electron with spin σ in the d-level is written in terms of fermionic operators f_σ and bosonic operators b as $d_\sigma^\dagger = f_\sigma^\dagger b$. This representation is exact, if the constraint that the total number operator of auxiliary fermions f_σ and bosons b is equal to unity is obeyed. $f_\sigma^\dagger$ and $b^\dagger$ may be envisaged as creating the three allowed states of the impurity: singly occupied with spin σ or empty.

In view of the possibility of both FL and NFL behavior in quantum impurity systems mentioned in the introduction it is useful to introduce M degenerate channels for the conduction electron operators $c_{\sigma\mu}^\dagger$, labeled $\mu = 1, 2, \ldots, M$, in such a way that in the limit of impurity occupation number $n_d \to 1$ (Kondo limit) the M-channel Kondo model is recovered, i.e. the model obeys an SU(M) channel symmetry. The slave bosons then form an SU(M) multiplet $b_{\bar\mu}$ which transforms according to the conjugate representation of SU(M), so that μ is a conserved quantum number. Generalizing, in addition, to arbitrary spin degeneracy N, $\sigma = 1, 2, \ldots, N$, one obtains the SU(N)×SU(M) Anderson impurity model in pseudoparticle representation

$$H = \sum_{\vec{k},\sigma,\mu} \varepsilon_{\vec{k}} c_{\vec{k}\mu\sigma}^\dagger c_{\vec{k}\mu\sigma} + E_d \sum_\sigma f_\sigma^\dagger f_\sigma + V \sum_{\vec{k},\sigma,\mu} (c_{\vec{k}\mu\sigma}^\dagger b_{\bar\mu}^\dagger f_\sigma + h.c.) , \qquad (1)$$

where the local operator constraint $\hat{Q} \equiv \sum_\sigma f_\sigma^\dagger f_\sigma + \sum_\mu b_{\bar\mu}^\dagger b_{\bar\mu} = 1$ must be fulfilled at all times.

2.2   GAUGE SYMMETRY AND PROJECTION ONTO THE PHYSICAL FOCK SPACE

The system described by the auxiliary particle Hamiltonian (1) is invariant under simultaneous, local $U(1)$ gauge transformations, $f_\sigma \to f_\sigma e^{i\phi(\tau)}$, $b_{\bar\mu} \to b_{\bar\mu} e^{i\phi(\tau)}$, with $\phi(\tau)$ an arbitrary, time dependent phase. While the gauge symmetry guarantees the conservation of the local, integer charge Q, it does not single out any particular Q, like $Q = 1$. In order to effect the projection onto the $Q = 1$ sector of Fock space, one may use the following procedure [6, 7]: Consider first the grand-canonical ensemble with respect to Q and the associated chemical potential $-\lambda$. The expectation value in the $Q = 1$ subspace of any physical operator $\hat{A}$ acting on the impurity states is then obtained as

$$\langle \hat{A} \rangle = \lim_{\lambda \to \infty} \frac{\frac{\partial}{\partial\zeta} \mathrm{tr}\left[\hat{A} e^{-\beta(H+\lambda Q)}\right]_G}{\frac{\partial}{\partial\zeta} \mathrm{tr}\left[e^{-\beta(H+\lambda Q)}\right]_G} = \lim_{\lambda \to \infty} \frac{\langle \hat{A} \rangle_G}{\langle Q \rangle_G}, \qquad (2)$$

where the index G denotes the grand canonical ensemble and ζ is the fugacity $\zeta = e^{-\beta\lambda}$. In the second equality of Eq. (2) we have used the fact that any physical operator $\hat{A}$ acting on the impurity is composed of the impurity electron operators d_σ, $d_\sigma^\dagger$, and thus annihilates the states in the $Q = 0$ sector, $\hat{A}|Q = 0\rangle = 0$. It is obvious that the grand-canonical expectation value involved in Eq. (2) may be factorized into auxiliary particle propagators using Wick's theorem, thus allowing for the application of standard diagrammatic techniques. The local U(1) gauge symmetry, which must not be broken according to Elitzur's theorem [8], precludes the existence of an auxiliary Bose condensate. The latter would imply a spurious phase transition at finite T in quantum impurity models and should, therefore, be avoided in any approximation.

It is important to note that, in general, λ plays the role of a time dependent gauge field. In Eq. (2) a time independent gauge for λ has been chosen. In this way, the projection is only performed at one instant of time, explicitly exploiting the conservation of the local charge Q. Thus, choosing the time independent gauge means that in the subsequent development of the theory, the Q conservation must be implemented exactly. This is achieved in a systematic way by means of conserving approximations [9], i.e. by deriving all self-energies and vertices by functional derivation from one common Luttinger-Ward functional Φ of the fully renormalized Green's functions,

$$\Sigma_{b,f,c} = \delta\Phi\{G_b, G_f, G_c\}/\delta G_{b,f,c}. \qquad (3)$$

This amounts to calculating all quantities of the theory in a self-consistent way, but has the great advantage that gauge field fluctuations need not be considered.

2.3   INFRARED THRESHOLD BEHAVIOR OF AUXILARY PROPAGATORS

The projection onto the physical subspace, Eq. (2), implies that the pseudo-fermion and slave boson Green's functions G_f, G_b are definied as the usual time-ordered, grand canonical expectation values of a pair of creation and annihilation operators, however evaluated in the limit $\lambda \to \infty$. It follows that the traces involved in G_f, G_b are taken purely over the the $Q = 0$ sector of Fock space, and thus the backward-in-time contribution to the auxiliary particle propagators vanishes. Consequently, the auxiliary particle propagators are formally identical to the core hole propagators appearing in the well-known X-ray problem [11], and the long-time behavior of G_f (G_b) is determined by the orthogonality catastrophe [12] of the overlap of the Fermi sea without impurity ($Q = 0$) and the fully interacting conduction electron sea in the presence of a pseudofermion (slave boson) ($Q = 1$). It may be shown that the auxiliary particle spectral functions have threshold behavior with vanishing spectral weight at $T = 0$ for energies ω below a threshold E_o, and power law behavior above E_o, $A_{f,b}(\omega) \propto \Theta(\omega - E_o)\omega^{-\alpha_{f,b}}$.

For the single-channel Anderson model, which is known to have a FL ground state, the threshold exponents may be deduced from an analysis in terms of scattering phase shifts, using the Friedel sum rule, since in the spin screened FL state the impurity acts as a pure potential scatterer [13, 14, 15, 16],

$$\alpha_f = \frac{2n_d - n_d^2}{N}, \qquad \alpha_b = 1 - \frac{n_d^2}{N} \qquad (N \geq 1, M = 1) \qquad (4)$$

These results have been confirmed by numerical renormalization group (NRG) calculations [17] and by use of the Bethe ansatz solution in connection with boundary conformal field theory (CFT) [18]. On the contrary, in the NFL case of the multi-channel Kondo model the threshold exponents have been deduced by a CFT solution [19] as

$$\alpha_f = \frac{M}{M + N}, \qquad \alpha_b = \frac{N}{M + N} \qquad (N \geq 2, M \geq N) \qquad (5)$$

Since the dependence of α_f, α_b on the impurity occupation number n_d shown above originates from pure potential scattering, it is characteristic for the FL case. The auxiliary particle threshold exponents are, therefore, indicators for FL or NFL behavior in quantum impurity models of the Anderson type.

2.4   SADDLEPOINT PROJECTION

Equivalently to the method shown in Eq. (2) the projection onto the $Q = 1$ subspace can be represented as an integration over the parameter λ,

$$\langle \hat{A} \rangle = \int \mathcal{D}[f, \bar{f}; b, \bar{b}; c, \bar{c}] \int \frac{d\lambda}{2\pi T} e^{-\beta[i\lambda(Q-1)+S\{f,\bar{f};b,\bar{b};c,\bar{c}\}]} \langle \hat{A} \rangle_G, \qquad (6)$$

where S is the action of the Anderson model corresponding to Eq. (1). This representation opens up the possibility of constructing an approximate projection scheme [10]: Performing the λ-integration in saddlepoint approximation (λSPA), i.e. evaluating Eq. (6) at the stationary point of the grand canonical action $S_G = S + i\lambda(Q - 1)$, is equivalent to determining $-i\lambda$ as a (real) thermodynamical chemical potential λ_o, fixing $\langle Q \rangle = 1$. The λSPA respects several of the exact properties discussed in sections 2.2 and 2.3: (1) Although the threshold property of the auxiliary spectral function no longer holds, their infrared powerlaw behavior is preserved. Within NCA (section 3.1) it can be shown explicitly that the λSPA does not change the exponents α_f, α_b, and the same may be expected to hold for higher order self-consistent approximations (section 3.3). (2) Since the boson field is treated as a purely fluctuating field, Bose condensation does not occur in the λSPA. Technically, this is achieved in that the boson spectral function $A_b(\omega)$, not restricted by the threshold property in this approximation, acquires weight at negative freqencies, which is negative because of the stability requirement $\omega A_b(\omega) \geq 0$ for bosonic functions. The latter is seen in Fig. 1. Obviously, the saddlepoint projection is immediately generalizable to lattice problems, since the *average* local charge $\langle Q \rangle$ and thus λ_o are space independent.

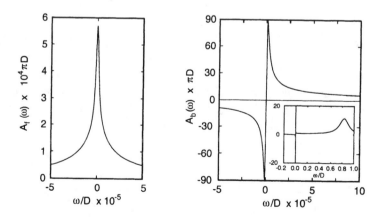

Figure 1 Auxiliary fermion and boson spectral functions $A_f(\omega)$, $A_b(\omega)$ near $\omega = 0$, evaluated within NCA in λ saddlepoint approximation for $T = 10^{-6}D$, $\varepsilon_d = -0.8D$, $\Gamma = 0.15D$, $T_K = 7 \cdot 10^{-5}D$; D = half band width. Power law behavior for $\omega \to 0$ and spectral weight at $\omega < 0$ are seen. The inset shows $A_b(\omega)$ at larger ω.

3.   CONSERVING SLAVE PARTICLE T-MATRIX APPROXIMATION

3.1   NON-CROSSING APPROXIMATION (NCA)

The conserving formulation discussed in section 2.2 precludes mean field approximations which break the $U(1)$ gauge symmetry, like slave boson mean field theory. Although the latter can in some cases successfully describe the low T behavior of models with a FL ground state, it leads to a spurious phase transition at finite T and, in particular, fails to describe NFL systems.

Rather, the approximation should be generated from a Luttinger-Ward functional Φ. Using the hybridization V as a small parameter, one may generate successively more complex approximations. The lowest order conserving approximation generated in this way is the Non-crossing Approximation (NCA) [20, 21], defined by the first diagram in Fig. 3, labeled "NCA". The NCA is successful in describing Anderson type models at temperatures above and around the Kondo temperature T_K, and even reproduces the threshold exponents Eq. (5) for the NFL case of the Anderson impurity model. However, it fails to describe the FL regime at low temperatures. This may be traced back to the failure to capture the spin-screened Kondo singlet ground state of the model, since coherent spin flip scattering is not included in NCA, as seen below.

3.2   DOMINANT LOW-ENERGY CONTRIBUTIONS

In order to eliminate the shortcomings of the NCA mentioned above, we may use as a guiding principle to look for contributions to the vertex functions which renormalize the auxiliary particle threshold exponents to their correct values, since this is a necessary condition for the description of FL and NFL behavior, as discussed in section 2.3. As shown by power counting arguments [22], there are no corrections to the NCA exponents in any finite order of perturbation theory. Thus, any renormalization of the NCA exponents must be due to singularities arising from an infinite resummation of terms. In general, the existence of collective excitations leads to a singular behavior of the corresponding two–particle vertex function. In view of the tendency of Kondo systems to form a collective spin singlet state, we expect a singularity in the spin singlet channel

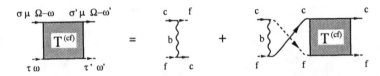

Figure 2   Diagrammatic representation of the Bethe–Salpeter equation defining the conduction electron–pseudofermion T-matrix $T^{(cf)}$.

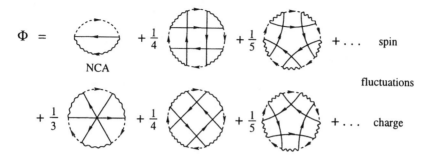

$$\Phi = \quad \text{NCA} \quad + \frac{1}{4} \quad + \frac{1}{5} \quad + \dots \quad \text{spin}$$

fluctuations

$$+ \frac{1}{3} \quad + \frac{1}{4} \quad + \frac{1}{5} \quad + \dots \quad \text{charge}$$

Figure 3 Diagrammatic representation of the Luttinger-Ward functional generating the CTMA. The terms with the conduction electron lines running clockwise (labelled "spin fluctuations") generate $T^{(cf)}$, while the terms with the conduction electron lines running counter-clockwise (labelled "charge fluctuations") generate $T^{(cb)}$. The two-loop diagram is excluded, because it is not a skeleton.

of the pseudofermion–conduction electron vertex function. It is then natural to perform a partial resummation of those contributions which, at each order in the hybridization V, contain the maximum number of spin flip processes. This amounts to calculating the conduction electron–pseudofermion vertex function in the "ladder" or T-matrix approximation, $T^{(cf)}$, where the irreducible vertex is given by $V^2 G_b$. The Bethe–Salpeter equation for $T^{(cf)}$ reads (Fig. 2),

$$T_{\sigma\tau,\sigma'\tau'}^{(cf)\,\mu}(i\omega_n, i\omega_n', i\Omega_n) = \quad + \quad V^2 G_{b\bar\mu}(i\omega_n + i\omega_n' - i\Omega_n)\delta_{\sigma\tau'}\delta_{\tau\sigma'}$$

$$- \quad V^2 T \sum_{\omega_n''} G_{b\bar\mu}(i\omega_n + i\omega_n'' - i\Omega_n) \times \qquad (7)$$

$$G_{f\sigma}(i\omega_n'')\, G_{c\mu\tau}^0(i\Omega_n - i\omega_n'')\, T_{\tau\sigma,\sigma'\tau'}^{(cf)\,\mu}(i\omega_n'', i\omega_n', i\Omega_n),$$

where σ, τ, σ', τ' represent spin indices and μ a channel index. A similar integral equation holds for the charge fluctuation T-matrix $T^{(cb)}$; it is obtained from $T^{(cf)}$ by interchanging $f_\sigma \leftrightarrow b_\mu$ and $c_{\sigma\mu} \leftrightarrow c_{\sigma\mu}^\dagger$. Inserting NCA Green's functions for the intermediate state propagators of Eq. (7), we find at low temperatures and in the Kondo regime ($n_d \gtrsim 0.7$) a pole in the singlet channel of $T^{(cf)}$ at an energy of the c-f pair which scales with T_K, indicating the onset of spin singlet formation.

3.3 SELF-CONSISTENT FORMULATION

In order to find a gauge invariant approximation incorporation multiple spin flip processes as discussed in the previous section, we have to find the Luttinger-Ward functional, which by second functional derivation w.r.t. G_c and G_f or G_b generates the T-matrices $T^{(cf)}$ and $T^{(cb)}$, respectively. This generating functional Φ is shown diagrammatically in Fig. 3. The auxiliary particle self-energies are obtained in the conserving scheme as the functional

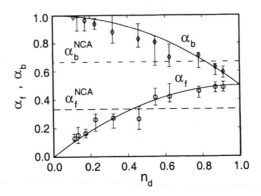

Figure 4   The fermion and boson threshold exponents α_f, α_b are shown for $N = 2$, $M = 1$ in dependence of the average impurity occupation n_d. Solid lines: exact values, Eq. (4); Symbols with error bars: CTMA; dashed lines: NCA.

derivatives of Φ with respect to G_f or G_b, respectively (Eq. (3)), and are, in turn, nonlinear functionals of the full, renormalized auxiliary particle propagators. This defines a set of self-consistency equations, which we term conserving T-matrix approximation (CTMA).

3.4   RESULTS

We have solved the CTMA equations numerically for a wide range of impurity occupation numbers n_d from the Kondo to the empty impurity regime both for the single-channel and for the two-channel Anderson model down to temperatures of the order of at least $10^{-2}T_K$.

The solution of the CTMA equations forces the T-matrices to have vanishing spectral weight at negative COM frequencies Ω. Indeed, the numerical evaluation shows that the poles of $T^{(cf)}$ and $T^{(cb)}$ are shifted to $\Omega = 0$ by self-consistency, where they merge with the continuous spectral weight present for $\Omega > 0$, and thus renormalize the threshold exponents of the auxiliary spectral functions. For $N = 2$, $M = 1$ the threshold exponents α_f, α_b extracted from the numerical solutions are shown in Fig. 4. In the Kondo limit of the multi–channel case ($N \geq 2$, $M = 2, 4$) the CTMA solutions are found not to alter the NCA values and reproduce the the correct threshold exponents, $\alpha_f = M/(M + N)$, $\alpha_b = N/(M + N)$.

The good agreement of the CTMA exponents with their exact values over the complete range of n_d for the single–channel model and in the Kondo regime of the multi–channel model may be taken as evidence that the T-matrix approximation correctly describes both the FL and the non–FL regimes of the SU(N)×SU(M) Anderson model (N=2, M=1,2,4). The static spin susceptibility χ of the single- and of the two-channel Anderson model in the Kondo regime calculated within CTMA as the derivative of the magnetization

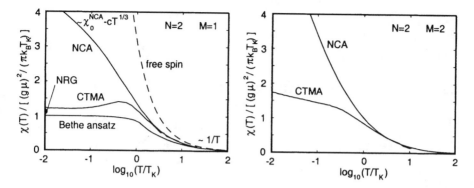

Figure 5 Static susceptibility of the single-channel ($N = 2$, $M = 1$) and the two-channel ($N = 2$, $M = 2$) Anderson impurity model in the Kondo regime ($E_d = -0.8D$, $\Gamma = 0.1D$, Landé factor $g = 2$). In the single-channel case, the CTMA and NCA results are compared to the NRG result ($T = 0$, same parameters) [23] and to the Bethe ansatz for the Kondo model [24]. The CTMA susceptibility obeys scaling behavior in accordance with the exact results (not shown).

$M = \frac{1}{2}g\mu_B\langle n_{f\uparrow} - n_{f\downarrow}\rangle$ with respect to a magnetic field H is shown in Fig. 5. Good quantitative agreement with exact solutions is found for $N = 2$, $M = 1$ (FL). For $N = 2$, $M = 2$ (NFL) CTMA correctly reproduces the exact [25] logarithmic temperature dependence below the Kondo scale T_K. In contrast, the NCA solution recovers the logarithmic behavior only far below T_K.

4. CONCLUSION

We have reviewed a novel technique to describe correlated quantum impurity systems with strong onsite repulsion, which is based on a conserving formulation of the auxiliary boson method. The conserving scheme allows to implement the conservation of the local charge Q without taking into account time dependent fluctuations of the gauge field λ. By including the leading infrared singular contributions (spin flip and charge fluctuation processes), physical quantities, like the magnetic susceptibility, are correctly described both in the Fermi and in the non–Fermi liquid regime, over the complete temperature range, including the crossover to the correlated many–body state at the lowest temperatures. As a standard diagram technique this method has the potential to be applicable to problems of correlated systems on a lattice as well as to mesoscopic systems out of equilibrium via the Keldysh technique.

We wish to thank S. Böcker, T.A. Costi, K. Haule, S. Kirchner, A. Rosch, A. Ruckenstein and Th. Schauerte for stimulating discussions. This work is supported by DFG through SFB 195 and by the Hochleistungsrechenzentrum Stuttgart.

References

[1] A. C. Hewson, *The Kondo Problem to Heavy Fermions* (C.U.P., Cambridge, 1993).

[2] For a comprehensive overview see D. L. Cox and A. Zawadowski, Adv. Phys. **47**, 599 (1998).

[3] W. Metzner and D. Vollhardt, Phys. Rev. Lett. **62**, 324 (1989).

[4] A. Georges et al., Rev. Mod. Phys. **68**, 13 (1996).

[5] S. E. Barnes, J. Phys. **F6**, 1375 (1976); **F7**, 2637 (1977).

[6] A. A. Abrikosov, Physics **2**, 21 (1965).

[7] P. Coleman, *Phys. Rev.* **B29**, 3035 (1984).

[8] S. Elitzur, Phys. Rev. D **12**, 3978 (1975).

[9] G. Baym and L.P. Kadanoff, Phys. Rev. **124**, 287 (1961); G. Baym, Phys. Rev. **127** 1391 (1962).

[10] J. Kroha, P. Hirschfeld, K. A. Muttalib, and P. Wölfle Solid State Comm. **83** (12), 1003 (1992).

[11] P. Nozières and C. T. De Dominicis, Phys. Rev. **178**, 1073; 1084; 1097 (1969).

[12] P. W. Anderson, Phys. Rev. Lett. **18**, 1049 (1967).

[13] K. D. Schotte and U. Schotte, Phys. Rev. **185**, 509 (1969).

[14] B. Menge and E. Müller-Hartmann, Z. Phys. **B73**, 225 (1988).

[15] J. Kroha, P. Wölfle and T. A. Costi, Phys. Rev. Lett. **79**, 261 (1997).

[16] For a more detailed discussion see J. Kroha and P. Wölfle, Acta Phys. Pol. B **29**, 3781 (1998); cond-mat# 9811074.

[17] T.A. Costi, P. Schmitteckert, J. Kroha and P. Wölfle, Phys. Rev. Lett. **73**, 1275 (1994); Physica (Amsterdam) **235-240C**, 2287 (1994).

[18] S. Fujimoto, N. Kawakami and S.K. Yang, J.Phys.Korea **29**, S136 (1996).

[19] I. Affleck and A.W.W. Ludwig, Nucl. Phys. **352**, 849 (1991); **B360**, 641 (1991); Phys. Rev. B **48**, 7297 (1993).

[20] N. Grewe and H. Keiter, Phys. Rev B **24**, 4420 (1981).

[21] Y. Kuramoto, Z. Phys. B **53**, 37 (1983); Y. Kuramoto and H. Kojima, *ibid.* **57**, 95 (1984); Y. Kuramoto, *ibid.* **65**, 29 (1986).

[22] D. L. Cox and A. E. Ruckenstein, Phys. Rev. Lett. **71**, 1613 (1993).

[23] We are grateful to T. A. Costi for providing the NRG data.

[24] N. Andrei, K. Furuya, J.H. Löwenstein, Rev.Mod.Phys. **55**, 331 (1983).

[25] N. Andrei, C. Destri, Phys. Rev. Lett. **52**, 364 (1984).

s+d MIXING IN CUPRATES:

Strong electron correlations and superconducting gap symmetry

N.M.Plakida and V.S. Oudovenko

Joint Institute for Nuclear Research, 141980 Dubna, Russia

Abstract   A microscopical theory of electronic spectrum and superconductivity is formulated within the two-dimensional anisotropic t-J model with $t_x \neq t_y$ and $J_x \neq J_y$. The Dyson equation is solved in the weak coupling approximation. The superconducting pairing mediated by spin-fluctuations has $d+s$ symmetry with the $s-$component being proportional to the asymmetry $|t_x - t_y|$.

1.   INTRODUCTION

Recently the d-wave symmetry of superconducting pairing in cuprates was unambiguously confirmed by observation of half-flux magnetic quanta (see, e.g., [1] and references therein). However, in orthorhombic phase of YBCO a clear s-wave component was observed in c-axis tunneling experiments (see, e.g., [2]) where a mixed $s+d$ symmetry of the superconducting order parameter should be expected from the symmetry consideration.

The problem of the mixed $s+d$ symmetry of the superconducting gap function in the orthorhombic phase within the conventional Fermi-liquid models was discussed in a number of papers (see, e.g., [3] and references therein). For models with strong electron correlations the s-wave component must be strongly suppressed due to on-site Coulomb correlations. For the 2D t-J model it follows from the constraint of no double occupancy on a single site given by the identity [4]:

$$\langle \hat{c}_{i,\sigma}\hat{c}_{i,-\sigma} \rangle = \frac{1}{N} \sum_{k_x,k_y} \langle \hat{c}_{\mathbf{k},\sigma}\hat{c}_{-\mathbf{k},-\sigma} \rangle = 0, \tag{1}$$

for the projected electron operators $\hat{c}_{i,\sigma} = c_{i,\sigma}(1 - n_{i,-\sigma})$. The anomalous correlation function $\langle \hat{c}_{\mathbf{k},\sigma}\hat{c}_{-k,-\sigma} \rangle$ has the symmetry of the gap function, $\Delta(k_x, k_y)$. Therefore to satisfy the condition (1) in the tetragonal phase with the Fermi surface (FS) symmetrical in respect to (k_x, k_y) permutation the gap function should have a lower symmetry, e.g. B_{1g}, "d-wave" symmetry: $\Delta_d(k_x, k_y) = -\Delta_d(k_y, k_x)$. In the orthorhombic phase the FS has no that

111

J. Bonča et al. (eds.), Open Problems in Strongly Correlated Electron Systems, 111–116.

permutation symmetry and the condition (1) can be satisfied for a gap function of the "mixed" ("$d+s$") symmetry (within the A_{1g} irreducible representation):

$$\Delta(k_x, k_y) = \Delta_d(k_x, k_y) + s \, \Delta_s(k_x, k_y). \tag{2}$$

where $\Delta_s(k_x, k_y) = \Delta_s(k_y, k_x)$ is "the extended s-wave" component.

   In this report we present results of calculations of superconducting T_c for the 2D t-t'-J model, within the theory developed by us in [5], both in tetragonal and orthorhombic phases. In the next Sect.2 we derive the Dyson equation and in Sect.3 formulate it in the weak coupling approximation. Numerical results and discussion are presented in Sect.4.

2.   MODEL AND DYSON EQUATION

   We consider the t-J model in the Hubbard operator (HO) representation for $\hat{c}_{i\sigma}^+ = X_i^{\sigma 0}$ and $\hat{c}_{j\sigma} = X_j^{0\sigma}$

$$H_{t-J} = - \sum_{i \neq j, \sigma} t_{ij} X_i^{\sigma 0} X_j^{0\sigma} + \frac{1}{4} \sum_{i \neq j, \sigma} J_{ij} \left(X_i^{\sigma\bar{\sigma}} X_j^{\bar{\sigma}\sigma} - X_i^{\sigma\sigma} X_j^{\bar{\sigma}\bar{\sigma}} \right), \tag{3}$$

where $\bar{\sigma} = -\sigma$. In the orthorhombic phase, with the lattice constants $a_x \neq a_y$, the electron hopping energy for the nearest neighbors, $a_1 = \pm a_x, \pm a_y$, can be written as $t_{ij} = t_x \delta_{j,i \pm a_x} + t_y \delta_{j,i \pm a_y}$, while for the second neighbors, $a_2 = \pm(a_x \pm a_y)$, it equals to $t_{ij} = t'$. The exchange interaction for the nearest neighbors is $J_{ij} = J_x \delta_{j,i \pm a_x} + J_y \delta_{j,i \pm a_y}$. The anisotropy parameters in the orthorhombic phase, $\alpha = (t_x - t_y)/(t_x + t_y)$ and $\beta = (J_x - J_y)/(J_x + J_y)$ are supposed to be small. In what follows, we will use the dimensionless wave numbers, $\tilde{k}_x = k_x a_x$, $\tilde{k}_y = k_y a_y$ so that in the first Brillouin zone $\pi \leq (\tilde{k}_x, \tilde{k}_x) \leq \pi$. Below we introduce also the chemical potential μ which is defined by the equation for the average number of electrons

$$n = \sum_\sigma \langle X_i^{\sigma\sigma} \rangle = \sum_\sigma \langle X_i^{\sigma 0} X_i^{0\sigma} \rangle. \tag{4}$$

To calculate the Green function we use the equation of motion for the HO:

$$i \frac{d}{dt} X_i^{0\sigma} = - \sum_l t_{il} B_{i\sigma\sigma'} X_l^{0\sigma'} + \frac{1}{2} \sum_l J_{il} (B_{l\sigma\sigma'} - \delta_{\sigma\sigma'}) X_i^{0\sigma'}, \tag{5}$$

where we introduced the operator $B_{i\sigma\sigma'} = (X_i^{00} + X_i^{\sigma\sigma})\delta_{\sigma'\sigma} + X_i^{\bar{\sigma}\sigma}\delta_{\sigma'\bar{\sigma}} = (1 - \frac{1}{2}N_i + \sigma S_i^z)\delta_{\sigma'\sigma} + S_i^{\bar{\sigma}}\delta_{\sigma'\bar{\sigma}}$. This Bose-like operator describes electron scattering on spin and charge fluctuations caused by the nonfermionic commutation relations for the HO (the first term in (5) – the kinematic interaction) and by the exchange spin-spin interaction (the second term in (5)).

By using equations of motion for the thermodynamic Green function $\hat{G}_{ij,\sigma}(t-t')=\langle\langle\Psi_{i\sigma}(t)|\Psi_{j\sigma}^{+}(t')\rangle\rangle$, in terms of the Nambu operators, $\Psi_{i\sigma}$, $\Psi_{i\sigma}^{+}=(X_{i}^{\sigma 0}\ X_{i}^{0\bar{\sigma}})$, we obtain the Dyson equation [5]:

$$\hat{G}^{\sigma}(k,\omega)=Q\tilde{G}^{\sigma}(k,\omega)=Q\{\omega\hat{\tau}_{0}-(E_{k}^{\sigma}-\mu)\hat{\tau}_{3}-\Delta_{k}^{\sigma}\hat{\tau}_{1}-\tilde{\Sigma}(k,\omega)\}^{-1} \quad (6)$$

where $\hat{\tau}_{0}$, $\hat{\tau}_{1}$, $\hat{\tau}_{3}$ are the Pauli matrices, $Q=\langle\{X_{i}^{0\sigma},X_{i}^{\sigma 0}\}\rangle=(1-n/2)$. The energy of the quasiparticles E_{k}^{σ} and the gap function Δ_{k}^{σ} are given by the k-representation the frequency matrix $\hat{E}_{ij\sigma}=\langle\{[\Psi_{i\sigma},H],\Psi_{j\sigma}^{+}\}\rangle Q^{-1}$ which is the mean-field approximation (MFA) within in the projection technique for the GF [5]. The self-energy operator $\tilde{\Sigma}(k,\omega)$ is calculated in the self-consistent Born approximation (SCBA), i.e. in the second order of effective electron-electron interaction induced by spin-charge fluctuations with the vertex $g(q,k-q)=t(q)-(1/2)J(k-q)$.

Numerical studies of the linearized system of the Dyson equations (6) in tetragonal phase for a model short-range AFM susceptibility were performed in [5]. The electron spectral density shows quasiparticle (QP) excitations at the Fermi surface crossing and a dispersive incoherent band. For low hole concentration the QP dispersion is small while the intensity of the incoherent band is quite large. With doping the QP band width strongly increases and the incoherent band is suppressed. The results for single-electron spectral functions were in general agreement with numerical studies for finite clusters [6]. The occupation numbers $N(\mathbf{k})$ have the characteristic behavior for strongly correlated systems. Being large throughout the BZ, due to the incoherent contribution, they show only a small drop at the FS. The volume of the FS at small doping is proportional to the hole concentration δ that does not obey the Luttinger theorem. The superconducting pairing due to the exchange and the kinematic interactions (in the second order) results in the d-wave pairing with high T_c in tetragonal phase. Below we consider modifications for T_c induced by orthorhombic distortion within the asymmetric t-J model (3).

3.   WEAK COUPLING APPROXIMATION

To study the symmetry of the order parameter in orthorhombic phase by taking into account the constraint equation (1) we consider the weak coupling approximation for the Dyson equation (6). Allowing for a renormalization of the coherent part of the spectral weight by Z we write the equation for the gap in the weak coupling approximation in the form:

$$\Phi_{k}^{\sigma}=\frac{1}{N}\sum_{q}\{2g(q,k-q)-\lambda(q,k-q)\}\frac{Z_{q}^{2}\ \Phi_{q}^{\sigma}}{2\Omega_{q}}\tanh\frac{\Omega_{q}}{2T}, \quad (7)$$

where $\Omega_k = [(E_k - \mu)^2 + |\Phi_k^\sigma|^2]^{1/2}$ is the QP energy in the supercon-
ducting state with the frequency independent gap function Φ_k^σ. The first term
comes from the frequency matrix (MFA) while the second one – from the
self-energy $\tilde{\Sigma}(k, \omega)$ evaluated in the weak-coupling approximation. It is given
by the effective interaction $\lambda(q, k - q) = g^2(q, k - q)\chi_s(k - q)$. Here we
take into account only spin-fluctuation coupling given by the static suscep-
tibility: $\chi_s(q) = \chi_0/[1 + \xi^2(1 + \gamma(q))]$, $\gamma(q) = (1/2)(\cos \tilde{q}_x + \cos \tilde{q}_y)$.
It is a periodic function in q-space with maxima at $\mathbf{Q} = (\pm\pi, \pm\pi)$ and the
width being defined by the AFM correlation length ξ. The intensity of the
interaction, $\chi_s(Q) = \chi_0$, is fixed by the normalization condition (see [5]):
$(1/N)\sum_i \langle \mathbf{S}_i \mathbf{S}_i \rangle = (3/4)n$.

To take into account a strong incoherent part of the spectrum defined by
the self-energy we used the following representation for the electronic spectral
density in the normal state:

$$A_{11}^\sigma(k, \omega) \simeq Z_k \delta(\omega + \mu - E_k^\sigma) + A_{11}^{inc}(\omega) . \tag{8}$$

The weight of the quasiparticle part Z_k and the incoherent part $A_{11}^{inc}(\omega)$ are
coupled by the sum rule for the spectral weight: $\int_{-\infty}^{+\infty} d\omega A_{11}^\sigma(k, \omega) = 1$.
By considering a coherent band in the range $-\Gamma \le \omega \le \Gamma$ and an incoherent
band in the range $-W \le \omega \le -\Gamma$ below the coherent band we get for
the incoherent part: $A_{11}^{inc} = (1 - Z_k)/(W - \Gamma)$. The chemical potential
μ is given by the equation for the average number of electrons (4): $n =$
$[(1 - n/2)/N]\sum_{k,\sigma} \int_{-\infty}^{+\infty} d\omega [\exp(\omega/T) + 1]^{-1} A_{11}^\sigma(k, \omega)$. To fix the value
of Z_k we assume that the Fermi surface for QP with the spectrum E_k obeys
the Luttinger theorem: $n = (1/N)\sum_{k,\sigma} \{\exp[(E_k^\sigma - \mu)/T] + 1\}^{-1}$. Now
by taking into account the above given equations we get for the QP weight:
$Z = (1 - n)/(1 - n/2)^2$, $A_{11}^{inc} = n^2/(W - \Gamma)(2 - n)^2$. It is remarkable
that in spite of our crude modelling of the spectral density, Eq. (8), the average
number of electrons, Eq. (4), as follows from the above given equation, obeys
the condition: $n/(1 - n/2) \le 2$,   or $n \le 1$.

Contrary to conventional superconductors, the equation for the gap (7) should
be solved by taking into account also the restriction imposed on the anomalous
correlation function by the constraint equation, Eq. (1), which in our approximation reads

$$\langle X_i^{0\sigma} X_i^{0\bar{\sigma}} \rangle = \frac{1}{N} \sum_q \frac{Z_q^2 \Phi_q^\sigma}{2\Omega_q} \tanh \frac{\Omega_q}{2T} = 0. \tag{9}$$

Below we present results for the superconducting T_c for different doping.

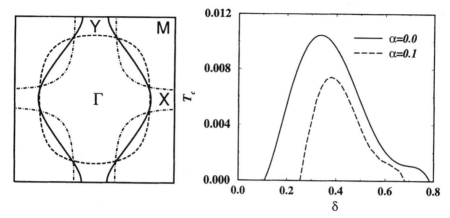

Figure 1 Fermi surface for $t' = 0.0$ (bold line), $t' = -0.2$ (dash-dotted line), and $t' = +0.2$ (dashed line), and hole concentration $\delta = 0.1$

Figure 2 $T_c(\delta)$ in tetragonal (bold line) and orthorhombic (dashed line) phases ($t' = 0.0$)

4.   NUMERICAL RESULTS AND DISCUSSION

In our previous studies [5] of the t-J model in tetragonal phase we have performed direct numerical solution of the linearized Dyson equation (6) to find T_c and the (q, ω)-dependent gap function. In that case, performing an exact diagonalization of the linear gap equation, we have found only the d-wave symmetry of the gap and the constraint equation (9) was fulfilled. In the present studies of orthorhombic phase we consider a simplified method based on the model representation to the gap function. We take the gap function in the form (2) which should obey to the constraint equation (9). Close to the superconducting temperature T_c we can solve Eq. (7) with the linearized QP energy, $\Omega_k = E_k - \mu$. Below we present results for T_c based on the direct numerical self-consistent solution of the gap equation (7) and the constraint equation (9). Numerical calculations were performed for several values of parameters for the t-J model: $J/t = 0.4,\ 0.6;\ \alpha = \beta = 0.0,\ 0.1;\ t' = 0.0,\ \pm 0.2$, the AFM correlation length $\xi = 2,\ 3$ in the model function $\chi_s(q)$, and the hole concentration $\delta = 0.1 - 0.4$. All the energies and temperature are measured in units of $t = (1/2)(t_x + t_y)$.

We observe quite a strong change of the QP dispersion depending on the n.n.n. hopping, t', and much larger effective bandwidth for large doping, $\delta = 0.4$. These changes of the QP spectrum strongly influence the Fermi surface (FS) as it is shown on Fig.1 for the parameters: $t' = 0.0$ (bold line), $t' = -0.2$ (dash-dotted line), and $t' = +0.2$ (dashed line), and the hole concentration $\delta = 0.1$.

The critical temperature $T_c(\delta)$ (in units of t) is shown on Fig.2 in the tetragonal, $\alpha = 0.0$, (bold line) and orthorhombic, $\alpha = 0.1$, (dashed line) phases

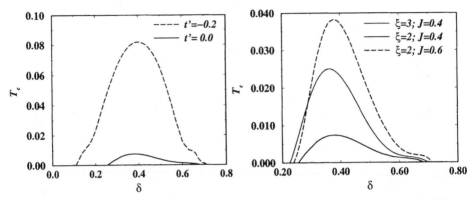

Figure 3 $T_c(\delta)$ in orthorhombic phase for $t' = 0.0$ (bold line), $t' = -0.2$ dot-dashed line) ($t' = 0.0$)

Figure 4 $T_c(\delta)$ in orthorhombic phase for $t' = 0.0$ and $J = 0.4$ with $\xi = 2$ (bold line), $\xi = 3$ (dotted line), and $J = 0.6$ with $\xi = 2$ (dashed line)

for $J = 0.4t$, $\xi = 2$, $t' = 0.0$. Suppression of T_c in the orthorhombic phase is due to a deformation of the FS resulting in a less favorable electron pairing by the AFM spin fluctuations. Its maximum at $\delta \simeq 0.35$ is due to an interplay between the shape of the FS (defined by the quasiparticle spectrum E_q) and the coherent spectral weight Z^2 in Eq.(7). Particularly, for $t'/t = -0.2 \ (+0.2)$ we observed a strong enhancement (suppression) of $T_c(\delta)$ due to change of the FS as shown on Fig.3 for $t'/t = -0.2$. Increasing of AFM interaction due to larger J or/and ξ strongly enhances T_c though it does not change the shape of the curve as shown on Fig.4.

The s-wave component in the orthorhombic phase in the gap function Eq.(2) depends on the parameters of the model and doping and can reach quite a large value in spite of small asymmetry parameters $\alpha \simeq \beta = 0.1$.

References

[1] J.R. Kirtley, C.C. Tsuei, K.A. Moler, *Science*, **285**, 1373 (1999)

[2] A.G. Kouznetsov, A.G. Sun, B. B. Chen, et al., *Phys. Rev. Let.*, **79**, 3050 (1997)

[3] C. O'Donovan, J.P. Carbotte, *Phys. Rev. B*, **52**, 4568 (1995)

[4] N.M. Plakida, V.Yu. Yushankhai, I.V. Stasyuk, *Physica C*, **160**, 80 (1989); V.Yu. Yushankhai, N.M. Plakida, P. Kalinay, *Physica C*, **174**, 401 (1991)

[5] N.M. Plakida, V.S. Oudovenko, *Phys. Rev. B*, **59**, 11949 (1999)

[6] J. Jaklič, P. Prelovšek, P., *Advances in Physics* **49**, 1-92 (2000)

III

CUPRATES:
STRIPE AND CHARGE
ORDERING

FERMI SURFACE, PSEUDOGAPS AND DYNAMICAL STRIPES IN LA$_{2-X}$SR$_X$CUO$_4$

A. Fujimori[1], A. Ino[2], T. Yoshida[1], T. Mizokawa[1]

[1]*Department of Physics and Department of Complexity Science and Engineering, University of Tokyo, Bunkyo-ku, Tokyo 113-0033, Japan*

[2]*Japan Atomic Energy Research Institute, SPring-8, Mikazuki, Sayo, Hyogo 679-5198, Japan*

Z.-X. Shen, C. Kim

Department of Applied Physics and Stanford Synchrotron Radiation Laboratory, Stanford University, Stanford, CA94305, U.S.A.

T. Kakeshita, H. Eisaki, S. Uchida

Department of Advanced Materials Science, University of Tokyo, Bunkyo-ku, Tokyo 113-0033, Japan

Abstract   Doping dependence of the electronic structure of La$_{2-x}$Sr$_x$CuO$_4$ (LSCO) has been systematically studied in a series of photoemission measurements. The unusual spectral features in the underdoped regime are attributed to the formation of dynamical stripes and the opening of large and small pseudogaps.

1.   INTRODUCTION

The most remarkable feature in the high-T_c cuprates is their characteristic phase diagram as a function of hole doping or band filling, which covers from the antiferromagnetic insulating phase near the undoped limit to the normal Fermi-liquid phase in the overdoped regime with the intervening superconducting phase. Furthermore, in the underdoped superconducting phase, "non-Fermi-liquid" properties such as pseudogap behaviors are observed.

In order to systematically understand the origin of the phase diagram and the nature of each phase, La$_{2-x}$Sr$_x$CuO$_4$ (LSCO) is a unique system in that it covers the whole range of the phase diagram in a single system. In addition,

119

J. Bonča et al. (eds.), Open Problems in Strongly Correlated Electron Systems, 119–128.
© 2001 *Kluwer Academic Publishers. Printed in the Netherlands.*

it has the simplest crystal structure with single CuO_2 layers and its hole concentration is rather accurately determined by the Sr concentration x (plus small oxygen non-stoichiometry). On the other hand, LSCO is complicated in that it undergoes a structural distortion from the high-temperature tetragonal (HTT) phase to the low-temperature orthorhombic (LTO) phase in the superconducting compositions and even has an inherent instability towards the low-temperature tetragonal (LTT) phase. The latter phase is realized in $La_{2-x-y}Nd_ySr_xCuO_4$ (LNSCO), accompanied by the ordering of charge and spins in a stripe form, especially around $x \sim 1/8$ [1]. Transport measurements of LNSCO have shown that the static stripes are indeed one-dimensional metals [2]. Recently, LNSCO with $y = 0.4$ and $x = 0.12$ was studied by angle-resolved photoemission spectroscopy (ARPES) by Zhou *et al.* [3]. They observed Fermi surface features characteristic of a (quarter-filled) one-dimensional metal, namely, confinement of spectral weight within $|k_x|, |k_y| < \pi/4$. In LSCO, the stripes are not static but are thought to remain dynamical fluctuations, as reflected on the incommensurate inelastic neutron peaks [4].

In this article, we summarize the results of our photoemission studies on LSCO, focusing on the systematic evolution of the electronic structure as a function of hole doping including the formation of the pseudogaps. Some unusual spectral features characteristic of LSCO are indeed attributed to the dynamical stripes.

2.   BAND DISPERSION AND FERMI SURFACE

The experimental band structure of optimally doped LSCO ($x \simeq 0.15$) studied by ARPES [5, 6] is similar to that of $Bi_2Sr_2CaCu_2O_8$ (BSCCO), which have been extensively studied by ARPES, as shown in Fig. 1. The band is flat around $\mathbf{k} = (\pi, 0)$ [especially along the $(\pi, 0)$ direction] while the band crossing the Fermi level (E_F) around $(\pi/2, \pi/2)$ is strongly dispersive. The "flat band" rises with x, crosses E_F at $x \sim 0.2$ and goes above E_F for $x > 0.2$ [6]. With decreasing x, the flat band is lowered, and around $x \sim 0.1$ it becomes as low as ~ 0.1 eV below E_F [7].

As the position of the flat band relative to E_F changes with x, the Fermi surface topology changes. For $x > 0.2$, Fermi surface crossing occurs on the $(0, 0) - (\pi, 0)$ line, giving rise to an electron-like Fermi surface centered at $(0, 0)$ [5], as shown in Fig. 2. For $x < 0.2$, Fermi-surface crossing occurs on the $(\pi, 0) - (\pi, \pi)$ line, resulting in a hole-like Fermi surface centered at (π, π) as in the other cuprates, as shown in the same figure. Real Fermi-surface crossing does not occur at the measurement temperatures (~ 10 K) for most of the samples, however, because a superconducting gap or a pseudogap is opened on the underlying "Fermi surface". In such a case, the Fermi surface can only be defined by a minimum gap locus [8]. As for the dispersive band around

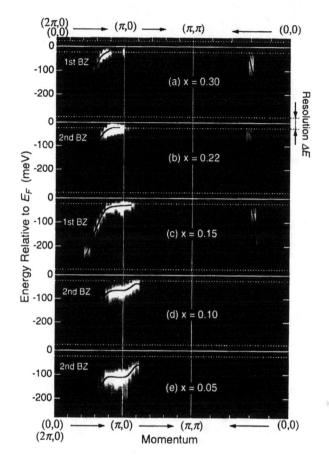

Figure 1   Band dispersions in LSCO [6].

$(\pi/2, \pi/2)$, the doping dependent shift is small. The doping dependence of the shift is strong for the band around $(\pi, 0)$ and weak for the band around $(\pi/2, \pi/2)$, as in the case for BSCCO [9].

Figure 2 also shows that, for the underdoped samples, the "Fermi surface" around $(\pi, 0)$ become rather straight. This suggests that a one-dimension-like electronic structure is realized, giving support to the formation of dynamical stripes. Very recently, the evolution of the Fermi surface described above was demonstrated in terms of two-dimensional intensity plots [10], too, as had been made for LNSCO [3]. Here, it should be noted that the spectral weight distribution around $(\pi, 0)$ is rather complicated, making the determination of the Fermi surface nontrivial. In addition to the intrinsic broadness of spectral features and the opening of the superconducting gap and pseudogap around $(\pi, 0)$, spectral weight remains finite at E_F around $(\pi, 0)$ even for $x = 0.3$, where the band is though to be located well above E_F.

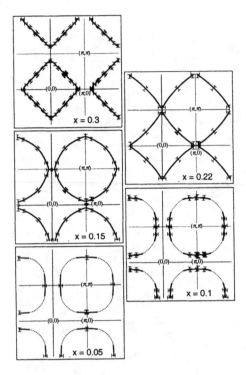

Figure 2   Fermi surface or minimum gap locus of LSCO determined by band dispersions in the ARPES spectra [6].

The intensity of the dispersive band around $(\pi/2, \pi/2)$ is unusually low in LSCO. In the underdoped samples ($x < 0.15$), most of its spectral weight is transferred to the higher binding energies of ~ 0.5 eV, where the insulating samples ($x \sim 0$) show the lower Hubbard band, as shown in Fig. 3 [6, 7]. Here, the insulating sample $x = 0.03$ shows essentially the same band dispersion as the parent Mott insulator such as $Sr_2CuO_2Cl_2$ [11]. The disappearance of the $(\pi/2, \pi/2)$ band or the "nodal" quasi-particle (QP) band can be intuitively understood as due to the presence of dynamical stripes, which extend along the vertical $(0, \pi)$ or horizontal $(\pi, 0)$ direction. That is, the propagation of QP along the diagonal direction would be always disturbed by the hole-poor region of the stripe phase, whereas the propagation along the Cu-O bond direction would be disturbed only with the probability of 50 %. In contrast to the nodal QP, the flat band around $(\pi, 0)$ is robust against the decrease of hole concentration. It persists down to $x \sim 0.05$, i.e., down to the superconductor-insulator boundary, and then spectral weight transfer to the lower Hubbard band at ~ 0.5 eV below E_F occurs.

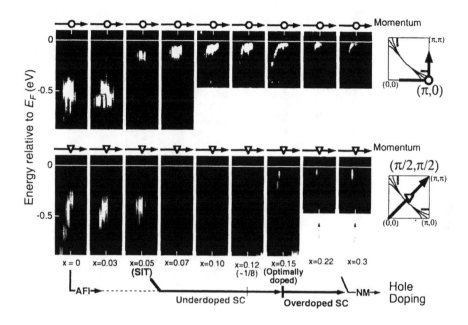

Figure 3 Doping dependence of the band dispersions near $(\pi, 0)$ and $(\pi/2, \pi/2)$ in LSCO [6].

2.1   CHEMICAL POTENTIAL SHIFT

According to the stripe picture, the hole density along each stripe is constant (0.5 per Cu-Cu distance, namely, each stripe is quarter-filled) whereas the density of the charge stripeschanges with hole doping. Therefore, as long as the separation between the stripes is sufficiently large so that the inter-stripe interaction is negligible, the energy of the system per hole remains almost constant. This means that the chemical potential of electron or hole remains unchanged with hole doping and hence apparently remains fixed. The pinning of the chemical potential has indeed been observed in LSCO for $x < 0.12$ [12].

Figure 3 demonstrates that ARPES data show remarkable changes around $x = 0.05$ [7]. The spectral change at $x \sim 0.05$ may reflect the superconductor-to-insulator transition at this composition or the change from the dynamical to static stripes because the coherent spectral weight near E_F, which remains appreciable around $(\pi, 0)$, disappears below this x. This may also reflect the change from the vertical to diagonal stripes at $x < 0.05$ as observed by a recent neutron scattering study [13] because the hole motion along the Cu-O bond direction would be hindered in this composition range. In the narrow composition range around $x = 0.05$, the spectrum at every **k** is modeled as a superposition of that of the insulator ($x \sim 0$) and that of the superconductor ($x \simeq 0.07$).

3.   LARGE PSEUDOGAP

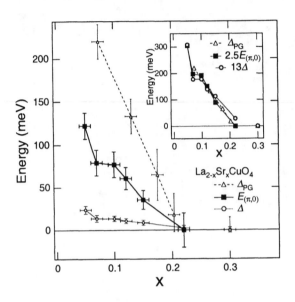

Figure 4   Magnitudes of the superconducting gap Δ, large pseudogap Δ_{PG} and the energy position of the $(\pi, 0)$ flat band $E_{(\pi,0)}$ as functions of hole doping x [6].

Angle-integrated photoemission (AIPES) studies of LSCO has revealed a reduction of the density of states (DOS) at E_F and around it below $x \sim 0.2$ [14]. This reduction tracks the decrease of the electronic specific heat [15, 16] and that of the Pauli component in the uniform magnetic susceptibility [18] with decreasing x, which occurs in the same composition range. These phenomena are consistent with the Fermi-surface crossing of the flat band at $x \simeq 0.2$ since the flat band is expected to naturally give a high DOS. The downward shift of the flat band away from E_F with decreasing x reduces the DOS at E_F. If we call the energy region around E_F with the reduced DOS a "large pseudogap", the large pseudogap expands from zero at $x \sim 0.2$ to ~ 0.1 eV at $x \sim 0.1$. As the magnitude Δ_{PG} of the large pseudogap increases with decreasing x (Fig. 4), the temperature $T_{\chi \max}$ at which the magnetic susceptibility takes the maximum increases [18], so that the relationship $\Delta_{PG}/k_B T_{\chi \max} \sim 3$ holds.

The appearance of a pseudogap below $x = 0.2$ and its nearly linear increase with decreasing hole concentration has been identified in recent specific heat measurements [17]. This observation together with the fact that Δ_{PG} is of the order of the in-plane super-exchange coupling constant J suggest that the antiferromagnetic coupling between the Cu spins is responsible for the formation of the large pseudogap. Such a scenario is consistent with the $t - J$ model calculation of photoemission spectra by Jaklič and Prelovšek [19].

4.   SMALL PSEUDOGAP AND SUPERCONDUCTING GAP

In addition to the large pseudogap, AIPES [14, 20] and ARPES [10] studies have revealed that a superconducting gap of several meV is opened at E_F. The gap is identified as a leading edge shift or a dip in the symmetrized spectra on the Fermi surface, $A(k_F, \omega) + A(k_F, -\omega)$. The magnitude of the gap reaches the maximum near $(\pi, 0)$ and satisfies the BCS relationship of a d-wave superconductor $2\Delta/k_B T_C \sim 4$ in the optimally-doped samples whereas $2\Delta/k_B T_C \gg 4$ in the underdoped samples. Is it found that Δ increases with decreasing x and that Δ_{PG}/Δ remains roughly constant in the underdoped regime (Fig. 4), implying a correlation between the large pseudogap and the superconducting gap, and hence a close connection between the antiferromagnetic correlations and the superconductivity. Although our measurements were made largely in the superconducting state, we expect that the superconducting gap of the underdoped samples, which is much larger than $4k_B T_C$, would not collapse above T_C and remains as a normal-state gap or a "small pseudogap" as in BSCCO [21].

5.   EVOLUTION OF ELECTRONIC STRUCTURE WITH HOLE CONCENTRATION

The photoemission results described have revealed the following characteristic hole concentration regions in LSCO:

(i) $x > 0.2$: Normal Fermi-liquid regime with an electron-like Fermi surface. The effective mass is enhanced with decreasing x. The flat band is located above E_F. The Wilson ratio approaches 2 in the heavily overdoped limit [18], indicating that the system is a strongly correlated two-dimensional Fermi liquid. The T_c becomes the highest in this crossover regime.

(ii) $0.12 < x < 0.2$: Crossover regime between the normal Fermi liquid and the dynamical stripe phase. The flat band is lowered below E_F and the Fermi surface becomes hole-like. The QP is weakened around the nodal point because of the stripe fluctuations. The pseudogaps start to form.

(iii) $0.05 < x < 0.12$: Metallic (superconducting) phase with dynamical stripes. The Fermi surface becomes straight lines around $(\pi, 0)$, implying one-dimensional character, and the nodal QP loses most its of spectral weight. The chemical potential is pinned associated with the stripe formation. The large pseudogaps are fully developed. The Wilson ratio becomes ~ 1 [18], suggesting that the spectral weight of low-energy fluctuations are depleted due to the opening of the large pseudogap.

(iv) $x < 0.05$: Insulating state with segregated holes.

One may ask the question of how the stripes and the pseudogaps are related with each other in the underdoped regime. Presumably, the hole-poor part of the stripe phase is more insulator-like and contributes to the gap-like DOS and hence to the large pseudogap. The hole-rich part contributes to the finite DOS at E_F and to the superconducting gap (and the small pseudogap).

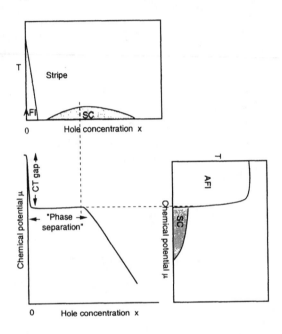

Figure 5   Phase diagram of LSCO plotted against the chemical potential μ [22].

Finally, let us come back to the question of whether the antiferromagnetic correlation helps the superconductivity or competes with it. The similar doping dependence of the large pseudogap and the superconducting gap may suggest that the antiferromagnetic correlation, which is likely to be the origin of the large pseudogap, is responsible for the Cooper pairing. On the other hand, if the antiferromagnetic correlation is the origin of the large pseudogap, it follows that the antiferromagnetic correlation disrupts the superconductivity through the reduction of the DOS at E_F. Then the antiferromagnetic correlation helps and disrupts the superconductivity simultaneously. This is analogous to the situation of phonon-mediated superconductivity in which too strong electron-phonon interaction causes lattice instability and thus disrupts the superconductivity. Now, it is illuminating to plot the phase diagram of LSCO against the chemical potential μ instead of hole concentration x, as shown in Fig. 5. Since the chemical potential does not move for $0 < x < 0.12$, in the

$\mu - T$ phase diagram, the underdoped region collapses into a phase boundary line between the superconducting phase and the antiferromagnetic insulating phase. As the chemical potential approaches that of the parent insulator, the T_c monotonously increases due to the increased pairing potential, but eventually the superconducting state becomes unstable and is taken over by the antiferro-magnetic insulating phase via a first-order phase transition. Such an instability would be inherent in strong coupling superconductors of magnetic origin [22]. Also, such a phase diagram implies the degeneracy of the antiferromagnetic and superconducting phases at the phase boundary, the situation to which SO(5) theory may be applicable [23].

Acknowledgments

We would like to thank K. Kishio, T. Kimura and K. Tamasaku for collaboration in the early stage of this work. This work is supported by a Grant-in-Aid for Scientific Research from the Ministry of Education, Science, Sports and Culture of Japan, the New Energy and Industrial Technology Development Organization (NEDO), the U. S. DOE's Office of Basic Energy Science, Division of Material Sciences. Experiments were performed at the Stanford Synchrotron Radiation Laboratory, which is operated by the Office's Division of Chemical Sciences.

References

[1] J. M. Tranquada *et al*, Nature **375**, 561 (1995).

[2] T. Noda, H. Eisaki and S. Uchida, Science **286**, 265 (1999).

[3] X. J. Zhou *et al.*, Science **286**, 268 (1999).

[4] K. Yamada *et al.*, Phys. Rev. B **57**, 6165 (1998).

[5] A. Ino *et al.*, J. Phys. Soc. Jpn. **68**, 1496 (1999).

[6] A. Ino *et al*, cond-mat/0005370.

[7] A. Ino *et al*, Phys. Rev. B **62**, 4137 (2000).

[8] H. Ding *et al.*, Phys. Rev. Lett. **78**, 2628 (1997).

[9] D. S. Marshall *et al.*, Phys. Rev. Lett. **76**, 4841 (1996).

[10] T. Yoshida *et al.*, unpublished.

[11] B. O. Wells *et al.*, Phys. Rev. Lett. **74**, 964 (1995).

[12] A. Ino *et al.*, Phys. Rev. Lett. **79**, 2101 (1997).

[13] S. Wakimoto *et al.*, Phys. Rev. B **62**, 3547 (IJ (B(2000).

[14] A. Ino *et al.*, Phys. Rev. Lett. **81**, 2124 (1998).

[15] J. W. Loram *et al.*, Physica C **162**, 498 (1989).

[16] N. Momono *et al.*, Physica C **233**, 395 (1994).

[17] J. W. Loram *et al.*, Physica C, in press.

[18] T. Nakano *et al.*, Phys. Rev. B **49**, 16000 (1994).

[19] J. Jaklič and P. Prelovšek, Phys. Rev. B **60**, 40 (1999).

[20] T. Sato *et al.*, Phys. Rev. Lett. **83**, 2254 (1999).

[21] A. G. Loeser *et al.*, Science **273**, 325 (1996).

[22] S. Tešanović, private communication.

[23] S-C. Zhang, Science **275**, 1089 (1997).

STRIPES AND NODAL FERMIONS AS TWO SIDES OF THE SAME COIN

J. Zaanen and Z. Nussinov

Instituut Lorentz for Theoretical Physics, Leiden University, P.O.B. 9506, 2300 RA Leiden, The Netherlands

Abstract

One of the central conceptual problems in High T_c superconductivity is to reconcile the abundant evidence for stripe-like physics at 'short' distances with the equally convincing evidence for BCS-like physics at large distance scales (the 'nodal fermions'). Our central hypothesis is that the duality notion applies: the superconductor should be viewed as a condensate of topological excitations associated with the fully ordered stripe phase. As we will argue, the latter are not only a form of 'straightforward' spin and charge order but also involve a form of 'hidden' or 'topological' long range order which is also responsible for the phenomenon of spin-charge separation in 1+1D. The topological excitation associated with the destruction of this hidden order is of the most unusual kind. We suggest that the associated disorder field theory has a geometrical, gravity like structure concurrent with topological phases with no precedent elsewhere.

1. THE PARADOX

Paradoxes are among the best weaponry available to a scientist. The paradox in science is associated with a flaw in the theoretical understanding on the most basic level. Recently an interview with Edward Witten was broadcasted on dutch TV. The interviewer tried to corner Witten, arguing that quantum-gravity is a shaky affair because it is not accessible by experimental means. Witten was prepared for this question, arguing that the situation is not that bad because quantum gravity is firmly rooted in a grand paradox. Einstein's theory of gravity and quantum mechanics are fundamentally incompatible. This is intimately linked to basic assumptions that are so self evident that they are not even explicitly formulated. The pursuit of string-theory should be considered as an attempt to lay bare these hidden assumptions, and in this sense progress is made.

J. Bonča et al. (eds.), Open Problems in Strongly Correlated Electron Systems, 129–140.

We want to suggest that high T_c superconductivity is in a similar state. The field revolves around a grand paradox, with the added merit that it has experimental physics on its side.

The arrival of a paradox is accompanied by raging controversy, and the controversy in high T_c superconductivity is not easy to overlook. The community has bifurcated in two schools of thought. The first school rests on the conceptual backbone of conventional BCS theory and has been quite successful in addressing the physics of the fully developed superconducting state [1]. Their stronghold are the quasiparticles associated with the d-wave order parameter. The other school refers to the growing body of empirical and theoretical evidence suggesting that the electrons have been eaten by dynamical stripes [1, 2]. The paradox is that stripes and nodal fermions are mutually exclusive.

The basic assumption underlying stripes is that the electrons are expelled from the magnetic domains, with the unavoidable consequence that the soft charge degrees of freedom are associated with the motions inside the stripes. Since the stripes are oriented along the $(1,0)/(0,1)$ lattice directions the low energy propagating electronic excitations should be found along the $\Gamma - X/Y$ directions in the Brillioun zone. Along $\Gamma - M$ the excitations have to traverse the insulating domains and this should cause a severe damping if not a complete gapping. Instead, photo-emission shows relatively sharp dispersive features along $\Gamma - M$ which are quite like the nodal fermions of a d-wave superconductor. The messy fermions are found along the stripe directions.

It seems a widespread reflex to postulate a 'two-fluid' picture: stripes and nodal fermions reflect separate universes, both governed by their own laws, which are for whatever reason completely disconnected. One has no other choice within the confines of BCS theory and the present understanding of dynamical stripes. However, this is no more than admitting defeat in the face of the paradox.

We want to pose the following question: can it be that stripes and nodal fermions are two manifestations of an underlying unity while they appear as dissimilar because of flawed hidden assumptions in the theory?

The remainder of this text will be a modest attempt to make the mind susceptible to the possibility that a positive answer exists for this question. A tactics will be followed which is not dissimilar to the habits in string theory. An alternative theory is suggested, of a highly speculative kind while its consequences are far from clear because of severe technical difficulties. However, it has the special merit that it does not suffer from the paradox, thereby shedding some light on what can be wrong in the current understanding.

The burden is on the stripe side. One has to get out of the narrow interpretations of textbook BCS theory to appreciate the nodal fermions on a sufficiently general level, and this work has already been done by others – see the next section. Our speculation is that the current way of viewing dynamical stripes is too classical. Instead, we assert that the superconducting- and stripe states are related by duality (section 3). Static stripes and superconductivity are competing orders and the duality principle of quantum field theory states that the competing phases can be viewed as 'two sides of the same coin' [3]: the disordered state (the superconductor) can be viewed as a condensate of the topological excitations (disorder fields) associated with the ordered state (the static stripes). The topological excitations of the stripe phase ('stripe disloca-tions') have such an unusual structure that it is a-priori unreasonable to assume that the associated disorder field-theory does not support nodal fermions.

2.   THE NODAL FERMIONS AS DIRAC SPINONS

All what is needed on this side of the coin is to obtain a sufficiently gen-eral view on the nature of what has to be demonstrated: the nodal fermions. Although controversial [4], we will take here the conservative position that BCS is correct as the fixed point theory. One has to be, however, aware of the over-interpretations associated with the weak-coupling treatments in the text-books. These are twofold: the ultraviolet is not governed by a non-interacting electron gas, even not to some degree of approximation [5]. Secondly, Bo-goliubov quasiparticles are in fact $S = 1/2$ excitations of the spin system (spinons) which acquire fully automatically a finite electron pole-strength in the superconducting state [6].

H_0 *is not a Sommerfeld gas.* The universality principle states that systems differing greatly at microscopic scales can nevertheless exhibit the same physics at macroscopic scales. BCS is a universal theory and its infrared structure can be deduced from a simple model. The standard textbook approach sets off by guessing a zero-th order Hamiltonian (H_0) which depicts the large energy scale physics. In systems such as Aluminum, the Fermi-liquid renormalizations are basically complete at T_c and H_0 is simply the Sommerfeld gas Hamiltonian. All one has to do is to add a small perturbation (the BCS-attractive interaction) which leaves the UV physics unaltered (the Fermi surface) while veering the system to the correct IR fixed point. Although the fixed point might still be the same, the way one gets there is entirely different in cuprates [5, 7]. There is no such thing as a close approach to the Fermi-liquids at short scales- and times as is the case in Al. Instead, the analysis of Shen and coworkers of the photo-emission suggests that at truly large energies the electrons move in stripes: the 'holy cross' [8, 9]. Upon descending in energy, the cross starts to deteriorate and the nodal fermions start to appear. It is as if the nodal fermions are a long wavelength phenomenon associated with the quantum disordering of stripes!

It is only at low temperatures, deep in the superconducting state that one finds features which behave like quantum-mechanically propagating particles (the 'quantum protection principle' [7]). The ramification is that it is not necessary to deduce a large, noninteracting Fermi-surface from quantum stripes. It is only necessary to demonstrate that the vacuum structure supports massless electron-like excitations living on Dirac cones: the nodal fermions.

Nodal fermions are spinons. As a next step, it is even not necessary to reinvent the electron. All that needs to be done is to find excitations carrying spin quantum number $S = 1/2$ living on the Dirac cones. The superconducting condensate will take care of connecting these to the electrons. For this purpose we only have to remind the reader of an insight by Kivelson and Rokshar [6], further elaborated by Fisher and coworkers [4]. According to the textbooks, the Bogoliubov quasiparticle is an electron because it has a pole-strength proportional to the square of a coherence factor. The finiteness of the pole-strength implies that the quantum numbers carried by the external electron can be attached to the excitations supported by the vacuum structure of the superconductor. Although spin- and momentum quantum numbers are sharply defined in the BCS state, there is a subtlety associated with the quantum of electrical charge: charge density is a fluctuating quantity in the superconductor and charge quanta can be added and removed at will from the condensate. Hence, the charge of the external electron can always be 'dumped' in the condensate.

To summarize, instead of reinventing aluminum, all what has to be done is to find out if a quantum disordered stripe phase can be constructed, which is superconducting while it carries $S = 1/2$ excitations with a nodal-fermion dispersion.

3.   STRIPE DUALITY

One of the quiet revolutions of mathematical physics is the discovery of the field-theoretic principle of duality. At first it appears as a mathematically rigorous procedure which can be carried through to the end in only a few simple cases (e.g., ref. [3]). However, it seems to reflect a physical principle of a far greater generality. Especially in the condensed matter context it has a stunning consequence: except for the critical state, the universality of duality seems to suggest that there are no truely disordered states at zero temperature. What appears as disorder is actually order of the disorder operators.

Duality can be formulated as an algorithm, with the following subroutines: (a) Characterize the order in the system in terms of an order parameter structure. (b) Enumerate the topological excitations, and link them to singular configurations of the order fields defined in (a). A single topological excitation suffices to destroy the order everywhere. (c) At a critical value of the coupling constant these topological excitations will proliferate, signalling the transition to the 'disordered' state. (d) The constituents of the disordered state are the topological excitations of its 'ordered' partner. As these objects interact this in turn defines a 'disorder' field theory describing the condensation of the disorder matter. The 'disordered' state corresponds with an ordered state in terms of the topological excitations of the 'ordered' state.

Why should this have anything to do with the cuprates? Static stripes and superconductivity are clearly competing forms of order. When stripe order sets in, superconductivity is suppressed and vice versa. Moreover, it appears that this competition is governed by a (near) continuous quantum phase transition [2]. This is not unimportant, since duality is only rigorously defined in continuum field theory and therefore the characteristic length scales should be large as compared to the lattice constant. Dynamical stripes seem to fulfill this condition at least in the underdoped regime. Finally, the ordered stripe phase and the superconductor appear to be very different states of matter, but this is not an a-priori problem. After all, the central notion of duality is that one is supposed to be the 'maximally disordered' version of the other, although at elevated energies they are bound to merge in a single critical regime. The remainder of this section is intended to illustrate the problems encountered in the duality construction which are so severe that it cannot be excluded that it is actually the correct way of viewing these matters.

According to the duality recipe, we have to start out specifying precisely what stripe order means. A stripe phase is a highly organized entity and characterized by a variety of distinct, coexisting orders:

(i) *The stripe phase is a Wigner-crystal.* This is obvious: the electrons form a crystal, breaking translational and rotational symmetry. We will adopt here the viewpoint that a fully ordered stripe state exists which can be used as reference state where translational symmetry is broken both parallel- and perpendicular to the stripes.

(ii) *The stripe phase is a Mott-insulator.* We use here 'Mott-insulator' in the general sense that the charge order discussed under (i) is commensurate with the underlying crystal structure [9], causing a full gap in the charge excitation spectrum. This is actually controversial, and not of central importance in the present context. It is merely helpful, because there is nothing mysterious about an insulating stripe phase. Specifically, we will associate a conserved charge of $2e$ to the stripe Mott-insulator, since with this choice the correct superconductivity emerges directly (see, however, [4]). The insulator would then correspond with a $2k_F$ on-stripe density wave [10].

(iii) *The stripe phase is a collinear antiferromagnet.* This is also obvious. Even when the charge order stays complete, the antiferromagnet can quantum disorder all by itself [11], and this is especially worth a consideration in the bilayer systems. However, we will ignore this possible complication since the focus here is in first instance on the 214 system where the charge ordered systems seem always to be Néel ordered as well.

(iv) *The stripe phase is 'topologically' ordered.* This is the novelty of the stripe phase: whenever stripes are observed in cuprates and nickelates the charges are localized on the antiphase domain walls in the Néel state. It is intuitively clear that this is a form of order, although of an unusual kind. In the fully disordered stripe phase this 'anti-phase boundarieness' must also be destroyed. Hence, the topological excitations of the topological order have to be considered and these are predominantly responsible for the unusual nature of the disorder theory.

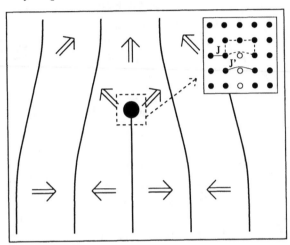

Figure 1   Sketch of the stripe dislocation. The lines indicate the stripes and the arrows the direction of the Néel order parameter in the vicinity of the dislocation in the classical limit ('π-vortex'.) The geometry of the 'curved' internal space seen by the spin system can be inferred from the exchange bonds indicated in the inset.

Given the complexity of stripe order, one anticipates a rather rich disorder-field theory. This is indeed the case. However, this structure can be built up starting from an elementary topological texture of a remarkable simplicity: the stripe dislocation as sketched in fig. 1. This is a stripe which is just ending somewhere in the middle of the sample. In the present context one should appreciate this object as a quantum particle, which can freely propagate through the lattice, occurring at a finite density in the quantum disordered stripe state.

The disorder fields are responsible for the fixed point physics in the disordered state, and these reflect the topological charges associated with the constituent topological excitations. What are the conserved charges associated with the stripe dislocations? Everything one needs to know for the charge sector is available, and the problems are associated with the topological- and spin sectors.

Let us first shortly discuss the charge sector – this will be discussed in detail elsewhere. The ordered reference state is assumed to be a Mott-insulator, characterized by local conservation of charge. The stripe dislocation destroys this local charge conservation and is thereby an electrically charge particle carrying the charge quantum of the insulator. Assuming this charge to be 2e and neglecting of the sign structure associated with the spin sector, the dislocations become hard-core bosons. Moreover, if the dislocations can move freely, then the infinitely long dislocation world lines of the dislocated state will wind around each other. The resulting entangled state is none other than a superconductor. This is just the inverse of the well known Abelian Higg's duality in 2+1 D [3]. This is not all, because the stripe phase is not just a featureless Mott-insulator but its charge sector also breaks translational- and rotational symmetry. The dislocation of fig. 1 is the topological defect associated with the restoration of translational symmetry, carrying a Burger's vector topological charge. Rotational symmetry is restored by a distinct topological excitation, and it is expected that these disclinations are initially suppressed. Although dislocations restore translational invariance, they leave the rotational symmetry breaking unaffected and this is the quantum-nematic state as introduced by Kivelson *et al.* [12] (see also Balents and Nelson [13]). As will be discussed elsewhere [14], instead of the single nematic of Kivelson *et al.* one finds actually a variety of physically distinct nematic like phases if one starts from a Mott-insulating stripe phase. For the present purposes all what matters is, however, that a state exists which is dislocated while the dislocations are subjected to 2+1D motions.

The stage is now set for the case we wish to make. The question is: what is the meaning of 'topological order' (or 'antiphase-boundarieness') and what does it mean to destroy this topological order? Our assertion is that the low energy effective theory associated with this 'order' is actually not an order parameter theory, but instead a *geometrical* theory. The spin system lives in a 'internal' space which is different from the space experienced by an external observer. In the absence of stripe dislocations this internal space is 'flat', but the dislocations are sources of 'curvature'. For the quantum antiferromagnet all that matters is the bipartiteness of the underlying lattice geometry. 'Curved' means that this bipartiteness is destroyed by the stripe dislocations. The spin system of the stripe-dual lives on a frustrating lattice which itself is fluctuating.

This can be discussed in a fairly rigorous setting, but given the space limitations let us just illustrate the main steps on an intuitive-geometrical level. What does anti-phase boundarieness mean? In fact, it does not make sense to call stripes domain walls in the spin system. Domain walls occur when a Z_2 symmetry is in charge and the (semiclassical) spin system is $O(3)$ invariant. Stripes are in this sense non-topological and some other principle is in charge, and this should be made explicit in order to construct the duality. This is a geometric principle and our claim is based on an exact result in 1+1D physics where 'antiphase boundarieness' is called 'spin-charge separation'.

Spin-charge separation has been demystified in a seminal contribution by Ogata and Shiba [15]. By inspecting the Bethe-Ansatz wave-function of the

Hubbard model in 1+1D in the large U limit they come up with a particular prescription for constructing the spin dynamics. Although it does not seem to be fully appreciated, this involves the notions of a geometric theory: on the most basic level it is similar to the Einstein theory of gravity. Their prescription is as follows: choose a particular distribution of holes on the lattice, and the amplitude of this configuration in the wave function will be entirely given by the configuration of the charges. Every given distribution of holes defines a pure spin problem which is indistinguishable at large U from the Heisenberg spin chain *after a redefinition of the lattice*. This is the 'squeezing' operation: take out the holes, together with the sites where they reside, and substitute an antiferromagnetic exchange bond between the spins neighbouring the hole for the taken out hole+site. In a geometrical language, the external observer (us) experiences the full chain. However, the internal observer (spin system) experiences a different space: the squeezed chain where the holes and their corresponding sites have been removed. Although the internal observer is 'blind' for the charge dynamics, it does matter for the external observer and this gives rise to a particular simple factorizable form for the spin-spin correlation function measured by the latter. Since this correlator is universal, the geometric structure which it reflects is also universal, and apparently even realized in the weak coupling (Tomonaga-Luttinger) limit.

Taking this geometrical principle as physical law, how does it generalize to a higher dimension? The only feature of the embedding space which matters for the quantum antiferromagnet is the bipartiteness of the lattice. There are two ways of dividing a bipartite lattice in two sublattices and this defines a sublattice parity p: $p = +1$ or $p = -1$ if the covering is $\cdots A - B - A - B - \cdots$ or $\cdots B - A - B - A - \cdots$, respectively. Divide now both the original- and the squeezed lattice in two sublattices: it is immediately seen that relative to the squeezed lattice the sublattice parity on the original lattice flips every time a hole is passed. The sublattice parity is the 'hidden' Z_2 symmetry!

This generalizes in a unique way to the D dimensional bipartite lattice. In order to keep the bipartiteness intact in the absence of the holes, the holes have to lie on D-1 dimensional manifolds. Hence, in 2+1D the holes are localized on 1+1D manifolds: the stripes. These manifolds can be of arbitrary shape in principle: the stripe fluctuations. Since the spin system on the higher dimensional squeezed lattice is unfrustrated, it will show long-range Néel order. When the hole manifolds order, this spin order will also become manifest.We claim that this prescription is consistent with all available experiments on stripes. We emphasize that we take here a phenomenological stand: the reason that this happens should be given by microscopic theory and this is far from settled. However, if the interest is in the long wavelength behavior one might as well pose the principle and take it for granted as long as it is consistent with the experiments.

In 2+1D the form of order described in the above can be destroyed in a way which is impossible in 1+1D: the stripe dislocation is the topological excitation of the 'topological' (sublattice parity) order in 2+1 D. Although it can be stated more precisely, it is already clear from Fig. 1: the sublattice parity order of the upper part of the figure cannot be matched with the lower part. More precisely, the space experienced by the spin system (the squeezed lattice) is no longer bipartite. This loss of bipartiteness, a frustration, is analogous to spatial curvature. In prior works [16], geometrically frustrated systems have been investigated on their own right- frustration was incurred by the noncommuting nature of the generators of translation. In the present context, the frustration inherent in the loss of bipartiteness- may be similarly reformulated in a geometrically precise manner [14]. Charge (stripe) dislocations destroy spin charge separation and act as gravitational sources for the spin texture.

The analogy with gravity becomes more literal in the classical limit. Consider $S \rightarrow \infty$ and static dislocations. The Néel order parameter texture is as indicated with the arrows in Fig. 1. This can be called a 'π-vortex' (see also [17]), since it looks like 'half' the topological excitation of a $O(2)$ system. However, the spin system is $O(3)$ invariant and the soliton of the $O(3)$ system in 2+1D is the skyrmion, corresponding with a texture where the plane in which the order parameter rotates in internal space depends on the direction one takes in the embedding space. The rotation in Fig. 1 is in a single plane (like a vortex) and therefore it does not carry a conserved topological spin charge. Also notice that it is distinguished from a $O(2)$ vortex because it carries a zero-mode. Ascribe the rotation as indicated in the figure to the equator of the $O(3)$ sphere. Keeping the order parameter fixed at left- and right infinity, degenerate configurations are obtained by canting the order parameter 'above' the dislocation in the direction of one of the poles.

Interestingly,the above is exactly reproduced by embedding a $O(3)$ sigma model in a 2+1D space with a metric given by Einstein theory in the presence of a mass source of strength $8Gm = 1$ (G is Newton's constant and m the mass). In this limit, stripe dislocations act like the famous 'conical singularities' of 2+1D gravity [18]. Unfortunately, the stripe dislocations are not Lorentz invariant, otherwise the semiclassical theory of quantum stripes would reduce to an exercise in 2+1D quantum gravity!

Although these textures are non-topological, they are clearly 'disorder operators' in the spin system and when the stripe dislocations are proliferated while their spin zero-modes are also disordered, they will destroy the Néel order completely, giving rise to a dynamical mass-gap. However, there is a next subtlety: even when $2e$ is chosen for the electrical charge quantum the theory can no longer be bosonic when free stripe dislocations are present. In order to see this, we have to go back to the lattice geometry. Take the Ogata-Shiba prescription and remove the charge-stripes, substituting an anti-ferromagnetic bond for the lattice sites where the stripe reside. The lattice geometry as seen by the spins is as indicated in the inset of Fig. 1. At the dislocation a 'pentagon' plaquette is found and this is directly recognized as a spin frustration event causing minus signs which cannot be transformed away.

The ground state wave function of a nearest-neighbor Heisenberg spin system on a bipartite lattice is nodeless. This is easily seen as follows. Keep the spin operators on the A sublattice fixed and regauge the spin-operators on the B sublattice according to $S^z \to S^z$ and $S^\pm \to -S^\pm$ which leaves the commutation relations unaffected. In the basis which is diagonal in the Ising term, all off-diagonal matrix elements become negative and this means that the ground state wave function only contains positive definite amplitudes. Repeating this on the squeezed spin lattice associated with the stripe dislocation, one finds a seam of positive bonds, starting at the dislocation and ending at infinity. The location of this seam is without physical meaning; it is easily checked that by repeatedly applying the gauge transformations [14, 19] the sign-string can be moved arbitrarily through the plane, and the locus of the string is therefore a gauge freedom. Elsewhere we will argue that the spin-system is also insensitive to the locus of the half-infinite stripe attached to the dislocation and this means that the stripe dislocation the stripe dislocation appears in the spin system as a quantum particle attached to infinity by the sign string.

In the presence of irreducible signs mathematical physics comes to a grinding halt, and we are not aware of a precedent for the above sign structure. All one can say in general is that deep in the semi-classical regime signs are not immediately detrimental. Studies of the $J1 - J2$ model show that the Néel state is robust against a substantial degree of geometrical frustration while the (spin-Peierls) physics found at optimum frustration can be understood without referral to Marshall signs [20]. However, also in the semi-classical case one encounters a problem with the above, which now takes the shape of a Wess-Zumino-Witten type Berry phase [21]. In the derivation of the semi-classical theory using the spin-coherent state path integral formalism one encounters imaginary terms in the Euclidean action which are proportional to the topological (Berry-) phase which takes care of the quantization of the microscopic spin. In a many-spin system it takes the form $S_{WZW} = S \sum_{\vec{r}} \int_0^1 d\tau \int_0^\beta dt \vec{n}(\vec{r}, t, \tau) \cdot \partial_t \vec{n}(\vec{r}, t, \tau) \times \partial_\tau \vec{n}(\vec{r}, t, \tau)$ (t is imaginary time). In the 1+1D case and for large S, this reduces to $2\pi S q$, where the integer q is the Skyrmion number associated with the order parameter texture in space-time. For half-integer spin this leads to alternating signs in the the quantum partition function and these are believed to be responsible for the collapse of the mass gap of the integer spin systems [22].

It was pointed out that in the large S limit these topological terms are inconsequential for the 2+1D quantum antiferromagnet on the bipartite lattice [21]. This lattice can be divided into even and odd 1+1D rows, and the topological phase associated with the even 'chains' exactly compensate those of the odd 'chains'. Consider now the stripe dislocation. Computing the topological phase for the 'conical' texture of Fig. 1, we find that the compensation is no longer complete. The texture can be smoothly deformed because the phase itself is topological, and it is easily demonstrated that it corresponds precisely with the 1+1D topological phase associated with the additional row in the lattice of half-infinite length, starting at the dislocation. Hence, even in the semi-classical case 'sign' problems remain although it is not at all clear to us what these imply.

4.   THE FAITH OF THE PARADOX

What did we accomplish? In fact very little. Following the duality algorithm to the letter, we found that in combination with our understanding of the 'antiphase-boundariness' of the stripes a novel problem is generated. We have no clue regarding the nature of the solution of this problem.

However, it is interesting to revisit the paradox discussed in the introduction. Its signature was that it was not possible to simultaneously take stripes and nodal fermions seriously. In this stripe-duality framework this is no longer true. The paradox has been resolved to yield a question: could it be that the stripe-disorder fields support nodal fermion excitations?

Let us first completely neglect the signs and in this case we know what to do. In the superconducting phase the world lines of the dislocations are winding around each other. To every world line a spin texture is attached of the kind as indicated in Fig. 1 – the signature of the spin system as it appears in the inelastic neutron scattering suggests that the spin system can be considered as semiclassical and since the spin-wave velocity is large it might well be that the spins can follow the charge motions nearly instantaneously. The 'π-vortices' are clearly disordering events in the spin system and, interestingly, they exert this disordering influence in the same way in all the directions in space. The 'π-vortex' covers the half-infinite plane 'above' (fig. 1) the dislocation and since the dislocation occurs at all 'vertical' positions the spin system is disordered identically in all directions. A quantum fluid of 'π-vortices' does not know about the directionally of stripes. This is a somewhat too rigorous resolution of the stripes-nodal fermion paradox: a dynamical mass gap should be generated in the spin sector and this gap should be rather uniform in momentum space, because of the isotropic disordering influence of the π-vortices.

Fortunately, there are the minus signs. Although little can be said in general, they do cause destructive interferences and have a reputation to diminish spin-gaps in favor of massless spinon excitations. The effective spin problem to be solved is that of a quantum-antiferromagnet living on a bipartite lattice pierced by the local frustration events associated with the stripe dislocations which themselves are moving around quantum mechanically. Is there any reason to exclude that this behaves like a d-wave superconductor?

Acknowledgments

We acknowledge stimulating discussions with S. A. Kivelson, H. V. Kruis, and O. Y. Osman. Financial support was provided by the Foundation for Fundamental Research on Matter (FOM), which is sponsored by the Netherlands Organization of Pure Research (NWO).

References

[1] J. Orenstein and A. J. Millis, Science **288**, 468 (2000).

[2] J. Zaanen, Nature **404**, 714 (2000), and ref's therein.

[3] H. Kleinert, "Gauge Fields in Condensed Matter", Vol I & II (World Scientific, Singapore, 1989).

[4] T. Senthil and M. P. A. Fisher, Physical Review B **60**, 6893 (1999).

[5] "The Theory of Superconductivity in the High Cuprates" (Princeton Univ. Press, Princeton, 1997).

[6] S. A. Kivelson and D. Rokshar, Phys. Rev. B **41**, 11693 (1990).

[7] R. B. Laughlin and D. Pines, Proc. Natl. Acad. Sci. USA **97**, 28 (2000).

[8] X. J. Zhou *et. al.*, Science **286**, 268 (1999); Z. X. Shen *et. al.*, unpublished.

[9] J. Zaanen, Science **286**, 251 (1999) and ref.'s therein.

[10] M. Bosch, W. van Saarloos and J. Zaanen, cond-mat/0003236.

[11] S. Sachdev, Science **288**, 475 (2000).

[12] S. Kivelson, E. Fradkin, and V. J. Emery, Nature **393**, 550 (1998)

[13] L. Balents and D. R. Nelson, Phys. Rev. B **52**, 12951 (1995).

[14] H. V. Kruis, Z. Nussinov, and J. Zaanen (in preparation)

[15] M. Ogata and H. Shiba, Phys. Rev. B **41**, 2326 (1990).

[16] J. P. Sethna, Physical Review B **31**, 6278 (1985); S. Sachdev and D. R. Nelson, Phys. Rev. B **32**, 1480 (1985); Z. Nussinov, J. Rudnick, S. A. Kivelson, and L. Chayes, Phys. Rev. Let. **83**, 472 (1999)

[17] O. Zachar, cond-mat/9911171.

[18] S. Deser, R. Jackiw and G. 't Hooft, Ann. Phys. **152**, 220 (1984)

[19] Z. Nussinov, A. Auerbach, and R. Budnik (in preparation)

[20] N. Read and S. Sachdev, Nucl. Phys. B **316**, 609 (1989); M. S. L. du Croo de Jongh, J. M. J. van Leeuwen and W. van Saarloos, Phys. Rev. B

[21] E. Fradkin, "Field Theories of Condensed Matter Physics" (Addison-Wesley, Redwood City, USA, 1991).

[22] F. D. M. Haldane, Phys. Rev. Lett. **50**, 1153 (1988).

DMRG STUDIES OF STRIPES AND PAIRING IN THE t-J MODEL

Steven R. White[1], and D.J. Scalapino[2]

[1] *Department of Physics and Astronomy, University of California, Irvine, CA 92697 USA*

[2] *Department of Physics, University of California, Santa Barbara, CA 93106 USA*

Abstract   We summarize recent density matrix renormalization group studies of stripes and pairing in the t-J model. These results imply that the ground state of this model is striped, with weak pairing. However, with the introduction of a next nearest neighbor hopping t', the stripes are destabilized, and a strongly paired state can arise.

1.   INTRODUCTION

It is clear from a variety of experiments [1] that stripes appear as important low-energy configurations in the underdoped cuprates. However, the basic questions of why stripes form and what role they play in superconductivity remain controversial. A decade ago Hartree-Fock solutions of the Hubbard model showed that stripes—one-dimensional domains of increased hole density forming anti-phase Néel boundaries—were present in mean field solutions of the Hubbard model [2]. Here the stability of the stripe structure arises from the reduction in kinetic energy that the holes experience in moving transverse to the stripes. However, the stripes in the Hartree-Fock solution are characterized by a filling of one hole per domain wall unit cell, while experiments on the cuprates at low doping find a filling of half this. In addition, within the Hartree-Fock framework, it is not clear how superconductivity enters.

An alternative view argues that stripe formation arises from competition between phase separation and long-range Coulomb interactions [3]. Central to this "frustrated phase separation" picture is the assumption that lightly doped $t-J$ or Hubbard models, with parameters in the relevant physical regime, will, in the absence of a long-range Coulomb interaction, globally phase separate into uniform hole-rich and undoped regions. In this approach, it is argued that the formation of stripes is governed by a larger charge energy scale and that pairing arises as a secondary effect associated with the transfer of a spin gap from the undoped regions to the stripes and a subsequent pair transfer between

J. Bonča et al. (eds.), Open Problems in Strongly Correlated Electron Systems, 141–150.

stripes, leading to a Josephson coupling and superconductivity. The fact that stripes act as antiferromagnetic domain walls is also a secondary "kinetic" effect in this picture.

Recently we have carried out numerical density matrix renormalization group (DMRG) calculations [4] on various $t - J$ systems which suggest a third view. These calculations show that low-lying striped states occur in the t-J model in the absence of long-range Coulomb interactions [5, 6, 7, 8]. Furthermore, unlike the Hartree-Fock solutions, the domain walls are characterized at low doping by a filling of one hole per two domain wall unit cells. The short-range structure of the domain wall contains strong antiferromagnetic singlet bond correlations crossing the holes. Just as in the case of the two-hole bound state [9], these spin correlations around and across the holes form in order to maximize the hopping overlap with other hole configurations, which lowers the kinetic energy, while at the same time minimizing the disturbance of the AF background. For this reason it is not surprising that the binding energy per hole of a domain wall is only slightly greater than that of a hole in a $d_{x^2-y^2}$ pair. When an additional next-nearest-neighbor, one-electron hopping term is added, the tendency to form stripes is weakened [10] and we find that the domain walls become unstable with respect to the $d_{x^2-y^2}$ pairing state [11].

In the following we will discuss numerical results for the t-J model which lead us to this view.

The Hamiltonian of the basic t-J model is

$$H = -t \sum_{\langle ij \rangle s} (c_{is}^\dagger c_{js} + \text{h.c.}) + J \sum_{\langle ij \rangle} (\mathbf{S}_i \cdot \mathbf{S}_j - \frac{n_i n_j}{4}), \tag{1}$$

where doubly occupied sites are explicitly excluded from the Hilbert space. Here $\langle ij \rangle$ are nearest-neighbor sites, and s is a spin index. The operator and $c_{is}^\dagger$ creates an electron of spin s on site i and $\vec{S}_i = \frac{1}{2} c_{is}^\dagger \vec{\sigma}_{ss'} c_{is'}$ and $n_i = c_{i\uparrow}^\dagger c_{i\uparrow} + c_{i\downarrow}^\dagger c_{i\downarrow}$ are the electron spin moment and charge density operators at site i. The nearest-neighbor hopping and exchange interactions are t and J, and the average site occupation $< n >= 1 - x$ is set by the hole doping parameter x.

2.   CONVERGENCE OF DMRG

An important issue in DMRG calculations for 2D clusters is the question of convergence, both in the number of sweeps required and in the cluster sized needed. In particular, these issues are central to the issue of the ground state of the 2D t-J model: is it uniform, striped, or phase separated? As an illustration of our calculations, we present detailed results for a DMRG calculation on a 16×6 system with $J/t = 0.35$ and cylindrical boundary conditions, with eight holes. This calculation shows that as our DMRG calculation progresses from a starting point with all the holes in a clump in the center of a long system, the stripes appear spontaneously long before the holes have any probability amplitude of being near the open ends of the system.

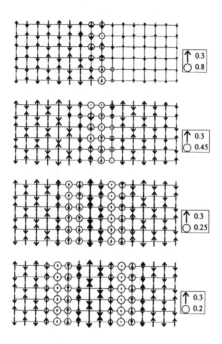

Figure 1   A 16 × 6 *t-J* system, with $J/t = 0.35$ and eight holes, with cylindrical boundary conditions (open in the x direction, periodic in y), is studied with DMRG. The plot shows the local hole density by the diameter of the circles, according to the scales shown. The local spin moment is shown by the length of the arrows. The four pictures represent the state of the system at the end of sweeps 1, 3, 6, and 15. The number of states kept per block was increased as the sweeps progressed, with 80, 200, 600, and 1000 states kept in these four sweeps, respectively. The energy steadily decreased, taking values -42.96, -49.68, -51.890, and -52.279, in the four sweeps. The quantum numbers during the DMRG warmup sweep (sweep "0") were manipulated to force all 8 holes onto the center two columns, in order to strongly favor a phase-separated state. However, the phase separated state is unstable, and splits into two four-hole stripes, which subsequently repel each other. Different style arrows are used to distinguish the two separate antiferromagnetic domains that form.

In Fig. (1), we show the results of our calculation as a function of the number of sweeps carried out so far. No external fields were applied. In the initial DMRG sweep, all the holes were forced onto the center two columns of sites. Subsequently, as the finite system DMRG sweeps are performed, the system moves the holes in order to decrease the energy of the wavefunction. Since hole density is locally conserved, the essentially local DMRG sweeps move the holes slowly. Néel order develops spontaneously in the z spin direction, since we have quantized the spins in the z basis. As the calculation converges, this spontaneously broken spin symmetry slowly disappears, corresponding to an averaging of the overall spin direction over all possible directions. (This reduction in the local spin moments is not yet visible in the sweeps shown in Fig. (1)) Substantially before the hole density has approached either end of the system, two stripes appear spontaneously. The π phase shift also appears spontaneously, and is visible in the local measurements because of the broken spin symmetry. As the calculation converges to the ground state, the π phase shift becomes visible only through spin-spin correlations. The two stripes repel, and continue to move slowly apart as the sweeps progress until they are roughly equidistant from each other and the open left and right ends.

In Fig. (2), we show the hole density as a function of the x coordinate for the same sweeps as in Fig. (1).

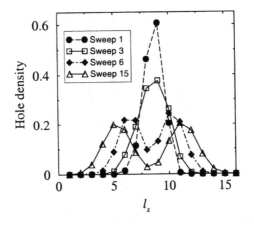

Figure 2   The hole density as a function of the x coordinate is shown for the same four sweeps as in the previous figure. Note that in sweep 6, where the striped pattern is clearly visible, the hole density on the left and right edges is still zero. Consequently, the stripes are not caused by the open boundaries. However, we find that whether the final stripe configurations are site centered or bond centered is influenced by the boundary conditions.

The calculation we have shown addresses the issue of stripe formation on $L \times 6$ systems. On a variety of other geometries we have also seen striping. In particular, in $L \times 8$ systems with cylindrical boundary conditions [5, 6], we see half-filled transverse stripes with a doping dependence remarkably similar to the behavior seen in $La_{1.6-x}Nd_{0.4}Sr_xCuO_4$.

As a further illustration of the robustness of the striped state, we have made an effort to stabilize *longitudinal* half-filled stripes in $L \times 8$ systems with cylindrical boundary conditions. This is somewhat difficult, because the transverse stripes appear to have slightly lower energy. Furthermore, we have found that domain walls do not like to end on open boundaries, which seem to

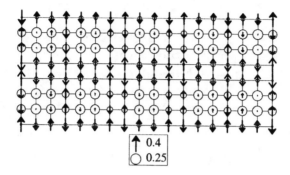

Figure 3   Hole density and spin moments showing longitudinal stripes on a 18 × 8 *t-J* lattice with cylindrical boundary conditions, $J/t = 0.35$, and 20 holes. The diameter of the circles is proportional to the hole density $1 - \langle n_i \rangle$ on the ith site and the length of the arrows is proportional to $\langle S_i^z \rangle$, according to the scales shown. Differently styled arrows are used to show the two different antiferromagetic domains. This structure depends on the boundary conditions as discussed in the text.

repel wall ends. However, we have found that we can stabilize the ends nicely by increasing the hopping slightly on a single edge link at which we wish the domain wall to end. In Fig. (3) we show the hole and spin densities in an 18 × 8 system with two longitudinal stripes. We used a hopping of $1.2t$ on the second and sixth vertical edge links on both the left and right edges, and we also applied staggered fields with a π phase shift built in on sites (1,1), (1,4), (1,5), and (1,8), and the equivalent sites on the right edge. To stabilize the stripe configuration it was also necessary to apply pinning fields throughout the system during the warmup sweep and first several finite system sweeps. Because of the mapping of the sites in the 2D system onto an effective 1D chain in DMRG, during these first sweeps the system is much better equilibrated in the y direction than the x-direction, and the system is unstable to the formation of transverse stripes. After these sweeps, all of the interior fields were turned off and about a dozen more sweeps were performed, with the final number of states kept per block equal to 1600. This calculation shows that with pinning terms applied only at the edges, a rather long cylinder supports longitudinal stripes. These stripes cannot be regarded as simple charge density oscillations induced by boundaries, as occurs in single chain system. Of course, on a long system the state with longitudinal stripes might have higher energy than a state with the bulk having transverse stripes. DMRG is unable to tunnel between two states which differ so much over large length scales. However, we believe that DMRG would have no trouble making the system shown in Fig. (3) uniform if the correct ground state was uniform, simply by smearing out the stripes in the central region.

3.   PAIRING

We consider now the issue of pairing correlations along a stripe. In Fig. (4), we have measured the pair field correlations along a longitudinal half-filled stripe in a 16×4 lattice. A π-phase shifted staggered magnetic field $0.1t$ was applied to the top and bottom edges to mimic the magnetic environment of a stripe in a 2D system. In Fig. (4) we show the pair-field correlations along the central two legs with

$$D_{yy}(\ell) = \left\langle \Delta_y(i + \ell) \Delta_y^\dagger(i) \right\rangle \tag{2}$$

and

$$D_{xy}(\ell) = \left\langle \Delta_x(i + \ell) \Delta_y^\dagger(i) \right\rangle \tag{3}$$

Here

$$\Delta_y^\dagger(i) = c_{i,2\uparrow}^\dagger c_{i,3\downarrow}^\dagger - c_{i,2\downarrow}^\dagger c_{i,3\uparrow}^\dagger \tag{4}$$

is an operator that creates a singlet pair on the ith rung between leg 2 and leg 3, and

$$\Delta_x(i) = c_{i+1,2\downarrow} c_{i,2\uparrow} - c_{i,2\downarrow} c_{i+1,2\uparrow} \tag{5}$$

destroys a singlet pair on leg 2 between the i and $i + 1$ rungs. The short range $d_{x^2-y^2}$-like structure of the pairing correlations is seen in the sign change between $D_{yy}(\ell)$ and $D_{xy}(\ell)$. The pair-field correlations are clearly suppressed at larger distances.

The suppression of pairing along a domain wall can be understood as arising from a suppression of charge fluctuations induced by the π-shifted antiferromagnetic background. Strong local charge fluctuations are essential for superconductivity. In a domain wall with $\rho_\ell = 0.5$, as two adjacent holes or hole pairs move away from each other, the resulting region in which the two π-phase shifted domains are in contact result in a restoring potential which grows linearly with the separation. This strongly suppresses such charge fluctuations, and leads to the decay of the longer-range pairing correlations. Further suppression of pairing correlations may come from a tendency for CDW formation at $\rho_\ell = 0.5$. We have found that a 2-leg ladder with a filling $x = 0.25$, corresponding to a linear filling $\rho_\ell = 0.5$, has a small charge gap and long range CDW order in its ground state.

We have recently studied systems with a next nearest neighbor hopping term t' added to Eq. (1) [11]. Although a variety of terms can be added to the basic t-J Hamiltonian to improve its applicability to experimental systems, t' is particularly interesting because it directly affects the competition between pairing and stripe formation. Fig. (5)(a) shows the hole and spin density for

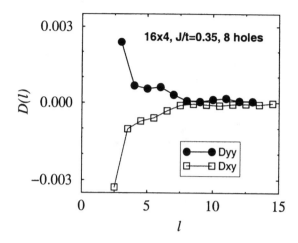

Figure 4   Pair field correlations along a domain wall running the length of a 16 × 4 lattice with $J/t = 0.35$ and 8 holes.

different values of t' for a 12 × 6 system with periodic boundary conditions in the y-direction. As t' increases, the static stripe structure is smeared out [10] and, as shown in Fig. (5)(b), the pairing correlations are enhanced [11]. We have measured the density-density CDW correlations for the lattices with the smeared out domain walls and find them to be negligible, implying that the smearing out of the charge density is not due to fluctuations of the domain walls, but rather a reduction in their ability to bind holes which eventually leads to the complete evaporation of the stripes into pairs. The effect of t' is to enhance the pair mobility, leading to a lowering of the stabilization energy of the domain walls. As this happens, the pairing correlations increase and the stripes disappear. For $t' = 0.3t$ the antiferromagnetic response driven by the staggered field at the open ends is peaked at (π, π) and $d_{x^2-y^2}$-pairing correlations are dominant. Fig. (5) clearly shows that the striped domain-wall state and the superconducting pairing state compete for $t' > 0$. However, there does appear to be an overlap region in which pairing is significant but weakly bound domain walls remain.

Thus, in the nearest-neighbor t-J model, domain walls are energetically favored over pairs and we see only weak pairing correlations. Turning on $t' > 0$ enhances the pair mobility, tipping the balance towards a $d_{x^2-y^2}$-pairing state. Phenomenologically, $t'/t > 0$ models the electron-doped materials, with $x = \langle n_i \rangle - 1$ the electron doping rather than the hole doping. Thus, one might have expected to see even stronger pairing correlations for $t'/t < 0$, corresponding to the hole doping case. However, as discussed in [11], in this

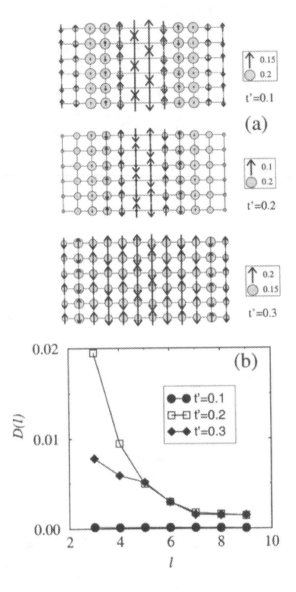

Figure 5 (a) Hole densities and spin moments on 12×6 systems with cylindrical boundary conditions. (b) d-wave pairing correlations for the same system with $t' = 0.1$, 0.2, and 0.3.

case the domain walls evaporate into quasi-particles and the $d_{x^2-y^2}$-pairing correlations remain weak.

4. SUMMARY

In summary, these results lead to the conclusions that in the t-J model, stripes and pair formation are driven by the same basic mechanism, the competition between kinetic and exchange energies, and that they compete with each other. In the nearest-neighbor t-J model, domain-wall/stripe formation is slightly favored over $d_{x^2-y^2}$ pairing. At low doping the stripes are characterized by a linear filling $\rho_\ell = 0.5$ and the repulsion between the stripes give a stripe spacing $d = (2x)^{-1}$. In the nearest-neighbor t-J model, domain walls are energetically favored over pairs, and the pairing correlations are weak. As the next-nearest-neighbor hopping t' is turned on, one goes continuously from a situation in which the stripe correlations dominate to one in which the pairing correlations are dominant. Similar effects should be seen for other changes in the model which enhance pair mobility or act to destabilize the stripes. The regime where the pairing is strongest is broad, and includes the case where the static stripes are completely absent as well as the case where weak, smeared-out stripes are still present. In this regime there is no evidence of stripe fluctuations in the density-density correlation function; rather, the stripes have completely or nearly evaporated into pairs.

Our conclusions differ from those of previous approaches in several respects. Contrary to the frustrated phase separation scenario, we do not have global phase separation; stripe formation is not driven by competition with long range coulomb interactions; and stripes and pairing compete, although there is a region in which both coexist. In this coexistence regime, the pairing correlations are two-dimensional rather than one-dimensional. A key difference between the view we have presented and the Hartree-Fock approach is that the local structure of the domain wall we have discussed involves short-range antiferromagnetic singlet bond correlations across the holes rather than the mean-field, single-particle correlations of the Hartree-Fock solution. Furthermore, we find domain walls with a linear filling of $\rho_\ell = 0.5$ rather than $\rho_\ell = 1$.

Acknowledgments

S.R. White acknowledges support from the NSF under grant #DMR98-70930 and D.J. Scalapino acknowledges support from the NSF under grant #DMR95-27304.

References

[1] "Conference on Spectroscopies in Novel Superconductors", *J. of Phys. & Chem.*, **59** (No.10-12) (1998).

[2] J. Zaanen and O. Gunnarsson, Phys. Rev. B **40**, 7391 (1989); D. Poilblanc and T.M. Rice, Phys. Rev. B **39**, 9749 (1989); H.J. Schulz, J. Physique, **50**, 2833 (1989); K. Machida, Physica C **158**, 192 (1989); K. Kato *et. al.*,

J. Phys. Soc. Jpn. 59, 1047 (1990); J.A. Vergés *et. al.*, Phys. Rev. B **43**, 6099 (1991); M. Inui and P.B. Littlewood, Phys. Rev. B **44**, 4415 (1991); J. Zaanen and A.M. Oles, Ann. Physik **5**, 224, (1996).

[3] S.A. Kivelson and V.J. Emery, p. 619 in *Proc. "Strongly Correlated Electronic Materials: The Los Alamos Symposium 1993,"* K.S. Bedell *et. al.*, eds. (Addison Wesley, Redwood City, Ca., 1994).

[4] S.R. White, Phys. Rev. Lett.**69**, 2863 (1992), Phys. Rev. B **48**, 10345 (1993).

[5] S.R. White and D.J. Scalapino, Phys. Rev. Lett.**80**, 1272 (1998).

[6] S.R. White and D.J. Scalapino, Phys. Rev. Lett.**81**, 3227 (1998).

[7] S.R. White and D.J. Scalapino, Phys. Rev. B **61**, 6320 (2000).

[8] Stefan Rommer, Steven R. White and D.J. Scalapino, Phys. Rev. B **61**, 13424 (2000).

[9] S.R. White and D.J. Scalapino, Phys. Rev. B **55**, 6504 (1997).

[10] T. Tohyama, *et. al.*, Phys. Rev. B **59**, R11649 (1999).

[11] S.R. White and D.J. Scalapino, Phys. Rev. B **60**, R753 (1999).

COEXISTENCE OF CHARGE AND SPIN-PEIERLS ORDERS IN THE 1/4-FILLED LADDER NaV$_2$O$_5$

D. Poilblanc[1] and J. Riera[1,2]

[1]*Laboratoire de Physique Quantique & UMR–CNRS 5626, Université Paul Sabatier, F-31062 Toulouse, France*

[2]*Instituto de Física Rosario, Consejo Nacional de Investigaciones Científicas y Técnicas, y Departamento de Física, Universidad Nacional de Rosario, Avenida Pellegrini 250, 2000-Rosario, Argentina*

Abstract   Charge and spin-Peierls instabilities in quarter-filled ($n = 1/2$) Hubbard and t-J ladders including local Holstein and/or Peierls couplings to the lattice are studied by numerical techniques. We show that, generically, these systems undergo instabilities towards the formation of Charge Density Waves, Bond Order Waves and (generalized) spin-Peierls modulated structures. Moderate electron-electron and electron-lattice couplings can lead to a coexistence of these three types of orders. In an isolated ladder, a zig-zag pattern is stabilized by the Holstein coupling and the nearest-neighbor Coulomb repulsion. Our results are applied to the NaV$_2$O$_5$ compound (trellis lattice) and various phases with coexisting charge disproportionation and spin-Peierls order are proposed and discussed in the context of recent experiments.

1.   INTRODUCTION

The vanadium inorganic compound NaV$_2$O$_5$ is believed to be a nearly perfect realization of a quarter-filled low dimensional system. Therefore, the nature of the SP phase [1, 2] below $T_{SP} \simeq 35K$ is expected to be quite different from the one occurring in the more conventional antiferromagnetic Heisenberg chain CuGeO$_3$ (Ref. [3]).

The NaV$_2$O$_5$ system is built from weakly coupled planes whose structure is shown in Fig. 1. It can be depicted as an array of parallel ladders (Fig. 1(b)) coupled in a trellis lattice. Oxygen atoms (not shown) are located at the center of the vertical and horizontal bonds of Fig. 1(a) and lead to effective hopping matrix elements and antiferromagnetic (AF) super-exchange interactions. LDA band structure calculations [4] and estimations based on empirical rules [5] lead to similar values of the hopping amplitudes along and perpendicular to

J. Bonča et al. (eds.), Open Problems in Strongly Correlated Electron Systems, 151–160.

the ladders, $t_\parallel \simeq 0.15\text{eV}$ and $t_\perp \simeq 0.35\text{eV}$ respectively. However, some controversy remains regarding the magnitude of the diagonal hopping (see Fig. 1(c)) t_{xy} with values ranging from 0.012eV (Ref. [4]) to 0.3eV (Ref. [5]).

Although the average valence of the vanadium in NaV_2O_5 is 4.5 (half an electron per vanadium d-orbital on average), the exact nature of the charge ordering is still under active debate. Recent structure refinement at room temperature [6] suggested that, in contrast to earlier reports [7], α'-NaV_2O_5 would have a centrosymmetric space group implying only one kind of vanadium site. Besides, below the transition temperature T_{SP}, joint neutron and X-ray diffraction experiments [8] reveal new superlattice reflections which can be ascribed to a lattice modulation associated to displacements of predominantly V atoms.

The insulating character of NaV_2O_5 could, in fact, be simply understood in the framework of the quarter-filled Hubbard or t–J ladders [5, 9, 10] without invoking any charge order mechanism (as it is expected at room temperature). A simple analytic picture, valid when the hopping along the legs is small [5, 9], gives a finite charge gap. The existence of a metal-insulator transition in quarter-filled t–J ladder has been confirmed by numerical calculations [9, 10]. In the insulating state, the ground state (GS) configuration corresponds to the single occupancy of the bonding states on each rung and it has been argued that, in this case, the low-energy processes can be described by an effective AF Heisenberg chain [5, 10], hence giving some relevance to earlier descriptions of this material [11, 12]. Furthermore, angle-resolved photoemission data at room temperature [13] are consistent both with a description in term of an effective half-filled t-J chain [11] or in terms of a quarter-filled t-J ladder [9].

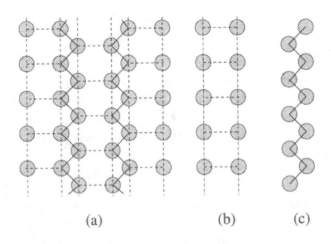

(a)   (b)   (c)

Figure 1   (a) Structure of the vanadium planes of NaV_2O_5 showing ladder (b) and chain (c) patterns. Each vanadium carries, on average, a charge of $+0.5$.

Recently, motivated by the experimental studies of NaV_2O_5 at low temperature, charge ordering has been investigated theoretically [15, 16, 10, 14]. It has been argued that repulsion between electrons on neighboring ladders can lead to charge disproportionations where the electrons are localized on a single leg of each ladder [10] (as in the original structure proposed in Ref. [7]) or form a zig-zag pattern within the ladder [15, 16, 14]. Although the Madelung energy, which includes also a long-range (LR) Coulomb interaction, slightly favors the chain charge ordering, it has been suggested [14] that, in contrast, the underlying coupling with the lattice could stabilize the zig-zag CDW. Experimental features in the optical conductivity at intermediate energies (0.6–2.5eV) [18] could be reproduced in a calculation of the t-J model on the trellis lattice [19] only by assuming an *ad-hoc* charge disproportionated GS, although low-energy charged magnons [18] could not be found.

Here, we investigate further, by numerical exact diagonalization (ED) techniques, the interplay between the electron-electron repulsion and the electron-phonon (or spin-phonon) couplings. Related work have been published elsewhere [17]. The likely smallness of the interladder as well as many of the experimental results point to the possibility that the charge or SP instabilities are driven primarily by the ladder physics. [4, 14] Thus, our analysis consists of the study of these structures by considering isolated quarter-filled t-J ladders including lattice-local charge couplings (Holstein) or modulations of the bond parameters (Peierls and SP couplings). Further details regarding the physics of the chains connecting the ladders can be found in Ref. [17]. Eventually, we shall consider the trellis lattice (Fig. 1(a)) of NaV_2O_5 where the most likely states will be determined based on the previous results obtained for the individual ladder.

2.   ISOLATED LADDER: CHARGE INSTABILITY

We shall first consider the quarter-filled anisotropic t-J ladder (Fig. 1(b)) in the presence of an Holstein-coupling,

$$H = H_{tJ} + H_V + H_H, \tag{1}$$

$$H_{tJ} = t_\parallel \sum_{i,\alpha,\sigma} (\tilde{c}^\dagger_{i,\alpha;\sigma} \tilde{c}_{i+1,\alpha;\sigma} + h.c.) + t_\perp \sum_{i,\sigma} (\tilde{c}^\dagger_{i,1;\sigma} \tilde{c}_{i,2;\sigma} + h.c.)$$

$$+ J_\parallel \sum_{i,\alpha} (\mathbf{S}_{i,\alpha} \cdot \mathbf{S}_{i+1,\alpha} - \frac{1}{4} n_{i,\alpha} n_{i+1,\alpha}) + J_\perp \sum_i (\mathbf{S}_{i,1} \cdot \mathbf{S}_{i,2} - \frac{1}{4} n_{i,1} n_{i,2}),$$

$$H_V = V \sum_i (n_{i,1} n_{i+1,1} + n_{i,2} n_{i+1,2} + n_{i,1} n_{i,2}),$$

$$H_H = \sum_{i,\alpha} n_{i,\alpha} \delta_{i,\alpha} + \frac{1}{2} K \sum_{i,\alpha} \delta_{i,\alpha}^2,$$

where $\tilde{c}_{i,\alpha;\sigma}^{\dagger} = c_{i,\alpha;-\sigma}(1 - n_{i,\alpha;\sigma})$ are *hole* Guzwiller projected creation operators (the large on-site Coulomb interaction prevents doubly-occupancy) and the index α stands for a chain index ($= 1, 2$). We consider *a priori* different fermion hopping amplitudes ($t_{\parallel}$, $t_{\perp}$) or magnetic exchange interactions ($J_{\parallel}$, $J_{\perp}$) along the legs and along the rungs. Hamiltonian (1) can be viewed as the strong coupling limit of a Hubbard ladder so that one can assume a relation of the form $J_{\parallel}/J_{\perp} = (t_{\parallel}/t_{\perp})^2$ between the parameters. A nearest neighbor (NN) Coulomb repulsion V has been included. Note that the electron-phonon coupling has been absorbed in the definition of the lattice displacement δ_i. The magnitude of the coupling to the lattice is then given by a single parameter namely the inverse of the lattice stiffness $1/K$. The phonons have been given an infinite mass since the charge and spin dynamics are assumed here to involve smaller time scales than lattice fluctuations (adiabatic approximation). Note that the on-site displacement δ_i corresponds in fact to an effective parameter which might combine several effects.

Before proceeding with the study of Hamiltonian (1), it is instructive to recall the properties of the t-J ladder in the absence of electron-phonon coupling and NN repulsion. The electronic properties of this model at quarter filling have been investigated previously [5, 9, 10] and the existence of a metal-insulator transition has been shown [9, 10]. It is believed that the system becomes insulating for (approximately) $t_{\perp} > 2t_{\parallel}$. Physically, this corresponds to the situation where the bonding and antibonding bands are completely separated, the lower band becoming effectively half-filled so that an arbitrary small repulsion opens a charge gap. This physical situation might be relevant for the insulating phase of the NaV_2O_5 material.

Although the lattice is considered here in the adiabatic approximation, no supercell order is assumed *a priori* and the lowest energy equilibrium lattice configuration is obtained through a self-consistent procedure. Indeed, the total energy functional $E(\{\delta_{i,\alpha}\})$ can be minimized with respect to the sets of distortions $\{\delta_{i,\alpha}\}$ by solving the non-linear set of local coupled equations, $K\delta_{i,\alpha} + \langle n_{i,\alpha} \rangle = 0$. Since the second term depends implicitly on the distortion pattern $\{\delta_{i,\alpha}\}$, it can be solved by a regular iterative procedure. [20]

The phase diagram of the Holstein-t-J ladder model is shown in Fig. 2(a) for a realistic set of parameters. These results have been obtained by studying 2×6 and 2×8 clusters. In the absence of NN repulsion V, a rapid transition occurs from a uniform phase (U) at small coupling (or equivalently large lattice rigidity) to a localized phase at large coupling. This strong coupling phase is characterized by a charge ordering with two types of (almost) completely empty or completely occupied sites arranged in some disordered patterns. More interestingly, a CDW phase with a "zig-zag" arrangement of the excess charge (which exists also for $V = 0$ only in a very narrow region around $1/K \sim 2.5$) is stabilized by the NN repulsion V. In contrast to the localized phase, the charge disproportionation in this zig-zag CDW state is not complete. Notice that there is a finite critical value of $1/K$ associated to the stability of the CDW phase. This feature might be due to the fact that the uniform ladder for $1/K = 0$ is, for an anisotropy ratio of $t_{\parallel}/t_{\perp} = 0.4$, already in an insulating state with a charge gap (see Ref. [9]).

3.   ISOLATED LADDER: COEXISTING CHARGE AND SP ORDERS

The possibility of a coexisting SP order in the previous CDW state can be studied by considering additional Peierls and SP couplings realized by making the following substitutions in Hamiltonian (1);

$$t_{\parallel} \rightarrow t_{\parallel}(1 + \delta^B_{i,\alpha}), \tag{2}$$

$$J_{\parallel} \rightarrow J_{\parallel}(1 + g\delta^B_{i,\alpha}), \tag{3}$$

and by adding a new elastic term $\frac{1}{2}K_B \sum_{i,\alpha}(\delta^B_{i,\alpha})^2$.

Note that the spin-Peierls order is, for a quarter-filled band, intrinsically linked to a bond order wave (BOW) characterized by a modulation of the hopping amplitudes [21] as given by Eq. (2). In other words, the expectation values $\langle c^{\dagger}_{i,\alpha;\sigma} c_{i+1,\alpha;\sigma} \rangle$ and $\langle S_{i,\alpha} \cdot S_{i+1,\alpha} \rangle$ should exhibit similar modulations along the legs. For simplicity, we shall assume here that the magnetic exchange interation $J_{\parallel}$ involves two virtual hops $t_{\parallel}$ so that one can take, in first approximation, $g \simeq 1$. Note that modulations of the rungs were not found.

We now turn to the investigation of the full Hamiltonian including both Holstein and Peierls-like lattice couplings, the amplitudes of each of them being characterized by the two independent coupling constants $1/K$ and $1/K_B$ respectively. The GS configuration for the site and bond deformations can be obtained exactly on small clusters by solving simultaneously, by the previous recursive method, the sets of non-linear equations. Our results, for the same set of parameters t_{μ}, J_{μ} as above, are summarized in the phase diagram of Fig. 2(c). In contrast to the previous model, no zig-zag CDW ordering appears at small Holstein coupling $1/K$. A uniform dimerized (D) phase is stable for large enough $1/K_B$. The dimerized zig-zag (D-ZZ) phase is found only in a narrow region located around $1/K \sim 2.5$ while a larger Holstein coupling stabilizes a localized phase (with no well-defined periodic pattern). In addition, these data suggest that the Peierls coupling rather tends to suppress the charge order and hence to destabilize the D-ZZ phase with respect to the D phase. However, the region of stability of the D-ZZ phase (of particular interest in the

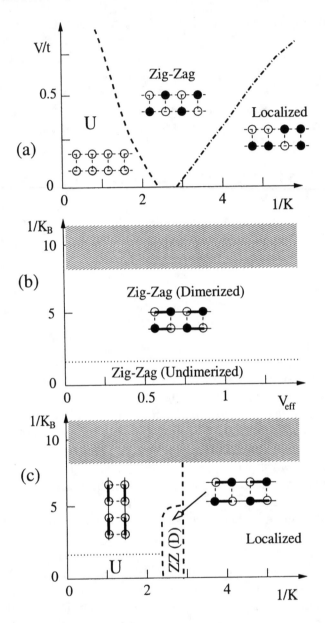

Figure 2   Typical phase diagrams of $\frac{1}{4}$-filled anisotropic t-J ladders (a) as a function of the NN Coulomb repulsion V and the on-site Holstein coupling of strength $1/K$; (b) as a function of the Peierls lattice coupling and an effective static checkerboard-like potential V_{eff}; (c) as a function of Holstein and Peierls lattice couplings. These results have been obtained by ED of small clusters with anisotropy ratios $t_{\parallel}/t_{\perp} = 0.4$ and $J_{\parallel}/J_{\perp} = 0.16$. Shaded regions are unphysical.

case of the NaV$_2$O$_5$ material) is greatly enhanced by a NN Coulomb repulsion (see Fig. 2(a)).

4.   TRELLIS LATTICE: CHARGE INSTABILITY

At this stage, it is interesting to apply our results (valid, strictly speaking, in the case of *isolated* ladders) to investigate the possibility of charge ordering in the 2D trellis lattice of NaV$_2$O$_5$ (see Fig. 1(a)). Obviously, the type of order the most likely to appear depends whether the chains or the ladders are the main structures of this compound. Since the interladder couplings, although still not well determined, are quite likely to be smaller than the couplings along the rungs of the ladders it is to be expected that the physics of the ladders is going to dominate the behavior of the trellis lattice. Note that a $4k_F$ charge order in the zig-zag chains [17] would imply the 2D formation of parallel chains of average charge $0.5 + \Delta n$ and $0.5 - \Delta n$. This corresponds to the structure originally proposed by Galy and coworkers [7] (assuming a complete disproportionation $\Delta n = 0.5$). However, experimental evidences accumulate in favor of a modulation of the charge also in the x direction of the ladder legs, hence more compatible with the $2k_F$-CDW orders in the zig-zag chains with NNN couplings [17]. Two typical patterns exhibiting bond-centered or site-centered CDW along the zig-zag chains are shown in Figs. 3(a) and 3(b) respectively. Note that only the structure of Fig. 3(a) is fully compatible with the zig-zag CDW ordering in *all* the ladders. However, some recent structure refinement [22] points rather towards the structure shown in Fig. 3(b) with an alternating sequence of zig-zag and uniform ladders. Based on our previous studies, we believe this structure should be stable in the 2D trellis lattice for intermediate on-site Coulomb and NNN repulsions U and V. Sizeable cross-bond J_{xy} and t_{xy} (corresponding to J and t in the chain model of Ref. [17]) might help to stabilize this structure further. Interestingly enough, in our picture, the charge ordering is not saturated ($\Delta n < 0.5$). However, a doubling of the periodicity occurs in the direction of the ladders consistently with recent experiments [8].

Based on our previous studies of the coexistence of CDW and SP orders in isolated ladders, we believe that similar generalized modulated structures are expected also for the 2D trellis lattice of NaV$_2$O$_5$ as argued in Ref. [17]. Examples of such structures are shown in Figs. 4.

5.   CONCLUSIONS

To summarize, the role of Holstein and Peierls electron-phonon couplings as well as magneto-phonon (SP) couplings has been investigated in the adiabatic approximation in quarter-filled ladder systems. A numerical method based on ED techniques supplemented by a self-consistent procedure has been used to

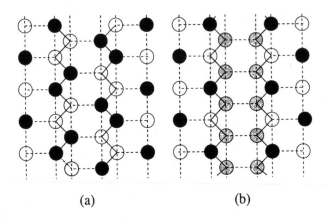

Figure 3 Typical patterns of charge ordering on the NaV_2O_5 lattice structure showing a supercell of 4 sites along the zig-zag chain. The filled (open) symbols correspond to an excess (depression) of charge compared to the average charge of 1/2 (grey sites have a density close to the average density). (a) Bond-centered charge density wave of period 4 ($Q = 2k_F$) with two types of non-equivalent sites. (b) Site-centered charge density wave of period 4 ($Q = 2k_F$) with three types of non-equivalent sites.

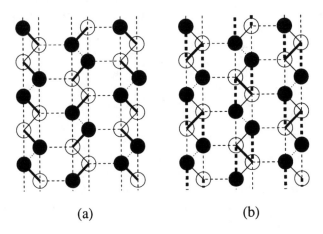

Figure 4 Typical patterns of coexisting lattice distortion and charge ordering on the NaV_2O_5 lattice structure showing a periodicity of 4 sites along the t-t'-J-J' chain. The filled (open) symbols correspond to an excess (depression) of charge compared to the average charge of 1/2. These patterns show a modulation of the bond exchange couplings along the zig-zag chain with three types of bonds and two types of sites ($2k_F$ order of the charge density). Thick, thin and dotted lines correspond to strong, intermediate and weak bonds respectively. (a) Mixed $2k_F$–$4k_F$ lattice distortion. (b) $2k_F$ lattice distortion.

determine various phase diagrams as a function of the strengths of the lattice couplings. We have shown that, generically, CDW, BOW and spin-Peierls orders coexist in such systems. Moreover, in many cases, the SP instability is enhanced by a charge disproportionation. Eventually, we have considered the case of the trellis lattice of NaV$_2$O$_5$. Based on the previous results for isolated chains and ladders, we have proposed mixed CDW-BOW-SP ground states with a zig-zag charge pattern in the ladders and a modulated bond structure.

Whether the charge ordering and the BOW-SP transition occur simultaneously at the same temperature is not yet clear but could be resolved experimentally by thermodynamical measurements. In addition, such transitions should have different spectroscopic signatures due to profound changes in the GS excitation spectrum. On one hand, charge ordering is expected to strongly enhance the charge gap (although the insulating character of the material at high temperature could be accounted for by a local on-site repulsion alone). On the other hand, the opening of a spin gap should be associated to the SP-BOW transition.

Very recent X-ray diffuse scattering studies of NaV$_2$O$_5$ [23] suggest that 3D structural fluctuations are important above the transition and, hence, that transverse interactions (other than magnetic) should be important to stabilize the ordered phase.

Acknowledgments

We thank IDRIS, Orsay (France) for allocation of CPU time on the supercomputers. We also acknowledge funding from the ECOS-SECyT A97E05 programme.

References

[1] M. Isobe and Y. Ueda, J. Phys. Soc. Jpn **65**, 1178 (1996); Y. Fujii et al. J. Phys. Soc. Jpn **66**, 326 (1997); M. Weiden et al., Z. Phys. B **103**, 1 (1997).

[2] For ultrasonic evidence of the SP transition in NaV$_2$O$_5$ see e.g. P. Fertey *et al.*, Phys. Rev. B **57**, 13698 (1998).

[3] For a review on CuGeO$_3$, see e.g. J. P. Boucher and L. P. Regnault, J. Phys. I (Paris) **6**, 1939 (1996).

[4] H. Smolinski, C. Gros, W. Weber, U. Peuchert, G. Roth, M. Weiden, C. Geibel, Phys. Rev. Lett. **80**, 5164 (1998).

[5] P. Horsch and F. Mack, Eur. Phys. J. B. **5**, 367 (1998).

[6] A. Meetsma, J. L. de Boer, A. Damascelli, T. T. M. Palstra, J. Jegoudez and A. Revcolevschi, Acta. Cryst., in press (1998); H.-G. von Schnering, R. Kremer, O. Jepsen, T. Chatterji and M. Weiden, preprint (1997).

[7] A. Carpy and J. Galy, Acta Cryst. B **31**, 1481 (1975).

[8] Tapan Chatterji *et al.*, Solid State Comm. **108**, 23 (1998). Note that the room temperature data of these authors indicate also a single *V* site in agreement with Ref. [6].

[9] J. Riera, D. Poilblanc and E. Dagotto, Eur. Phys. J. B. **7**, 53 (1999).

[10] S. Nishimoto and Y. Ohta, preprint cond-mat/9805336.

[11] D. Augier, D. Poilblanc, S. Haas, A. Delia and E. Dagotto, Phys. Rev. B **56**, R5732 (1997).

[12] David Augier and Didier Poilblanc, Eur. Phys. J. B **1**, 19 (1998); D. Smirnov *et al.*, Phys. Rev. B 57, R11035 (1998); For treatment of non-adiabatic phonons see also D. Augier, D. Poilblanc, E. Sorensen, I. Affleck, Phys. Rev. B **58**, 9110 (1998).

[13] K. Kobayashi *et al.*, Phys. Rev. Lett. **80**, 3121 (1998).

[14] M. V. Mostovoy and D. I. Khomskii, preprint cond-mat/9806215.

[15] H. Seo and H. Fukuyama, preprint cond-mat/9805185.

[16] P. Thalmeier and P. Fulde, preprint cond-mat/9805230.

[17] More details on this issue can be found in J. Riera and D. Poilblanc, Phys. Rev. B. **59**, 2667 (1999).

[18] A. Damascelli *et al.*, Phys. Rev. Lett. **81**, 918 (1998).

[19] S. Nishimoto and Y. Ohta, preprint cond-mat/9808061; See also T. Ohama, H. Yasuoka, M. Isobe and Y. Ueda, preprint (1998).

[20] A. Dobry and J. Riera, Phys. Rev. B **56**, 2912 (1997).

[21] Coexistence between BOW and CDW has been discussed, in the context of charge transfer salts, in e.g. K. C. Ung, S. Mazumdar and D. K. Campbell, Solid St. Commun., **85**, 917 (1993); K. C. Ung, S. Mazumdar and D. Toussaint, Phys. Rev. Lett. **73**, 2603 (1994).

[22] C. L. de Boer et al., preprint (2000).

[23] S. Ravy, J. Jegoulez and A. Revcolevschi, preprint cond-mat/9808313.

IV

CUPRATES:
NUMERICAL METHODS AND
QUANTUM HALL EFFECT

NORMAL STATE PROPERTIES OF CUPRATES:
t-J MODEL VS. EXPERIMENT

P. Prelovšek[1,2]

[1] *J. Stefan Institute, SI-1000 Ljubljana, Slovenia*

[2] *Faculty of Mathematics and Physics, University of Ljubljana, SI-1000 Ljubljana, Slovenia*

Abstract   We discuss some recent results for the properties of doped antiferromagnets, obtained within the planar t-J model mainly by the finite-temperature Lanczos method, with the emphasis on the comparison with experimental results in cuprates. Among the thermodynamic properties the chemical potential and entropy are considered, as well as their relation to the thermoelectric power. At the intermediate doping model results for the optical conductivity, the dynamical spin structure factor and spectral functions reveal a marginal Fermi-liquid behaviour, close to experimental findings. It is shown that the universal form of the optical conductivity follows quite generally from the overdamped character of single-particle excitations.

1.   INTRODUCTION

It is by now quite evident through numerous experiments on electronic properties that cuprates, being superconductors at high temperatures, are also strange metals in the normal phase [1]. On the other hand it also appears that most features can be well represented by prototype single-band models of correlated electrons, as the Hubbard model and the t-J model. In spite of their apparent simplicity these models are notoriously difficult to treat analytically, in particular in the most interesting regime of strong correlations. This has led to intensive efforts towards numerical approaches [2], mostly using quantum Monte Carlo (QMC) and the exact diagonalization (ED) methods.

The subject of this contribution is the planar t-J model (Hubbard model is expected to show similar behaviour in the strong correlation regime), which represents layered cuprates as doped antiferromagnets (AFM) and within a single band both mobile charges and spin degrees of freedom,

$$H = -t \sum_{\langle ij \rangle s} (\tilde{c}_{js}^{\dagger} \tilde{c}_{is} + \text{H.c.}) + J \sum_{\langle ij \rangle} (\mathbf{S}_i \cdot \mathbf{S}_j - \frac{1}{4} n_i n_j). \qquad (1)$$

163

J. Bonča et al. (eds.), Open Problems in Strongly Correlated Electron Systems, 163–172.

Strong correlations are here imposed by strictly forbidding doubly occupied sites. So far most calculations were performed for the ground state at $T = 0$, where the standard Lanczos algorithm offers an efficient exact-diagonalization analysis of small systems [2]. More recently a novel numerical method, finite-temperature Lanczos method (FTLM) [3, 4], has been introduced combining the Lanczos method with a random sampling, which allows for an analogous treatment of static and dynamic properties at $T > 0$.

One of most important conclusion of experimental efforts in the last decade is the realization that the electronic phase diagram in the parameter space of planar hole concentration c_h and temperature T is quite universal. Materials are usually classified as underdoped, optimally doped and overdoped, with respect to the highest T_c in a given class. In our analysis we cannot establish the superconductivity, so we will use the highest entropy (at low T) as a criterion for the optimum doping. In fact both criteria are quite close for real cuprates [5] and one could conjecture that this relation is not accidental. In particular since in thermodynamic quantities at the same doping also the pseudogap scale disappears.

In this contribution we mainly discuss two topics related to the normal-state properties of cuprates. In Sec.II we deal with the thermodynamic quantities: entropy, chemical potential and closely related thermoelectric power, all of them in the full range of c_h. In Sec.III we concentrate on the appearance and the relation between different manifestations of the marginal-Fermi-liquid (MFL) behaviour, observed in the optical conductivity, dynamic spin susceptibility and spectral functions at the intermediate (optimum) doping.

2.   THERMODYNAMICS

Let us first consider thermodynamic quantities, which can be directly derived from the grand-canonical sum: free energy density $\mathcal{F}$, chemical potential μ and entropy density s. For these the FTLM is particularly simple to implement [6, 4], since it requires only a minor generalization of the usual Lanczos method. Results presented below were obtained mostly for $N = 20$ sites and parameters $J/t = 0.3$, corresponding to the situation in cuprates (where $t \sim 0.4\text{eV}$). Note that in small systems only results above certain (size dependent) T are meaningful, i.e. in cases below typically $T > T_{fs} \sim 0.1\,t$.

Let us first discuss results for s, shown in Fig.(1) for various T. It seems quite generic feature of such a system that $s(c_h)$ exhibits a (rather broad) maximum at the intermediate doping $c_h \sim c_h^*$. The increase of s on doping can be plausibly related to the frustration between the AFM exchange $\propto J$ and the hole kinetic energy $\propto c_h t$ preferring the FM configuration. This naturally leads to a most frustrated situation at $c_h^* \sim J/t$. It is quite fortunate that the FTLM works best, i.e. T_{fs} is lowest, just in the cases with large s and large

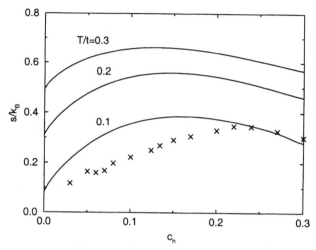

Figure 1 s vs. c_h at several T [4]. For comparison also experimental results for LSCO [5] at highest $T = 320 \ K \sim 0.07 \ t$ are shown.

frustration, while other methods have difficulties in such situation. E.g. QMC is plagued with the minus-sign problem which seems to be intimately related to fermionic frustration.

Even more surprising fact is the magnitude of s at $T < J$. E.g. at $T = 0.1 \ t$ at c_h^* we get 40% of $s(T = \infty)$. Clearly we are dealing with a system which has very low degeneracy (Fermi) temperature $T_{deg} < J$, far below the free fermion value $T_{deg}^0 \sim 8t$. Such a conclusion is in agreement with experiments in cuprates. In recent years s has been measured in YBCO and LSCO (also presented in Fig.(1) in a wide doping regime [5] and our results show good quantitative agreement.

For $\mu_h(T)$, presented in Fig.(2), we mostly do not find a T^2 dependence of μ_h at low T, as expected for a normal Fermi liquid, except within the extremely overdoped regime $c_h \geq 0.3$. In particular, in a broad range $0.05 < c_h < 0.3$ we find a roughly linear variation $\mu_h(T) = \mu_h(T = 0) + \alpha k_B T$, whereby the slope α changes the sign at $c_h = c_h^* \sim 0.15$. The variation $\mu(c_h)$ at low T has been recently deduced experimentally from the shift of photoemission spectroscopy spectra in LSCO [7], and the agreement with our results is quite satisfactory [4]. From photoemission results as well as from our Fig.(2) it is also evident that $\mu(c_h < c_h^*)$ is very flat which would indicate that at $T \rightarrow 0$ and $c_h < c_h^*$ the compressibility $\kappa \propto -dc_h/d\mu$ is very large or even diverging, as would e.g. follow from the phase separation scenario [8] or the singular $c_h \rightarrow 0$ limit [1]. It should be however stressed that the distinction between these scenarios could be relevant only at very low $T \ll J$, since both experiment and numerics

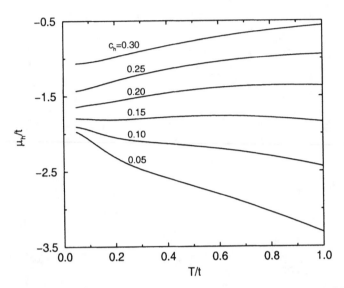

Figure 2   Hole chemical potential μ_h vs. T at several dopings c_h [4].

indicate on quite large s, i.e. a distribution over a wide spectrum of states, even at $T \sim J/10$.

It is quite helpful to realize that the free energy density $\mathcal{F}(c_h, T)$ relates the variation of $s = \partial\mathcal{F}/\partial T$ and $\mu = \partial\mathcal{F}/\partial c_h$, i.e.

$$\left.\frac{\partial s}{\partial c_h}\right|_T = -\left.\frac{\partial \mu_h}{\partial T}\right|_{c_h} = \frac{\partial^2 \mathcal{F}}{\partial c_h \partial T}. \tag{2}$$

This connects the maximum $s(c_h^*)$ with the change in slope $d\mu_h(c_h^*)/dT = 0$. Moreover Eq.(2) allows us to discuss more confidently the slope $d\mu_h/dT = \alpha k_B$ for which we find in the underdoped regime $\alpha \sim 2$. Although the latter has not been so far verified directly for cuprates, one can extract in the same regime from the measured $\partial s/\partial c_h$ for LSCO and YBCO at $T > 100\ K$ similar values $\alpha > 1$ [5]. It is quite evident that at $\alpha > 1$ we are not dealing with a degenerate Fermi liquid but rather with the nondegenerate doped carriers, which is a situation typical for a doped (nondegenerate) semiconductor. One should just recall the standard expression for μ_h in p-type semiconductor,

$$c_h = P_v e^{-(\epsilon_v - \mu_h)/k_B T} \implies -\frac{\partial \mu_h}{\partial T} = k_B \ln\frac{P_v}{c_h} > k_B, \tag{3}$$

where in our notation $P_v \sim 1$. The constant slope $d\mu_h/dT$ observed in our calculations down to $T < 0.1\ t$ is a confirmation of such a picture. In experimental results for s one should however notice a reduction of α with T,

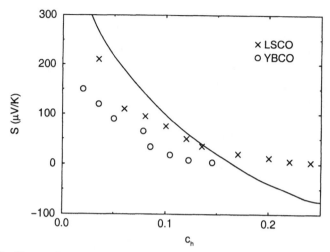

Figure 3 Thermoelectric power S vs. c_h for $T/t = 0.1$ [4]. Experimental result for LSCO and oxygen deficient YBCO are taken from Ref. [5].

but even at $T \sim T_c$ the system is not evidently a normal Fermi liquid with $\alpha < 1$.

Another consequence of such a semiconductor picture is an expression for the thermopower S,

$$S \sim \frac{\bar{\epsilon} - \mu_h(T)}{e_0 T} \sim \frac{\mu_h(T=0) - \mu_h(T)}{e_0 T} \sim \alpha S_0, \qquad (4)$$

where $S_0 = k_B/e_0 = 86 \mu V/K$. The validity of this approximation we have verified within the t-J model also directly by evaluating the mixed current-energy current correlation function and observing that they are proportional to the current-current correlation $C_{j_E j}(\omega) = \mu_h(0)C(\omega)$. The result is in Fig.(3) is good agreement with the general experimental observation in cuprates [9] of a large and rather T-independent S at low doping. In fact instead of the usual semiconductor expression for α at $c_h \ll 1$, Eq.(4), in a strongly correlated system it is more appropriate to use the proper statistics for the t-J model leading to $\alpha \sim \ln[2(1 - c_h)/c_h]$, which even predicts the change of sign at $c_h \sim 0.3$.

3.   DYNAMICS AT OPTIMUM DOPING

It has been quite early established from experiments that cuprates in the normal state do not follow the behaviour consistent with the normal Fermi liquid. In contrast several static and dynamic quantities at optimum doping can be quite well accounted for within the marginal Fermi liquid (MFL) concept

[10]. Most evident example is the dynamic conductivity $\sigma(\omega)$ which does not obey the usual Drude form with a constant rate $1/\tau$ but can be well fitted in a broad range of ω, T with the generalized MFL form $\tau^{-1}(\omega, T) = \tilde{\lambda}(|\omega| + \eta T)$, describing also the well established linear resistivity law $\rho \propto T$. It has been natural to postulate an analogous MFL behaviour for quasiparticle (QP) relaxation in spectral functions as e.g. measured by the angle-resolved photoemission spectroscopy (ARPES). Only recently, however, the high resolution ARPES experiments on BSCCO [11] seem to be in position to confirm beyond doubt this behaviour, obeyed in the optimum-doped materials surprisingly even at $T < T_c$ for QP along the nodal direction in the Brillouin zone. Most evident indication that also spin dynamics follows the MFL concept is the observed anomalous NMR and NQR spin-lattice relaxation rate $1/T_1(T) \sim const.$ [12] instead of usual Korringa law in metals.

By calculating using FTLM several related quantities, desciribing charge and spin dynamics within the t-J model, we established that the MFL concept applies well in a broad range of intermediate hole doping $0.1 < c_h < 0.3$. We discuss here in particular the dynamical conductivity $\sigma(\omega)$, the local spin susceptibility $\chi_L(\omega)$ [13] and the QP relaxation rate as obtained from the analysis of spectral functions $A(\mathbf{k}, \omega)$ [14]. Moreover, $\sigma(\omega)$ has been been found close to a universal form [13],

$$\sigma(\omega) = C_0 \frac{1 - e^{-\omega/k_B T}}{\omega}, \tag{5}$$

in a remarkably broad frequency regime $0 < \omega < \omega^* \sim 2t$, while C_0 is essentially T-independent for $T < J$. Resulting $\sigma(\omega < \omega^*)$ is clearly governed by T only. Evidently, Eq.(5) reproduces the linear resistivity law $\rho = T/C_0$ and is consistent with the MFL scenario for $\tau^{-1}(\omega, T)$, however in a very restrictive way since both MFL parameters are essentially fixed. A reasonable overall fit can be e.g. achieved by $\tilde{\lambda} \sim 0.6$ and $\eta \sim 2.7$. When optical experiments on $\sigma(\omega)$ in cuprates are analysed within the MFL framework quite close values for λ, η are in fact reported [15, 16]. In addition, the model results reproduce well also the absolute value of $\sigma(\omega)$ and $\rho(T)$ [4].

Analogous universality has been found also in the spin dynamics, in particular when looking at the local spin susceptibility $\chi_L(\omega)$ and related spin correlation function $S_L(\omega)$,

$$\chi_L''(\omega) = \frac{1}{\pi} \tanh\left(\frac{\omega}{2T}\right) \bar{S}_L(\omega), \tag{6}$$

where $\bar{S}_L(\omega) = S_L(\omega) + S_L(-\omega)$ is the symmetrized function, having a fixed sum rule

$$\int_0^\infty \bar{S}(\omega) d\omega = \langle (S_i^z)^2 \rangle = \frac{1}{4}(1 - c_h). \tag{7}$$

The most important message of numerical results on spin dynamics at intermediate doping is that $\bar{S}_L(\omega)$ is nearly T-independent in a broad range of T, in particular for $T < J$. Moreover $\bar{S}_L(\omega)$ is only weakly doping dependent consistent with the sum rule. So we have a conclusion that at intermediate doping the more fundamental and universal quantity is the correlation function $S_L(\omega)$ and not the susceptibility $\chi_L(\omega)$, which is the analogy to the relation between $C(\omega)$ and $\sigma(\omega)$ in Eq.(5). For the spin dynamics this is also the message of the anomalous NMR $1/T_1$ in cuprates [12]. As a result $\chi_L''(\omega)$, Eq.(6), follows the MFL behaviour i.e. at $\omega < T$ one gets anomalous T dependence $\chi_L''(\omega) \propto \omega/T$.

A MFL-type QP relaxation is extracted within the t-J model also from the analysis of the spectral functions $A(\mathbf{k}, \omega)$ near the optimum doping [14, 4]. For the characterization of QP properties the self energy $\Sigma(\mathbf{k}, \omega)$ is crucial. On the other hand the same information can be also expressed in terms of QP parameters $Z_\mathbf{k}, \Gamma_\mathbf{k}, \epsilon_\mathbf{k}$. Both definitions are related as

$$A(\mathbf{k}, \omega) = -\frac{1}{\pi} \mathrm{Im} \frac{1}{\omega - \Sigma(\mathbf{k}, \omega)} = \frac{1}{\pi} \frac{Z_\mathbf{k} \Gamma_\mathbf{k}}{(\omega - \epsilon_\mathbf{k})^2 + \Gamma_\mathbf{k}^2}. \tag{8}$$

For QP near the Fermi surface the hole-part self energy $\omega < 0$ is found to be of the MFL form, i.e. $\mathrm{Im}\Sigma \sim -\tilde{\gamma}(\omega + \xi T)$ with $\tilde{\gamma} \sim 1.4$ and $\xi \sim 3.5$. $\tilde{\gamma} > 1$ means an overdamped character of QP, since the the full width at half maximum $\Delta \sim 2\Gamma(\epsilon) > \epsilon$ is larger than the QP (binding) energy ϵ. This should be contrasted with the electron-like regime $\omega > 0$ where the damping is found to be essentially smaller and consequently QP can be underdamped.

Here we comment on the relation of our results to recent ARPES results in BSCCO. The analysis for hole-like excitations in the nodal direction $(0,0) - (\pi, \pi)$ shows the MFL form with the QP damping $\Gamma \sim 0.75\omega$ for $\omega > T$ and $\Gamma \sim 2.5T$ for $\omega < T$ [11]. This again means an overdamped character of hole excitations, since $2\Gamma(\epsilon) > \epsilon$. In making the comparison one should take into account that $\Gamma = Z|\mathrm{Im}\Sigma|$. Since at the peak position we find $Z \sim 0.5$ experimental and model values appear reasonably close.

Let us finally discuss the relation of $\sigma(\omega)$ and the associated relaxation rate $1/\tau$ to the QP damping Γ [17]. In the case of weak scattering one finds $1/\tau \sim 2\Gamma$. In cuprates as well in the t-J model we are apparently dealing with overdamped QP, so the relation is at least questionable. Also, the conductivity form Eq.(5) appears to be universal, while the QP damping does not seem to be parameter free.

One approach is to approximate the current-current correlation function $C(\omega)$, which in general replaces C_0 in Eq.(5), by a decoupling in terms of spectral functions $A(\mathbf{k}, \omega)$ neglecting possible vertex corrections, i.e.

$$C(\omega) = \frac{2\pi e_0^2}{N} \sum_{\mathbf{k}} (v_{\mathbf{k}}^\alpha)^2 \int d\omega' f(-\omega') f(\omega' - \omega) A(\mathbf{k}, \omega') A(\mathbf{k}, \omega' - \omega). \quad (9)$$

In order to reproduce the MFL form of $\sigma(\omega)$ one has to assume the MFL form for the spectral function, Eq.(8), i.e. the QP damping of the form $\Gamma = \gamma(|\omega| + \xi T)$. We neglect also the $\mathbf{k}$ dependence of Γ and Z. In fact it is enough to assume that $\Gamma_{\mathbf{k}}(\omega)$ is independent of deviations $\Delta\mathbf{k}_\perp$ perpendicular to the Fermi surface. The latter is just what is observed in recent ARPES studies of BSCCO [11]. Replacing in Eq.(9) the $\mathbf{k}$ summation with an integral over ϵ with a slowly varying density of states we can derive

$$\int d\epsilon A(\epsilon, \omega') A(\epsilon, \omega' - \omega) = \frac{Z^2}{\pi} \frac{\bar{\Gamma}(\omega, \omega')}{\omega^2 + \bar{\Gamma}(\omega, \omega')^2}, \quad (10)$$

where $\bar{\Gamma}(\omega, \omega') = \Gamma(\omega') + \Gamma(\omega' - \omega)$. We are thus dealing with a function $C(\omega)$ depending only on the ratio ω/T, and on MFL parameters γ, ξ. For $\gamma \ll 1$ we recover via such an analysis $C(\omega)$ strongly peaked at $\omega = 0$ and consequently MFL-type $\sigma(\omega)$ with $1/\tau(\omega) = 2\Gamma(\omega/2)$ invoked in connection with the MFL concept [10]. For the regime of overdamped QP excitations $\gamma \sim 1$ or more appropriate $\gamma\xi \sim 1$ we present in Fig.(4) $C(\omega)$ for several γ fixing $\xi = \pi$. For $\gamma < 0.2$ still a pronounced peak shows up at $\omega \sim 0$, on the other hand $C(\omega)$ becomes for $\gamma > 0.3$ nearly constant or very slowly varying in a broad range of ω/T.

The main message of the above simple analysis is that for systems with overdamped QP excitations the universal form (5) describes quite well $\sigma(\omega)$ for a wide range of parameters. It should be stressed that nearly constant

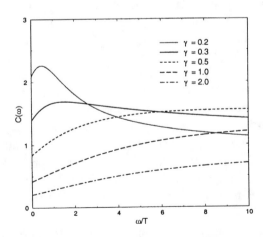

Figure 4   Current-current correlation spectra $C(\omega)$ vs. ω/T for various γ at fixed $\xi = \pi$.

$C(\omega < \omega^*)$ also means that the current relaxation rate $1/\tau^*$ is very large, $1/\tau^* \sim \omega^* \gg 1/\tau$, i.e. much larger than the conductivity relaxation scale apparent from Eq.(5) where $1/\tau \propto T$ follows solely from thermodynamics.

One should also be aware of the upper cutoff scale ω^* for the validity of the MFL-like QP damping. In the problem considered here it appears that the cutoff is directly related to the current relaxation rate $\omega^* \sim 1/\tau^*$ found in the t-J model to be extremely high at the intermediate doping, i.e. $\omega^* \sim 2t$. The latter allows for an effective mean free path l^* of only few cells, essentially independent of T. Such a short l^* can be plausibly explained by assuming that charge carriers - holes entirely loose the phase coherence in collisions with each other due to the randomizing effect of an incoherent spin background. Note again that the short correlation length (even at $T < T_c$) appears also from the analysis of ARPES spectral functions $A(\mathbf{k}, \omega)$ varying $\Delta \mathbf{k}_{\parallel}$ along the Fermi surface [11].

4.   DISCUSSION

Cuprates in their metallic phase are anomalous in several respects. One important conclusion at least for theoreticians is that most of anomalous properties are quite well reproduced also in the prototype t-J model. The analysis of this model has been so far restricted to numerical calculations of small systems, nevertheless in the $T > T_{fs}$ window where macroscopic relevance of FTLM results is expected the agreement with experiments is even quantitative, without any adjustable parameters. Since the behaviour found experimentally is quite generic and universal down to lowest $T \sim T_c$ there is no reason to doubt in the generality of model results.

Nevertheless there are open questions of the existence and the origin of low energy scales in cuprates as well as in the t-J model. In the underdoped or weakly doped regime FTLM shows the indication for the pseudogap scale T^*, in particular in the uniform susceptibility χ_0 and in the density of states [4]. This scale seems to be related to the onset of short range AFM correlations, hence $T^* \propto J$. Still for $T < T^*$ the entropy remains large as manifested by experiments and our results. The electron liquid is thus closer to a nondegenerate system of composite particles than to a degenerate Fermi gas. Only at $T \to 0$ the entropy is low enough to make the discussion of possible orderings or instabilities relevant.

The origin of the MFL behaviour of several dynamic quantities and of the universal form of $\sigma(\omega)$ and $\chi_L(\omega)$ in the intermediate doping has to be intimately related to the large degeneracy in this regime. It has been shown that QP are essentially overdamped down to lowest $T \sim T_c$. This can be only explained by the scattering on spin fluctuations, which mainly contribute to the entropy. On the other hand spins are just strongly perturbed by holes introduced by doping, so a self-consistent enhancement seems to be the mechanism which dominates the relevant physics. Only at low $T \sim T_c$ apparently this behaviour breaks down by an emergence of coherence and new energy scales.

The author acknowledges the support of the Swiss National Fund and the Institute for Theoretical Physics, ETH Zürich, where part of this work has been done, as well as the support of the Japan Society for Promotion of Science and the hospitality of the Institute for Materials Research, Tohoku University, Sendai, where the manuscript has been prepared.

References

[1] for a review see M. Imada, A. Fujimori, and Y. Tokura, Rev. Mod. Phys. **70**, 1039 (1998).

[2] E. Dagotto, Rev. Mod. Phys. **66**, 763 (1994).

[3] J. Jaklič and P. Prelovšek, Phys. Rev. B **49**, 5065 (1994).

[4] for a review see J. Jaklič and P. Prelovšek, Adv. Phys. **49**, 1 (2000).

[5] J.R. Cooper and J.W. Loram, J. Phys. I France **6**, 2237 (1996); J.W. Loram, K.A. Mirza, J.R. Cooper, and J.L. Talllon, J. Phys. Chem. Solids **59**, 2091 (1998).

[6] J. Jaklič and P. Prelovšek, Phys. Rev. Lett. **77**, 892 (1996).

[7] A. Ino *et al.*, *Phys. Rev. Lett.* **79**, 2101 (1997).

[8] V.J. Emery, S.A. Kivelson, and H.Q. Lin, *Phys. Rev. Lett.* **64**, 475 (1990).

[9] for a review see, A.B. Kaiser and C. Uher, in *Studies of High Temperature Superconductors*, Vol.7, ed. A. V. Narlikar (Nova Science Publishers, New York), p. 353 (1991).

[10] C.M. Varma *et al.*, Phys. Rev. Lett. **63**, 1996 (1989); P.B. Littlewood and C. M. Varma, J. Appl. Phys. **69**, 4979 (1991); E. Abrahams, J. Phys. France I **6**, 2192 (1996).

[11] T. Valla *et al.*, Science **285**, 2110 (1999); A. Kaminski *et al.*, Phys. Rev. Lett. **84**, 1788 (2000).

[12] T. Imai, C.P. Slichter, K. Yoshimura, and K. Kasuge, Phys. Rev. Lett. **70**, 1002 (1993).

[13] J. Jaklič and P. Prelovšek, Phys. Rev. Lett. **74**, 3411 (1995); Phys. Rev. B **52**, 6903 (1995).

[14] J. Jaklič and P. Prelovšek, Phys. Rev. B **55**, R7307 (1997).

[15] Z. Schlesinger *et al.*, Phys. Rev. Lett. **65**, 801 (1990).

[16] A. El-Azrak *et al.*, Phys. Rev. **49**, 9846 1994; C. Baraduc, A. El-Azrak, and N. Bontemps, J. Supercond. **8**, 1 (1995).

[17] P. Prelovšek, cond-mat/0005330.

STABILITY OF d-WAVE SUPERCONDUCTIVITY IN THE $t-J$ MODEL

F. Becca, L. Capriotti and S. Sorella

Istituto Nazionale per la Fisica della Materia (INFM) and International School for Advanced Studies (SISSA), Via Beirut 4, I-34013 Trieste, Italy

Abstract   We use a recently developed technique, which allows to perform few Lanczos steps on a given wavefunction even for large system sizes, to investigate the $t-J$ model in the physical parameter region and to check the stability of the BCS d-wave variational wavefunction [1]. Our statistical Lanczos algorithm, which extends and improves the one Lanczos step proposed in Ref. [2], has been extensively tested on the small $L = 18$ sites cluster where many Lanczos iterations can be performed exactly. In this case, at doping $\delta \sim 10\%$ the BCS wavefunction represents a very good initial state to achieve extremely accurate energies and correlation functions with few Lanczos iterations. For large sizes ($L \leq 98$) the behavior is similar: the low-energy d-wave order parameter P_d is weakly affected by a couple of Lanczos iterations in the low doping $\delta \sim 10\%$ region, whereas the energy is considerably lowered. As a further test of our calculation we have computed the variance of the Hamiltonian $\Delta E_p = (\langle \hat{H}^2 \rangle - \langle \hat{H} \rangle^2)/L^2$ on the BCS wavefunction with $p = 0, 1, 2$ Lanczos steps. For large p, when the Lanczos algorithm converges to the exact ground state, the variance vanishes exponentially with increasing p. The remarkable reduction of the variance, observed for $p = 1, 2$ Lanczos steps even for the largest lattice size considered, suggests a smooth and rapid convergence to the exact ground state. These results support the existence of off-diagonal long-range d-wave superconducting order in the two-dimensional $t-J$ model.

1.   INTRODUCTION

One of the most important question raised after the discovery of high-Tc superconductivity is whether a simple model of strongly correlated electrons can capture the low-energy physics of real materials. In particular it is still a very much debated issue whether a purely repulsive electronic interaction can give rise to a d-wave superconducting ground state by doping an antiferromagnetic Mott insulator with a small amount of holes. The simplest model that has been proposed immediately after the discovery of high-Tc superconductors is the

J. Bonča et al. (eds.), Open Problems in Strongly Correlated Electron Systems, 173–185.

$t-J$ model [3, 4]

$$\hat{H} = J \sum_{\langle i,j \rangle} \left(\hat{\mathbf{S}}_i \cdot \hat{\mathbf{S}}_j - \frac{1}{4}\hat{n}_i\hat{n}_j \right) - t \sum_{\langle i,j \rangle,\sigma} \tilde{c}_{i,\sigma}^\dagger \tilde{c}_{j,\sigma}, \qquad (1)$$

where $\tilde{c}_{i,\sigma}^\dagger = \hat{c}_{i,\sigma}^\dagger (1 - \hat{n}_{i,\bar{\sigma}})$, $\hat{n}_i = \sum_\sigma \hat{n}_{i,\sigma}$ is the electron density on site i, $\hat{\mathbf{S}}_i = \sum_{\sigma,\sigma'} \tilde{c}_{i,\sigma}^\dagger \tau_{\sigma,\sigma'} \tilde{c}_{i,\sigma'}$ is the spin operator and $\tau_{\sigma,\sigma'}$ are Pauli matrices. In the following we put $t = 1$.

After many years of intense numerical and theoretical efforts there is no general consensus on the properties of this simple Hamiltonian and of the related Hubbard model. From the numerical point of view, the density matrix renormalization group (DMRG) [5] predicts that a charge density wave instability [6], nowadays called *striped-phase* for its one dimensional character, is strongly competing with pairing and superconductivity [7]. Within DMRG it appears therefore difficult to explain Copper-Oxide superconductors with a simple one-band model, especially because, in order to be consistent with photoemission experiments [8], a negative next-nearest-neighbor hopping amplitude t' has to be included in the model. In fact, the negative t' suppresses even further superconductivity, so that the $t-J$ model becomes unrealistic to describe the low-energy physics of high-Tc superconductors. However, the DMRG results, though quite accurate, are not exact in two dimensions. Moreover, for technical reasons it is possible to consider only particular boundary conditions (open in one direction and periodic in the other), which certainly make the DMRG calculation still far to be representative of the thermodynamic limit.

Quantum Monte Carlo (QMC) is an appealing alternative numerical approach. This numerical method is still severely limited for two-dimensional fermionic systems by the well-known *sign problem* and is consequently biased by the initial guess of the ground state used to control this numerical instability [9, 10, 11]. However this technique has the important advantage to work very well with periodic boundary conditions since translation invariance can be explicitly used to improve the efficiency of QMC algorithms. In particular, an approximate ground state can be obtained starting from a translation invariant wavefunction $|\psi_G\rangle$ by applying exactly few powers of the Hamiltonian $(-\hat{H})^p$ [12] or many approximate ones (by using for instance the *Fixed-Node* FN approximation [9]). However, within various QMC schemes the situation is still controversial. First Heeb [2] and Khono [13] have found d-wave superconductivity in a reasonable parameter range. Later Shih and *co-workers* [14], using a very similar method, have excluded drastically this possibility. The latter results were obtained within the monotonic-behavior assumption of the off-diagonal superconducting order parameter as a function of the number p of Hamiltonian powers applied to the initial wavefunction. This assumption, although reasonable, is highly questionable. By contrast, in a recent QMC work [15], a clear tendency to d-wave superconductivity in the $t-J$ model was found. Moreover, very recently, an almost realistic phase diagram with a corresponding high-Tc d-wave superconducting transition has been obtained for the Hubbard model within the Dynamical Mean Field Approximation [16, 17]. Furthermore, in the contest of the Hubbard model, also a weak-coupling renormalization group approach [18, 19] gives rise to a d-wave order parameter in a large region of the phase diagram. The latter results strongly support the relevance of a single band model for the explanation of high-Tc superconductivity.

2.   NUMERICAL METHOD

In this work we make a further attempt to clarify the controversial numerical findings on the issue of d-wave superconductivity in the $t-J$ model, using the statistical *few Lanczos-step technique* (FLST), efficiently implemented by means of the stochastic reconfiguration (SR) [10, 11]. Within the latter scheme all kind of correlation functions can be computed efficiently without any *mixed average* [11] approximation: an enormous advantage compared to the FN or to the original SR technique [10, 11]. In these cases this bias can be removed at the price of adding a small field coupled to the desired correlation function [15]. However, it turns out that it is extremely difficult and computationally demanding to work in the small field limit when unbiased correlation functions can be obtained. FLST is instead a very good compromise that solves efficiently this numerical problem of QMC methods.

The wavefunctions that we are able to sample statistically read:

$$|\psi_p\rangle = \left(1 + \sum_{k=1}^{p} \alpha_k \hat{H}^p\right)|\psi_G\rangle \qquad (2)$$

with parameters $\{\alpha_k\}$ for $k = 1, \cdots, p$ minimizing the energy expectation value $\langle\psi_p|\hat{H}|\psi_p\rangle/\langle\psi_p|\psi_p\rangle$. For any p it is simple to show that the wavefunction (2) corresponds exactly to apply p Lanczos step iterations to the initial wavefunction $|\psi_G\rangle$. This wavefunction is sampled statistically with the SR technique, by using in the reconfiguration scheme the first p powers of the Hamiltonian. Unlike the previous method [10, 11] the reference wavefunction $|\psi^f\rangle$ is not evolved during the Markov iteration. $|\psi^f\rangle$ is instead kept statistically equal to the initial wavefunction $|\psi_G\rangle$ using the variational scheme proposed by Hellberg and Manousakis [20], which highly reduce the statistical fluctuations related to the SR technique. The equivalence of FLST to the standard Lanczos algorithm will be discussed in a forthcoming paper [21].

The initial wavefunction to which FLST will be applied can be written as follows [1]:

$$|\psi_G\rangle = |\psi_{p=0}\rangle = \hat{P}_0 \, \hat{P}_N \hat{J}|D\rangle. \qquad (3)$$

where $|D\rangle$ is a BCS wavefunction, which is an exact eigenstate of the following Hamiltonian:

$$\hat{H}_{BCS} = \hat{H}_0 + \frac{\Delta_{BCS}}{2}(\hat{\Delta}^\dagger + \hat{\Delta}) \tag{4}$$

$$\hat{\Delta}^\dagger = \sum_{\langle i,j\rangle} M_{i,j}\left(\tilde{c}_{i,\uparrow}^\dagger \tilde{c}_{j,\downarrow}^\dagger + \tilde{c}_{j,\uparrow}^\dagger \tilde{c}_{i,\downarrow}^\dagger\right) \tag{5}$$

where $\hat{H}_0 = \sum_{k,\sigma} \epsilon_k \tilde{c}_{k,\sigma}^\dagger \tilde{c}_{k,\sigma}$ is the free electron tight binding nearest-neighbor Hamiltonian, $\epsilon_k = -2(\cos k_x + \cos k_y) - \mu$, μ is the free-electron chemical potential and $\hat{\Delta}^\dagger$ creates all possible nearest-neighbor singlet bonds with d-wave symmetry being $M_{i,j}$ +1 or −1 if the bond $\langle i,j\rangle$ is in the x or y direction, respectively. $\hat{P}_N$ and $\hat{P}_0$ are the projectors over the subspaces with a fixed number N of particles and no doubly occupied states. Finally the Jastrow factor $\hat{J} = \exp\left(\gamma/2 \sum_{i,j} v(i-j)\hat{n}_i\hat{n}_j\right)$ couples the holes via the density operators $\hat{n}_i$ and contains another variational parameter $\gamma \sim 1$ which scales an exact analytic form, obtained by approximating the holes with hard-core bosons at the same density, and applying the spin-wave theory to the corresponding XY model [22]. We note here that by performing a particle-hole transformation on the spin down $\tilde{c}_{i,\downarrow}^\dagger \to (-1)^i \tilde{c}_{i,\downarrow}$, the ground state of the BCS Hamiltonian is just a Slater-determinant with $N = L$ particles [23]. This is the reason why this variational wavefunction can be considered of the generic Jastrow-Slater form, a standard variational wavefunction used in QMC. Using the particle-hole transformation, it is also possible to control exactly the spurious finite system divergences related to the nodes of the d-wave order parameter.

3.   NUMERICAL TESTS

In this section we show the accuracy of FLST applied to the BCS wavefunction (3) on a small 18-site cluster, where exact results are available.

Our main task is to compute the order parameter at finite system size

$$P_d = \frac{1}{L}\langle \psi_p^{N+2}|\hat{\Delta}^\dagger|\psi_p^N\rangle, \tag{6}$$

where $|\psi_p^N\rangle$ and $|\psi_p^{N+2}\rangle$ are the states with N and $N+2$ particles, respectively. If P_d is finite in the thermodynamic limit this necessarily implies off-diagonal long-range order in the ground state. Following Ref. [15], it is convenient with an approximate technique to calculate a short-range quantity like P_d, instead of the more conventional long-range expectation value $\langle \psi_p^N|\hat{\Delta}\hat{\Delta}^\dagger|\psi_p^N\rangle/L^2$. In Table 1 we show a comparison between FLST and the exact results for 18 and 16 electrons at $J = 0.4$.

In Table 2 we show P_d as a function of the number of Lanczos step iterations for the 18-site cluster at $J = 0.4$. In the same Table we have computed also

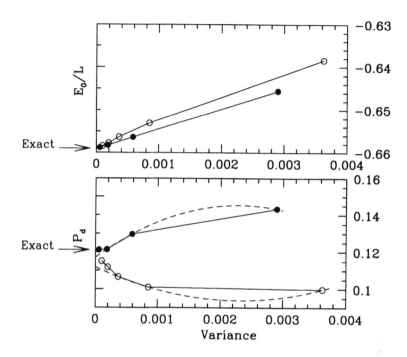

Figure 1   Ground-state energy per site E_0/L and d-wave order parameter P_d as a function of the variance for $N = 16$, $L = 18$, $J = 0.4$: full dots (optimal Δ_{BCS}), empty dots ($\Delta_{BCS} \rightarrow 0$). Dashed lines are quadratic fits of the estimates with $p = 0, 1, 2$.

the variance $\Delta E_p = (\langle \psi_p | \hat{H}^2 | \psi_p \rangle - \langle \psi_p | \hat{H} | \psi_p \rangle^2)/L^2$, the overlap squared $Z_p = |\langle \psi_p | \psi_0 \rangle|^2$ of the FLST wavefunction with the true ground state $|\psi_0\rangle$, and the average sign of the FLST wavefunction:

$$\langle S_p \rangle = \sum_x \langle x | \psi_0 \rangle^2 \text{Sgn} \left(\langle x | \psi_p \rangle \langle x | \psi_0 \rangle \right) , \qquad (7)$$

where $|x\rangle$ denotes configurations with definite electron positions and spins. For an exact calculation, namely $p \gg 1$, both $Z_p \rightarrow 1$ and $S_p \rightarrow 1$, whereas $\Delta E_p \rightarrow 0$. The variance thus represents a very important tool to estimate the 'distance' from the exact ground state when the latter one is not known. In particular whenever $Z_p \simeq 1$ the energy approaches the exact result linearly with the variance ΔE_p, allowing us to estimate the error in the variational energy.

Table 1 Comparison between the estimates of the ground-state energy per site E_0/L and of the d-wave order parameter P_d obtained with the exact and the statistical (FLST) application of p Lanczos steps on the variational wavefunction of Eq. (3). $L = 18$, $N = 16, 18$ and $J = 0.4$.

N	Δ_{BCS}	p	E_0/L (FLST)	P_d (FLST)	E_0/L (Exact)	P_d (Exact)
18	0.55	1	-0.4765(1)		-0.47668	
18	0.55	2	-0.4775(1)		-0.47749	
16	0.20	1	-0.6541(1)	0.1074(4)	-0.65420	0.10730
16	0.55	2	-0.6583(1)	0.122(1)	-0.65826	0.12135

Table 2 Average sign $\langle S_p \rangle$, overlap squared on the exact ground state Z_p and variance times the volume squared $\Delta E_p \times L^2$ obtained applying exactly p Lanczos steps on the variational wavefunction of Eq. (3). $L = 18$, $N = 16, 18$ and $J = 0.4$.

N	Δ_{BCS}	p	$\langle S_p \rangle$	Z_p	$\Delta E_p \times L^2$	E_0/L
18	0.00	0	1.0000	0.6898	1.194	-0.43833
16	0.00	0	0.9656	0.8306	1.174	-0.63847
18	0.80	0	1.0000	0.8850	0.335	-0.46639
18	0.80	1	1.0000	0.9915	0.042	-0.47662
18	0.80	2	1.0000	0.9995	0.004	-0.47752
18	0.80	3	1.0000	1.0000	0.0003	-0.47759
18	0.80	4	1.0000	1.0000	0.00002	-0.47759
18	0.80	∞	1.0000	1.0000	0.0	-0.47759
16	0.55	0	0.9891	0.9260	0.940	-0.64559
16	0.55	1	0.9988	0.9814	0.191	-0.65638
16	0.55	2	0.9999	0.9942	0.060	-0.65826
16	0.55	3	1.0000	0.9983	0.018	-0.65882
16	0.55	4	1.0000	0.9995	0.005	-0.65898
16	0.55	5	1.0000	0.9999	0.002	-0.65902
16	0.55	6	1.0000	0.9999	0.0005	-0.65904
16	0.55	7	1.0000	1.0000	0.0001	-0.65904
16	0.55	∞	1.0000	1.0000	0.0	-0.65904

This can be achieved by plotting the variational energies E_p as a function of the corresponding variance ΔE_p, and performing a very stable linear or quadratic fit to the $\Delta E_p = 0$ exact limit (see Fig. 1). Similar fits can be attempted for correlation functions though, in this case, also a term $\propto \sqrt{\Delta E_p}$ is expected for $\Delta E_p \to 0$. This term is however negligible for quantities like P_d that are averaged bulk correlation functions in a large system size (see the Appendix). In practice even in the small 18-site cluster the non-linear term turns out to be negligible (see Fig. 1). We believe that, being the convergence of the Lanczos algorithm particularly well behaved and certainly unbiased, the variance extrapolation method is in this case particularly useful and reliable. However for bad initial wavefunction (e.g., randomly generated) or very large sizes the approach to zero of the variance may behave rather wildly, requiring many Lanczos steps to reach the regime where the extrapolation is possible.

As shown in Table 2 the quality of the variational BCS wavefunction (2) is *exceptionally good*, especially in the doped $N = 16$ case. Here Z_p is larger than 0.9 even at the simplest $p = 0$ variational level, and is drastically improved with really few Lanczos step iterations. Remarkable is also the behavior of the average sign S_p which measures directly the accuracy of the BCS wavefunction phases, without caring about the amplitudes. In the undoped case the signs of the BCS wavefunction $\langle S_0 \rangle$ can be proven to be exact, i.e., $\langle S_0 \rangle = 1$, having the BCS state the well-known Marshall signs, i.e., the phases of the exact ground state of Heisenberg model. For the two-hole case, the BCS nodes change in a non trivial way. Nevertheless, $\langle S_0 \rangle$ remains very close to 1 and it is much higher than the average sign of the corresponding Gutzwiller wavefunction ($\Delta_{BCS} \to 0$), also shown in the table for comparison.

These results suggest that there is a tendency to d-wave BCS pairing in the $t-J$ model at $\sim 10\%$ doping and $J \sim 0.4$, and that the BCS wavefunction is a particularly accurate wavefunction to describe the small and even zero doping region of the $t-J$ model.

4. LARGER SIZE CALCULATIONS

Though few Lanczos steps may appear inadequate for large system size, this simple scheme is instead providing us very good variational energies up to $L \sim 100$ sites in the $t-J$ model, even when this variational energies are compared with more complicated schemes like the FN. In order to show that the Lanczos scheme remains effective for larger sizes it is useful to consider first a relevant case where a numerically exact solution is possible: the zero doping limit of the $t-J$ model, i.e., the Heisenberg model. In Fig. 2 we plot the energy results calculated for few Lanczos steps as a function of the variance. The fact that the energy is approaching the exact result smoothly with the variance both in the 50- and 98-site clusters (with a slightly larger curvature in the latter case) indicates that the projected BCS wavefunction should have a substantially large overlap squared Z_p with the exact ground state of the Heisenberg model even on these large system sizes. This is also confirmed by the behavior of the square

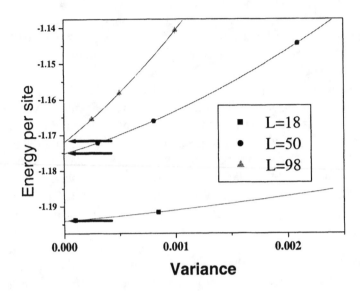

Figure 2 Energy per site of the finite-size Heisenberg model. Comparison of exact results (indicated by arrows) and the approximate $p = 0, 1, 2$ Lanczos step iterations over the projected d-wave wavefunction. Continuous lines are quadratic fit of the data.

order parameter (Fig. 3), which is considerably improved by few Lanczos steps. This result is particularly important since the starting BCS wavefunction has not antiferromagnetic long-range order, whereas the two-dimensional Heisenberg model is widely believed to be antiferromagnetically ordered.

The above test represents also a further strong evidence that the ground-state wavefunction of the Heisenberg model is smoothly connected to a d-wave BCS superconducting wavefunction. This circumstance represents a very interesting numerical fact that clearly supports the experimental observation of high-Tc d-wave superconductivity coming just upon a small doping of a quantum antiferromagnet.

Within the Lanczos approximate states $|\psi_p\rangle$ ($p = 0, 1, 2$) acting on the best variational BCS wavefunction, we have performed a finite-size scaling of the order parameter P_d as defined in Eq. (6), at a fixed doping $\delta = 13.3\%$ (corresponding to $N = 84$ on the largest size $L = 98$). Since P_d is computed between two states with N and $N + 2$ particle numbers, we assume that it refers to the intermediate doping $\delta = 1 - (N + 1)/L$. For smaller sizes the doping $\delta = 13.3\%$ is not possible and we have interpolated linearly P_d between the two fillings closest to this doping. The main results of this paper is then shown in the Fig. 4. Here the size scaling for $p = 0, 1, 2$ Lanczos steps and for their zero-variance extrapolation clearly indicates a finite P_d in the thermodynamic limit. The stability of the BCS variational wavefunction is evident from this figure: P_d remains much larger than the corresponding value of the Gutzwiller ($\Delta_{BCS} \to 0$) metallic state. Instead, as shown in Fig. 5 and similarly to the undoped case, the energy is very much lowered by FLST, suggesting that the method remains effective even for large sizes and finite doping.

5.   CONCLUSIONS

In conclusion we have provided evidences in favor of the stability of d-wave pairing correlations in the $t-J$ model. Our result disagrees with a recent QMC calculation [14], for which, however, the self-consistent assumption mentioned in the introduction leads to a very poor, high-energy variational state. Furthermore, also the DMRG results provide a qualitatively different scenario, namely that stripes suppress d-wave superconductivity [7]. In order to clarify this disagreement, we have performed a simulation at $J = 0.4$ for a cluster with $L = 6 \times 12$, 8 holes, and open boundary conditions on the long direction, as usually done within DMRG. Our accuracy is in this case

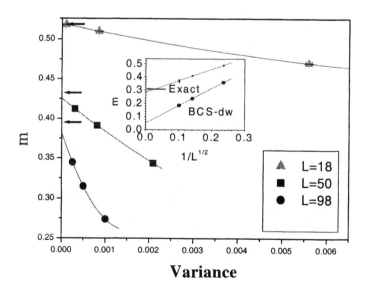

Figure 3   Order parameter $m = \sqrt{S(\pi, \pi)/L}$ in the finite-size Heisenberg model ($S(\pi, \pi)$ being the spin isotropic antiferromagnetic structure factor). Comparison of exact results (indicated by arrows) and the approximate $p = 0, 1, 2$ Lanczos step iterations over the projected d-wave wavefunction. Continuous lines are quadratic fit of the data. Inset: finite-size scaling with the variational (BCS d-wave) wavefunction and with the variance extrapolated one.

worse than with periodic boundary condition, especially for quantities that are not bulk average correlation functions (see the Appendix). For P_d instead we have found that, even at the simplest variational level open boundary conditions strongly suppress P_d by more than 30%, much more than the difference found in Fig. (4). Thus we find that boundary conditions play a very important role and may strongly destabilize the uniform BCS d-wave ground state, as obtained within DMRG. This is consistent with experimental findings [24], showing that any small asymmetry in the CuO planes, acting in our point of view as a distorted boundary condition, may enhance the tendency to stripe formation and then suppressing superconductivity. Our results instead support the recent numerical works [16, 17, 18, 19], indicating that d-wave superconductivity can be obtained in a one band model with repulsive electron interaction.

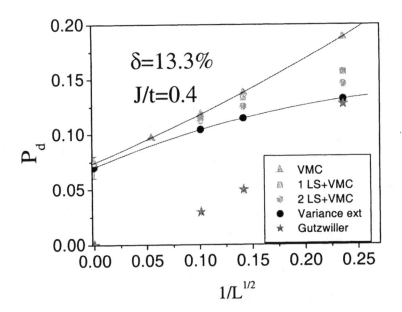

Figure 4 Superconducting d-wave order parameter P_d in the $t-J$ model as defined in the text in Eq. (6). These results were obtained starting by the variational wavefunction (triangles) defined in Eq. (3) and by applying to it one (squares) or two (dots) Lanczos steps. Black dots are obtained by quadratic extrapolation to the zero variance exact limit in order to estimate the size dependent error of the approximate variational calculations. The stars refer to the Gutzwiller wavefunction with $\Delta_{BCS} \to 0$ and $\gamma = 1$ in Eq. (3). Continuous lines are quadratic fit of the data.

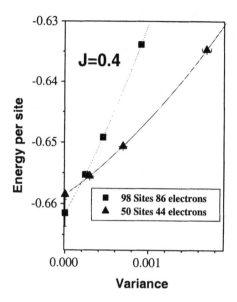

Figure 5 Energy per site of the $t - J$ model for 44 electrons on 50 sites (triangles) and 86 electrons on 98 sites (squares), $J = 0.4$, for the approximate $p = 0, 1, 2$ Lanczos step iterations over the projected d-wave wavefunction Eq. (3). Continuous lines are quadratic fit of the data.

Appendix: Variance estimate of the error on bulk correlation functions

In this Appendix we estimate the error on correlation functions assuming that the ground state $|\psi_0\rangle$ is approximated with the wavefunction $|\psi_p\rangle$:

$$|\psi_0\rangle = |\psi_p\rangle + \epsilon_p|\psi'\rangle \tag{1}$$

where $\langle\psi_p|\psi_p\rangle = \langle\psi'|\psi'\rangle = 1$, and $|\psi'\rangle$ represents a normalized wavefunction orthogonal to the exact one, $\langle\psi_0|\psi'\rangle = 0$. We restrict our analysis to thermodynamically averaged correlation functions $\hat{O}$, the ones which can be written as a bulk average of local operators $\hat{O}_i$: $\hat{O} = \sum_i \hat{O}_i/L$. This class of operators includes for instance the average kinetic or potential energy or the spin-spin correlation function at a given distance d, namely $\hat{O}_i = \hat{\mathbf{S}}_i \cdot \hat{\mathbf{S}}_{i+d}$. If we use periodic boundary conditions the expectation value of $\hat{O}_i$ on a state with given momentum does not depend on i and the bulk average does not represents an approximation

$$\frac{\langle\psi_0|\hat{O}_i|\psi_0\rangle}{\langle\psi_0|\psi_0\rangle} = \frac{\langle\psi_0|\hat{O}|\psi_0\rangle}{\langle\psi_0|\psi_0\rangle} = C. \tag{2}$$

We show here that the expectation value of bulk-averaged operators $\hat{O}$ on the approximate state $|\psi_p\rangle$ satisfy the following relation:

$$\langle\psi_p|\hat{O}|\psi_p\rangle = C + O(\epsilon_p^2, \epsilon_p/\sqrt{L}), \tag{3}$$

thus implying that for large enough size the expectation value (3) approaches the exact correlation function C linearly with the variance. The validity of the above statement is very simple to show under very general grounds. In fact by definition:

$$\langle\psi_p|\hat{O}|\psi_p\rangle = C + 2\epsilon_p\langle\psi'|\hat{O}|\psi_0\rangle + \epsilon_p^2\langle\psi'|\hat{O}|\psi'\rangle. \tag{4}$$

The term proportional to ϵ_p in the above equation can be easily bounded by use of the Schwartz inequality:

$$|\langle\psi'|\hat{O}|\psi_0\rangle|^2 = |\langle\psi'|\hat{O} - C|\psi_0\rangle|^2 \le \langle\psi_0|(\hat{O} - C)^2|\psi_0\rangle . \tag{5}$$

The final term in the latter inequality can be estimated under the general assumption that correlation functions $C(d) = \langle\psi_0|(\hat{O}_i - C)(\hat{O}_{i+d} - C)|\psi_0\rangle/\langle\psi_0|\psi_0\rangle$ decay sufficiently fast with the distance $|d|$, as a consequence of the cluster property:

$$\langle\psi_0|(\hat{O} - C)^2|\psi_0\rangle = (1 + \epsilon_p^2)\frac{1}{L}\sum_d C(d).$$

This concludes the proof of the statement of this Appendix, provided $\sum_d C(d)$ is finite for $L \to \infty$.

Acknowledgments

This work was partially supported by MURST (COFIN99) and INFM. Useful discussions with T.M. Rice, M. Troyer, C. di Castro, A. Parola, M. Calandra, and M. Capone are gratefully acknowledged.

References

[1] C. Gros, Phys. Rev. B **38**, 931 (1988).

[2] E.S. Heeb and T.M. Rice, Europhys. Lett. **27**, 673 (1994).

[3] G. Baskaran, Z. Zou, and P.W. Anderson, Solid State Comm. **63**, 973 (1987).

[4] F.C. Zhang and T.M. Rice, Phys. Rev. B **37**, 3759 (1988).

[5] S.R. White, Phys. Rev. Lett. **69**, 2863 (1992).

[6] S.R. White and D. Scalapino, Phys. Rev. Lett. **80**, 1272 (1998).

[7] S.R. White and D. Scalapino, Phys. Rev. B **60**, 753 (1999); S.R. White and D. Scalapino, e-print cond-mat/0006071.

[8] D.S. Dessau, Z.X. Shen, D.M. King, D.S. Marshall, L.W. Lombardo, P.H. Dickinson, A.G. Loeser, J. Di Carlo, C.H. Park, A. Kapitulnik, and W.E. Spicer, Phys. Rev. Lett. **71**, 2781 (1993); D.S. Marshall, D.S. Dessau, A.G. Loeser, C.H. Park, A.Y. Matsuura, J.N. Eckstein, I. Bosovic, P. Fournier, A. Kapitulnik, W.E. Spicer, and Z.X. Shen, Phys. Rev. Lett. **76**, 4841 (1996).

[9] H.J.M. van Bemmel, D.F.B. ten Haaf, W. van Saarloos, J.M.J. van Leeuwen, and G. An, Phys. Rev. Lett. **72**, 2442 (1994); D.F.B. ten Haaf, H.J.M. van Bemmel, J.M.J. van Leeuwen, W. van Saarloos, and D.M. Ceperley, Phys. Rev B **51**, 13039 (1995).

[10] S. Sorella, Phys. Rev. Lett. **80**, 4558 (1998).

[11] S. Sorella and L. Capriotti, Phys. Rev. B **61**, 2599 (2000).

[12] Y.C. Chen and T.K. Lee, Phys. Rev. B **51**, 6723 (1995).

[13] M. Kohno, Phys. Rev. B **55**, 1435 (1997).

[14] C.T. Shih, Y.L. Chen, H.Q. Lin, and T.K. Lee, Phys. Rev. Lett **98**, 1294 (1998).

[15] M. Calandra and S. Sorella, Phys. Rev. B **61**, 11894 (2000).

[16] A.L. Lichtenstein and M.I. Katsnelson, e-print cond-mat/9911320.

[17] Th. Maier, M. Jarrell, Th. Pruschke, and J. Keller, e-print cond-mat/0002352.

[18] N. Furukawa, T.M. Rice, and M. Salmhofer, Phys. Rev. Lett. **81**, 3195 (1998).

[19] C.J. Halboth and W. Metzner, Phys. Rev. B **61**, 7364 (2000).

[20] C.S. Hellberg and E. Manousakis, Phys. Rev. B **61**, 11787 (2000).

[21] S. Sorella, in preparation.

[22] F. Franjic and S. Sorella, Mod. Phys. Lett. B **10**, 873 (1996).

[23] H. Yokoyama and H. Shiba, J. Phys. Soc. Jpn. **57**, 2482 (1988).

[24] J. M. Tranquada, B.J. Sternlieb, J.D. Axe, Y. Nakamura, and S. Uchida, Nature **375**, 561 (1995).

A NEW SIMULATION METHOD FOR INFINITE SIZE LATTICES

H.G. Evertz and W. von der Linden

Inst. f. Theor. Physik, Techn. Univ. Graz, 8010 Graz, Austria

Abstract   We introduce a Monte Carlo method, as a modification of existing cluster algorithms, which allows simulations directly on systems of infinite size, and, for quantum models, also at $\beta = \infty$. All two-point functions can be obtained, including dynamical information. When the number of iterations is increased, correlation functions at larger distances become available. Limits $q \to 0$ and $\omega \to 0$ can be approached directly.

1.   INTRODUCTION

Standard Monte Carlo simulations are limited to systems of finite size. Physical results for infinite systems have to be obtained by finite size scaling, assuming that one knows the correct scaling laws, and assuming that the data are already in a suitable asymptotic regime. It is therefore very desirable to obtain results also *directly* at infinite system size. We will show how to do so, with only a small modification of existing cluster algorithms, by using the cluster representation of the models to calculate two-point functions. In the quantum case we can then also simulate directly at $\beta = \infty$ and calculate correlation functions and dynamical greens functions. As examples we will show calculations for the classical Ising model and for quantum Heisenberg ladder systems.

2.   CLUSTER METHODS

The *Swendsen Wang cluster method* [1] for the classical Ising model is based on the Fortuin-Kasteleyn representation

$$e^{\beta J(s_i s_j - 1)} = \sum_{b_{ij}=0,1} p\, \delta_{s_i s_j}\, \delta_{b_{ij},1} + (1-p)\, \delta_{b_{ij},0} =: \sum_{b_{ij}=0,1} W_{ij}(s_i, s_j, b_{ij})$$

(1)

for the Boltzmann weight of a spin-pair, with $p = 1 - e^{-2\beta J}$, which enlarges the phase space of spin variables s_i by additional bond variables b_{ij}. The

J. Bonča et al. (eds.), Open Problems in Strongly Correlated Electron Systems, 187–192.

partition function $Z = \sum_{\{s_i\}} \sum_{\{b_{ij}\}} W(s,b)$ with total weight $W(s,b) = \prod_{\langle ij \rangle} W_{ij}(s_i, s_j, b_{ij})$ is then simulated efficiently by switching between representations: given a spin-configuration $s := \{s_i\}$, one generates a bond configuration $b := \{b_{ij}\}$ with the conditional probability $p(b|s) \sim W(s,b)$, thus creating a configuration of *clusters* of sites connected by bonds $b_{ij} = 1$. Given a bond configuration, a new spin configuration is generated with probability $p(s|b) \sim W(s|b)$. Because of the factor $\delta_{s_i s_j} \delta_{b_{ij},1}$ in W_{ij}, this amounts to setting all spins of each cluster randomly to a common new value, independent of other clusters.

Observables $\hat{O}$ can be computed either in spin-representation as $O(s)$ or in bond representation as $O(b) = \left(\sum_s O(s) W(s,b) \right) / \left(\sum_s W(s,b) \right)$ (so called "improved estimators") [2]. For two-point functions $O(s) = s_i s_j$ the bond representation is particularly simple:

$$O(b) = \delta(\text{sites } i \text{ and } j \text{ are in the same cluster}) . \tag{2}$$

Thus two-point functions and derived quantities (including susceptibility, energy, specific heat) can be computed from the properties of individual clusters.

A variant of the Swendsen-Wang method is Wolff's *single cluster method* [2]. Given a spin configuration, only one of the bond-clusters is constructed, namely the cluster which contains a randomly chosen initial starting site x_0. All spins in this cluster are then flipped. One advantage is that the single cluster will on average be larger than the average Swendsen-Wang cluster (so that its flip results in a bigger move in phase space). Correcting for this fact, one obtains [2]

$$\langle s_i \, s_j \rangle = \langle \frac{V}{V_{cl}} \, \delta(\text{sites } i \text{ and } j \text{ are inside the single cluster}) \rangle , \tag{3}$$

where V_{cl} is the number of sites in the single cluster.

3.   NEW METHOD

Our new *infinite system method* now employs the single cluster method, except that it starts each cluster not from a randomly chose site, but always from the *same* site x_0 (e.g. the origin of the coordinate system). For the ferromagnetic Ising model, we begin with an initial staggered spin configuration of unlimited size. (Only a finite part if this configuration will have to be stored). We iterate the following two steps: (1) For the current spin configuration, a cluster containing site x_0 is constructed with Swendsen-Wang bond-probabilities, and (2) all spins in this cluster are flipped, resulting in a new spin configuration.

After a sufficient total number $N(r)$ of iterations, this process will "equilibrate" all spins within a radius r around x_0, and two-point functions can then be measured within this area. With increasing number of iterations, the cluster will occasionally reach larger and larger distances. Because of eq. (3), the probability to do so is (roughly) proportional to the the correlation function $C(r)$. A region of the lattice at distance r will become "equilibrated" after a sufficient number n_{eq} of times that the cluster has reached that region, thus $N(r) \sim \frac{n_{eq}}{C(r)}$. Since the cluster can grow without bounds, we are calculating correlation functions for the *infinite* lattice, while we only need to store the spin configuration within the area actually reached by a cluster during a finite run.

3.1   EXAMPLE: CLASSICAL ISING MODEL

Fig. 1 shows as an example the correlation function for the two dimensional Ising model at $\beta = 0.42 < \beta_c = 0.44068...$

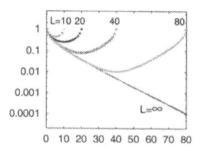

Figure 1   Correlation functions of the $d = 2$ Ising model at $\beta = 0.42$.

For comparison we show results for finite size lattices, with periodic boundary conditions. Usually, the infinite lattice result can only be obtained as the extrapolation of the finite lattice ones. In contrast, with our new method we can obtain the infinite lattice result in a single simulation, producing the lower straight line in fig. 1. The computational effort per sweep is about the same as for the $L = 80$ system. We checked that the method works just as well in three dimensions.

The calculation of error bars for $C(r)$ needs special care. For each distance $|r|$, we demand that the cluster contains sites with that distance $|r|^{d-1} \times n_{eq}$ times, i.e. it reaches each site at that distance $O(n_{eq})$ $(= O(10))$ times, before we consider data at that distance thermalized. Only then do we start to accumulate measurements for $C(r)$ according to eq. (3) for all sites i, j in the cluster. Alternative similar procedures are possible.

3.2   RESTRICTIONS

When will the new method work ? The main constraint is that the computational effort, proportional to the average cluster size, must be finite. There must not be a finite probability for a percolating cluster, thus we need $\beta < \beta_c$.

If we assume $C(r) \sim e^{-r/\xi}/r^{d-2+\eta}$, then the average cluster size is finite for either $\xi < \infty$, or $\xi = \infty$ and η sufficiently large. The new method will thus work if the unsubtracted correlation function decays sufficiently rapidly.

4.   QUANTUM CASE

For nonrelativistic quantum systems, the loop algorithm [3] provides a cluster method. It is based on an enlarged representation in terms of the original spin operators and additional loop operators, similar in spirit to the Fortuin-Kasteleyn representation, eq. (1). We will in the following specialize to the spin $\frac{1}{2}$ quantum Heisenberg model

$$\mathcal{H} = J \sum_{\langle ij \rangle} \frac{1}{2} \left(S_i^+ S_j^- + S_i^- S_j^+ \right) + \lambda S_i^z S_j^z . \tag{4}$$

The two-point function for the single cluster version of the loop algorithm then reads

$$\begin{aligned} \left\langle \frac{1}{2} \left(S_i^+ S_j^- + S_i^- S_j^+ \right) \right\rangle &= \left\langle \frac{V}{V_{cl}} \delta(\text{i and j on the loop}) \right\rangle \\ \left\langle S_i^z S_j^z \right\rangle &= \left\langle \frac{V}{V_{cl}} \delta(\text{i and j on the loop}) \, S_i^z S_j^z \right\rangle \end{aligned}, \tag{5}$$

where V_{cl} is the number of sites on the loop. Therefore our approach to simulate infinite size systems can be used in the same way as for the Ising model. "Infinite size" here can be applied to the spatial directions as well as, independently, to the direction of imaginary time, yielding simulations directly at $\beta = \infty$, while retaining all dynamical information. Note that our method is not a projector method. We can simulate directly *at* $L = \infty$ and $\beta = \infty$. This also enables us to obtain the limits $q \to 0$ and $\omega \to 0$ directly from the simulations. Contributions from $q \equiv 0$ or $\omega \equiv 0$, which are in general different from those limits, and are present in normal simulations with periodic boundary conditions, are completely avoided in our approach. As statistics increases, clusters will grow to larger distances in spatial and/or imaginary time direction, thereby improving the available resolution in q and ω.

Again, our present method will be applicable provided that the unsubtracted correlation function drops off sufficiently quickly (for which finite correlation length and/or finite gap are sufficient but not necessary).

4.1   EXAMPLE: SPIN LADDERS

As an example we studied spin ladder systems [4] with $N = 2$ and 4 legs for the isotropic antiferromagnet ($\lambda = 1$). We used the discrete time version of the loop algorithm ($\Delta \tau = 1/16$). The method can be applied just as well in continuous time [6].

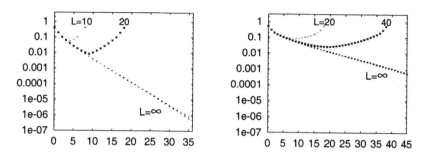

Figure 2 Equal time staggered spatial correlation function at $q_\perp = \pi$ for N=2 (left) and N=4 (right)

Fig. 2 shows results for the equal time staggered spatial correlation functions along the chains, which behave similarly to the Ising case. A fit to the infinite lattice result gives $\xi = 2.93(2)$ for $N = 2$ and $\xi = 8.2(1)$ for $N = 4$. Fig. 3 shows greens functions for $L = \infty$, the infinite size system. Whereas

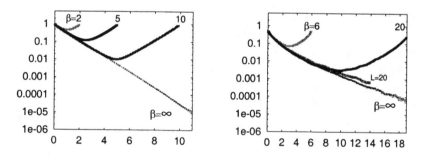

Figure 3 Greens functions $\langle S(\vec{q}, 0) \, S(\vec{q}, t) \rangle$ at $\vec{q} = (\pi, \pi)$ for $L = \infty$ and N=2 (left) and N=4 (right)

finite temperature calculations give results periodic in imaginary time, which have to be extrapolated, the new approach provides the $\beta = \infty$ $(T = 0)$ result directly. With increased number of iterations, the infinite system results become available at larger distances both in spatial and in imaginary time direction. A fit to the exponential decay $G(\tau) \sim e^{-\tau\Delta}$ directly gives the gaps $\Delta = 0.5059(4)$ for $N = 2$ and $\Delta = 0.19(1)$ for $N = 4$, consistent with previous results [5]. We also show results for $L = 20$ and $\beta = \infty$ to exemplify the effect of finite size systems.

Continuing the imaginary time greens function to real frequencies by maximum entropy provides the spectra of fig. 4, in which the gaps, the single magnon peaks, and higher excitations for $N = 4$ are clearly visible. This appears to be the first time that the spectrum for $N = 4$ has been calculated.

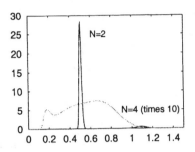

Figure 4   Spectrum $S(\vec{q}, \omega)$ at $\vec{q} = (\pi, \pi)$ for $L = \infty$ and $\beta = \infty$

5.   DISCUSSION

Let us compare our method to other approaches. Exact diagonalization provides ground state results but is limited to small systems. Projector methods are also limited in system size and rely critically on proper convergence to the ground state. DMRG can achieve $\beta = \infty$ (mostly without dynamical information) for fairly small systems, or infinite size for large temperatures and finite $\Delta\tau$. The most powerful method to extrapolate to infinite system size is the Finite Size Scaling method of Kim [7] and Caracciolo et al. [8], which allows extrapolation at correlation lengths far larger than the system size, but does require knowledge about scaling and corrections to scaling of the model.

In summary, we have introduced a new method, as a small modification of existing cluster methods, to simulate both classical and quantum systems at infinite system size and/or at zero temperature, while obtaining all two-point functions. Larger distances (in space and/or imaginary time) are accessed for longer simulations. The method in its present form is applicable as long as the unsubtracted two-point correlation function decays sufficiently quickly.

References

[1] R.H. Swendsen and J.S. Wang, Phys. Rev. Lett. 58, 86 (1987).

[2] U. Wolff, Phys. Rev. Lett. 62, 361 (1989), Nucl. Phys. B334, 581 (1990).

[3] H.G. Evertz, G. Lana, and M. Marcu, Phys. Rev. Lett. 70, 875 (1993), for a review see H.G. Evertz, cond-mat/9707221 (2nd ed. 2000)

[4] See e.g. E. Dagotto and M. Rice, Science 271, G18 (1996).

[5] B. Frischmuth et al., Phys. Rev. B54, R3714, (1996); S.R. White et al., Phys. Rev. Lett. 73, 886 (1994).

[6] B.B. Beard and U.J. Wiese, Phys. Rev. Lett. 77, 5130 (1996).

[7] J.K. Kim, Phys. Rev. Lett. 70, 1735 (1993).

[8] S. Caracciolo et al., Phys. Rev. Lett. 74, 2969 (1994).

UNIVERSALITY IN TWO-DIMENSIONAL
QUANTUM HEISENBERG ANTIFERROMAGNETS

Matthias Troyer

Theoretische Physik, Eidgenössische Technische Hochschule Zürich, 8093 Zürich, Switzerland

Abstract   The quantum Monte Carlo loop algorithm has led to a breakthrough in the numeric simulation of quantum magnets. It enables us to perform simulations on lattices with up to millions of quantum spins. This has made possible the accurate investigation of the low temperature universal scaling regime in quantum magnets and the properties of quantum phase transitions. I will review theoretical and numerical work for the phase diagram of two-dimensional quantum Heisenberg antiferromagnets, emphasizing quantum phase transitions and crossovers as well as applications to quantum Hall bilayer systems.

1.   INTRODUCTION

Recent progress in quantum Monte Carlo algorithms [1, 2] and the availability of massively parallel computers now enable the accurate simulation of non-frustrated quantum magnets on huge lattices. The quantum Monte Carlo loop algorithm is a generalization of classical cluster Monte Carlo algorithms [3, 4] to quantum systems. It solves the problem of critical slowing down, making the computational complexity of the Monte Carlo simulation scale optimally, namely linearly with the spatial volume times the inverse temperature. Furthermore it can be formulated directly in continuous imaginary time [2], removing any discretization error arising from finite time steps.

This algorithmic advance in combination with parallel implementations [5] now allow us to study not only the qualitative properties of these models, but to critically check predictions made for the universal critical behavior near quantum phase transitions [6] and finite temperature phase transitions [7, 8] in quantum systems, as well as the low-temperature scaling and crossovers away from criticality [9, 10, 11, 12, 13].

Furthermore accurate quantitative comparisons of quantum Monte Carlo simulations to experimental measurements allow the determination of microscopic coupling parameters and the explanation of unusual magnetic properties of low-dimensional quantum magnets [14, 15, 16, 17, 18].

193

J. Bonča et al. (eds.), Open Problems in Strongly Correlated Electron Systems, 193–202.

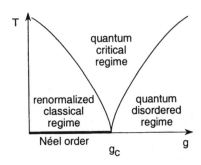

Figure 1 Generic phase diagram of a two dimensional quantum Heisenberg antiferromagnet, as discussed by Chakravarty, Halperin and Nelson. As a function of a control parameter g two different ground states can be observed. A long range ordered Néel state and a gapped quantum disordered state, separated by a quantum critical point. At finite temperatures three regimes can be observed, a renormalized classical regime, a quantum critical regime, and a quantum disordered regime, separated from each other by smooth crossovers.

Analytic investigations of these phase diagram [19, 20, 21] are based on a mapping of the quantum Heisenberg antiferromagnet to a (2+1)-dimensional nonlinear sigma model (NLσM), ignoring the Berry phase terms:

$$S_{eff} = \frac{2}{g} \int_0^\beta d\tau \int d^2x \left(|\nabla \Omega|^2 + \frac{1}{c_0^2} |\partial \Omega / \partial \tau|^2 \right).$$ (1)

Here Ω is a three-component unit vector. g is the coupling constant and c_0 the bare spin wave velocity. Detailed predictions have been made for the various phases and regimes sketched in Fig. 1. In the following I will review numeric simulations of various quantum Heisenberg antiferromagnets using the loop algorithm and compare the simulation results to these analytic predictions.

2.   THE ORDERED GROUND STATE

For weak quantum fluctuations the ground state has long range Néel order which is destroyed immediately at any finite temperature by the Goldstone modes, long-wave length gapless spin wave excitations.

Until about ten years ago the existence of long range order in the ground state of the 2D square lattice quantum Heisenberg antiferromagnet was uncertain. Numerical simulations provided strong evidence [23]. A very elegant numerical "proof" are the quantum Monte Carlo simulations by Wiese and Ying [24]. They show that the low-temperature finite size results of the uniform and staggered susceptibility and the energy fit perfectly to the corresponding scaling forms derived by Hasenfratz and Niedermayer [22] for a Lorentz invariant effective field theory with $O(3)$ symmetry that is broken to $O(2)$ in the ground state. Up to second order in temperature and inverse system size these

Table 1   Estimates of ρ_s, c, and M from quantum Monte Carlo simulations.

	ρ_s	c	M
Kim and Troyer [12]	0.1784^{+0024}_{-0015}	$1.649^{+0.015}_{-0.007}$	–
Wiese and Ying [24]	$0.185(2)$	$1.68(1)$	$0.3074(4)$
Sandvik [25]	$0.175(2)$	$1.673(7)$	$0.3070(3),$
Calandra Buonaura and Sorella [26]	–	$1.64(8)$	$0.3075(2)$

asymptotic scaling functions depend are characterized by three constants: the staggered magnetization M, the spin wave velocity c and the spin stiffness ρ_s. Numerical estimates for these three parameters are summarized in Table 1.

3.   THE RENORMALIZED CLASSICAL REGIME

In the renormalized classical regime, at finite temperatures above the ordered ground state the correlation length diverges exponentially [19, 20, 22]. The behavior is that of a classical magnet but with renormalized parameters. The most accurate low-temperature result was obtained by Hasenfratz and Niedermayer, who got not only the exact prefactor but also the first correction term: [22]

$$\xi^{RC}(T) = \frac{e}{8}\frac{c}{2\pi\rho_s}\exp\left(\frac{2\pi\rho_s}{T}\right)\left[1 - \frac{T}{4\pi\rho_s} + \mathcal{O}\left(\frac{T}{2\pi\rho_s}\right)^2\right] \qquad (2)$$

The behavior at low temperature is thus completely characterized by the ground state stiffness ρ_s and the ground state spin wave velocity c. Comparing Eq. (2) to experimental measurements on copper oxides (spin $S = 1/2$) showed good agreement [27], while for nickel oxides large discrepancies were found [27, 28].

Numerical simulations were first done for spin $S = 1/2$ by Makivic and Ding [9] on small systems and by Elstner *et al.* for all $\leq 5/2$ by high temperature series expansion [29]. Later Beard *et al.* [11] and we [12] did simulation for $S = 1/2$ on larger lattices of up to one million spins and down to lower temperatures. $S = 1$ was studied by us in Ref. [30]. The simulations confirmed the discrepancies seen in experiments but also pointed out the reason [12, 29, 30, 31] Eq. (2) is valid only once the lattice spacing a is much smaller than the correlation length. *Additionally* the extent in imaginary time β has to be much larger than the lattice spacing mapped to imaginary time a/c. Thus

$$T \ll 2\pi\rho_s \simeq S^2 \qquad \text{and} \qquad T \ll c/a \simeq S. \qquad (3)$$

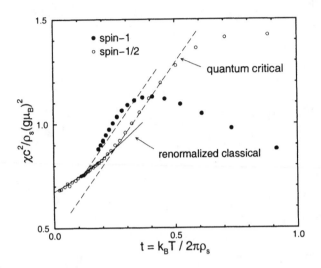

Figure 2   The uniform susceptibility χ as a function of temperature t for both spin $S = 1/2$ [12]) and $S = 1$ [30]. The solid line is the predicted low temperature form of the renormalized classical regime. The dashed lines have the universal slope expected for the quantum critical regime. As expected, contrary to $S = 1/2$, no extended quantum critical regime exists for $S = 1$.

Thus for larger spin S the second criterion is more restrictive and the validity of the asymptotic expression (2) is restricted to larger correlation lengths. This effect of the lattice cutoff on the renormalization in the imaginary time direction was recently quantified in leading order by Hasenfratz [32] in the form of an additional multiplicative correction which was seen to reduce the discrepancies [13, 32].

It also follows that for temperatures $T \gg c/a$ quantum effects should become weak and a semi-classical description be appropriate. It was indeed found by Cuccoli *et al.* that the pure-quantum self-consistent harmonic approximation (PQSCHA), a semi-classical treatment can describe square lattice quantum Heisenberg antiferromagnets with $S \geq 1$ very well. [30, 33]

The uniform susceptibility χ_u is also known analytically up to second order in temperature [20, 22]:

$$\chi_u^{RC}(T) = \frac{2}{3}\frac{\rho_s}{c^2}\left[1 + \frac{T}{2\pi\rho_s} + \left(\frac{T}{2\pi\rho_s}\right)^2\right], \tag{4}$$

As finite size effects are small for χ_u we could obtain it down to very low temperatures and extract accurate values for ρ_s and c (see Tab. 1 and Fig. 2).

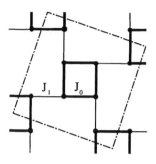

Figure 3 Lattice structure of the 1/5-th depleted square lattice of CaV_4O_9. The dashed square indicates the eight spin unit cell used in the calculations of the critical exponents in Ref. [6].

Table 2 Critical exponents β, ν, η and z for the quantum phase transition in a planar Heisenberg antiferromagnet as calculated in Ref. [6]. For comparison the exponents of the 3D classical Heisenberg ($O(3)$) model are listed.

model	ν	β	η	z
2D QAFM	0.685(35)	0.345(25)	0.015(20)	1.018(20)
3D O(3) [38]	0.7048(30)	0.3639(35)	0.034(5)	—

4. THE QUANTUM PHASE TRANSITION

Increasing quantum fluctuations destroy the long range order at a quantum critical point g_c where the staggered magnetization M vanishes. Based on the mapping to the NLσM this quantum critical point at zero temperature should be in the same universality class as the 3D classical NLσM, which is the continuum field theory for the classical 3D Heisenberg ferromagnet. It was however argued by Chakravarty [34] that the neglected Berry phase terms could be relevant at this quantum phase transitions and change the universality class. First numerical simulations on a bilayer quantum magnet by Sandvik and Scalapino [35] and on a dimerized square lattice by Katoh and Imada [36] as well as by Sandvik and Vekic [37] were contradictory, because the system sizes (up to 200 spins) were not large enough to accurately determine the critical behavior.

Using the loop algorithm we could study up to 20000 spins at low temperatures and accurately determine the critical exponents for the quantum phase transitions in a 1/5-depleted square lattice shown in Fig. 3. This lattice shows two quantum phase transitions as the ratio of couplings $\delta = J_0/J_1$ (which determines the effective coupling g in the continuum field theory) is changed. >From finite size scaling at the critical point we could determine the dynamical critical exponent $z = 1.018 \pm 0.02$ and the correlation exponent η. This is perfectly consistent with $z = 1$, expected from a Lorentz invariant effective field theory. We calculated M and ρ_s using the finite size scaling equations of Hasenfratz and Niedermayer [22]. Fitting the results for various δ to a power law in $\delta - \delta_c$ allowed the determination of the exponents ν for the correlation length and β for the staggered magnetization. The results, summarized in Tab. 2.

The good agreement of the exponents with those of the classical Heisenberg model is a strong support for the conjecture of Chubukov *et al.* that the Berry phase terms are dangerously irrelevant and do not change the universality of the quantum phase transition [20].

5.   THE QUANTUM DISORDERED REGIME

At large quantum fluctuations the ground state is quantum disordered, with exponentially decaying correlations and a finite gap Δ for all magnetic excitations. This spin gap Δ controls the low-temperature behavior in the quantum disordered regime at temperatures $T \ll \Delta$, where the uniform susceptibility is exponentially suppressed:

$$\chi^{QD} \propto \exp(-\Delta/T) \tag{5}$$

A typical example are weakly coupled spin ladders [40]. In this case the loop algorithm has not only confirmed the existence of a spin gap [41, 42] but even allows accurate quantitative fits of simulation to experiments and the determination of microscopic coupling constants. Examples are $SrCu_2O_3$ [14, 18], CaV_2O_5 and MgV_2O_5 [15, 16, 18], NaV_2O_5 [17] and CaV_4O_9 [16].

6.   THE QUANTUM CRITICAL REGIME

The quantum critical regime exists in the vicinity of the quantum critical point at finite temperatures as long as $T \gg \Delta$ or $T \gg 2\pi\rho_s$ respectively but the quantum renormalization is not yet cut off by lattice effects ($T \ll c/a$). This regime widens as the temperature increases (see the phase diagram Fig. 1).

The quantum critical regime exhibits remarkable universal behavior. Not only the exponents for temperature dependencies but even amplitudes can be universal. For example the uniform susceptibility is

$$\chi_u^{QC} = A_\chi \left(\frac{g\mu_B}{c}\right)^2 T. \tag{6}$$

and the inverse correlation length is

$$\xi^{-1} = A_\xi \frac{T}{c}. \tag{7}$$

where A_χ and A_ξ are dimensionless universal numbers [19, 20]. The c-independent combination of the amplitudes $A_\xi/\sqrt{A_\chi}$ was calculated by Sandvik *et al.* [39] and $A_\chi = 0.26(1)$ was calculated by us [6]. The excellent agreement with a $1/N$ expansion for the NLσM [20] again confirms the NLσM as the effective field theory at the quantum critical point.

Away from criticality the linear T-dependence with universal slope remains but a finite offset different from zero appears in these quantities.

7.   CROSSOVERS

The possibility of observing a crossover from the quantum critical to the renormalized classical regime in square lattice materials was a heavily debated subject. [9, 19, 20, 29] and could be investigated using the loop algorithm [12, 13, 30].

As first pointed out by Chubukov *et al.* [20] we found convincing evidence for such a crossover in the uniform susceptibility of the spin-1/2 square lattice in the regime $0.3J < T < 0.5J$ [12] (see Fig. 2). As expected [29] we did not find any crossover in the $S = 1$ case [30] where lattice effects become important before the crossover temperature is reached (as $2\pi\rho_s > c/a$ in this case).

For the correlation length of the square lattice ξ on the other hand we can find no evidence for a quantum critical region or a crossover [12, 13]. There logarithmic corrections to the asymptotic scaling do not cancel like for the uniform susceptibility and hide any quantum critical region and crossover.

8.   BILAYER ANTIFERROMAGNET IN A MAGNETIC FIELD AND QUANTUM HALL BILAYERS

Recently the bilayer quantum Heisenberg antiferromagnet in a *magnetic field* has attracted interest since it is in the same universality class as certain quantum Hall bilayer systems. The bilayer antiferromagnet has the Hamiltonian

$$\mathcal{H} = \sum_i \left[J_\perp \hat{\mathbf{S}}_{1i} \cdot \hat{\mathbf{S}}_{2i} - \mathbf{H} \cdot \left(\hat{\mathbf{S}}_{1i} + \hat{\mathbf{S}}_{2i} \right) \right] + \sum_{<ij>} J \left[\hat{\mathbf{S}}_{1i} \cdot \hat{\mathbf{S}}_{1j} + \hat{\mathbf{S}}_{2i} \cdot \hat{\mathbf{S}}_{2j} \right],$$

$$\tag{8}$$

where $\hat{\mathbf{S}}_{ai}$ are quantum spin-1/2 operators in 'layers' $a = 1, 2$ residing on the sites, i, of a two-dimensional square lattice, $\mathbf{H} = (0, 0, H)$ is the external magnetic field, and $J_\perp > 0$, J are intra- and inter-layer exchange constants respectively. The phase diagram in zero field has been extensively studied

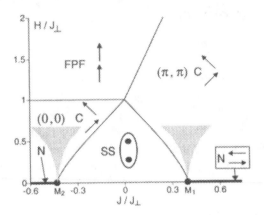

Figure 4   Ground states of $\mathcal{H}$. The arrows denote the mean orientations of the spins in the two layers, for the case where H points vertically upwards; in the SS phase the spins have no definite orientation. The in-plane ordering wave vectors of the x, y components of the spins in the C phase are indicated. There is a Kosterlitz Thouless transition at non-zero T above the C phases, and the result (9) applies above the shaded regions.

before (see References in [7]). The phase diagram in nonzero field as studied in a collaboration with Sachdev [7] is shown in Fig. 4. For large $J_\perp$ the ground state consists of singlets on the "rungs" between the layers and is quantum disordered with a finite spin gap which survives even in magnetic fields up to $H = \Delta$. At smaller values of $J_\perp$ the ground state shows antiferromagnetic long range order. Applying a magnetic field in that range leads to a canted spin state with broken rotational order. The $O(2)$ order of this canted state has a finite temperature phase transition of Kosterlitz Thouless type. Most interesting are the regions above the two critical points (the shaded regions in Fig. 4). There we have a universal relation for the Kosterlitz Thouless transition temperature

$$T_{KT} = \kappa H, \tag{9}$$

with a universal constant $\kappa = 0.38 \pm 0.06$ determined in a quantum Monte Carlo simulation [7].

This model is of special interest since it is in the same universality class as quantum Hall bilayers at total filling fractions $\nu = 2/n$ with n an odd integer [43]. Two recent experiments [44, 45] have found magnetic transitions in bilayer quantum Hall systems at total filling fraction $\nu = 2$. This gives the possibility to confirm the universality of Eq. (9) in experiments on quantum Hall systems.

9. CONCLUSIONS

The quantum Monte Carlo loop algorithm enables for the first time accurate simulations of the universal scaling of quantum systems at low temperatures and close to quantum phase transitions. This allows detailed comparisons with both theoretical prediction and experimental measurements. Quantum Monte Carlo simulations provide for stringent checks on the validity of theories and allow the determination of universal critical exponents and amplitudes. We can now study the interesting properties of quantum phase transitions with accuracies that until recently was possible only for classical phase transitions. It will be of great interest to apply these techniques also to phase transitions in other systems that do not suffer from the sign problem, such as bosonic systems.

The author acknowledges support by the Swiss National Science Foundation and is thankful to the organizers of the workshop for their hospitality.

References

[1] H. G. Evertz, M. Marcu, and G. Lana, Phys. Rev. Lett. **70**, 875 (1993).

[2] B. B. Beard and U.-J. Wiese, Phys. Rev. Lett. **77**, 5130 (1996).

[3] R.H. Swendsen and J-S. Wang, Phys. Rev. Lett. **58**, 86 (1987).

[4] U. Wolff, Phys. Rev. Lett. **62**, 361 (1989).

[5] M. Troyer, *et al.*, Lecture Notes in Computer Science **1505**, 191 (1998).

[6] M. Troyer, *et al.*, J. Phys. Soc. Jpn **66**, 2957 (1997); M. Troyer and M. Imada in *Computer Simulations in Condensed Matter Physics X*, ed. D.P. Landau *et al.*, (Springer Verlag, Heidelberg, 1997).

[7] M. Troyer and S. Sachdev, Phys. Rev. Lett. **81**, 5418 (1998).

[8] K. Harada and N. Kawashima, Phys. Rev. B **55**, R11949 (1997)

[9] M. S. Makivic and H.-Q. Ding, Phys. Rev. B **43**, 3562 (1991).

[10] J.-K. Kim, D. P. Landau, and M. Troyer, Phys. Rev. Lett. **79**, 1583 (1997).

[11] B.B. Beard *et al.*, Phys. Rev. Lett. **80**, 1742 (1998).

[12] J.-K. Kim and M. Troyer, Phys. Rev. Lett. **80**, 2705 (1998).

[13] J.-K. Kim and M. Troyer, in preparation.

[14] D. C. Johnston, Phys. Rev. B **54**, 13 009 (1996).

[15] S. Miyahara *et al.*, J. Phys. Soc. Jpn. **67**, 3918 (1998).

[16] M. A. Korotin *et al.*, Phys. Rev. Lett. **83**, 1387 (1999).

[17] D. C. Johnston *et al.*, hys. Rev. B **61**, 9558 (2000).

[18] D. C. Johnston *et al.*, cond-mat/0001147

[19] S. Chakravarty *et al.*, Phys. Rev. B **39**, 2344 (1989)

[20] A. V. Chubukov, S. Sachdev, and J. Ye, Phys. Rev. B **49**, 11919 (1994).

[21] For a review see S. Sachdev, *Quantum Phase Transitions*, Cambridge University Press (1999).

[22] P. Hasenfratz and F. Niedermayer, Z. Phys. B **92**, 91 (1993).

[23] E. Manousakis, Rev. Mod. Phys. **63**, 1 (1991).

[24] U. J. Wiese and H. P. Ying, Z. Phys B **93**, 147 (1994).

[25] A.W. Sandvik, Phys. Rev. B **56**, 11678 (1997).

[26] M. Calandra Buonaura and S. Sorella, Phys. Rev. B **57**, 11446 (1998).

[27] M. Greven *et al.*, Z. Phys. B **96**, 465 (1995).

[28] K. Nakajima *et al.*, Z. Phys. B **96**, 479 (1995).

[29] N. Elstner *et al*, Phys. Rev. Lett. **75**, 938 (1995).

[30] K. Harada *et al.*, J. Phys. Soc. Jpn. 67, 1130 (1998).

[31] S. Ty, B.I. Halperin and S. Chakravarty, Phys. Rev. Lett. **62**, 835 (1989).

[32] P. Hasenfratz, preprint, cond-mat/9901355.

[33] A. Cuccoli *et al.*, Phys. Rev. Lett. **77**, 3439 (1996); Phys. Rev. B 58, 14151 (1998).

[34] S. Chakravarty in *Random magnetism and high temperature superconductivity*, ed. by W.P. Beyermann, N.L. Huang-Liu and D.E. MacLaughlin, World Scientific (Singapore 1993).

[35] A. W. Sandvik and D. J. Scalapino, Phys. Rev. Lett. **72**, 2777 (1994).

[36] N. Katoh and M. Imada, J. Phys. Soc. Jpn. **63**, 4529 (1994).

[37] A. W. Sandvik and M. Vekić, J. Low. Temp. Phys. **99**, 367 (1995).

[38] K. Chen *et al.*, Phys. Rev. B **48**, 3249 (1993).

[39] A.W. Sandvik *et al.*, Phys. Rev. B**51**, 16483 (1995)

[40] E. Dagotto and T. M. Rice, Science **271**, 618 (1996)

[41] B. Frischmuth, B. Ammon, and M. Troyer, Phys. Rev. B **54**, R3714 (1996)

[42] M. Greven *et al.*, Phys. Rev. Lett. **77**, 1865 (1996).

[43] S. Das Sarma, *et al.*, Phys. Rev. Lett. **79**, 917 (1997); Phys. Rev. B **58**, 4672 (1998).

[44] V. Pellegrini *et al.*, Science **281**, Aug 7 (1998).

[45] A. Sawada *et al.*, Phys. Rev. Lett. **80**, 4534 (1998).

STRIPES AND PAIRING
IN THE $\nu = 5/2$ QUANTUM HALL EFFECT

F. D. M. Haldane

Dept. of Physics, Princeton University, Princeton NJ 08544-0708, USA.

Abstract   The observed quantum Hall state at $\nu = 5/2$ is believed to be an "exotic" quantum Hall states that can be described as a BCS analog state with paired electrons. It is sensitive to tilting the magnetic field direction which appears to destroy the quantum Hall effect. I will describe studies that show how the nature of the ground state of electrons at $\nu = 5/2$ changes between composite Fermi liquid, (non-Abelian) BCS-like paring, and compressible anisotropic (smectic) striped phases, as the structure of the interaction potential is varied. In this system, at least, it seems that stripes are a competing phase that is unrelated to the paired state, and the transition between them appears to be first order.

1.   INTRODUCTION

The issue of whether the "stripes" (one-dimensional charge density wave ordering) seen in some of the high-temperature superconducting materials is deeply related to their superconductivity, or whether it is a competing phase that is unrelated to superconductivity, is today a topical question [1]. It is interesting to note that a rather different system, the two dimensional electron gas (2DEG) at high magnetic fields, such that the Landau level filling factor $\nu = 5/2$, also appears to exhibit different phases, which are respectively related to stripes and superconductivity (of "composite fermions") [2, 3]. In addition, the paired quantum Hall state appears to exhibit some very exotic properties, the so-called "non-Abelian statistics", that make it potentially one of the most interesting of the quantum Hall states.

2.   ABELIAN AND NON-ABELIAN QUANTUM HALL STATES

The general classification of Quantum Hall states has progressed from the integer quantum Hall states, with Slater-determinant-type wavefunctions, and Hall conductance $\sigma_H = ne^2/h$, with integer n, to the now-familiar "Abelian" fractional quantum Hall states, with correlated wavefunctions of the Laughlin

J. Bonča et al. (eds.), Open Problems in Strongly Correlated Electron Systems, 203–213.

type, which have fractional Hall conductances $\sigma_H = \nu e^2/h$, with rational $\nu = p/q$. Finally, with the recent proposed identification of the $\nu = 5/2$ state as a spin-polarized state [4, 5] related to the Moore-Read state [6] it appears that "non-Abelian" fractional quantum Hall states may be physically realized.

2.1   ABELIAN QUANTUM HALL EDGE STATES

A key feature of quantum Hall states is that they exhibit a type of "topological order", which is manifested by the existence of chiral edge states, which necessarily occur at the boundary between the quantum Hall state and "normal matter", or at the boundary between topologically distinct quantum Hall states, which are the only places that low-energy excitations that propagate can be made. The existence of a minimal set of gapless edges states is protected by the topological order - an example of the "quantum protectorate" concept [7].

In the case of Abelian quantum Hall states, these edge states can be described by a chiral, multicomponent generalization [8, 9] of the "bosonized" description of the 1D Fermi gas:

$$H_{\text{edge}}^{\text{eff}} = \oint \frac{ds}{4\pi} \sum_{ij} V_{ij}(s) : \partial_s \varphi_i \partial_s \varphi_j :, \tag{1}$$

where $V_{ij}(s)$ is a real, positive-definite symmetric matrix, defined at each point s on the edge. The fields $\varphi_i(s)$ obey

$$[\varphi_i(s), \varphi_j(s')] = i\pi K_{ij}(\text{sgn}(s - s') + \text{sgn}(i - j)), \tag{2}$$

where K_{ij} is a non-singular real symmetric matrix with integer elements. (An arbitrary ordering of the fields is introduced.) ($\mathbf{K} = \pm 1$, with rank n, corresponds to the integer Hall effect.) Since the commutator (2) is a c-number, this is a harmonic oscillator system, and $: \ldots :$ means normal-ordering of oscillator variables with respect to the ground state of H^{eff}. The momentum (generator of translations along the edge) is

$$P_{\text{edge}} = \oint \frac{ds}{4\pi} \sum_{ij} K_{ij}^{-1} : \partial_s \varphi_i \partial_s \varphi_j :, \tag{3}$$

The creation operators (or "vertex operators")

$$\Psi_i^\dagger(s) \propto \eta_i : e^{i\varphi_i(s)} : \tag{4}$$

add integer charge eQ_i to the edge at position s, where η_i are Klein factors that ensure correct anticommutation of odd-Q_i operators:

$$\eta_i = \prod_{j<i}(-1)^{Q_i Q_j N_j}, \quad N_i = \oint \frac{ds}{2\pi} \partial_s \varphi_i, \tag{5}$$

where N_i are integer winding numbers of the fields around the edge of the Hall fluid. The total charge enclosed by the edge is

$$Q = e \sum_{ij} Q_i K_{ij}^{-1} N_j, \quad (-1)^{Q_i} = (-1)^{K_{ii}}, \tag{6}$$

where the second condition guarantees that odd-charge creation-operators are fermions, and even-charge creation-operators are bosons.

Despite its simplicity, this "chiral Luttinger liquid" effective theory gives a rich account of the physics of quantum Hall state edges, where the low-energy response is concentrated. The key results are that the electrical Hall conductance (dissipationless charge current perpendicular to an electric field) is given by the quadratic form

$$\sigma_H = \frac{e^2}{h}(\mathbf{Q}, \mathbf{K}^{-1}\mathbf{Q}), \tag{7}$$

and that the *thermal* Hall conductance (dissipationless heat flow perpendicular to a temperature gradient, the *Leduc-Righi* coefficient) is given by

$$\kappa_H = \frac{\pi^2}{3}\frac{k_B^2 T}{h}\gamma_{LR}, \quad \gamma_{LR} = \text{Sign}(\mathbf{K}) \tag{8}$$

where $\text{Sign}(\mathbf{K})$ is the integer-valued *signature* of the matrix (number of positive minus number of negative eigenvalues), essentially the net conformal anomaly, if the description is phrased in the language of conformal field theory.

The topological excitations of the theory ("quasi-particles" or "vortices") correspond to excitations in the interior of the Hall fluid that correspond to changes $\{N_i\} \rightarrow \{N_i + m_i\}$ of the set of winding numbers of its edge modes. These carry *fractional charge* $q(\mathbf{m}) = e(\mathbf{Q}, \mathbf{K}^{-1}\mathbf{m})$ and accumulate a *fractional statistics* phase $\Theta(\mathbf{m}, \mathbf{m}') = \pi(\mathbf{m}, \mathbf{K}^{-1}\mathbf{m};)$ as they are exchanged.

The topological structure of the Abelian quantum Hall states is classified by the integer matrix $\mathbf{K}$ and vector $\mathbf{Q}$ up to integer similarity transforms. Various models for multicomponent quantum Hall states (*e.g.*, the "hierarchy" [10] and "composite fermion" [3] pictures) suggest apparently different $(\mathbf{K}, \mathbf{Q})$ pairs which are in fact equivalent under such transformations. There is a body of work [11, 12] devoted to developing the unique classification of the distinct topological structures as "integer lattices".

2.2   NON-ABELIAN QUANTUM HALL STATES

The *non-Abelian* Hall states take fractionalization one degree further, and correspond to *fractional quantization of the Leduc-Righi coefficient* with rational, non-integer γ_{LR}. Evidently, they cannot be described by a simple bosonization formalism, and require more of the machinery of conformal field theory.

With a single exception, all the fractional quantum Hall states discovered to date are believed to correspond to Abelian theories. The exception is the enigmatic even-denominator $\nu = 5/2$ state, the *only* even-denominator state known in a single-layer 2D electron gas (even-denominator states can easily occur in multi-layer structures). This is now believed to be a fully spin-polarized paired Hall state with $\gamma_{LR} = 3/2$, essentially described by a wavefunction first written down by Moore and Read [6], and also studied by Greiter et al. [13]. Its conformal field theory edge-mode description requires one bosonic field $\varphi(s)$ and an additional neutral Majorana (real) fermion field $\chi(s)$, related to the Bogoliubov quasiparticle of a spin-polarized BCS state, which contributes the half-integer part of the Leduc-Righi quantization.

One of the remarkable properties of this system is that as vortices are injected into the interior of the droplet of this quantum Hall fluid, they each detach from the edge, and carry with them, a single Majorana fermion mode such that a system of $2N$ vortices is associated with a 2^N-fold degenerate (or quasi-degenerate) manifold of low-energy states localized at the vortex cores in the interior of the fluid. Recall that a Majorana fermion mode is half of a standard (complex) Dirac fermion mode, so the number of possible states gets a factor of 2 ("empty" + "filled") for each *pair* of Majorana modes (*i.e.*, for each effective Dirac mode); if the number of vortices is odd, the missing "partner" for the last Majorana mode is found at the edge of the system, as a zero-energy edge mode.) More details can be found in [15]. In contrast, there are no low-energy states in the interior of Abelian quantum Hall fluids.

Adiabatic (or quasi-adiabatic) motion of the vortex configuration along a trajectory that winds vortices around each other and finally that returns the vortex coordinates to their initial configuration may leave the internal state of the Majorana degrees of freedom changed. This so-called "non-Abelian statistics" contrasts with the Abelian character of the fraction statistic excitations of the usual Abelian quantum Hall states, where the adiabatic transport of vortices that returns to the initial configuration reproduces the original state up to a non-trivial Berry's phase factor.

2.3   ANALOGIES WITH CHIRAL T-BREAKING SUPERCONDUCTORS

This "non-Abelian statistics" feature first showed up in the context of the Moore-Read quantum Hall fluid wavefunction, which is the exact ground state of a particular three-body interaction potential that preserves the exact degeneracy of manifold of Majorana fermion states, but is a somewhat abstract and difficult model to explore.

However, a very interesting recent development is the recognition [15, 16] that essentially the *same* physics can be found in the Bogoliubov-de Gennes

(BdG) equation as applied to exotic 2D superconducting states with chirality, and broken time reversal symmetry (like "$p_x + ip_y$" spin-polarized $S_z = 1$ pairing). The Majorana fermions show up as the Bogoliubov quasiparticles, which are gapless at the edge of the superconductor, as well as having a zero-energy state in each isolated vortex core; the bosonic modes are there as the collective modes of the superconducting order parameter, and differ from the quantum Hall case, by being the Goldstone modes of a true broken symmetry of the ground state, and are gapless in the bulk of the sample. However, the simple addition of a Chern-Simons term to the effective Ginzburg-Landau theory would suffice to destroy the superconducting order, and banish the gapless charge excitations from the bulk regions, leaving behind only the gapless charge edge modes of a topologically-ordered quantum Hall state.

The superconducting analogy would also exhibit a two-fold Ising-like order, with domains of "$p_x + ip_y$" ($L_z = +1$) and "$p_x - ip_y$" ($L_z = -1$) pairing. The maximal chirality breaking by confinement to the lowest Landau level in quantum Hall systems selects one handedness only.

One feature that emerges clearly from the BdG theory is that the ground state degeneracy of the Majorana fermion modes will be lifted by mixing between states on different vortex cores, essentially a tunneling process. This effect will remain exponentially small if the inter-vortex distance is much larger than the coherence length, but means that "adiabatic" processes must be carried out quasi-adiabatically - slowly but not too slowly - to leave the system in a final state at the end of a periodic cycle that is different from the initial state. In contrast, the exact Hamiltonian for which the Moore-Read wavefunction (plus vortices) is exact, strictly maintains exact ground state degeneracies as a function of vortex coordinates. Presumably adding other residual interactions to this model quantum hall system will also lift the degeneracies in analogy with the BdG equation results.

2.4   EXPERIMENTAL AND THEORETICAL CHALLENGES

The potential identification of a physically-realized non-Abelian quantum Hall state is an exciting prospect for entering a new chapter in the quantum Hall story. However, a key question is what new phenomenology should be looked for by experimentalists as a sign of the non-Abelian nature of the state.

Apart from the unusual even-denominator Hall conductance, the $\nu = 5/2$ state outwardly appears rather unexceptional. In principle, a measurement of the Leduc-Righi conductance would reveal its non-integer quantization: but while the value would be unusual, the phenomenology seems prosaic. In any case, the Leduc-Righi quantization is so far a theorist's preoccupation, as it has not yet proved possible to be experimentally measured in any quantum Hall system.

There are interesting proposals [17] that such a system of vortices with non-Abelian statistics could in principle become the basis of a quantum computing device, with the information encoded in the Majorana variables. However, the issues of how the initial pure quantum state could be prepared to encode the input, and the final state probed to extract the output, appear to be quite open questions, even in principle.

3.   PAIRED QUANTUM HALL STATES

Paired quantum Hall states are analogs of BCS states, where the quantum Hall fluid can be thought of as made of pairs of electrons. While the Moore-Read paired state is a "non-Abelian" quantum Hall state, paired states are not in general non-Abelian.

The idea of a paired Hall state was proposed early on as a natural variant of Laughlin's state, but only some time later was a paired state - the "Haldane-Rezayi" (HR) spin singlet state - discovered to be the exact zero-energy ground state of an explicit Hamiltonian, with the "hollow-core" two-body lowest-Landau level interaction pseudopotential. This state proved to be a puzzle, as it seemed to be a counter-example to the emerging idea that all quantum Hall edge states should be described by some unitary conformal field theory. The recent work [15] has resolved this long-standing paradox, and now there is no longer any conflict with the identification of quantum Hall edge states in general with unitary conformal field theories, whether Abelian or not.

As a function of the matrix elements, the determinant of an antisymmetric matrix $\mathbf{A}$ is a perfect square, the square of a function $\mathrm{Pf}(\mathbf{A})$ known as the *"Pfaffian"* of the matrix. At fixed particle number (as opposed to in the indeterminate-particle-number broken-gauge-symmetry formulation), *all* BCS states can be written as a Pfaffian wavefunction. The HR state can be written in this way as

$$\Psi_{HR} = \Pf_{ij} \left(\frac{\alpha_i \beta_j - \beta_i \alpha_j}{(z_i - z_j)^2} \right) \Psi_L^{1/2}, \tag{9}$$

where $\Psi_L^{1/2}$ is the symmetric Laughlin wavefunctions for $\nu = 1/2$ spinless bosons, z_i are the complex 2D spatial coordinates, and $(\alpha, \beta) = (\uparrow, \downarrow)$ are the spin coordinates. The spin-singlet HR state was not originally written in Pfaffian form, but inspired Moore and Read (MR) to formulate [6] their spin-triplet pairing state:

$$\Psi_{MR} = \Pf_{ij} \left(\frac{\alpha_i \alpha_j}{(z_i - z_j)} \right) \Psi_L^{1/2}. \tag{10}$$

This state is often referred to as "the Pfaffian wavefunction", but since *all* paired states can be written as Pfaffians, the terminology "Moore-Read state" is to be preferred. This is again the exact zero-energy eigenstate of a model Hamiltonian, this time one with a three-body interaction. (This is a significant detail: spin-polarized eigenstates of a two-body interaction at $\nu = 1/2$ (projected into a single Landau level) are (or can be chosen) particle-hole symmetric in their Landau level: the MR state does *not* have this required property of a physical state, as it is the eigenstate of a three-body, not a two-body interaction.)

The simplest picture of a paired Hall state at $\nu = 1/2$ is a $\nu^{\text{eff}} = 1/8$ Laughlin state of charge-$2e$ bosons, which would only occur when the number of electrons, N_e, is even, and would have an eight-fold ground-state degeneracy with periodic boundary conditions (*i.e.*, on a Torus). Results from numerical diagonalization of the model Hamiltonians for which these states are exact ground states, were available for a while [14] but not interpreted until recently: the HR Hamiltonian had 10 zero-energy $S = 0$ ground states for even N_e, and 1 zero-energy $S = 1/2$ multiplet for odd N_e; in contrast, the MR state had 6 zero-energy ground states for N_e even, and 2 for N_e odd.

The recent analysis by Read and Green [15], based on the BdG equation for superconducting quasiparticle excitations, has explained this structure: "strong-pairing" states which can be thought of a condensates of charge-$2e$ bosons, and in the quantum Hall context, would have 8-fold ground state degeneracy at even N_e only. But if the pairing is chiral, with a node in the BCS order parameter at the center ($\mathbf{k}=0$) of the 2D Fermi surface, there is a necessary transition to a weak-pairing phase where the Bogoliubov quasi-particle spectrum becomes gapless at $\mathbf{k} = 0$. As the system enters the weak-coupling phase, this gapless quasiparticle excitation becomes gapped again, but survives as gapless fermionic edge modes that are mandated by the extra topological order in the weak-coupling phase. In the case of the MR state, 2 of the 8 states switch from occurring at even N_e to odd N_e, clearly identifying the MR state with the weak-coupling phase. γ_{RL} changes from $\gamma_{RL} = 1$ in the strong pairing phase to $\gamma_{RL} = 3/2$ in the weak pairing MR state.

In contrast, this analysis indicates the the HR state is *exactly* at the critical point, and has gapless bulk $S = 1/2$ excitations. This finally explained why the analysis of its edges, and pattern of zero modes in disc geometries, *etc.*, could not be reconciled with any conformal field theory, which can only describe a system where the only gapless modes are edge modes! The HR model appears to be at the critical point at which extra topological order appears in the weak coupling phase, with an additional set of *electrically neutral $S=1/2$* Dirac fermion edge modes (carrying a conserved "pseudocharge" as well as spin). (γ_{RL} changes from $\gamma_{RL} = 1$ in the strong pairing phase to $\gamma_{RL} = 3$ in the weak pairing phase). The edge modes of the weak-pairing phase turn out [6] to be described by a three-component Abelian quantum Hall state with det K

$= 8$:

$$\mathbf{K} = \begin{pmatrix} 3 & 1 & 1 \\ 1 & 2 & 0 \\ 1 & 0 & 2 \end{pmatrix}, \quad \mathbf{Q} = \begin{pmatrix} 1 \\ 0 \\ 0 \end{pmatrix}, \quad \mathbf{V} = \begin{pmatrix} a & b & b \\ b & c & d \\ b & d & c \end{pmatrix}, \quad (11)$$

which has a $SU(2)$ symmetry in the neutral-mode sector. The strong-pairing phase is the one-component, $K = 8$, $Q = 2$ Abelian state.

The existence of the zero energy states on the torus for *odd N* is a key confirmation of this analysis of the HR model; for even N, the exact 10-fold ground-state degeneracy becomes the expected 8-fold quasi-degeneracy if the model is perturbed away from the special critical coupling at which the HR state is exact (det K is the quasi-degeneracy of Abelian Hall states). While Read [15] comments that this analysis shows the HR state is "non-generic" as a paired quantum Hall state, its new status as a *critical state* with a gapless neutral $S = 1/2$ bulk fermion excitation (apparently dispersing as $\Delta E(k) \propto k^2$ at long wavelengths) would seem to still leave it as an interesting system for further study, as an example of a continuous transition at which topological order changes.

4.   TRANSITION TO THE STRIPED PHASE

A rather different state of matter of the 2D electron gas in high magnetic fields is the striped phase. Early Numerical studies of the the 2DEG showed that as interactions were varied, the incompressible fractional quantum Hall states could collapse to leave the system in a gapless compressible state. Recently, work by Fogler et. al. [18] and Moessner and Chalker [19] should that Hartree-Fock theory predicts a highly anisotropic charge-density wave (CDW) state near half filling in higher Landau levels.

This stripe phase occurs at a wave vector selected by the most favorable distance between 1D edges (from empty to filled regions) that maximizes the Fock exchange energy. Given that there is a "best" inter-wall distance, it can be seen that if all walls are at this distance from their neighbors, the width of the stripes of empty regions and filled regions must be the same, so the filling factor must be 1/2.

Away from this filling this effect is partially frustrated, and eventually, more isotropic crystalline phases are preferred [18]. Hartree-Fock theory also predicts that a purely 1D stripe-ordered ground state is unstable against the formation of a weak secondary CDW along the stripes; whether this weak effect can be destroyed by quantum fluctuations has been as subject of controversy [20, 22, 23], but may be an academic issue, as above very low temperatures, any true long range translational symmetry breaking will be destroyed, leading to a 2D "nematic liquid crystal" phase [20]. Experimental observation [24, 25] of the spontaneous appearance of highly-anisotropic conductivity in systems where higher Landau levels are partially-filled appear to be naturally explained by striped states.

Surprisingly, this type of state can be found very nicely in numerical finite size diagonalization studies [21], carried out on the torus (i.e., with periodic boundary conditions). The shape of the periodically-repeated cell determines a grid of allowed center-of-mass momentum states on a many-electron system, and translational symmetry-breaking is seen by the development of large and characteristic quasi-degeneracy of the ground state on a subset of the momenta that define a Bragg superlattice. The shape of the cell is varied to optimize this, and hence determine the physically-preferred charge-ordering pattern [21].

When the magnetic field direction is tilted away from the normal to the plane of the 2D electron gas, the ν = 5/2 quantum Hall state is destroyed, and replaced by a compressible state. As tilting the field at fixed filling factor increases the Zeeman energy, this effect was for many years interpreted as evidence that the state was spin- unpolarized, or partially polarized, and hence destroyed by increased Zeeman energy. However, numerical diagonalization seemed not to find any clear evidence for a quantum Hall state at physically-realistic interaction parameters.

Recently, Morf [4] reexamined the issue, and found evidence that the 5/2 ground state was fully spin polarized, and suggested that a paired quantum Hall state could be stabilized by varying the interaction away from the Coulomb potential. These studies were carried out using the widely-used spherical finite-size geometry, but while this is very good for studying incompressible states, it has technical difficulties for comparing different states at the same nominal filling factor, or for characterizing compressible state.

Subsequently, with E. Rezayi [5], I extended our similar lowest-Landau-level studies of the crossover between the composite fermion liquid and paired Hall states as the interaction potential is varied, to the spin-polarized second Landau level. What clearly emerges, is that the system with a unmodified pure $(1/r)$ Coulomb interaction potential in the second Landau level is *almost exactly at a transition point between a paired Hall state and a compressible state*, which we identified as similar to the striped phases we saw in the third and fourth Landau levels [21]. The reader is referred to Ref. [5] for the evidence. The proximity to what appears to be a first order transition means that the excitation spectra of the two phases will be mixed up together in the finite size systems, but tuning the potential away from this point by changing the relative strength of its short-range and long-range components separates them, and reveals the transition.

The real effective potential is influenced by the detailed shape of the quantum well that confines the 2D electron gas to a plane, and tilting the magnetic field makes it rotationally-anisotropic. While a quantitative demonstration that tilting the field induces the transition has not yet been achieved, the scenario that field-tilting (which introduces anisotropy) stabilizes a competing striped phase is very appealing.

In the light of the current interest in the high-T_c superconductivity field in the relation between paired and striped phases, their apparent proximity at Landau level filling 5/2 is very intriguing. The evidence suggests that the transition between these states is strongly first-order, and hence that, in this system, they are unrelated, competing phases.

Often, when finite-size numerical diagonalization is used to compare the ground state to the "correct" trial wavefunction, spectacularly good results (e.g., 99% projections) can be achieved, even for quite large symmetry-reduced Hilbert spaces (say, a few hundred states when *all* unitary symmetries have been accounted for). Such results are certainly significant vindication of a trial wavefunction such as the Laughlin state. Initially, we found that the projection of the motional paired ground state on the MR state was not so spectacular. However, the MR state does not have the particle-hole symmetry that any homogeneous spin-polarized liquid state must have, if the Hamiltonian only has two-particle interactions projected into a Landau level. (The MR state is an eigenstate of a three-body interaction, which does not have this property.)

The resolution of this was that the *particle-hole-symmetrized* MR state *does* have very high projection on the paired ground state in the region of parameter space identified as incompressible. In general, particle-hole transformations in a magnetic field are anti-unitary operations (requiring field reversal), so are not associated with a quantum number; however, reflection in a line in the plane is also anti-unitary, and for rectangular periodic boundary conditions (which have reflection symmetry), the two operations combine into a unitary symmetry that splits the Hilbert space into even and odd sectors. Projection of the MR state into one of these sectors produced the "correct" state.

5.   SUMMARY

The $\nu = 5/2$ quantum Hall state seems to be plausibly identified as the first experimental realization of the class of "non-Abelian" paired quantum Hall states, with potentially novel features. It also competes with a charge-ordered "striped" phase, somewhat reminiscent of high-T_c superconductivity. It is perhaps one of the most interesting current problems in the quantum Hall effect.

Acknowledgments

This work was supported in part by NSF DMR-MRSEC-9809483 at the Princeton Center for Complex Materials. I also wish to thank the organizers of the NATO ASI; my ancestral connections to Begunje and Koroška made it very gratifying to attend the meeting at Bled.

References

[1] V. J. Emery, S. A. Kivelson, and J. M. Tranquada, Proc. Natl. Acad. Sci. (USA) **96**, 8814 (1999).

[2] B. I. Halperin, P. A. Lee, and N. Read, Phys. Rev . B **47**, 7312 (1993).

[3] J. K. Jain, Phys. Rev. Lett **63**, 199 (1989); Phys. Rev B **40**, 8079 (1989).

[4] R. Morf, Phys. Rev. Lett. **80**, 1505 (1998).

[5] E. H. Rezayi and F. D. M. Haldane, Phys. Rev. Lett. **84**, 4685 (2000).

[6] G. Moore and N. Read, Nucl. Phys. B**360**, 362 (1991).

[7] R. B. Laughlin and D. Pines, Proc. Natl. Acad. Sci. (USA) **97**, 28 (2000).

[8] X. G. Wen, Phys. Rev. B **41**, 12838 (1990).

[9] J. Fröhlich and A. Zee, Nucl. Phys. B **364**, 517 (1991). 1502

[10] F. D. M. Haldane Phys. Rev. Lett. **51**, 605 (1983).

[11] N. Read, Phys. Rev. Lett. **65**, 1502 (1990).

[12] J. Fröhlich, U. M. Studer and E. Thiran, J. Stat. Phys. **86**, 821 (1997).

[13] M. Greiter, X.-G.Wen, and F. Wilczek, Phys. Rev. Lett . **66**, 3205 (1991); Nucl. Phys. B**374**, 567 (1992).

[14] F. D. M. Haldane and E. H. Rezayi, unpublished.

[15] N. Read and D. Green, Phys. Rev. B **61**, 10267 (2000).

[16] D. A. Ivanov, cond-mat/0005069.

[17] A. Kitaev, quant-ph/9707021 ; S. Bravyi and A. Kitaev, quant-ph/0003137

[18] M. M. Fogler, A. A. Koulakov, and B. I. Shklovskii, Phys. Rev. B. **54**, 1853 (1996).

[19] R. Moessner and J. T. Chalker, Phys. Rev. B **54**, 5006 (1996).

[20] E. Fradkin and S. A. Kivelson, Phys. Rev. B **59**, 8065 (1999).

[21] E, H. Rezayi, F. D. M. Haldane, and K. Yang, Phys. Rev. Lett. **83**, 1219 (1999).

[22] A. H. MacDonald and Matthew P. A. Fisher, Phys. Rev. B **61**, 5724 (2000).

[23] V. J. Emery, E. Fradkin, S. A. Kivelson, and T. C. Lubensky, Phys. Rev. Lett. **85**, 2160 (2000),

[24] M. P. Lilly, K. B. Cooper, J. P.Eisenstein, L. N. Pfeiffer. and K. W. West, Phys. Rev. Lett **82**, 394 (1999).

[25] R. R. Du, D. C. Tsui, H. L. Stormer, L. N. Pfeiffer, K. W. Baldwin, and K. W. West, Solid State Commun. **109**, 389 (1999).

V
MANGANITES, ORBITAL DEGENERACY

THEORY OF MANGANITES:
THE KEY ROLE OF PHASE SEGREGATION

E. Dagotto, A. Feiguin, and A. Moreo

National High Magnetic Field Lab, Florida State University, Tallahassee, FL 32306, USA

Abstract

Recent computational and mean-field studies of models for manganese oxides have revealed a rich phase diagram, not anticipated in early calculations in this context. In particular, the transition between the antiferromagnetic insulator of the hole-undoped limit and the ferromagnetic metal at finite hole-density was found to occur through a mixed-phase process, with coexisting nanometer size clusters. More recently, the influence of disorder on the first-order transitions present in non-disordered manganite models has been shown to lead to giant cluster coexistence, as reported in several experiments. The size of the clusters can be very large since the two competing phases have the same electronic density. The results are illustrated using the random field Ising model. A plethora of experimental data for manganites and other materials are consistent with the phase segregation scenario. Overall it is concluded that inhomogeneities are the key ingredient regulating the physics of manganites. The ideas are general and should apply not only to manganites but to other compounds as well.

The recent wide interest in the study of manganese oxides has been triggered by the discovery of the so-called "colossal" magnetoresistance (CMR) effect [1], where a drastic reduction in the d.c. resistivity ρ_{dc} of most manganites occurs upon the application of magnetic fields of ~ 10 Teslas. In some compounds fields of this value, very small in natural electronic units, can produce changes in the resistivity of several orders of magnitude, inducing a metal-insulator transition. The effect is correlated with the appearance of a magnetic moment in the sample, namely the metallic phase has a ferromagnetic (FM) character. In addition to this curious magneto-transport behavior, the phase diagram of manganites is very rich, with many phases in competition, including charge-ordered regions, antiferromagnetic and ferromagnetic phases, and several regimes whose characteristics are unusual [2]. Thus, independently of its potential applications it is very important to understand the behavior of these materials as a basic science problem.

The theoretical study of models for manganites started several decades ago, when these materials were synthesized for the first time. In those days the

J. Bonča et al. (eds.), Open Problems in Strongly Correlated Electron Systems, 217–226.

emphasis was focused on the origin of the ferromagnetic phase, and the concept of double exchange was coined [3]. To improve their kinetic energy the electrons in the e_g-band of the Mn network induce the localized t_{2g}-spins into a polarized state, and the conductivity is improved by the generation of ferromagnetism. The essence of this idea is still believed to be basically correct. However, certainly these double-exchange concepts are incomplete to explain the manganites, with its complicated phase diagram and exotic behavior upon the application of magnetic fields.

Improving over those early studies, in recent years the computational analysis of manganite models has started [4,5,6,7]. The first model analyzed involved only one e_g-orbital. This model contains a hopping term for the e_g-fermions regulated by an amplitude t, a Hund coupling J_H ferromagnetically linking the mobile and localized spins at a given Mn-ion, and a nearest-neighbors antiferromagnetic (AF) coupling J_{AF} among the localized spins (assumed classical). In this context it was realized that when unbiased methods such as a Monte Carlo simulation were used to handle the model, a curious phase diagram was observed. Its most novel feature was the presence of regions of unstable densities at low temperatures [5]. These unstable densities manifest as a discontinuity in the electronic density vs the chemical potential, especially at large and realistic values of J_H. These discontinuities are a signal of *phase-separation* between two phases, each with the physical characteristics at the extreme of the density discontinuity. In the one-orbital model these phases are: AF insulating with e_g-density equal to one, and FM metal with e_g-density less than one [5,7]. Phase separation tendencies are expected to lead to a mixed-phase state with a complicated pattern involving clusters of the two phases.

Similar results were found when a two-orbital model was studied numerically. In this case the mobile electrons are coupled to the Jahn-Teller modes of oscillation Q_2 and Q_3 through a coupling λ of order 1 [6]. The e_g-electrons occupy two orbitals per ion, and they move using a nontrivial 2×2 hopping matrix which, aside from geometrical factors, it is regulated by a hopping parameter t. A Hund coupling J_H links mobile and localized electrons, the latter being antiferromagnetically coupled among themselves by an exchange J_{AF}. Under the assumption that the phonons are classical, Monte Carlo simulations without sign problems are possible. The results of those simulations [6] and others [8,9] showed two interesting facts. First, the complicated phases found in experiments are present in the simple two-orbital model: (i) FM metallic regime, an (ii) A-type AF orbital-ordered insulating phase at zero hole density, an (iii) A-type orbital-ordered metallic phase at x=0.5, a (iv) charge-ordered CE-type insulating phase at x=0.5, a (v) G-type AF insulating state at x=1.0, and others. The existence of the CE-type phase, which appears prominently in experiments, can be deduced from topological arguments [10]. Second, large regions of parameter space are electronically phase-separated similarly as found in the one-orbital model. The phase separation in some cases involves phases with FM and AF spin order, but in other cases involves phases that differ only in their orbital arrangement [6].

Then, after considerable investigations it is established that electronic phase separation in a variety of forms appears in models of manganites [7]. This includes both systems where electrons interact through phonons or even in cases where electrons interact through on-site Hubbard interactions [11]. However, in the studies described above the (1/r) tail of the repulsive Coulomb interactions has not been included. It is expected that this interaction may have an important effect since the electronic phase separation described thus far involves phases with *different* electronic densities. If the separation is macroscopic, a large energy penalization is expected due to the fact that the phases have macroscopically different charges. For this reason it is widely believed that the macroscopic phase separation will be reduced to a separation at a microscopic scale. In particular, estimations locate this size in the range of the nanometer range. Actually, explicit calculations in one dimensional models showed that the large clusters in phase-separated regimes break down into pieces of a few lattice spacings in size upon the introduction of a nearest-neighbor charge repulsion [12]. This repulsion tends to arrange the charge in an ordered pattern. Evidence of coexisting small clusters has been discussed in many experiments, as recently reviewed [7]. For lack of space it is not possible to reproduce here the vast experimental literature that has reported evidence of inhomogeneities in manganites (for one of the first reports see [13]). Note that the inhomogeneities are *intrinsic*, i.e. not caused by polycrystal physics but they appear in single crystals as well. Also other theoretical investigations have arrived to conclusions similar to those of the computational studies described above. For a longer list of references the reader should consult Refs.[7] and [4-12].

In spite of the success of the previously described simulations, they do not provide the whole picture for manganites. In fact, recently important new experimental information about the microscopic properties of manganites has been reported, and these results are difficult to understand with the theory described thus far. Electronic diffraction and transport techniques have been applied to the compound $La_{5/8-y}Pr_yCa_{3/8}MnO_3$, material that changes from a ferromagnetic (FM) metal to a charge-ordered (CO) insulating state at y~0.35 [14]. At low temperatures in this regime the unexpected coexistence of giant clusters of FM and CO phases was observed [14]. Similar results were also reported using scanning tunneling spectroscopy applied to $La_{0.7}Ca_{0.3}MnO_3$ [15]. The clusters found in both these experiments were as large as 0.1 μm=1000Å~250a, with a~4Å the Mn-Mn distance. The metal-insulator FM-CO transition occurs percolatively, varying either temperature or magnetic fields. These results rule out theories based on homogeneously distributed small polarons to describe manganites in the CMR region.

The discovery of such huge coexisting FM-CO clusters in manganite single-crystals is puzzling. The only theoretical framework with similar characteristics is the electronic phase separation scenario involving phases with different electronic densities described in previous paragraphs [7]. The phase-separation ideas could be successful in describing manganites at some special densities [7,16]. However, the micrometer clusters at intermediate densities [14,15] require an alternative explanation since the energy cost of charged μm-size domains is too large to keep the structure stable. A novel framework involving *equal-density* large clusters is needed to rationalize these challenging experimental results.

The main purpose of this paper is to discuss new theoretical developments [17] where a novel proposal has been formulated to explain the presence of coexisting giant clusters of FM and CO phases found experimentally. The new approach [17] combines (i) strong coupling interactions, necessary to produce the ordered phases, and (ii) quenched disorder. The latter is caused by the random chemical replacement of ions, such as La and Pr, with different sizes. This replacement affects the hopping of e_g-electrons and the exchange J_{AF} between the t_{2g}-spins due to the buckling of the Mn-O-Mn bonds near Pr [2]. The main idea of the new proposal can be easily described. Suppose that in a given system, manganite or any other, two phases are in competition. Suppose these two phases are sufficiently different such that the transition from one to the other must necessarily proceed abruptly in a first-order transition. Recent computer simulations have shown that this is indeed the case for the FM metallic and CO insulating phases in manganites [9]. This happens not only for the two-orbital case but also for the one-orbital model where the insulating phases are not charge-ordered but they have a complicated spin arrangement [18]. In Fig.1a, some illustrations of the ideas of Ref.[17] are provided, namely the energy obtained in a Monte Carlo simulation of the two-orbital model at hole-density x=0.5 is shown as a function of J_{AF} in units of t=1, for fixed values of the other parameters in the problem. A clear crossing of levels is observed at $J_{AF}{\sim}0.21$, between phases with FM and AF characteristics, according to the measured spin correlations (not shown).

Suppose now that one works in the vicinity of the first-order transition and introduces *disorder* in the parameter as a function of which the transition occurs, in this case J_{AF}. In other words, now J_{AF} is no longer uniform but it carries a bond index and its values are randomly selected from a uniform distribution centered at a value common to all bonds and with a width δ. The same occurs for the hopping amplitudes since they are also located at bonds and the chemical replacement of ions in their vicinity may alter their corresponding values. Intuitively it is clear that a complex ground state is to be expected by

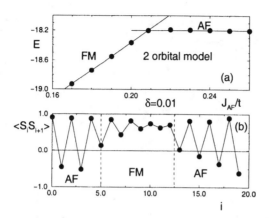

Figure 1 Figure illustrating the generation of large coexisting clusters of equal density in realistic models for manganites. Shown are Monte Carlo results for the two-orbital model with $\langle n \rangle = 0.5$, $T=1/100$, $J_H = \infty$, $\lambda = 1.2$, $t=1$, periodic boundary conditions, and using a chain of $L=20$ sites. (a) is the energy per site vs J_{AF}/t for the non-disordered model. Note the level crossing between FM and AF regions, corresponding to a first-order transition. (b) MC averaged nearest-neighbor t_{2g}-spins correlations vs position along the chain (denoted by i) for one set of random t_{ab}^{α} and J_{AF} couplings (J_{AF}/t at every site is between $0.21-\delta$ and $0.21+\delta$, with $\delta=0.01$). FM and AF regions are shown. Large clusters are generated, much larger than the lattice spacing.

this procedure since couplings J_{AF} slightly above the critical value will favor the AF phase, while slightly below will favor the FM phase. Thus, coexisting clusters will likely emerge from the influence of disorder. However, since the random component of J_{AF} is independent from site to site (the disorder is uncorrelated in the bond index), naively one would expect that the size of these clusters is of the order of the lattice spacing. This would be too costly in energy due to the interfaces. Then, due to the competing tendencies of having small clusters from the random disorder and large clusters to avoid the surface energy, an intermediate situation can be stabilized, as discovered by Moreo et al. [17]. In Fig.1b, results for a one dimensional system are presented in the form of nearest-neighbor spin correlations. It is clear that *large* clusters can be formed, in this case of the order of 10 lattice spacings out of the effect of disorder which is uncorrelated from site to site. Actually, changing the width δ the size of the clusters can change. As δ decreases, with the uniform component of J_{AF} fixed at the critical value, the size of the clusters *increase*.

The universality of the Monte Carlo simulation results suggest that there is a general principle behind. To understand this effect let us briefly review the phenomenology of the random-field Ising model (RFIM) [19] defined by $H = -J\sum_{\langle ij \rangle} S_i S_j - \sum_i h_i S_i$, where $S_i = \pm 1$, and the rest of the notation is standard. The random fields $\{h_i\}$ have the properties $[h_i]_{av}=0$ and $[h_i^2]_{av}=h^2$, where h is the width of the distribution, and $[...]_{av}$ is the average over the fields. In manganites, $S_i=\pm 1$ would represent the competing metallic and insulating states on a small region of space centered at i. The random field mimics the t_{ab}^α and J_{AF} fluctuations locally favoring one state over the other. Without disorder, the T=0 Ising model has a first-order transition located at zero magnetic field between the two fully-ordered states, analogous to the first-order transitions of non-disordered manganite models. However, at $h \neq 0$ the Ising transition is drastically modified [19]. Key arguments guiding RFIM investigations [20] can be restated for manganites. Working close to a first-order transition, consider that in a region of phase-I (either AF or FM), a phase-II bubble of radius R is created. The energy cost R^{d-1} is proportional to the domain wall area. To stabilize the bubble an energy compensation originating in the (t, J_{AF}) disorder is needed. Consider the average hopping inside the bubble using $S_R=\sum_l t_l$, where l labels bonds and t_l is the hopping deviation at bond l from its non-disordered value, the latter fixed at the critical coupling of the first-order transition without disorder. Although the random hopping deviations mostly cancel inside the bubble, important *fluctuations* survive. In particular, the S_R standard deviation is $\sigma_{S_R}=(\Delta t)R^{d/2}$ (d the spatial dimension) since $[t_l t_{l'}]_{av}=(\Delta t)^2\delta_{ll'}$, with (Δt) the width of the random hopping distribution about the non-disordered value. Similar expressions hold for the J_{AF} fluctuations. Even though individual random deviations t_l cannot exceed a (small number) Δt, the strength of the overall fluctuations can be large.

To provide qualitative guidance to manganite experts, standard MC simulations of the RFIM were performed. In Fig.2a, low temperature results are shown for one representative set $\{h_i\}$, individually taken from $[-W, +W]$ with W=3.0 (J=1,W=$\sqrt{3}$h). The formation of large coexisting clusters is clear, even with uncorrelated random fields at neighboring sites. These clusters can be made as large as those found in experiments (250a) adjusting W. This figure also illustrates the influence of an external field $H_{ext}\sum_i S_i$. As H_{ext} grows, the region most affected by the field is the surface of the clusters. This suppresses the narrowest portions of the spin-down regions (Fig.2a), inducing a connection among separated spin-up domains . Then, as H_{ext} increases, a *percolative* transition is to be expected, as described in a variety of experiments [13,14,15]. Similar results were here found at zero field, decreasing the temperature, and in three dimensional systems (Fig.2b). Based on the RFIM-manganite analogy, the picture described here predicts a metal-insulator percolative transition in manganites as chemical compositions, temperatures, or magnetic fields are varied near first-order transitions. Note, however, that even though manganite models and the RFIM have qualitatively similar behavior, subtle manganite properties such as critical exponents at the metal-insulator transition cannot be easily predicted.

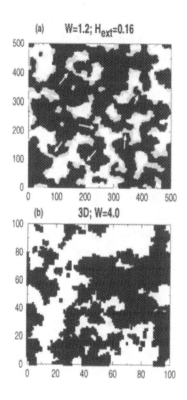

Figure 2   Results of a Monte Carlo simulation of the random-field Ising model at T=0.4 (J=1), with periodic boundary conditions. The dark (white) small squares represent spins up (down). At T=0.4 the thermal fluctuations appear negligible and the results shown are those of the lowest energy configuration. (a) Results using a 500×500 cluster with W=1.2 and for one configuration of random fields. The dark regions are spins up in the H_{ext}=0 case, the grey regions are spins down at zero field that have flipped to up at H_{ext}=0.16, while the white regions have spins down with and without the field. Note the (i) huge clusters created by disorder, and (ii) the percolative-like features of the giant clusters induced by the magnetic field. Special places are arrow-marked where narrow spin-down regions have flipped linking spin-up domains. (b) Results for the RFIM on a 100^3 cluster with disorder. A representative slice at constant z-coordinate is shown, illustrating the presence of giant coexisting clusters even in three dimensions.

The picture of coexisting clusters emerging from these calculations inspires a simple effective model to understand the behavior of ρ_{dc} in the manganites. Let us assume that the system is divided into metallic and insulating phases, with a fraction "p" between 0 and 1 regulating the amount of metallic phase. The real compound is mimicked by a cubic lattice where electrons move with hopping t_m in metallic regions and t_i in insulating regions. The lattice spacing of this lattice should not be considered the true lattice spacing but a much larger scale characteristic of the cluster size in manganites. The cubic grid is used only for simplicity to discretize the problem and carry out the calculations. A considerable literature on percolation corresponding to electrons moving in a system of these characteristics reveals a critical value p_c of about 0.5, where a metal-insulator transition occurs. Note that this percolation is quantum, as opposed to classical, namely in addition to the geometrical percolation threshold for conductance to be finite it is necessary to avoid the Anderson localization. The results observed in the study are qualitatively so clear that it is unlikely that the classical vs quantum character of the percolation matters. To simulate the effect of temperature, t_m starts at low temperature at 1, but decreases as the temperature increases. This is natural and for, example, in the FM region it occurs due to the deviations from a perfect ferromagnet induced by thermal effects. Regarding t_i, at low temperature it must be 0 but at finite temperature it can be allowed to be finite. As a matter of fact, the experimentally measured ρ_{dc} at room temperature is not too different between the metallic and insulating regimes. Thus, it can be assumed that t_m and t_i converge to a common value as the temperature reaches room temperature.

To understand the experimental plots of ρ_{dc} vs temperature the following arguments applies. Let us work slightly above p_c, i.e. after the percolation has occurred. Here the lattice contains metallic filaments going from one side to the other of the sample, with a complex (fractal-like) shape. Near room temperature since the hopping in the insulating regions is not too different from those in the metallic regions, the movement of electrons across the sample is done mainly through the dominant insulating part of the lattice. Then, ρ_{dc} increases as the temperature decreases. However, the conductivity of the insulating region rapidly degenerates (i.e. the experimental ρ_{dc} rapidly grows) as the temperature is reduced, reproduced by a rapid change of t_i with temperature. Then, at some intermediate temperature clearly the insulating regions can no longer carry charge and the conductivity comes from the metallic filaments. As the temperature keeps on going down, the conductivity in the filament improves since the ferromagnetism becomes more and more saturated. At zero temperature a finite and usually large ρ_{dc} is reached due to the complex form of the conducting filaments. Then, ρ_{dc} has all the features of the experimentally observed ρ_{dc}. This expected behavior is neatly reproduced through a computer simulation (see Fig.3) using an algorithm prepared by J. Verges [21].

Summarizing, the giant clusters in manganites are conjectured to be caused by disorder in the couplings, hopping amplitudes and J_{AF}, induced by chemical substitution, and which affect transitions that otherwise would be of first-order without disorder. The origin of the disorder is *intrinsic* to the compound, namely it is caused by the different ionic sizes in the chemical composition. The mixed-phase state involves clusters with equal electronic density, comple-

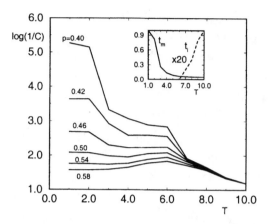

Figure 3 Inverse of the conductivity C (see text) versus temperature (T) in arbitrary units (T=10 is approximately room temperature). The model is a cubic network where electrons move with a hopping amplitude either "metallic" or "insulating". The metallic (insulating) amplitude decreases (increases) from one (zero) as T grows (see inset). This is the indirect way T is here considered. "p" is the fraction of metallic bonds, related with, e.g., the amount of Pr in $La_{5/8-y}Pr_yCa_{3/8}MnO_3$ [14]. Note the presence of features similar to those found in experiments, namely as T decreases from its largest value, 1/C first increases following the insulating behavior of the low-p region, reaches a maximum and then decreases following a metallic behavior. More details can be found in the text.

menting the electronic phase-separation scenario [7]. Although non-disordered models remain crucial to determining the competing tendencies in manganites, disordering effects appear necessary to reproduce the subtle percolative nature of the metal-insulator transition and the conspicuous presence of μm domains in these compounds. The physics described does not depend on whether the phases are generated by Jahn-Teller or Hubbard interactions, as long as a competition metal vs insulator exists. The present observations are very general, and the formation of coexisting giant clusters when two states are in competition through first-order transitions should be a phenomenon frequently present in transition-metal-oxides and related compounds. The theoretical prediction for the ρ_{dc} against temperature (Fig.3) is in excellent agreement with experiments.

The overall conclusion is that the physics of manganites appears governed by intrinsic inhomogeneities which are caused by electronic phase separation tendencies and the influence of disorder on the first-order FM-CO transition.

E.D. and A.M. acknowledges the support of NSF under grant NSF-DMR-9814350. A.F. acknowledges the support of the Antorchas Foundation and the National High Magnetic Field Lab.

References

[1] S. Jin *et al.*, Science **264**, 413 (1994). Y. Tokura et al., J. Appl. Phys. **79**, 5288 (1996); A. P. Ramirez, J. Phys.: Condens. Matter **9**, 8171 (1997).

[2] S-W. Cheong and H. Y. Hwang, in *Colossal Magnetoresistance Oxides*, ed. Y. Tokura, Gordon & Breach, Monographs in Cond. Matt. Science.

[3] C. Zener, Phys. Rev. **82**, 403 (1951).

[4] J. Riera, K. Hallberg, and E. Dagotto, Phys. Rev. Lett. **79**, 713 (1997).

[5] S. Yunoki, J. Hu, A. Malvezzi, A. Moreo, N. Furukawa, and E. Dagotto, Phys. Rev. Lett. **80**, 845 (1998).

[6] S. Yunoki, A. Moreo, and E. Dagotto, Phys. Rev. Lett. **81**, 5612 (1998).

[7] A. Moreo, S. Yunoki and E. Dagotto, Science **283**, 2034 (1999).

[8] T. Hotta, S. Yunoki, M. Mayr, and E. Dagotto, Phys. Rev. B**60**, R15009 (1999).

[9] S. Yunoki, T. Hotta, and E. Dagotto, cond-mat/9909254. Accepted in PRL.

[10] T. Hotta, Y. Takada, H. Koizumi, and E. Dagotto, Phys. Rev. Lett. **84**, 2477 (2000).

[11] T. Hotta, A. Malvezzi, and E. Dagotto, submitted to Phys. Rev. B. cond-mat/0003056.

[12] A. L. Malvezzi, S. Yunoki, and E. Dagotto, Phys. Rev. **B 59**, 7033 (1999).

[13] M. R. Ibarra and J. M. De Teresa, contribution to *Colossal Magnetore-sistance, Charge Ordering and Related Properties of Manganese Oxides*, edited by C. N. R. Rao and B. Raveau, World Scientific, 1998c.

[14] M. Uehara, S. Mori, C. H. Chen, and S.-W. Cheong, Nature **399**, 560 (1999).

[15] M. Fäth, S. Freisem, A. A. Menovsky, Y. Tomioka, J. Aarts, and J. A. Mydosh, Science **285**, 1540 (1999).

[16] J. J. Neumeier and J. L. Cohn, preprint.

[17] A. Moreo, M. Mayr, A. Feiguin, S. Yunoki and E. Dagotto, 1999 preprint, cond-mat/9911448, to appear in Phys. Rev. Letters.

[18] S. Yunoki and A. Moreo, Phys. Rev. B**58**, 6403 (1998).

[19] See *Spin Glasses and Random Fields*, ed. A. P. Young, World Scientific.

[20] Y. Imry and S. K. Ma, Phys. Rev. Lett. **35**, 1399 (1975).

[21] J. Verges, preprint, cond-mat/9905235.

MAGNETIC AND ORBITAL ORDERING IN MANGANITES

Andrzej M. Oleś[1] and Louis Felix Feiner[2,3]

[1]*Institute of Physics, Jagellonian University, Reymonta 4, PL-30059 Kraków, Poland*

[2]*Utrecht University, Princetonplein 5, NL-3584 CC Utrecht, The Netherlands*

[3]*Philips Research Laboratories, Prof. Holstlaan 4, NL-5656 AA Eindhoven, The Netherlands*

Abstract   We present an analysis of superexchange and double exchange interactions in colossal magnetoresistance manganites. The superexchange explains the A-type antiferromagnetic and orbital ordering observed in $LaMnO_3$, and the doping dependence of the anisotropic exchange constants $J_{(a,b)}$ and J_c in the polaronic phase for doping by $x \leq 0.1$ holes. In the ferromagnetic metallic phase ($x > 0.15$ holes) we analyze various possible orbital states, and demonstrate that the ordered orbital states are unstable against the orbital liquid of disordered complex orbitals. The double exchange in the latter state reproduces the observed cubically symmetric magnon dispersion in the ferromagnetic manganites.

1.   SUPEREXCHANGE IN $La_{1-x}Ca_xMnO_3$

The multiple magnetic phase transitions and complex phase diagrams of the colossal magnetoresistance (CMR) manganites have attracted a lot of attention recently. Although these compounds are known since fifty years, the various observed phase transitions and the CMR phenomenon itself are not yet fully understood. The anisotropic A-type antiferromagnetic (AF) order in $LaMnO_3$ changes to a ferromagnetic (FM) phase at some finite doping, which may be either metallic or insulating. It has been argued that the A-AF state can be qualitatitively explained by superexchange (SE) in a doubly-degenerate e_g band, with FM interactions that arise along two crystallographic directions due to a particular staggering of the occupied e_g orbitals [1]. The second (FM) state follows qualitatively from the double exchange (DE) model [2]. However, this model is highly unsatisfactory, as it cannot explain the value of the transition temperature T_c to the FM state, and the temperature dependence of the resistivity [3]. A complete theory has to include the orbital degree of freedom which allows for orbital ordering (OO), observed both in undoped and doped manganites [4]. Therefore, we address here the problem of the magnetic

J. Bonča et al. (eds.), Open Problems in Strongly Correlated Electron Systems, 227–236.

properties of manganites in the whole doping range, resulting from the interplay of the SE and DE interactions for electrons in degenerate e_g orbitals.

We start with the undoped insulating compound $LaMnO_3$ which has to be understood before the problem of more complicated magnetic interactions in doped materials is addressed. A unique feature of the correlated oxides with partially filled e_g orbitals is the presence of an orbital degree of freedom in addition to the usual spin problem. Therefore, the magnetic ordering and OO and their changes under hole doping have to be considered on equal footing. In the simplest case of Cu^{2+} (d^9) ions this leads to a competition between different types of magnetic ordering and magnetic disordered (spin-liquid) phases [5]. Examples of ordered phases are known since long in cubic and layered cuprates, such as $KCuF_3$ and K_2CuF_4, respectively, which are Mott insulators with a two-sublattice OO in FM planes [1], while the nature and the physical examples of a spin liquid state are still under investigation. The case of manganites is more classical as the spins are larger. Although the magnetic couplings between e_g electrons of Mn^{3+} $[d^4(t_{2g}^3 e_g)]$ ions in $LaMnO_3$ might look similar to those in $KCuF_3$ [1], an important difference is the local Hund's rule coupling J_H between an e_g electron and a t_{2g} $S_t = 3/2$ core, leading to large $S = 2$ spins of Mn^{3+} ions coupled by SE interactions [6, 7].

The collective behavior of the e_g electrons follows from their local interactions: Coulomb $U \simeq 7.3$ eV and exchange $J_H = 0.69$ eV, as estimated from spectroscopic data [7]. The Coulomb interaction is the dominating energy scale that leads to an effective low-energy Hamiltonian, where spin and orbital degrees of freedom are interrelated. The SE interactions are obtained by virtual processes which involve either $d_i^4 d_j^4 \rightleftharpoons d_i^3(t_{2g}^3) d_j^5(t_{2g}^3 e_g^2)$ transitions by an e_g electron, or $d_i^4 d_j^4 \rightleftharpoons d_i^3(t_{2g}^2 e_g) d_j^5(t_{2g}^4 e_g)$ transitions by one of $=t_{2g}$ electrons. This yields the superexchange Hamiltonian, $H_J = H_J^e + H_J^t$, between spins $S = 2$ after the derived excitations are decomposed into the local high-spin ($S = 5/2$) and low-spin ($S = 3/2$) states of Mn^{2+} (d^5) ions. For the processes promoted by hopping of e_g electrons between two Mn^{3+} ions ($n_i = 1$), with the characteristic energy $J = t^2/U = 23$ meV, one finds [7]:

$$H_J^e = \frac{1}{16} \sum_{\langle ij \rangle} n_i n_j \left\{ -\frac{8}{5} \frac{t^2}{\varepsilon(^6A_1)} \left(\vec{S}_i \cdot \vec{S}_j + 6 \right) \mathcal{P}_{\langle ij \rangle}^{\zeta\xi} + \left(\vec{S}_i \cdot \vec{S}_j - 4 \right) \right.$$

$$\left. \times \left[\left(\frac{t^2}{\varepsilon(^4E)} + \frac{3}{5} \frac{t^2}{\varepsilon(^4A_1)} \right) \mathcal{P}_{\langle ij \rangle}^{\zeta\xi} + \left(\frac{t^2}{\varepsilon(^4E)} + \frac{t^2}{\varepsilon(^4A_2)} \right) \mathcal{P}_{\langle ij \rangle}^{\zeta\zeta} \right] \right\}, \quad (1)$$

where $^6A_1$ is the high-spin state, and $^4A_1$, ^4E and $^4A_2$ are the low-spin states, and $\varepsilon(^6A_1) = U - 5J_H, \varepsilon(^4A_1) = U, \varepsilon(^4E) = U + \frac{2}{3}J_H, \varepsilon(^4A_2) = U + \frac{10}{3}J_H$, are the respective excitation energies [7]. The hopping elements result from two-step processes which go via the oxygen orbitals and depend on the pair of e_g orbitals involved in the hopping process; $t \simeq 0.41$ eV in Eq. (1) stands for the largest element between two directional orbitals $|\zeta\rangle$ parallel to the direction of the bond $\langle ij \rangle$. The hopping to/from the orbitals $|\xi\rangle$ perpendicular to the bonds vanishes due to compensating orbital phases. Thus, the SE on the bond $\langle ij \rangle$ involves only either the pairs of two different orbitals, $|\zeta\rangle$ and $|\xi\rangle$, or the pairs of directional orbitals $|\zeta\rangle$, as expressed by the projection operators:

$$\mathcal{P}^{\zeta\xi}_{\langle ij \rangle} = P_{i\zeta}P_{j\xi} + P_{i\xi}P_{j\zeta}, \qquad \mathcal{P}^{\zeta\zeta}_{\langle ij \rangle} = 2P_{i\zeta}P_{j\zeta}, \tag{2}$$

where $P_{i\zeta} = \frac{1}{2} - \tau_i^\alpha$ and $P_{j\xi} = \frac{1}{2} + \tau_j^\alpha$ are the respective local projection operators on orbital $|\zeta\rangle$ and $|\xi\rangle$ at site i. They are represented by the orbital operators τ_i^α associated with the three cubic axes ($\alpha = a$, b, or c),

$$\tau_i^{a(b)} = \frac{1}{4}(-\sigma_i^z \pm \sqrt{3}\sigma_i^x), \qquad \tau_i^c = \frac{1}{2}\sigma_i^z, \tag{3}$$

where the σ's are Pauli matrices acting on the orbital pseudospins: $|x\rangle = \begin{pmatrix} 1 \\ 0 \end{pmatrix}$ and $|z\rangle = \begin{pmatrix} 0 \\ 1 \end{pmatrix}$, corresponding to the usual basis in the e_g subspace defined by $|x\rangle \equiv |x^2 - y^2\rangle$ and $|z\rangle \equiv |3z^2 - r^2\rangle$ orbitals.

The t_{2g}-hopping leads to an approximately isotropic SE [8],

$$H_J^t = \sum_{\langle ij \rangle} \left\{ \frac{1}{4}J_t n_i n_j \left(\vec{S}_i \cdot \vec{S}_j - 4 \right) + \frac{4}{9}\hat{J}_t(1 - n_i)(1 - n_j) \left(\vec{S}_i \cdot \vec{S}_j - \frac{9}{4} \right) \right.$$

$$\left. + \frac{1}{3}\bar{J}_t \left[n_i(1 - n_j) + (1 - n_i)n_j \right] \left(\vec{S}_i \cdot \vec{S}_j - 3 \right) \right\}, \tag{4}$$

with SE constants (as estimated using the spectroscopic data [7, 8]): $J_t = 2.1$ meV, $\hat{J}_t = 4.6$ meV, and $\bar{J}_t = 5.5$ meV for the pairs of Mn^{3+}–Mn^{3+}, Mn^{4+}–Mn^{4+}, and Mn^{3+}–Mn^{4+} ions, respectively, and $\vec{S}_i$ is an $S = 2$ ($S = 3/2$) spin operator for a Mn^{3+} (Mn^{4+}) ion with $n_i = 1$ ($n_i = 0$) e_g electrons.

The intersite Jahn-Teller (JT) interaction leads to purely orbital interactions [9] which favor orbital alternation, and one finds [7]:

$$H_{JT} = \kappa \sum_{\langle ij \rangle} n_i n_j \left(\mathcal{P}^{\zeta\zeta}_{\langle ij \rangle} - 2\mathcal{P}^{\zeta\xi}_{\langle ij \rangle} + \mathcal{P}^{\xi\xi}_{\langle ij \rangle} \right) - E_z \sum_i \tau_i^c n_i, \tag{5}$$

where $\mathcal{P}^{\xi\xi}_{\langle ij \rangle} = 2P_{i\xi}P_{j\xi}$, and $\kappa \simeq 11$ meV is determined from the temperature of the structural phase transition $T_s = 750$ K [7]. The tetragonal crystal-field splitting $\propto E_z$ acts as a magnetic field in the pseudospin space.

The SE interaction due to e_g electrons (1) depends on the Hund's rule coupling J_H/U and on the type of occupied orbitals. As in the d^9 model [5], the e_g interactions are frustrated in the $J_H/U \to 0$ limit at $E_z = 0$, but the t_{2g} SE removes this degeneracy and stabilizes the G-type AF phase for small J_H/U, while for the actual values of the interaction parameters the A-AF phase is stable [Fig. 1(a)]. Due to the particular stability of the high-spin $^6A_1$ state of Mn^{2+} ions, the system is quite close to the FM insulating (FI) state, and one might expect that a transition to this state occurs under doping.

In the weakly doped regime the holes are localized and form Mn^{4+} lattice polarons with energy E_p. They interact with the surrounding Mn^{3+} ions by FM SE $\propto J_p = K_p/(1 + E_p/2J_H)$, where $K_p = t^2/2E_p$ stands for the effective charge interaction in which the polaronic energy is lost at both sites [8],

$$H_{pol} = -E_p \sum_i (1 - n_i) - \sum_{\langle ij \rangle} (1 - n_i) \left[K_p + \frac{1}{8} J_p \left(\vec{S}_i \cdot \vec{S}_j - 3 \right) \right] P_{j\zeta} n_j. \tag{6}$$

Taking the Hamiltonian $\mathcal{H} = H_J^e + H_J^t + H_{JT} + H_{pol}$, we determined the total exchange constants on the bonds within the FM (a, b) planes $J_{(a,b)}$, and along the c axis J_c, assuming a random distribution of polarons over the lattice. In addition, we weighted the polaron energy E_p by the number of holes at the first neighbors of the polaron, $\sum_j (1 - n_j)/6$, which suppresses partly the polaronic energy when doped holes are close to each other and the lattice cannot distort. The (average) AF interaction J_c is influenced much stronger by doping than the FM $J_{(a,b)}$, and the order changes to the FI state at $x \simeq 0.11$. The model reproduces well the values of $J_{(a,b)}$ and J_c in $LaMnO_3$ [7], being -1.15 and 0.88 meV (experiment gives -0.83 and 0.58 meV, respectively), and the observed doping dependence of $J_{(a,b)}$ and J_c in $La_{1-x}Ca_xMnO_3$ [10], but only if a rigid OO is assumed [Fig. 1(b)]. This suggests that the lattice plays an important role in the manganites, and prevents the formation of orbital polarons [11].

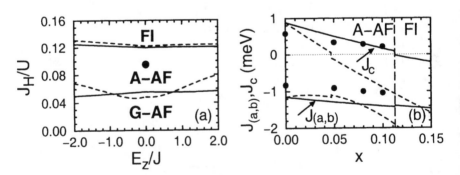

Figure 1 (a) Mean-field phase diagram of $LaMnO_3$ obtained using SE interactions H_J^e and H_J^t (dashed lines), and both SE and cooperative JT effect H_{JT} (full lines); the parameters of $LaMnO_3$ are indicated by a dot [8]. (b) Exchange interactions $J_{(a,b)}$ and J_c as functions of doping x for a hole in the OO state (full lines), and for an orbital polaron (dashed lines). Experimental points for $La_{1-x}Ca_xMnO_3$ [10] are shown by dots.

2. ORBITAL LIQUID WITH COMPLEX ORBITALS

The FM states observed in manganites at doping $x > 0.2$ are frequently metallic and have *isotropic* magnetic properties, with very similar stiffness constants in magnon dispersion. In contrast, models which combine SE interactions with the kinetic energy of a partially filled e_g band give typically *anisotropic* FM states with OO [12]. Such states arise from a competition between the SE and the kinetic energy of the e_g electrons — the resulting stable states are those with the lowest kinetic energy, as for instance $|\Phi_x\rangle = \prod_i c_{ix}^\dagger |0\rangle$ or $|\Phi_z\rangle = \prod_i c_{iz}^\dagger |0\rangle$, with either $|x\rangle$ or $|z\rangle$ orbitals occupied at every Mn^{3+} site. Nagaosa proposed that the fluctuations between the regions of $|x\rangle$-like orbitals give an *orbital liquid* state [13]. Here we show that a lower energy is obtained when *orbitals with complex coefficients* are disordered.

A natural route to determine the most stable orbital pattern in doped manganites is to consider first the one- and two-sublattice OO given by the eigenstates of the Pauli matrices: σ^x, σ^y, and σ^z. The eigenstates of σ^z, $|x\rangle$ and $|z\rangle$, form the usual basis in the e_g subspace, with the kinetic energy in the FM state given by spinless orbital fermions,

$$H_t = -\frac{t}{4} \sum_{\langle ij \rangle \| (a,b)} \left[3c_{ix}^\dagger c_{jx} + c_{iz}^\dagger c_{jz} \pm \sqrt{3}(c_{ix}^\dagger c_{jz} + c_{iz}^\dagger c_{jx}) \right] - t \sum_{\langle ij \rangle \| c} c_{iz}^\dagger c_{jz}.$$

(7)

The kinetic energy in the correlated e_g band has to be determined using a formalism which includes explicitly the constraints coming from the large local Coulomb interaction U. We investigated the kinetic energy in the OO phases within the slave-boson formalism, where the fermion creation operators are replaced by a simultaneous creation of a boson $b_{i\eta}^\dagger$ and a charge excitation $f_i^\dagger$, $c_{i\eta}^\dagger = b_{i\eta}^\dagger f_i^\dagger$, with $\eta = x, z$. Assuming OO, only the hopping between the occupied states contributes to the kinetic energy, while the unoccupied orbitals are at high energy and give no contribution. The e_g SE (1) favors a two-sublattice ordering which however results in a reduced bandwidth of the effective band and therefore increased kinetic energy, both in the case of $|x\rangle/|z\rangle$ and $(|x\rangle + |z\rangle)/(|x\rangle - |z\rangle)$ alternation. The uniform states, such as $|\Phi_x\rangle$ and $|\Phi_z\rangle$, are different — the bands have a full bandwidth of $6t$, as in the case of an uncorrelated e_g band structure [see Fig. 2(a)]. However, all these OO states are highly anisotropic and thus cannot serve as an explanation of the observed FM isotropic phase. Moreover, they are expected to be highly renormalized by quantum corrections as the pseudospins are not conserved.

Yet, another possibility to investigate H_t is to use instead the complex orbitals which are the eigenstates of σ^y with eigenvalues ± 1, $c_{i+}^\dagger = (c_{iz}^\dagger - ic_{ix}^\dagger)/\sqrt{2}$ and $c_{i+}^\dagger = (c_{iz}^\dagger + ic_{ix}^\dagger)/\sqrt{2}$, as a basis to write the kinetic energy in FM states,

$$H_t = -\frac{t}{2} \sum_{\langle ij \rangle} \left(c_{i+}^\dagger c_{j+} + c_{i-}^\dagger c_{j-} + e^{i\chi_\alpha} c_{i+}^\dagger c_{j-} + e^{-i\chi_\alpha} c_{i-}^\dagger c_{j+} \right), \qquad (8)$$

where the angle χ_α depends on the bond direction: $\chi_\alpha = \pm 2\pi/3$ for $\alpha = a, b$, and $\chi_\alpha = 0$ for $\alpha = c$. Because $\langle \sigma_i^z \rangle = \langle \sigma_i^x \rangle = 0$ in these complex orbitals, OO states built from such orbitals will retain cubic symmetry and will cause no lattice distortion, as also pointed out by Khomskii [14] for the uniform state $|\Phi_+\rangle = \prod_i c_{i+}^\dagger |0\rangle$. The kinetic energy in this state, $\varepsilon_+(\mathbf{k}) = -t[2\gamma_+(\mathbf{k}) + \gamma_z(\mathbf{k})]$, where $\gamma_\pm(\mathbf{k}) = (\cos k_x \pm \cos k_y)/2$ and $\gamma_z(\mathbf{k}) = \cos k_z$, is *isotropic* and has the same minimum of the band at $\varepsilon_+(\mathbf{k} = 0) = -3t$ as for $|\Phi_x\rangle$ and $|\Phi_z\rangle$. However, the density of states (DOS) is different for all these states — the $|\Phi_x\rangle$ and $|\Phi_z\rangle$ states are effectively lower-dimensional, and therefore the DOS has a larger second moment than for the cubic $|\Phi_+\rangle$ state. As a result, $|\Phi_+\rangle$ is *unstable* against $|\Phi_x\rangle$ and $|\Phi_z\rangle$ *at all doping concentrations* x [Fig. 2(b)]. We note also that in contrast to the corresponding spin model, the orbital polarized states, $|\Phi_+\rangle$ and $|\Phi_-\rangle$, *are not eigenstates of* H_t.

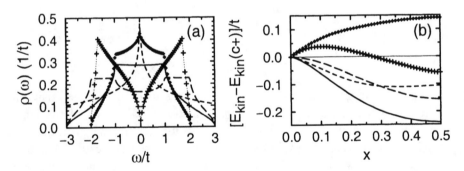

Figure 2 (a) Density of states $N(\omega)$, and (b) kinetic energy calculated with respect to the orbital ordered uniform $|\Phi_+\rangle$ state [14], $[E_{\mathrm{kin}} - E_{\mathrm{kin}}(c+)]/t$ for FM states with OO: uniform $|x\rangle$ (dashed lines) and $|z\rangle$ (long-dashed lines), two-sublattice $(|x\rangle + |z\rangle)/(|x\rangle - |z\rangle)$ (diamonds), two-sublattice $|c+\rangle/|c-\rangle$ (pluses). Full line in (b) shows the energy of the orbital disordered state with KR renormalization.

The two-sublattice $|\Phi_\pm\rangle = \prod_{i \in A} c_{i+}^\dagger \prod_{i \in B} c_{i-}^\dagger |0\rangle$ state has also an isotropic band structure, $\varepsilon_\pm(\mathbf{k}) = \pm t[(\gamma_+(\mathbf{k}) - \gamma_z(\mathbf{k}))^2 + 3\gamma_-^2(\mathbf{k})]^{1/2}$, but with band minimum at $-2t$. Thus, one would expect that such a state is never stable. However, this is not the case, as the DOS has distinct maxima close to the band edges, [Fig. 2(a)], and the $|\Phi_\pm\rangle$ state becomes more stable than $|\Phi_+\rangle$ for $x > 0.27$. The instability of $|\Phi_+\rangle$ against this kind of orbital alternation was found recently also at quarter filling in weak coupling [15]. Nevertheless, also $|\Phi_\pm\rangle$ is unstable against the states $|\Phi_{x(z)}\rangle$ in the interesting regime of $x \simeq 0.3$ even after the SE and JT energies are included.

Next we show that the *orbital disordered state* has the lowest energy in the metallic regime. In the presence of strong correlations the kinetic energy may be formally obtained by including slave bosons $\{b_{i+}^{\dagger}, b_{i-}^{\dagger}\}$ for the complex orbitals, and a slave boson $b_i^{\dagger}$ for an empty orbital, with the physical space defined by the constraint $b_{i+}^{\dagger} b_{i+} + b_{i-}^{\dagger} b_{i-} + b_i^{\dagger} b_i = 1$. The electron operators are replaced by: $c_{i+}^{\dagger} = z_{i+}^{\dagger} f_{i+}^{\dagger}$ and $c_{i-}^{\dagger} = z_{i-}^{\dagger} f_{i-}^{\dagger}$, where the bosonic factors $z_{i\pm}^{\dagger}$ are selected following Kotliar and Ruckenstein [16]. In the mean-field approximation we find the renormalization factors $\sqrt{q_{i\pm}} = \langle z_{i\pm} \rangle = \sqrt{x/[1 - \langle b_{i\pm}^{\dagger} b_{i\pm} \rangle]}$. After minimizing the average energy one finds that the minimum corresponds to a disordered state with *equal* density of $|c+\rangle$ and $|c-\rangle$ fermions, and the same renormalization factor $q = q_{i\pm} = 2x/(1 + x)$ for both states. We note that the bands found in this correlated disordered state,

$$\varepsilon_{\pm}(\mathbf{k}) = -tq \left\{ 2\gamma_{+}(\mathbf{k}) + \gamma_{z}(\mathbf{k}) \pm [(\gamma_{+}(\mathbf{k}) - \gamma_{z}(\mathbf{k}))^2 + 3\gamma_{-}^2(\mathbf{k})]^{1/2} \right\}, \quad (9)$$

are isotropic and interpolate correctly to the limit of the empty band ($x \to 1$), where the correlations are not important. The OO states ($|\Phi_{+}\rangle, |\Phi_{\pm}\rangle, |\Phi_{x}\rangle$, and $|\Phi_{z}\rangle$) can be obtained within this formalism by implementing the constraints by Lagrange multipliers, yielding results equivalent to those discussed above and corresponding to a single slave fermion representing the occupied orbital states.

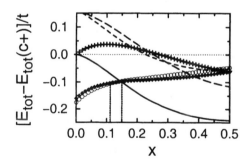

Figure 3 Total energy calculated with respect to the $|\Phi_{+}\rangle$ state, $[E_{\text{tot}} - E_{\text{tot}}(c+)]/t$ for FM states with OO, and for the orbital disordered state with KR renormalization. The A-AF state is marked by circles; the other lines and symbols have the same meaning as in Fig. 2(b). Two phase transitions are indicated by vertical dotted lines.

The total energies per site in the FM phases, $E_{\text{tot}} = \langle H_t + H_J^e + H_{\text{JT}} \rangle / N$ (with $\langle H_J^t \rangle = 0$), were determined by taking the expectation values of the orbital operators in the SE and JT terms. No SE energy is gained in the $|\Phi_x\rangle$ and $|\Phi_z\rangle$ phases, while the JT part gives a positive energy as the orbitals are uniformly polarized. In contrast, in $|\Phi_{+}\rangle, |\Phi_{\pm}\rangle$, and the disordered phase, the same e_g SE energy is gained due to $\langle \mathcal{P}_{\langle ij\rangle}^{\zeta\xi} \rangle = 1/2$, and the JT contribution vanishes. However, these two contributions do not suffice to make the energies of $|\Phi_+\rangle$ and $|\Phi_\pm\rangle$ lower than those of the $|\Phi_{x(z)}\rangle$ phases. The orbital liquid state with disordered orbitals $|c+\rangle$ and $|c-\rangle$ has the lowest energy, except at low doping, where the polaronic energies $\langle H_{\mathrm{pol}}\rangle$ contribute in place of the kinetic energy term $\langle H_t\rangle$, and the polaronic phases are stable (Fig. 3): the A-AF phase for $x < 0.11$, and the FI phase in the intermediate regime of $0.11 < x < 0.15$.

3.   MAGNONS IN FERROMAGNETIC MANGANITES

The magnetic excitations are derived using the first order terms in Schwinger bosons $a_{i\sigma}^\dagger$. First, the kinetic energy H_t has to be written for both spins, and the electron operators are decomposed as $c_{i\pm,\sigma}^\dagger = a_{i\sigma}^\dagger z_{i\pm}^\dagger f_{i\pm}^\dagger$ [17]. We separate the charge dynamics from the spin dynamics and expand the Schwinger boson terms around the $\uparrow$-spin FM ground state. The constraint $\sum_\sigma a_{i\sigma}^\dagger a_{i\sigma} + b_i^\dagger b_i = 2S$, with $S = 2$, gives $a_{i\uparrow} \simeq \sqrt{2\bar{S}}\sqrt{1 - a_{i\downarrow}^\dagger a_{i\downarrow}/2\bar{S}}$ and leads then to

$$a_{i\uparrow}^\dagger a_{j\uparrow} + a_{i\downarrow}^\dagger a_{j\downarrow} \simeq 2\bar{S} - \frac{1}{2}\left(a_{i\downarrow}^\dagger a_{i\downarrow} + a_{j\downarrow}^\dagger a_{j\downarrow} - 2a_{i\downarrow}^\dagger a_{j\downarrow}\right), \qquad (10)$$

with $2\bar{S} = 4 - x$ standing for the average number of Schwinger bosons in the doped system. After inserting Eq. (10) into the hopping Hamiltonian, one finds from the zeroth order term $\propto 2\bar{S}$ the kinetic energy considered above, while the first order term gives the magnon excitations with the energies $\omega_\mathbf{q} = (\langle H_t^{(0)}\rangle/z\bar{S})[3 - 2\gamma_+(\mathbf{q}) - \gamma_z(\mathbf{q})]$. As observed experimentally [18, 19, 20], the magnon dispersion is isotropic (i.e., has cubic symmetry), and the effective DE constant is $J_{\mathrm{DE}} = \langle H_t^{(0)}\rangle/2z\bar{S}^2$. Thus, the DE contribution to the magnon dispersion vanishes in the $x \to 0$ limit, and increases with increasing x, while electronic structure calculations that ignore electron correlations incorrectly give the largest FM interactions at $x = 0$ [21].

Whereas only the FM part of the SE contributes to the ground state energy, both FM and AF parts are needed to derive the SE contribution to the magnon dispersion [Fig. 4(a)]. Using the expansion similar to Eq. (10) of both SE terms, H_J^e and H_J^t, we derived an isotropic reduction of the magnon dispersion obtained from the DE mechanism. The total effective FM exchange interaction is therefore $J_{\mathrm{eff}} = J_{\mathrm{DE}} + J_{\mathrm{SE}}$, where $J_{\mathrm{SE}} < 0$. For the numerical evaluation we adopted the realistic parameters of $LaMnO_3$ as given in Ref. [7]. One finds that H_J^t gives the larger contribution, because the FM and AF terms in H_J^e almost compensate each other in the orbital liquid state with $\langle \mathcal{P}_{\langle ij\rangle}^{\zeta\xi}\rangle = \langle \mathcal{P}_{\langle ij\rangle}^{\zeta\xi}\rangle = 1/2$, and give a weak net AF interaction between pairs of Mn^{3+} ions. The Mn^{3+}–Mn^{4+} pairs feel a stronger e_g AF SE due to excitations to the low-spin ^3E states, $\propto (\bar{J}_e/3)(1 - n_i)n_j\left(\vec{S}_i \cdot \vec{S}_j - 3\right)P_{j\zeta}$, with $\bar{J}_e = 22.7$ meV.

The total magnon width W increases with x and almost saturates close to $x = 0.5$. The calculated result is somewhat below the experimental points,

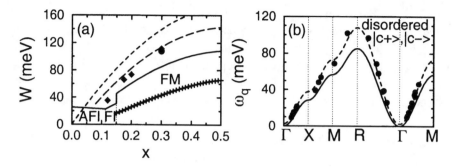

Figure 4 Magnons in the FM metallic phase with disordered complex orbitals: (a) width W of the magnon band as a function of doping x (full line), its DE part (dashed line), and DE together with SE from t_{2g} electrons (long-dashed line); (b) magnon dispersion at $x = 0.30$ (full line) compared with the experimental points for $La_{0.7}Pb_{0.3}MnO_3$ (circles and dashed line) [18]. The result found with a two-sublattice $|c+\rangle/|c-\rangle$ state is shown by pluses in (a); experimental points correspond to: $La_{1-x}Sr_xMnO_3$ [19] (diamonds) and $La_{0.7}Pb_{0.3}MnO_3$ [18] (filled circle).

but reproduces well the observed increase of W [19] with increasing x in the FM metallic regime of $x > 0.15$ [Fig. 4(a)]. At small doping $x \leq 0.15$ we show instead W for two polaronic phases: A-AF and FI. The magnon width *decreases* somewhat with increasing x within the A-AF phase which agrees qualitatively with the observed decrease of the Néel temperature under doping in $La_{1-x}Ca_xMnO_3$ [10]. The magnon dispersion found at $x = 0.3$, with $J_{eff}\bar{S} = 7.10$ meV, is *isotropic* and reproduces well the experimental points for $La_{0.7}Pb_{0.3}MnO_3$ [Fig. 4(b)], and the value of $J_{exp}\bar{S} = 8.79$ meV deduced by Perring *et al.* [18]. Note that no fit is used, and all the parameters follow from the spectroscopic data.

4.   SUMMARY

In conclusion, the exchange constants in the A-AF phase, stable in $LaMnO_3$ and in the polaronic regime, are well explained by the anisotropic SE due to e_g electrons. The t_{2g} SE is much weaker and gives only quantitative corrections in $LaMnO_3$. The quality of the derived SE interactions (1) and (4) may be best appreciated by comparing the calculated values of Néel temperatures T_N (using mean-field theory and the appropriate reduction of T_N due to quantum fluctuations) of 106, 95, and 124 K for $LaMnO_3$, $La_{0.92}Ca_{0.08}MnO_3$, and $CaMnO_3$, with the experimental values of 136, 122, and 110 K, respectively.

The magnon dispersion derived from DE for degenerate e_g orbitals supplemented by smaller SE terms agrees well with the experimental findings in FM metallic manganites. The latter result provides another argument in favor of disordered complex orbital state $\{|c+\rangle, |c-\rangle\}$ in the doped regime, while any form of polarized states, including $|\Phi_+\rangle$ and $|\Phi_\pm\rangle$ phases, is unstable. We believe that an even better agreement between theory and experiment could be obtained by including an incoherent part of the kinetic energy which would increase the DE contribution to the magnon dispersion $\propto J_{DE}$. A better understanding of the metal-insulator transition between the FI and FM phases [22], and a quantitative description of differences between magnon spectra of various FM compounds, remain still challenging questions in the theory of manganites.

Acknowledgments

We thank G. Aeppli, F. Moussa, P. Wölfle, and J. Zaanen for valuable discussions, and acknowledge the support by KBN of Poland, Project No. 2 P03B 175 14.

References

[1] K.I. Kugel and D.I. Khomskii, Sov. Phys. Usp. **25**, 232 (1982).

[2] P.W. Anderson and H. Hasegawa, Phys. Rev. **100**, 675 (1955).

[3] A.J. Millis *et al.*, Phys. Rev. Lett. **74**, 5144 (1995).

[4] Y. Murakami *et al.*, Phys. Rev. Lett. **80**, 1932 (1998); **81**, 582 (1998).

[5] L.F. Feiner, A.M. Oleś, and J. Zaanen, Phys. Rev. Lett. **78**, 2799 (1997); A.M. Oleś, L.F. Feiner, and J. Zaanen, Phys. Rev. B **61**, 6257 (2000).

[6] R. Shiina, T. Nishitani, and H. Shiba, J. Phys. Soc. Jpn. **66**, 3159 (1997).

[7] L.F. Feiner and A.M. Oleś, Phys. Rev. B **59**, 3295 (1999).

[8] L.F. Feiner and A.M. Oleś, Physica B **259-261**, 796 (1999).

[9] A. J. Millis, Phys. Rev. B **53**, 8434 (1996).

[10] F. Moussa *et al.*, Phys. Rev. B **60**, 12299 (1999).

[11] R. Kilian and G. Khaliullin, Phys. Rev. B **60**, 13458 (1999).

[12] S. Okamoto, S. Ishihara, and S. Maekawa, Phys. Rev. B **61**, 451 (2000).

[13] S. Ishihara, M. Yamanaka, and N. Nagaosa, Phys. Rev. B **56**, 686 (1997).

[14] D.I. Khomskii, unpublished and these Proceedings (2000).

[15] A. Takahashi and H. Shiba, unpublished (2000).

[16] G. Kotliar and A.E. Ruckenstein, Phys. Rev. Lett. **57**, 1362 (1986).

[17] A.M. Oleś and L.F. Feiner, Acta Phys. Polon. A **97**, 193 (2000).

[18] T.G. Perring *et al.*, Phys. Rev. Lett. **77**, 711 (1996).

[19] Y. Endoh and K. Hirota, J. Phys. Soc. Jpn. **66**, 2264 (1997).

[20] J.A. Fernandez-Baca *et al.*, Phys. Rev. Lett. **80**, 4012 (1998).

[21] I.V. Solovyev and K. Terakura, Phys. Rev. Lett. **82**, 2959 (1999).

[22] L.F. Feiner and A.M. Oleś, unpublished and these Proceedings (2000).

ORBITAL DYNAMICS:
THE ORIGIN OF ANOMALOUS
MAGNON SOFTENING
IN FERROMAGNETIC MANGANITES

G. Khaliullin[1] and R. Kilian[2]

[1]*Max-Planck-Institut für Festkörperforschung, Heisenbergstrasse 1, D-70569 Stuttgart, Germany*

[2]*Max-Planck-Institut für Physik komplexer Systeme, Nöthnitzer Strasse 38, D-01187 Dresden, Germany*

Abstract   We study the renormalization of magnons by charge and coupled orbital-lattice fluctuations in colossal magnetoresistance compounds. The model considered is an orbitally degenerate double-exchange system coupled to Jahn-Teller active phonons. The modulation of ferromagnetic bonds by low-energy orbital fluctuations is identified as the main origin of the unusual softening of the zone-boundary magnons observed experimentally in manganites.

1.   INTRODUCTION

Recently, distinct new features in the spin dynamics of the ferromagnetic manganese oxide compound $Pr_{0.63}Sr_{0.37}MnO_3$ have been reported [1]. Striking deviations from the predictions of the canonical double-exchange (DE) theory [2, 3, 4] were observed. In particular, an anomalous softening of magnons at the zone boundary was found even well below the Curie temperature T_C. It is noted that these anomalies are closely related to the reduced values of T_C: For higher-T_C compounds no considerable deviations from a simple cosine dispersion is observed [5].

These experimental findings seem to be of high importance. They in fact indicate that some very specific features of magnetism in colossal magnetoresistance manganites have still to be identified. In this paper we propose a mechanism which might explain the above experimental observations. Our basic idea is the following: The strength of ferromagnetic interaction in a given bond strongly depends on which orbital is occupied by an e_g electron (Fig. 1). Suppose that orbitals and Jahn-Teller (JT) distortions are disordered in the

237

ferromagnetic phase. It is then evident that temporal fluctuations of orbitals affect the short-wavelength magnons through a strong modulation of exchange bonds. Quantitatively, the effect is expected to be controlled by the characteristic time scale of orbital fluctuations: A slowing-down of the dynamics of the coupled system of orbitals and JT phonons should lead to a stronger magnon renormalization. In other words, the observed zone-boundary magnon softening is interpreted in this picture as a precursor effect in the proximity of static orbital-lattice ordering.

More specifically, we calculate the dispersion of one-magnon excitations at zero temperature. First, we map the ferromagnetic Kondo-lattice model onto a Hamiltonian of interacting magnons and spinless fermions. Yet the fermions carry orbital quantum number, the fluctuations of which the magnon can scatter on. Second, we show that as the orbitals are strongly coupled to the lattice there is an indirect coupling of magnons to JT phonons via the orbital sector. Put another way, the orbital-lattice coupling produces a low-energy component in the orbital fluctuation spectrum at phonon frequencies. We then calculate the self-energy corrections to the magnon dispersion perturbatively (employing a $1/S$ expansion). Charge fluctuations are found to produce a moderate softening of magnons throughout the Brillouin zone. The effect of orbital and lattice fluctuations on the magnon dispersion is more pronounced and becomes dramatic as static order in the orbital-lattice sector is approached.

2.   THE MODEL

We start with a model describing the ferromagnetic coupling of double-degenerate e_g-band electrons to an otherwise noninteracting system of localized spins S_t on a cubic lattice:

$$
\begin{aligned}
H = \ & -\sum_{\langle ij\rangle_\gamma} t_\gamma^{\alpha\beta}\left(e_{i\sigma\alpha}^\dagger e_{j\sigma\beta} + \text{H.c.}\right) - J_H S_{it} s_{ie} \\
& + \sum_\alpha U n_{i\uparrow\alpha} n_{i\downarrow\alpha} + \sum_{\alpha\neq\beta}' \left(U' - J_H \hat{P}\right) n_{i\alpha} n_{i\beta}
\end{aligned}
\tag{1}
$$

with $\gamma \in \{x, y, z\}$ and $\hat{P} = (s_{i\alpha}s_{i\beta} + \frac{3}{4})$. Hereafter, the indices α/β and σ stand for orbital and spin quantum numbers of e_g electrons, respectively; summation over repeated indices is implied and double counting is excluded from the primed sum. The spin operator $s_{i\alpha}$ acts on orbital α and $s_{ie} = \sum_\alpha s_{i\alpha}$. The last two terms in Eq. (1) describe intra- (inter-) orbital Coulomb interaction U (U') and Hund's coupling between e_g electrons in doubly occupied states. Correlations are assumed to be strong: $U, U', (U' - J_H) \gg t$. The important point is the peculiar orbital and bond dependencies of the electron hopping matrix elements. In orbital basis $\alpha \in \{|3z^2 - r^2\rangle, |x^2 - y^2\rangle\}$ (see Fig. 1) one has [6]:

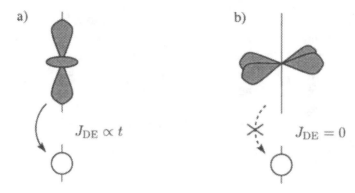

Figure 1 The e_g-electron transfer amplitude which controls the double-exchange interaction J_{DE} strongly depends on orbital orientation.

$$t_{x/y}^{\alpha\beta} = t \begin{pmatrix} 1/4 & \mp\sqrt{3}/4 \\ \mp\sqrt{3}/4 & 3/4 \end{pmatrix}, \quad t_z^{\alpha\beta} = t \begin{pmatrix} 1 & 0 \\ 0 & 0 \end{pmatrix}.$$

We work in the limit $J_H \to \infty$. Then the combined action of the on-site Hund's "ferromagnetism" and the spin-diagonal nature of electron hopping results in a global ferromagnetic ground state. Two different contributions to the spin stiffness should be noticed: Conventional double exchange due to metallic charge motion and superexchange which accounts for the kinetic energy gain by virtual hoppings of e_g electrons. We now derive the effective Hamiltonian describing the spin excitations at $T \ll T_C$.

3. DOUBLE EXCHANGE

At $J_H \to \infty$, $\mathbf{S}_t$ and the e_g spin $\mathbf{s}_e$ are not independent anymore: $\mathbf{s}_{ie} = n_{ie}\mathbf{S}_t/2S_t$. They form a total on-site spin $\mathbf{S}_i$ with spin value $S_t + \frac{1}{2}$ if an e_g electron is present, $n_{ie} = 1$, or simply S_t otherwise. The unification of band and local spin subspaces implies *spin-charge separation*. That is, the spin component $b_{i\sigma}$ of the e_g electron $e_{i\sigma\alpha} = c_{i\alpha}b_{i\sigma}$ is decoupled and "absorbed" by the total spin, and we are left with a charged fermion $c_{i\alpha}$ carrying the orbital index. At $T \ll T_C$, the Schwinger boson $b_{i\downarrow}$ simply becomes a part of the magnon operator b_i, namely $b_{i\downarrow} = b_i/(2S)^{1/2}$ and $e_{i\downarrow\alpha} = c_{i\alpha}b_i/(2S)^{1/2}$, while $b_{i\uparrow}$ is almost condensed, so $e_{i\uparrow\alpha} = c_{i\alpha}(1 - n_{i\downarrow})^{1/2} \approx c_{i\alpha}(1 - b_i^\dagger b_i/4S)$. Hereafter, $S = S_t + \frac{1}{2}$. Further, we assume $S \gg 1$ and keep the relevant leading $1/S$ terms only. The kinetic energy then reads

$$
\begin{aligned}
H_{\text{kin}} &= -\sum_{\langle ij\rangle_\gamma} t_\gamma^{\alpha\beta} c_{i\alpha}^\dagger c_{j\beta} + \frac{1}{2S}\sum_{\langle ij\rangle_\gamma} t_\gamma^{\alpha\beta} \\
&\quad \times c_{i\alpha}^\dagger c_{j\beta}\left(\frac{1}{2}b_i^\dagger b_i + \frac{1}{2}b_j^\dagger b_j - b_i^\dagger b_j\right) + \text{H.c.}
\end{aligned}
\tag{2}
$$

The first term here describes the fermionic motion in a ferromagnetic background (yet this motion is strongly correlated in the presence of orbital disorder [7, 8]), while the second term controls the spin dynamics and spin-fermion interaction. The latter term would come out from a spin-wave expansion of an effective Heisenberg model $J_{DE}(\mathbf{S}_i \mathbf{S}_j)$ with $J_{DE} = (2S^2)^{-1} t_\gamma^{\alpha\beta} \langle c_{i\alpha}^\dagger c_{j\beta} \rangle$ if one considers the fermionic sector on *average* as in mean-field treatments of the DE model. However, the bond variable $c_{i\alpha}^\dagger c_{j\beta}$ is a fluctuating complex quantity and the spin structure of Eq. (2) is therefore not of Heisenberg form. Explicitly $(T \ll T_C)$:

$$
\begin{aligned}
H_{\text{kin}} &= -\sum_{\langle ij \rangle_\gamma} t_\gamma^{\alpha\beta} c_{i\alpha}^\dagger c_{j\beta} \\
&\quad \times \left[\frac{3}{4} + \frac{1}{4S^2} \left(S_i^z S_j^z + S_i^- S_j^+ \right) \right] + \text{H.c.} \tag{3}
\end{aligned}
$$

In the classical limit for spins, one recovers from Eq. (3) the effective fermionic model with phase-dependent hopping [9, 10, 11]. In addition to the above peculiarities, the band/local duality of spin in the DE system further results in the following important point: Going to momentum representation in Eq. (2) one immediately realizes that the magnon-fermion interaction vertex does not vanish at zero-momentum transfer. What is wrong? The boson b_i should not be regarded as a physical magnon. The true Goldstone particle of the DE model is the following object with both local spin and itinerant features:

$$
\begin{aligned}
B_i &= b_i \left[n_{ic} + \sqrt{\frac{2S-1}{2S}} (1 - n_{ic}) \right] \\
&\approx b_i - \frac{1}{4S} (1 - n_{ic}) b_i. \tag{4}
\end{aligned}
$$

The composite character of the physical magnon B_i is the price one has to pay for spin-charge separation and for the rearrangement of the original Hilbert space. The itinerant component of B_i is of order $1/S$ only. However, the spin stiffness is itself of the same order, and the $1/S$ correction in Eq. (4) is of crucial importance to ensure spin dynamics consistent with the Goldstone theorem. Now, commuting Eq. (4) with Eq. (2) one finds the mean-field magnon dispersion $\omega_{\mathbf{p}}$ and the correct momentum structure of the magnon-fermion scattering vertex:

$$
[B_{\mathbf{p}}, H] = \omega_{\mathbf{p}} B_{\mathbf{p}} + \frac{t}{2S} \sum_{\mathbf{q}} A_{\mathbf{p}}^{\alpha\beta}(\mathbf{k}) c_{\mathbf{k}\alpha}^\dagger c_{\mathbf{k}-\mathbf{q},\beta} B_{\mathbf{p}+\mathbf{q}}, \tag{5}
$$

$$
A_{\mathbf{p}}^{\alpha\beta}(\mathbf{k}) = \gamma_{\mathbf{k}}^{\alpha\beta} - \gamma_{\mathbf{k}+\mathbf{p}}^{\alpha\beta}.
$$

Here, $\omega_{\mathbf{p}} = z D_s (1 - \gamma_{\mathbf{p}})$, $D_s = J_{DE} S$, J_{DE} is defined above, $z = 6$, and the form factors $\gamma_{\mathbf{k}} = z^{-1} \sum_\delta \exp(i\mathbf{k}\delta)$, $\gamma_{\mathbf{k}}^{\alpha\beta} = (zt)^{-1} \sum_\delta t_\delta^{\alpha\beta} \exp(i\mathbf{k}\delta)$. The physics behind the second term in Eq. (5) is the temporal fluctuations of the "exchange constant" due to the charge and orbital dynamics. From now on, the conventional diagrammatic method with bare magnon-fermion vertex given in Eq. (5) can be used.

Now we turn to the correlations in the fermionic band. We separate the *charge* and *orbital* degrees of freedom by the parameterization $c_{i\alpha} = h_i^\dagger f_{i\alpha}$, where the slave boson h_i and the fermion $f_{i\alpha}$ represent the density and orbital fluctuations, respectively [8]. Further, approximating $\delta(c_{i\alpha}^\dagger c_{j\beta}) = \langle h_j^\dagger h_i \rangle f_{i\alpha}^\dagger f_{j\beta} + \langle f_{i\alpha}^\dagger f_{j\beta} \rangle h_j^\dagger h_i$, one finds from Eq. (5) two different contributions to the magnon self energy.

4. SUPEREXCHANGE

At low dopings, the virtual charge transfer across the Hubbard gap becomes of importance. In limit of large J_H, transitions to the high-spin intermediate state with energy $U_1 = (U' - J_H) \ll U', U$ dominate. The corresponding superexchange Hamiltonian is then obtained to be:

$$H_{SE} = -J_{SE} \sum_{\langle ij \rangle_\gamma} [\mathbf{S}_i \mathbf{S}_j + S(S+1)] \left(1 - \tau_i^\gamma \tau_j^\gamma \right). \tag{6}$$

Here, $J_{SE} = (t^2/U_1)[2S(2S+1)]^{-1}$ and $\tau^{x/y} = \left(\sigma^z \pm \sqrt{3}\sigma^x \right)/2$, $\tau^z = \sigma^z$ with Pauli matrices $\sigma^{x/z}$ acting on orbital subspace f_α. Eq. (6) is the generalization of orbitally degenerate superexchange models [6, 12] for arbitrary values of spin. Superexchange is of *ferromagnetic* nature because the $J_H \to \infty$ limit is assumed, and the exchange strength depends on orbital orientations [13]. In an orbitally disordered state one can represent the orbital part of Eq. (6) in terms of fluctuating bond operators $f_{i\alpha}^\dagger f_{j\beta}$ [8]. Then the effect of superexchange is simply to add a new term

$$A_\mathbf{p}^{\alpha\beta}(\mathbf{k}, \mathbf{q}) = x_0 \left(\gamma_\mathbf{k}^{\alpha\beta} + \gamma_{\mathbf{k}-\mathbf{q}}^{\alpha\beta} - \gamma_{\mathbf{k}+\mathbf{p}}^{\alpha\beta} - \gamma_{\mathbf{k}-\mathbf{q}-\mathbf{p}}^{\alpha\beta} \right) \tag{7}$$

to the scattering amplitude in Eq. (5) and to renormalize $D_s \to (J_{DE}+J_{SE})S = t\chi_0(x + x_0)/2S$. Here the constant $x_0 = 2\chi_0 t/U_1$ with mean-field parameter $\chi_0 = \langle f_{i+c}^\dagger f_i \rangle \approx \frac{1}{2}(1 - x)$ defines the effective doping level x below which the superexchange contribution becomes of importance.

To sum up to now: We have developed a quantum theory of the double-exchange model at $T < T_C$ which also includes the correlation effects present in transition metal oxides. The obvious advantage of this approach over earlier treatments [3, 4] is that the theory has a transparent structure adjusted to describe the low-energy spin dynamics, and the large atomic scale J_H does not appear in the calculations, either. Yet the corrections t/J_H can be accounted for perturbatively, if necessary.

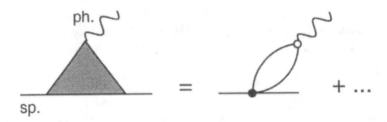

Figure 2   Effective spin-phonon coupling. Perturbatively, it is controlled by the orbital susceptibility (fermion bubble), Jahn-Teller (open dot $\propto g_0$) and double-exchange (filled dot $\propto t$) interaction vertices.

5.   INDIRECT MAGNON-PHONON COUPLING VIA THE ORBITAL SECTOR

An important piece of physics relevant to the experiment [1] is still missed in model (1), that is the Jahn-Teller orbital-lattice coupling [14]. This can be written as:

$$H_{JT} = -g_0 \left(Q_{2i}\sigma_i^x + Q_{3i}\sigma_i^z \right), \qquad (8)$$

where Q_2 and Q_3 are the distortions of appropriate symmetry [6]. The deformation energy including intersite correlations is given by

$$H_{ph} = \frac{1}{2}K \sum_i \mathbf{Q}_i^2 + K_1 \sum_{\langle ij \rangle_\gamma} \tilde{Q}_{3i}^\gamma \tilde{Q}_{3j}^\gamma, \qquad (9)$$

with $\tilde{Q}_3^{x/y} = (Q_3 \pm \sqrt{3}Q_2)/2$, $\tilde{Q}_3^z = Q_3$, and $\mathbf{Q} = (Q_2, Q_3)$. In general, the JT interaction strongly mixes the orbital and lattice dynamics, leading to a fluctuation of exchange bonds at low (phonon) frequencies. Here, we treat this problem only perturbatively as shown in Fig. 2. We arrive at the following effective spin-phonon Hamiltonian:

$$H_{s-ph} = -\sum_{pq} (\mathbf{g_{pq}Q_{-q}}) B_p^\dagger B_{p+q}. \qquad (10)$$

The coupling constants $g_{pq}^{(\alpha)} = g_0 a_0 (\eta_q^{(\alpha)} - \eta_p^{(\alpha)} - \eta_{p+q}^{(\alpha)})/S$, where $\eta_q^{(2)} = -\sqrt{3}(c_x - c_y)/2$, $\eta_q^{(3)} = c_z - \frac{1}{2}c_x - \frac{1}{2}c_y$, $c_\alpha \equiv \cos q_\alpha$, and the parameter $a_0 = t(x + x_0)\langle (f_{i+c}^\dagger f_i)(\sigma_i^z) \rangle_{\omega=0}$.

6.   MAGNON SELF ENERGIES

We are now in the position to calculate the renormalization of the magnon dispersion. The leading $1/S^2$ corrections are shown in Fig. 3. Charge and

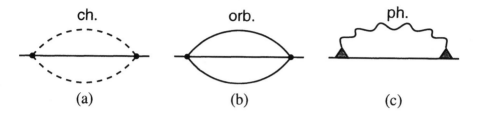

ch. orb. ph.

(a) (b) (c)

Figure 3 Magnon self energies.

orbital susceptibilities in Figs. 3(a) and 3(b), respectively, are calculated using mean-field Green's functions in slave boson h and fermion f subspaces. For the spectral density of JT phonons in Fig. 3(c) we use

$$\rho_\pm^{ph}(\omega, \mathbf{q}) = \frac{1}{\pi} \frac{\omega}{\omega_{\mathbf{q}}^\pm} \frac{\Gamma}{(\omega - \omega_{\mathbf{q}}^\pm)^2 + \Gamma^2} \qquad (11)$$

accounting phenomenologically for the damping Γ due to coupling to orbital fluctuations. The phonon dispersions $\omega_{\mathbf{q}}^\pm = \omega_0^{ph}[\kappa_{1\mathbf{q}} \pm (\kappa_{2\mathbf{q}}^2 + \kappa_{3\mathbf{q}}^2)^{1/2}]^{1/2}$ with $\omega_0^{ph} = (K/M)^{1/2}$ follow from Eq. (9). Here, $\kappa_{1\mathbf{q}} = 1 + k_1(c_x + c_y + c_z)$, $\kappa_{2\mathbf{q}} = k_1\eta_{\mathbf{q}}^{(2)}$, $\kappa_{3\mathbf{q}} = k_1\eta_{\mathbf{q}}^{(3)}$, and $k_1 = K_1/K$. The expressions obtained from the diagrams in Fig. 3 contain summations over momentum space which we perform numerically. We find the effect of charge fluctuations on the magnon spectrum to be quite featureless and moderate (see Fig. 4), which is due to the fact that the spectral density of charge fluctuations lies well above the magnon band. On contrary, the relatively low-energy fluctuations of the orbital and lattice degrees of freedom are found to affect the spin-wave dispersion in a peculiar way, particularly in $(0, 0, q)$ and $(0, q, q)$ directions. Fig. 4 shows the magnon dispersion. The hopping amplitude $t = 0.4$ eV is chosen to fit the spin stiffness in $Pr_{0.63}Sr_{0.37}MnO_3$; further we use $U_1 = 4$ eV [12]. The phonon contribution depends on the quantity $(g_0 a_0)^2/2K \equiv E_{JT}a_0^2$. We set $E_{JT}a_0^2 = 0.004$ eV [15], $\omega_0^{ph} = 0.08$ eV [16], $\Gamma = 0.04$ eV. Our key observation is the crucial effect of intersite correlations of JT distortions, controlled by k_1, on the magnon dispersion (see Fig. 5). To explain the experimental data [1] we are forced to assume *ferro-type* correlations ($k_1 < 0$). We interpret this surprising result in the following way: Conventionally $k_1 > 0$ corresponding to AF order of JT distortions and orbitals is expected in undoped compounds [6]. However, in the doped case charge mobility prefers a ferro-type local orientation of orbitals which minimizes the kinetic energy. This competition between JT and kinetic energies can be simulated by tuning k_1. At large enough doping, ferro-type orbital correlations are expected to prevail, hence effectively $k_1 < 0$. In fact, a ferro-type ordering of orbitals leading to a layered AFM spin structure is experimentally observed [17, 18] at dopings

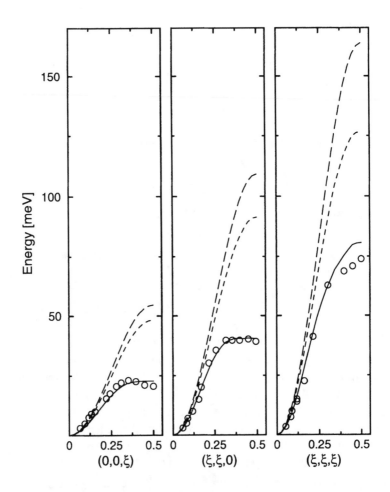

Figure 4   Magnon dispersion along $(0, 0, \xi)$, $(\xi, \xi, 0)$, and (ξ, ξ, ξ) directions, where $\xi = 0.5$ at the zone boundary. Solid lines represent the theoretical result including charge, orbital, and lattice effects and are fitted to experimental data [1] denoted by circles. For comparison the bare dispersion and the one including only charge effects are indicated by long-dashed and dashed lines, respectively. $k_1 = -0.33$ is chosen.

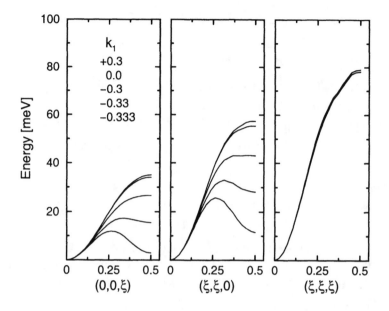

Figure 5 Magnon dispersion for different values of k_1. $E_{JT}a_0^2 = 0.006$ eV is used. The softening enhances as $k_1 \to -\frac{1}{3}$ corresponding to ferro-type orbital-lattice order.

about $x = 0.5$. As this instability is approached, low-energy fluctuations of exchange bonds develop, yielding a magnon evolution as shown in Fig. 5. Remarkably, the small-q spin stiffness is not affected by this physics, while a strong reduction of T_C by soft magnons at the zone boundary is predicted. This explains the origin of the anomalous enhancement of the D_s/T_C ratio in low-T_C manganites [19].

7.   SUMMARY

In conclusion, we have presented a theory of spin dynamics in a model relevant to manganites, emphasizing particularly the interplay between double-exchange physics and orbital-lattice dynamics. The unusual magnon dispersion in lower-T_C compounds is explained as being due to the proximity to orbital-lattice ordering. Apparently, strongly correlated orbital fluctuations play a crucial role in the physics of manganites.

References

[1] H. Y. Hwang *et al.*, Phys. Rev. Lett. **80**, 1316 (1998).

[2] C. Zener, Phys. Rev. **82**, 403 (1951); P. W. Anderson and H. Hasegawa, *ibid.* **100**, 675 (1955); P.-G. de Gennes, *ibid.* **118**, 141 (1960).

[3] K. Kubo and N. Ohata, J. Phys. Soc. Jpn. **33**, 21 (1972).

[4] N. Furukawa, J. Phys. Soc. Jpn. **65**, 1174 (1996).

[5] T. G. Perring *et al.*, Phys. Rev. Lett. **77**, 711 (1996).

[6] K. I. Kugel and D. I. Khomskii, Sov. Phys.-Usp. **25**, 231 (1982).

[7] S. Ishihara, M. Yamanaka, and N. Nagaosa, Phys. Rev. B **56**, 686 (1997).

[8] R. Kilian and G. Khaliullin, Phys. Rev. B **58**, R11 841 (1998).

[9] E. Müller-Hartmann and E. Dagotto, Phys. Rev. B **54**, R6819 (1996).

[10] A. J. Millis, P. B. Littlewood, and B. I. Shraiman, Phys. Rev. Lett. **74**, 5144 (1995).

[11] It is noticed that the phase-dependent part of the fermionic hopping is only of order $1/S$ (see Eq. (2)).

[12] L. F. Feiner and A. M. Oleś, Phys. Rev. B **59**, 3295 (1999).

[13] The effect of superexchange-bond fluctuations on the magnon spectrum in the antiferromagnetic Kugel-Khomskii model was recently studied in G. Khaliullin and V. Oudovenko, Phys. Rev. B **56**, R14 243 (1997).

[14] A. J. Millis, b. I. Shraiman, and R. Mueller, Phys. Rev. Lett. **77**, 175 (1996).

[15] A mean-field calculation gives $a_0 \approx 0.1$. Our fitting then implies a reasonable Jahn-Teller binding energy $E_{JT} \approx 0.4$ eV.

[16] This number is consistent with optical reflectivity data in Y. Okimoto *et al.*, Phys. Rev. Lett. **75**, 109 (1995).

[17] H. Kawano *et al.*, Phys. Rev. Lett. **78**, 4253 (1997).

[18] Y. Moritomo *et al.*, Phys. Rev. B **58**, 5544 (1998).

[19] J. A. Fernandez-Baca *et al.*, Phys. Rev. Lett. **80**, 4012 (1998).

FIELD INDUCED METAL-INSULATOR TRANSITION IN (PR:CA:SR)MnO₃

J. Hemberger, M. Paraskevoupolos, J. Sichelschmidt, M. Brando, R. Wehn, F. Mayr, K. Pucher, P. Lunkenheimer, and A. Loidl

Experimentalphysik V, Elektronische Korrelation und Magnetismus, Universität Augsburg, D - 86159Augsburg, Germany

A.A. Mukhin[1] and A.M. Balbashov[2]

[1]*Moscow Power Engineering Institute, 105835 Moscow, Russia*
[2]*General Physics Institute of the Russian Acad. Sci., 117942 Moscow, Russia*

Abstract   In the system $Pr_{0.65}Ca_{0.28}Sr_{0.07}MnO_3$ (PCSM) the field induced metal-insulator transition exhibits changes of the resistivity up to 11 orders in magnitude. In order to study this colossal magnetoresistance effect (CMR) we carried out measurements on a PCSM single crystal, applying a broad range of experimental methods. Resistivity, magnetization and magnetic AC-susceptibility were measured in magnetic fields up to 140 kOe and in the temperature range 1.5 $K < T < 800$ K. With decreasing temperature and in zero magnetic field PCSM undergoes a sequence of phase transitions from a paramagnetic into a charge ordered insulating and finally antiferromagnetic phase. The conductivity in these phases is dominated by variable range hopping processes. By applying magnetic fields $H \geq 10$ kOe a ferromagnetic metallic phase is induced. This transition exhibits an extremely marked hysteresis behavior. Below $T \approx 100$ K metastable states can be generated, in which ferromagnetic metallic and insulating phases coexist.

1.   INTRODUCTION

At present in the field of transition-metal oxides a lot of attention is payed to materials exhibiting the colossal magnetoresistance effect (CMR), stimulated by the observation of this effect in manganite films [1]. The system (Pr:Ca)MnO₃ shows very pronounced CMR features. In this system the metal-insulator transition (MIT) not only can be driven by a magnetic field, but the transformation of the insulating (I) into a metallic (M) and ferromagnetic (FM) phase can also be x-ray, photo- or electric field induced [2]. The system

J. Bonča et al. (eds.), Open Problems in Strongly Correlated Electron Systems, 247–252.

$Pr_{1-x}Ca_xMnO_3$ has been investigated in detail with respect to its structural, magnetic and transport properties [3, 4]. The (x, T)-phase diagram shows a orthorhombic (O) phase at high temperatures. For $x < 0.3$ a Jahn-Teller distorted O' phase develops at lower temperatures, while for concentrations $0.3 < x < 0.9$ the structural groundstate is tetragonal. For $x > 0.9$ the system stays orthorombic in the whole temperature range. Between $0.3 < x < 0.7$ an charge- and orbitally ordered state has been proposed [5]. For $x > 0.3$ the system is insulating (I) and charge order (CO) transition is followed by an antiferromagnetic (AFM) transition while for $x < 0.3$ a ferromagnetic (FM) metallic (M) phase is established [4]. The magnetic field induced transition from the AFM/CO/I state into the FM/M state is strongly of first order and shows large hysteresis effects. This MIT also can be induced by internal chemical pressure substituting Ca by Sr [6]. The reported structural and electronic properties for $Pr_{0.7}Ca_{0.3}MnO_3$ are as follows [7]: The CO transition at $T \approx 200$ K is followed by an AFM transition ($T_N \approx 150$ K). At lower temperatures the evolving of a FM component is reported, which was ascribed to a canting of the AFM order but equally well can be described assuming electronic phase separation into AFM and FM clusters, which is believed to play an important role in manganite CMR compounds [9].

To optimize the system with respect to its CMR properties, we tried to tune the Sr:Ca ratio to get as close as possible towards the border-line between the FM/M state and AFM/CO/I state in zero external magnetic field. In the following we discuss the results obtained from investigations of single crystals of the $Pr_{0.65}Ca_{0.28}Sr_{0.07}MnO_3$ (PCSM).

2.   RESULTS AND DISCUSSION

The PCSM single crystals were grown by the floating-zone method, details are described elsewhere [8]. The magnetic susceptibility and the magnetization were measured using an Oxford AC-susceptometer in an Oxford cryostat for fields up to 140 kOe. The electrical resistivity was measured employing a four probe technique in the same cryostat.

In Fig. (1) we show the $B(T)$-phasediagram of PCSM, determined from resistivity and magnetization measurements, which are illustrated in part in Fig. 2. In zero magnetic field the system undergoes a sequence of phase-transitions: At $T_{CO} \approx 210$ K a transition between two PM/I states takes place. Evidence for the inset of charge order is reported in similar systems [6]. At the same time structural changes are reflected in a clear change of the phonon-spectra [8]. At $T_N = 160$ K AFM order sets in and below $T_{irr} \approx 100$ K the system gets strongly irreversible and hysteresis phenomena with respect to field and temperature appear. In this regime a metastable FM metallic state can be induced by applying an external magnetic field. As pointed out later it can be

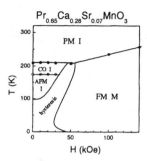

Figure 1 $B(T)$-phasediagram for $Pr_{0.65}Ca_{0.28}Sr_{0.07}MnO_3$. The 'hysteresis'-area reveals metastable behaviour.

assumed that these irreversibilities are introduced by phase separation into FM clusters within an AFM background.

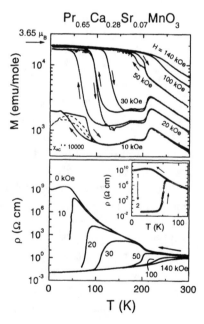

Figure 2   Resistivity and magnetization as a function of temperature for various fields up to 14 T. The dashed line (upper frame) displays the AC-susceptibility. The inset sows the thermoremanent behaviour of the resistivity. The data was taken in zero field. Between (1) and (2) a field of 5 T was switched on for several minutes.

Fig. 2 displays the temperature dependence of the resistivity (lower frame, only cooling) and the magnetization (upper frame) for various fields up to

140 kOe and in addition the AC-susceptibility (dashed line in the upper frame). The CO transition can well be identified in the resistivity as well as the magnetization curves. For fields below 50 kOe the resistivity rises on cooling across T_c while the magnetization drops. A more detailed analysis of the DC-resistivity in zero magnetic field reveals that the underlying transport processes can be ascribed to variable range hopping (VRH) for temperatures 50 K $< T <$ 200 K and is dominated by the hopping of adiabatic small polarons for temperatures 220 K $< T <$ 650 K above T_{CO} [8].

The AFM transition only can be detected in the magnetization and susceptibility curves. Applying fields $H \geq 10$ kOe a ferromagnetic component is induced. This phenomenon is accompanied by large hysteresis effects and a drop of the resistivity up to 11 orders of magnitude. The temperature of induced FM increases with increasing magnetic field and for $H > 50$ kOe it is shifted above T_{CO}. It is interesting to note, that for intermediate fields ($H = 20$ kOe, $H = 30$ kOe) a small hysteresis of the magnetization extends up to 200 K. The saturation value of the magnetization in applied magnetic field is found to be close to the expected value of $M_S = 3.65$ μ_B for $Mn_{0.65}^{3+}$ and $Mn_{0.35}^{4+}$. No additional contribution of the Pr^{3+} ions can be detected. At the same time the effective moment of $\mu_{eff} = 5.35$ μ_B evaluated from the reciprocal susceptibility data up to 800 K (not shown) is clearly enhanced compared with the manganese-only expectation ($\mu_{eff}(Mn) = 4.57$ μ_B). This fact implies that the Pr^{3+} moments order antiferromagnetically at low temperatures (see the anomaly in χ_{AC} and $M(10$ kOe) near $T \approx 25$ K).

The inset of Fig. 2 shows the resistivity measured in zero external field on cooling and heating. Between the points (1) and (2) the FM/M state was induced by applying a magnetic field of $H = 50$ kOe. After switching off the field the system remains metallic up to approximately 70 K. One should mention that in spite of the persistence of the conductivity at low temperatures the detected macroscopic magnetization returns to zero after switching off the external field in an analogous experiment (not shown). This can be explained by domain effects according to the soft magnetic character of the ferromagnetism in this system. The metastable conducting state below this temperature decays with time, as it is illustrated in Fig. 2. Fig. 2 shows the magnetic AC-susceptibility (upper frame) and resistivity (lower frame) versus time after switching off the magnetic field $H = 50$ kOe at $T = 66$ K. With increasing time χ_{AC} decays smooth and continuously. At the same time in the resistivity sharp steplike jumps appear. Such behavior can not be explained in the framework of a homogeneous canted AFM state but has to be considered as clear evidence for a inhomogeneous, phase separated state with FM. The susceptibility of the FM state is higher than that of the AFM state. The decay of $\chi_{ac}(t)|_{H=0}$ documents the vanishing fraction of ferromagnetic sample volume and can be described as stretched exponential behavior $\chi(t) = \chi_{FM}\left(1 - \exp\left(-\left(t/\tau\right)^\beta\right)\right) + \chi_{AFM}$

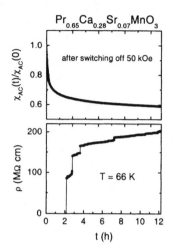

Figure 3   Time dependence of the normalized AC-susceptibility and the resistivity at $T = 66$ K after switching off a magnetic field of $H = 50$ kOe.

with $\beta \approx 0.3$. The steps in the resistivity denote the breakdown of conducting percolation paths. Phase separation scenarios in manganites are documented in literature experimentally and theoretically [8]. In the present case it stays unclear, if the groundstate of the system is intrinsically phase separated or if the observed phenomena are only due to induced metastable inhomogeneities with respect to the nearly degenerate AFM/CO/I and FM/M states of the system.

3.   SUMMARY

In the system $Pr_{0.65}Ca_{0.28}Sr_{0.07}MnO_3$, which was optimized with respect to the CMR properties, investigations of the DC-resistivity, the magnetic AC-susceptibility and the magnetization were carried out in fields up to 140 kOe. This system is very close to the boundary between a FM metallic and a AFM charge-ordered insulating phase. In this system metal insulator-transitions of more than 11 orders of magnitude in resistivity can be induced by applying an external magnetic field of $H \approx 10$ kOe. At low temperatures ($T < 100$ K) the system becomes strongly irreversible and metastable transport and magnetization phenomena appear. Those nonergodic states are related to the coexistence of nearly degenerate FM/M and AFM/CO/I phases. In this regime the nature of the transport properties appears to be determined by percolation processes.

Acknowledgments

This work was supported in part by the Sonderforschungsbereich 484 of the Deutsche Forschungsgemeinschaft.

References

[1] R. von Helmolt *et al.*, Phys. Rev. Lett. **71**, 2331 (1993); S. Jin *et al.*, Science **264**, 413 (1994).

[2] Y. Tomioka *et al.*, J. Phys. Soc. Jpn. **64**, 3626 (1995); V. Kiryukhin *et al.*, Nature **386**, 813 (1997); K. Miyano *et al.*, Phys. Rev. Lett. **78**, 4257 (1997); A. Asamitsu *et al.*, Nature **388**, 50 (1997)

[3] Z. Jirak *et al.*, J. Magn. Magn. Mat. **53**, 153 (1985).

[4] Y. Tomioka *et al.*, Phys. Rev. B **53**, R1689 (1996).

[5] E.O. Wollan and W.C. Koehler, Phys. Rev. **100**, 545 (1955).

[6] H. Yoshizawa *et al.*, Phys. Rev. B **55**, 2729 (1997).

[7] H. Yoshizawa *et al.*, Phys. Rev. B **52**, R13145 (1995); D.E. Cox *et al.*, Phys. Rev. B **57**, 3305 (1998).

[8] J. Sichelschmidt *et al.*, to be published

[9] A. Moreo *et al.*, Science **283**, 2034 (1999); Uehara *et al.*, Nature **399**, 560 (1999); I.F. Voloshin *et al.*, 71, 106 (2000)

TRIPLET PAIRING VIA LOCAL EXCHANGE IN CORRELATED SYSTEMS

J. Spałek

Institute of Physics, Jagiellonian University, Reymonta 4, 30-059 Kraków, Poland

Abstract   The role of local (Hund's-rule) exchange is discussed in stabilizing spin-triplet superconducting state in an orbitally degenerate and correlated fermion system. Both gapfull and gapless modes appear even for the isotropic pairing potential if only the bands are degenerate. The Bogolyubov-De Gennes equation has 4-component nature for a doubly degenerate band case.

1.   INTRODUCTION

The question posed here is whether there is a connection between the ferromagnetic exchange and the spin-triplet superconductivity. Historically, this type of pairing mediated by paramagnons has been proposed [1] as a possibility of such superconductivity to occur in almost ferromagnetic systems. Such concepts were particularly popular after the discovery of superfluidity in helium-3 and its subsequent microscopic explanation [2]. A renewed hope for the existence of the spin-triplet superconductivity is associated with the superconducting Sr_2RuO_4 [3], which is close to both ferromagnetism and the Mott-Hubbard insulating state [4]. The existence of strong ferromagnetic interactions is possible only in orbitally degenerate systems, where the intraatomic (Hund's rule) exchange appears. However, the strong Hund's rule coupling accompanies even stronger Coulomb repulsive interaction. Therefore, we can analyse the Hund's rule role in the pairing only if we treat properly the Coulomb correlations.

In this paper we introduce *real space* pairing induced by the Hund's rule exchange. We consider a model doubly degenerate band system leaving the analysis in a realistic band situation to a separate paper [5]. Such a simplified analysis allows us to single out the principal features of the problem in analytic terms. The real space pairing seems to be operative if electrons are strongly correlated, i.e. when the kinetic energy is comparable to the interactions among correlated particles. Here we present only the results for an effective Fermi-

J. Bonča et al. (eds.), Open Problems in Strongly Correlated Electron Systems, 253–259.

liquid model, which is based on the model of correlated fermions devised previously [6].

2.   NAMBU-DE GENNES METHOD FOR TRIPLET PAIRING IN 2-BAND CASE

Nambu-De Gennes Method for the Triplet Pairing in the Two-band Case

We consider first a two-band band model of almost-localized Fermi liquid with a local form of the triplet pairing. The corresponding effective Hamiltonian is of the simple form

$$\mathcal{H} = \sum_{\mathbf{k}\sigma l=1,2} E_{\mathbf{k}l} a^{\dagger}_{\mathbf{k}l\sigma} a_{\mathbf{k}l\sigma} - 2\tilde{J} \sum_{im} A^{\dagger}_{im} A_{im}, \tag{1}$$

where $E_{\mathbf{k}l}$ are the quasiparticle energies with enhanced masses by the band narrowing factor q^{-1} (calculated self-consistently [6]) in the bands $l = 1, 2$, $\tilde{J} \sim Jt^2$ is the effective Hund's rule coupling (the local interorbital exchange), and t^2 is the probability of having interorbital local spin-triplet configurations, characterized by the creation operators $A^{\dagger}_1 = a^{\dagger}_{il\uparrow} a^{\dagger}_{il'\uparrow}$, $A^{\dagger}_{-1} = a^{\dagger}_{il\downarrow} a^{\dagger}_{il'\downarrow}$, and $A^{\dagger}_0 = \frac{1}{\sqrt{2}}(a^{\dagger}_{il\uparrow} a^{\dagger}_{il'\downarrow} + a^{\dagger}_{il\downarrow} a^{\dagger}_{il'\uparrow})$ for $l \neq l'$. The local exchange origin of the second term derives from the exact relation between the pairing operators in real space and the full exchange operator projecting the corresponding two-particle state onto the spin-triplet configuration, which has the form

$$\sum_{m=-1}^{1} A^{\dagger}_{im} A_{im} = \mathbf{S}_{il} \cdot \mathbf{S}_{il'} + \frac{3}{4} n_{il} n_{il'}, \tag{2}$$

where $\mathbf{S}_{il}$ and n_{il} are respectively the spin and the particle number operators for electron on site i and orbital l. Explicitly $n_{il} = \sum_{\sigma} n_{il\sigma}$, $n_{il\sigma} = a^{\dagger}_{il\sigma} a_{il\sigma}$, whereas the spin operators $\mathbf{S}_{il} \equiv (S^+_{il}, S^-_{il}, S^z_{il}) \equiv (a^{\dagger}_{il\uparrow} a_{il\downarrow}, a^{\dagger}_{il\downarrow} a_{il\uparrow}, (1/2)(n_{il\uparrow} - n_{il\downarrow}))$. The right-hand side of (2) represents thus the full exchange operator in the Dirac sense [7].

After making the BCS-type approximation in the local form [6], we can cast Hamiltonian (1) into the four-component form, which in the reciprocal (**k**) space takes the form

$$\mathcal{H}_{BCS} = \sum_{\mathbf{k}} \mathbf{f}^{\dagger}_{\mathbf{k}} \mathbf{H}_{\mathbf{k}} \mathbf{f}_{\mathbf{k}} + \sum_{\mathbf{k}} E_{\mathbf{k}2}, \tag{3}$$

where the corresponding Nambu operators take the form: $\mathbf{f}^{\dagger}_{\mathbf{k}} = (f^{\dagger}_{\mathbf{k}1\uparrow}, f_{\mathbf{k}1\downarrow}, f_{-\mathbf{k}2\uparrow}, f_{-\mathbf{k}2\downarrow})$, $\mathbf{f}_{\mathbf{k}} = (\mathbf{f}^{\dagger}_{\mathbf{k}})^{\dagger}$, and the Hamiltonian matrix for selected **k** state reads

$$\mathbf{H_k} = \begin{pmatrix} E_{k1} - \mu, & 0, & \Delta_1, & \Delta_0 \\ 0, & E_{k1} - \mu, & \Delta_0, & \Delta_{-1} \\ \Delta_1^*, & \Delta_0^*, & -E_{k2} + \mu, & 0 \\ \Delta_0^*, & \Delta_{-1}^*, & 0, & -E_{k2} + \mu \end{pmatrix} \equiv \begin{pmatrix} E_{k1}\hat{\sigma}_0, & \hat{\Delta} \\ \hat{\Delta}^*, & -E_{k2}\hat{\sigma}_0 \end{pmatrix},$$

(4)

where $\hat{\sigma}_0 \equiv 1$ is the unit 2×2 matrix, and μ is the chemical potential. The superconducting gap is parametrized as $\Delta_m \equiv -2\tilde{J}\sum_k < f_{k1\sigma}^\dagger f_{-k2\sigma'}^\dagger >$, with $m = (\sigma + \sigma')/2$, and $\sigma, \sigma' = \pm 1$. The 2×2 matrix $\hat{\Delta}$ is parametrized in the usual form [7, 8]

$$\hat{\Delta} = i(\mathbf{d} \cdot \tilde{\sigma})\sigma_y = \begin{pmatrix} -d_x + id_y, & d_z \\ d_z, & d_x + id_y \end{pmatrix},$$

(5)

where $\tilde{\sigma}$ is composed of the three Pauli matrices, whereas the vector $\mathbf{d}$ in spin space has the components $d_x = (\Delta_{-1} - \Delta_1)/2$, $d_y = (\Delta_{-1} + \Delta_1)/2$, and $d_z = \Delta_0$. The form (4) is a generalization of the Nambu representation to the triplet case with three, in general different, gaps Δ_m.

It is straightforward to introduce the 4×4 Dirac matrices

$$\tilde{\beta} \equiv \begin{pmatrix} 1, & 0 \\ 0, & -1 \end{pmatrix} \quad \text{and} \quad \tilde{\alpha}_i = \begin{pmatrix} 0, & \sigma_i \\ \sigma_i, & 0 \end{pmatrix},$$

and then rewrite (4) for the strictly degenerate case $E_{k1} = E_{k2}$, and for $\Delta_m = \Delta_m^*$, in the form

$$\mathbf{H_k} = \tilde{\beta}(E_k - \mu) + i(\mathbf{d} \cdot \tilde{\alpha})\Sigma_2,$$

(6)

where Σ_2 is the y component of the relativistic spin operator. We discuss in detail the simple situation of degenerate electrons $(E_{k1} = E_{k2})$ with a real gap Δ_m in the next section.

One can also look at the approach from a different prospective. Let us introduce the four component wave function for a single quasiparticle in the superconducting phase propagating in the real space as follows

$$\hat{\Psi}(\mathbf{x}, t) = \frac{1}{\sqrt{N}} \sum_k \begin{pmatrix} \psi_{1k} f_{k1\uparrow} \\ \psi_{2k} f_{k1\downarrow} \\ \psi_{3k} f_{-k2\uparrow}^\dagger \\ \psi_{4k} f_{-k2\downarrow}^\dagger \end{pmatrix} \exp\left[i\left(\mathbf{k} \cdot \mathbf{x} - \frac{E_k}{\hbar}t\right)\right],$$

(7)

where $\psi_{\mu k}$ are the quasiparticle amplitudes which are determined for each eigenstate (see below). In this representation the Bogolyubov-De Gennes equation for a single quasiparticle in the superconducting states reads:

$$i\hbar\partial_t\hat{\Psi} = \tilde{\beta}\{E_k(\mathbf{k} \Rightarrow \frac{\nabla}{i}) - \mu\}\hat{\Psi} + i(\mathbf{d} \cdot \tilde{\alpha})\Sigma_2\hat{\Psi},$$

(8)

where $E_k(\mathbf{k} \Rightarrow \frac{\nabla}{i})$ represents now the differential operator $(1/i)\nabla$ replacing the wave vector $\mathbf{k}$ in the dispersion relation E_k for quasiparticles. In the effective-mass approximation and in the stationary case this wave equation for quasiparticles in the superconducting phase has the following form

$$\lambda \begin{pmatrix} \psi_1 \\ \psi_2 \\ \psi_3 \\ \psi_4 \end{pmatrix} = -\left(\frac{\hbar^2}{2m^*}\nabla^2 + \mu \right) \begin{pmatrix} \psi_1 \\ \psi_2 \\ -\psi_3 \\ -\psi_4 \end{pmatrix} + \begin{pmatrix} \Delta_1\psi_3 + \Delta_0\psi_4 \\ \Delta_0\psi_3 + \Delta_{-1}\psi_4 \\ \Delta_1\psi_1 + \Delta_0\psi_2 \\ \Delta_0\psi_1 + \Delta_{-1}\psi_2 \end{pmatrix},$$

(9)

where $\psi_\mu \equiv \psi_\mu(\mathbf{x})$ and λ is an eigenvalue of quasiparticle state in the super-conducting state with the above 4-component wave function (Δ_m are regarded as real). The validity of this equation goes beyond the simple solution (7), as one can include the magnetic and electric fields and other inhomogeneities if they appear on the mesoscopic or macroscopic scale. In the next section we will use explicitly the momentum representation of Eqs.(9), as we will discuss exclusively homogeneous superconducting states. One should also note that finding the eigenvalues for Hamiltonian in the forms (3) or (6) can be achieved by diagonalizing of the matrix 4×4 in general case, as discussed elsewhere [9].

We now discuss three principal solutions of Eq.(9) by taking $\psi_\mu(\mathbf{x}) = \psi_\mu \exp(i\mathbf{k} \cdot \mathbf{x})/\sqrt{V}$, where V is the system volume. We also assume that $\Delta_\mu = \Delta_\mu^*$, (e.g. neglect the applied magnetic fields).

2.1   ISOTROPIC PAIRING: $\Delta_0 = \Delta_{-1} = \Delta_1 \equiv \Delta$

We have the modes with a gap of the form

$$\lambda = \lambda_{k1,2} = \pm\sqrt{(E_k - \mu)^2 + 4\Delta^2} \equiv \pm\lambda_k.$$

(10)

For those two modes the eigenstates are characterized by the following quasi-particle operators

$$\alpha_k = u_k \frac{1}{\sqrt{2}} (f_{k1\uparrow} + f_{k1\downarrow}) - v_k \frac{1}{\sqrt{2}} \left(f_{-k2\uparrow}^\dagger + f_{-k2\downarrow} \right),$$

(11)

and

$$\beta_{-k}^\dagger = v_k \frac{1}{\sqrt{2}} (f_{k1\uparrow} + f_{k1\downarrow}) + u_k \frac{1}{\sqrt{2}} \left(f_{-k2\uparrow}^\dagger + f_{-k2\downarrow} \right),$$

(12)

with the usual Bogolyubov coherence factors. The wave function is symmetric with respect to particle-spin interchange ($\uparrow \leftrightarrow \downarrow$).

We obtain also the gapless modes of the form

$$\lambda = \lambda_{k3,4} = \pm(E_k - \mu),$$

(13)

which correspond to the eigenstates characterized by the operators

$$\gamma_k = \frac{1}{\sqrt{2}}\left(f_{k1\uparrow} - f_{k1\downarrow}\right), \quad \text{and} \quad \delta^\dagger_{-k} = \frac{1}{\sqrt{2}}\left(f^\dagger_{-k2\uparrow} - f^\dagger_{-k2\downarrow}\right) \quad (14)$$

and constitute the spin antisymmetric operators. These gapless modes disappear when the gap components are not equal. One should note that the gapless modes appear even though the superconducting gap here is k-independent.

Combining the solutions we can express the original ("old") particle operators in terms of quasiparticle ("new") operators in the following manner

$$\begin{pmatrix} f_{k1\uparrow} \\ f_{k1\downarrow} \\ f^\dagger_{-k2\uparrow} \\ f^\dagger_{-k2\downarrow} \end{pmatrix} = \frac{1}{\sqrt{2}} \begin{pmatrix} u_k, & v_k, & 1, & 0 \\ u_k, & v_k, & -1, & 0 \\ -v_k, & u_k, & 0, & 1 \\ -v_k, & u_k, & 0, & -1 \end{pmatrix} \begin{pmatrix} \alpha_k \\ \beta^\dagger_{-k} \\ \gamma_k \\ \delta^\dagger_{-k} \end{pmatrix}. \quad (15)$$

The equation for the gap e.g. $\Delta_1 = < f^\dagger_{k1\uparrow} f^\dagger_{-k2\uparrow} >$ is obtained by substituting the relevant transformed operators in (15) to Δ_1. In effect, we obtain the usual BCS form $(E_k \equiv E_k - \mu)$.

2.2 EQUAL-SPIN PAIRING: $\Delta_0 \equiv 0$

The eigenvalues take now the form

$$\lambda \equiv \lambda_{k1...4} = \pm\sqrt{(E_k - \mu)^2 + \Delta_\sigma^2} \equiv \pm\lambda_k^{(\sigma)}, \quad (16)$$

where for each spin orientation $\sigma = \pm1$ of the quasiparticles we have two solutions with the gap $\pm\sqrt{(E_k - \mu)^2 + \Delta_\sigma^2}$. The quasiparticle operators $(\alpha_{k\sigma}, \beta^\dagger_{-k\sigma})$ diagonalizing Hamiltonian (3) in this case are:

$$\alpha_{k\sigma} = u_k^{(\sigma)} \frac{1}{\sqrt{2}}\left(f_{k1\sigma} + f^\dagger_{-k2\sigma}\right) - v_k^{(\sigma)} \frac{1}{\sqrt{2}}\left(f_{k1\sigma} - f^\dagger_{-k2\sigma}\right), \quad (17)$$

and

$$\beta^\dagger_{-k\sigma} = -v_k^{(\sigma)} \frac{1}{\sqrt{2}}\left(f_{k1\sigma} + f^\dagger_{-k2\sigma}\right) + u_k^{(\sigma)} \frac{1}{\sqrt{2}}\left(f_{k1\sigma} - f^\dagger_{-k2\sigma}\right), \quad (18)$$

with the coherence factors $u_k^{(\sigma)}$ and $v_k^{(\sigma)}$ determined by Δ_σ and $\lambda_k^{(\sigma)}$.

2.3 SPIN-POLARIZED PHASE: $\Delta_0 = \Delta_\downarrow = 0$

In this limit the system is totally spin polarized, i.e. is a *spin superconductor*. In that limit we recover again the spectrum both with and without gap, i.e. $\lambda_{k1,2} = \pm\sqrt{(E_k - \mu)^2 + \Delta_\uparrow^2}$, $\lambda_{k3,4} = \pm(E_k - \mu)$. Thus paired and unpaired states coexist also in this phase, as can be easily seen from Eqs.(17) - (18), which yield the form written there for $\sigma = \uparrow$ and $\alpha_{\mathbf{k}\downarrow} = f_{\mathbf{k}1\downarrow}$ and $\beta^{\dagger}_{-\mathbf{k}\downarrow} = -f^{\dagger}_{\mathbf{k}2\downarrow}$.

Summarizing cases A-C, the lowest energy will have the homogeneous state with $\Delta_{\uparrow} = \Delta_{\downarrow} = \Delta_0$ so that the effective gap is equal to 2Δ. The most interesting feature of the results is that the gapless modes coexist in cases A and C and represent half of the spectrum. Also, the condensed phases described by the cases A-C above correspond respectively to the solutions for superfluid ^3He, which are labelled B, A, and A1. However, under the present circumstances we have momentum *independent* gaps, since the pairing is of the local (intrasite, but interorbital) nature.

3.   FINAL REMARKS

Two problems should be tackled next within this simple model situation. First, the analysis of the Meissner effect, since in the present situation the orbital diamagnetism will compete with the ferromagnetic spin polarization (particularly, if the triplet superconducting and ferromagnetic phases can coexist [10]). An intriguing question here is: Can we reach the limiting superconducting phase (corresponding to A1 phase in the case of superfluid ^3He), the critical temperature T_c of which can be enhanced by the applied magnetic field?

Second, one should derive microscopically the Ginzburg-Landau equation for the condensed pairs. Note that the De Gennes equation (8) or (9) is useful in describing the quasiparticle tunneling, whereas the Ginzburg-Landau equation is useful when considering the Josephson (pair) tunneling. Here an intriguing question to what extent the gapless quasiparticles influence the tunneling between the spin-singlet and the spin-triplet superconductors or between the triplet superconductor and normal metal. We should be able to see the progress in answering those questions in a near future.

Obviously, the present calculations do not include the anisotropic nature of the bands (we may have only pockets of the Fermi-surface coincidence for $E_{\mathbf{k}1} \neq E_{\mathbf{k}2}$). Therefore, they are *not* directly applicable to the quantitative analysis of the properties of Sr_2RuO_4 (see [5] for more detailed analysis).

Acknowledgments

The work was supported by KBN, Grant No. 2P03B 092 18. I am also grateful to Włodek Wójcik and Robert Podsiadły for discussions and the technical help.

References

[1] A. Layzer and D. Fay, Int. J. Magn. **1**, 135 (1971); Solid State Commun. **15**, 599 (1974); D. Fay and J. Appel, Phys. Rev. **B16**, 2325 (1977); I.I. Mazin and D.J.Singh, Phys. Rev. Lett. **79**, 733 (1997).

[2] P.W. Anderson and W.F. Brinkman, Phys. Rev. Lett. **30**, 1108 (1973); also in: *The Physics of Liquid and Solid Helium*, Part II, ed. K.H. Bennemann and J.B. Ketterson (Wiley, New York, 1978), p.177.

[3] Y. Maeno, H. Hashimoto, K. Yoshida, S. NishiZaki, S. Fujita, and J.G. Bednorz, Nature **372**, 532 (1994); T. Akima, S. NishiZaki, and Y. Maeno, J. Phys. Soc. Japan, **68**, 694 (1999).

[4] P.B. Allen *et al.*, Phys. Rev. **B53**, 4393 (1996); G. Cao, S. McCall, M. Shepard, J.E. Crow, and R.P. Guertin, Phys. Rev. **B56**, R2916 (1997); S. Nakatsuji, S. Ikeda, and Y. Maeno, J. Phys. Soc. Japan **66**, 1868 (1997).

[5] J. Spałek and W. Wójcik, in preparation.

[6] A. Klejnberg and J. Spałek, J. Phys.: Condensed Matter **11**, 6553 (1999).

[7] The role of the Hund's rule was raised qualitatively in G. Baskaran, Physica **B223-224**, 498 (1996); the role of Coulomb interaction was also considered in: D. Agterberg, T.M. Rice, and M. Sigrist, Phys. Rev. Lett. **78**, 3374 (1997).

[8] A.J. Leggett, Rev. Mod. Phys. **47**, 331 (1975); M. Sigrist and K. Ueda, Rev. Mod. Phys. **63**, 239 (1991).

[9] J. Spałek , submitted to Phys. Rev. B.

[10] A. Klejnberg and J. Spałek, Phys. Rev. **B61**, No.23 (2000).

VI

LOW DIMENSIONAL
SYSTEMS AND
TRANSPORT

DIMENSIONAL CROSSOVER, ELECTRONIC CONFINEMENT AND CHARGE LOCALIZATION IN ORGANIC METALS

G. Mihály[1,2], F. Zámborszky[1], I. Kézsmárki[1,2], and L. Forró[2]

[1]*Department of Physics, Institute of Physics, Budapest University*
of Technology and Economics, H-1111 Budapest, Hungary
[2]*IGA, Ecole Politechnique Federale de Lausanne, CH-1015 Lausanne,*
Switzerland

Abstract   The unusual "normal phase" behavior of the Bechgaard salts are discussed in view of some recent results on intra- and interchain transport. Application of hydrostatic pressure allowed to investigate the process of electron delocalization along the chains, then the progressive development of coherent 2 dimensional transport. Reconsideration of the p-T phase diagram is suggested, conflicting optical and Hall results are discussed. The experimental results reveal the importance of the strong electron-electron correlations, but do not support the Luttinger liquid picture.

1.   INTRODUCTION

The π-electrons of the Bechgaard salts are generally described by a highly anisotropic band structure. In the $(TMTSF)_2X$ family the transfer integrals are in the order of $t_a : t_b : t_c \approx 3000 : 300 : 10$ K while in their sulfur analog, in the somewhat less conducting $(TMTTF)_2 X$ compounds, these ratios are typically $t_a : t_b : t_c \approx 1000 : 100 : 3$. The overlap of the molecular orbits can be changed by anion substitution ($X = PF_6$, Br, ClO_4, etc.) and can also be tuned continuously by application of hydrostatic pressure [1, 2].

The strong electron-electron interaction, characteristic to these salts, leads not only to various interesting ground states but also to exotic electrical properties at higher temperatures. The quasi-one-dimensional nature of the electron system suggests that the unusual "normal-phase" features may be related to Luttinger-liquid behavior, where the quasi-particle excitations of a Fermi-liquid are replaced by collective states.

J. Bonča et al. (eds.), Open Problems in Strongly Correlated Electron Systems, 263–271.

The Luttinger liquid description was first proposed for the TMTTF compounds and later it was extended to some regions of the normal phase in the TMTSF salts [2, 3, 4]. On the other hand the Fermi liquid picture seems adequate in the metallic Se compounds, at least at low temperatures. The best example for this is that the metal-SDW transition in $(TMTSF)_2PF_6$ is well understood in terms of the Fermi surface instability of a 2-dimensional electron system [5].

In this paper the normal phase properties of $(TMTTF)_2Br$ and $(TMTSF)_2PF_6$ are characterized in terms of the results of recent experiments on the anisotropic electron transport [6, 7, 8]. In the Br compound, where the transfer integrals are small, the charge carriers are localized by Coulomb interactions. In contrast, $(TMTSF)_2PF_6$ is metallic with extended electronic states. We discuss the crossover from the diffuse to the coherent transport on the basis of the pressure dependence of the resistivity in $(TMTTF)_2Br$, where metallic behavior - similar to that of $(TMTSF)_2PF_6$ - is achieved under high pressures. In the metallic range the strong electron-electron correlations are still present and are well reflected in the unusual optical response of $(TMTSF)_2PF_6$ [4, 9, 10]. The analysis of these optical data is compared to the results of Hall measurements.

2.   PRESSURE INDUCED DELOCALIZATION

In first approximation both $(TMTTF)_2Br$ and $(TMTSF)_2PF_6$ have molecular chains forming a one-dimensional, quarter-filled hole band. The slightly alternating molecular distance may split this band into two non-overlapping bands, however, for noninteracting carriers a metallic character is still expected. The insulating behavior observed in the TMTTF salts is generally attributed to strong Coulomb interactions. It was suggested that the gap in the excitation spectrum may arise either from the one-site repulsion (if the band splitting due to dimerization is relevant) or from nearest neighbor interactions [4].

The upper panel in Fig. 1 shows the temperature dependence of the chain direction (a-direction) resistivity in $(TMTTF)_2Br$ at ambient pressure. At low temperatures the resistivity is activated, with an activation energy of about $\Delta_\rho = 120$ K. The resistivities measured along the other crystallographic directions (b' and c*) show very similar variation in the whole temperature range investigated (Fig. 1 lower panel). This implies a hopping process where the temperature dependence is solely determined by a molecular on-site interaction, while the magnitude of the conductivity measured along the various directions differs only due to the differences in the inter-molecular overlaps. Above $T_{min} \approx 100$ K the thermal energy exceeds the localization energy, $k_BT > \Delta_\rho$, and the conduction gradually changes from thermal activation to Einstein-diffusion. Note that this crossover happens simultaneously for all the three crystallographic directions. Even though this diffuse conduction leads

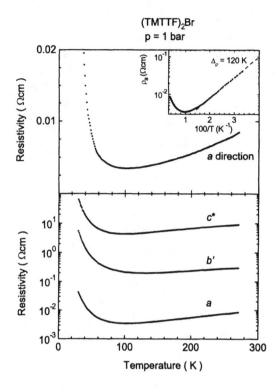

Figure 1 Temperature dependence of the resistivity in $(TMTTF)_2Br$. The inset show the Arrhenius plot of the chain direction resistivity. a, b' and $c*$ denotes the various crystallographic directions.

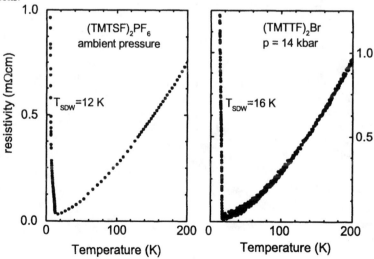

Figure 2 Temperature dependence of the resistivity in $(TMTSF)_2PF_6$ at ambient pressure and in $(TMTTF)_2Br$ at $p = 14$ kbar.

to a "metallic" slope, $d\rho/dT > 0$, it is clear that no phase coherent transport occurs above T_{min} in any direction.

Under hydrostatic pressures of about $p > 10$ kbar (TMTTF)$_2$Br becomes a real metal exhibiting an incommensurate SDW ground state arising from Fermi surface instability [11]. Figure 2 compares the temperature dependence of the resistivity measured at ambient pressure in (TMTSF)$_2$PF$_6$ and at $p = 14$ kbar in (TMTTF)$_2$Br. The correspondence is obvious: not only the qualitative features are similar both, but even the numerical values of quantities like ρ_{RT} or T_{SDW} are very close to each other.

(TMTSF)$_2$PF$_6$ is an anisotropic metal with 2-dimensional electron gas, at least at temperatures as low as about 20 K . At this temperature the mean free path in the ab' plane is estimated to be larger than the lattice constant along both directions, $l_a > a$ and $l_b > b$. On the other hand, with increasing temperature l_b is expected to be reduced below the interchain distance. It was suggested that this reduction is to be considered as a 2D $\longrightarrow$ 1D dimensional crossover and the confinement of the carriers on the 1D molecular chains results in a Luttinger liquid at high temperatures [2, 3] . Investigation of the $a - b'$ anisotropy, however, have not shown any signature of this sort of transition in the 50 - 100 K temperature range where $l_b \approx b$. This is shown in Fig. 3 and discussed in Ref.[6].

As already mentioned, when hydrostatic pressure is applied on (TMTTF)$_2$Br at $p = 14$ kbar the chain direction conductivity achieves the value measured in

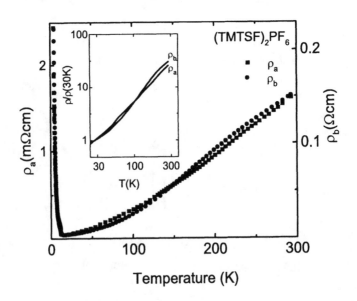

Figure 3   Temperature dependence of the resistivity in (TMTSF)$_2$PF$_6$ measured along the a and b' directions.

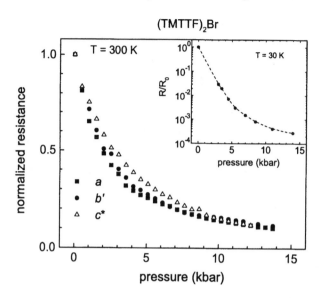

Figure 4   Pressure dependence of the resistivity in (TMTTF)$_2$Br measured along the various directions. The inset shows the pressure induced suppression of the resistivity at $T = 30$ K on logarithmic scale.

(TMTSF)$_2$PF$_6$. Figure 4 demonstrates that the resistivities measured along the different directions decrease by the same factor. The overlap of the molecular orbitals is enhanced simultaneously in each directions and at room temperature this results in an order of magnitude drop in the resistivities. The pressure induced change is smooth in spite of the fact that the thermal diffusion of the carriers (at $p = 0$ kbar) transforms to metallic conduction of extended states (at $p = 14$ kbar).

According to the temperature variations shown in Figs. 1 and 2, the lower the temperature the larger the difference between the $\rho(T)$ curves measured at $p = 0$ and 14 kbar. The inset in Fig. 4 shows that at $T = 30$ K four orders of magnitude change of the resistivity occurs without any sign of a phase transition. We emphasize that the low pressure side exhibits an activated temperature dependence of localized carriers, while at the high pressure side there is a well defined 2D Fermi surface. Detailed $\rho(p, T)$ experiments on (TMTTF)$_2$Br are presented in Ref[7].

In a Mott-Hubbard picture the pressure increases the overlap integrals, while, in first approximation, the strength of the Coulomb interaction is unchanged. In the experiments the t/U ratio has been continuously tuned from the insulator limit to a state which is already metallic. The question if such a transition should be a continuous crossover or a sharp phase transition is still debated theoretically [12, 13]. Our results suggest, that in the above organic conductors the electrons can be continuously transformed from states localized on molecules to band states exhibiting well established Fermi surface. This transition is smooth in any part of the p-T phase diagram, even in that region where the coherent transport extends from the 1D chains to the 2D ab' planes.

The pressure and temperature dependence of the anisotropic electrical conduction do not require Luttinger liquid phase boundaries (see also Refs. [6,7]). Experimental results on magnetic susceptibility [14] and NMR relaxation [15] reveal the importance of the strong electron-electron correlations, but in our view there is no firm evidence for the Luttinger liquid behavior [8].

3.   ELECTRON CORRELATIONS

The metallic state stabilized in $(TMTTF)_2Br$ by pressure or developing in $(TMTSF)_2PF_6$ and in $(TMTSF)_2ClO_4$ at ambient pressure still carries the signature of strong electron correlations. This is well established by recent low-temperature optical data [4, 9, 10]. The plasma edges observed in the optical reflectivity for electric field polarized both along the a and b' directions confirm the 2d nature of the electron gas, however the optical conductivity deviates from a simple metallic response: there is a gapped high frequency mode and a narrow low frequency peak (Fig. 5). The oscillator strength, $D = \int \sigma(\omega)d\,\omega$, associated with this narrow mode implies that only about 1 % of the carriers participate in the d.c. transport. The width of this mode corresponds to unusually long relaxation time; $\tau \approx 2*10^{-11}$s [$(TMTSF)_2PF_6$

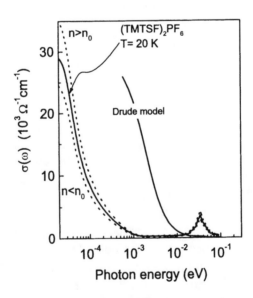

Figure 5 The frequency dependence of the conductivity in $(TMTSF)_2PF_6$ (schematic representation of the data of Ref[4]). The low frequency peak carries only 1% of the oscillator strength.

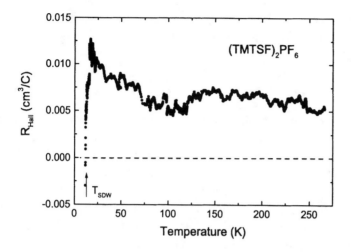

Figure 6 The temperature dependence of the Hall resistance in the normal phase of (TMTSF)$_2$PF$_6$.

at $T = 20$ K] and $\tau \approx 10^{-10}$ s [(TMTSF)$_2$ClO$_4$ at $T = 10$ K]. Accordingly, this means anomalously long mean free paths along the chains, $l_{PF_6} > 10$ μ and $l_{ClO_4} > 50$ μ, and may suggest that at low temperature the ballistic regime is reached.

The concentration of the carriers participating in the d.c. transport can also be tested by measuring the Hall-effect. Figure 6 shows the result obtained for (TMTSF)$_2$PF$_6$ [6] in a $B \parallel a$ configuration with current applied along the c^*-direction and Hall-voltage measured along b'. This is the only arrangement which is free from systematic error due to inhomogeneous current injection (inhomogeneity may mix to the Hall voltage the strongly temperature dependent longitudinal voltage). In the metallic phase we found a Hall resistance which is temperature independent within the experimental accuracy ($\pm 30\%$). Note that independent measurement performed in a $B \parallel c^*$, $I \parallel a$ and $U_H \parallel b'$ configuration [16] gave essentially the same result (within $\pm 30\%$). In both cases the magnitude of the Hall constant is close to that expected in a naive Drude picture for one hole/unit cell [$R_H = -1/ne$, and $n = 2.5 \times 10^{21}$ cm^{-3} corresponds to $R_H = 4 \times 10^{-3}$ cm^3/C].

In a simple Drude picture

$$\sigma(\omega) = \frac{ne^2}{m} \frac{\tau}{1 + (\omega\tau)^2}, \quad D = \int \sigma(\omega)d\omega = \frac{\pi ne^2}{2m} \quad (1)$$

and the reduction of the oscillator strength found in the optical response should lead to a factor of 100 enhancement in R_H. This has not been observed. On the other hand the structure of the optical conductivity clearly deviates from the prediction of the Drude model, and call for novel interpretation of the Hall data. Recently Zotos et al. [17] introduced a general formalism to describe the Hall constant for arbitrary many particle interactions. For the $T \longrightarrow 0$ limit they derived

$$R_H = -\frac{1}{D}\frac{\partial D(n)}{\partial n} \tag{2}$$

where $D(n)$ is concentration dependence of the oscillator strength. This implies that not only the Drude expression, but any shape of $\sigma(\omega)$ which results in $\int \sigma(\omega)d\omega = An$ is consistent with a Hall constant of $R_H = -1/ne$. This does not depend on the actual value of the prefactor A, and allows the small oscillator strength observed in the experiments for the low frequency mode. In Fig. 5 dashed lines indicate the possible concentration dependence of $\sigma(\omega)$ which restores the consistency of the optical and the Hall results. The experimental verification of this possibility remains to be done.

4.   SUMMARY

Pressure studies on (TMTTF)$_2$Br show continuous transformation of the localized carriers to band states. In the anisotropic structure the different transfer integrals increase simultaneously and the electrons extend from the molecular orbitals to conducting 1D chains, or (at low temperatures) to 2D Fermi surface. The resulting metal is still strongly correlated, as reflected in the optical response. The Hall effect in this correlated system can be analyzed in terms of the recent model of Zotos et al.

Acknowledgments

The experiments presented in this study were performed on single crystals prepared by B. Alavi and L.K. Montgomery. This work was supported by Hungarian Research Funds OTKA T015552 and FKFP 0355-B10.

References

[1] for a recent review see: C. Bourbonnais and D. Jerome, in "Advances in Synthetic Metals, Twenty years of Progress in Science and Technology", edited by P. Bernier, S. Lefrant, and G. Bidan (Elsevier, New York, 1999), pp. 206-301.

[2] C. Bourbonnais and D. Jerome, Science **281**, 1155 (1998).

[3] J. Moser et al, Eur.Phys. J. B **1**, 39 (1998).

[4] V. Vescoli et al., Science **281**, 1181 (1998).

[5] K. Yamaji, J. Phys. Soc. Jpn. **51**, 2787 (1982); G. Montambaux, Phys. Rev. B **38**, 4788 (1988); G. Mihály et al. Phys. Rev B **55**, 13465 (1997).

[6] G. Mihály et al, Phys. Rev. Lett. **84**, 2670 (2000).

[7] F. Zámborszky et al, to be published.

[8] G. Mihály et al, to be published.

[9] M. Dressel et al., Phys. Rev. Lett. **77**, 398 (1996).

[10] A. Schwartz et al., Phys. Rev. B **58**, 1261 (1998).

[11] B.J. Klemme et al., Phys. Rev. Lett. **75**, 2408 (1995).

[12] R.M. Noak and F. Gebhard, Phys. Rev. Lett. **82**, 1915 (1999).

[13] Ph. Nozieres, Eur. Phys. J. B **6**, 447 (1998), M.J. Rosenberg et al., Phys. Rev. Lett. **83**, 3498 (1999).

[14] M. Dumm et al., Phys. Rev. **61**, 511 (2000).

[15] P. Wzietek et al., J. Phys. I France **3**, 171 (1993), C. Bourbonnais et al., Phys. Rev. Lett. **62**, 1532 (1989).

[16] J. Moser et al, Phys. Rev. Lett. **84**, 2674(2000).

[17] X. Zotos et a., cond-mat/0002449 (2000).

DRUDE WEIGHT, INTEGRABLE SYSTEMS AND THE REACTIVE HALL CONSTANT

X. Zotos, and F. Naef

Institut Romand de Recherche Numérique en Physique des Matériaux (IRRMA), PPH-Ecublens, CH-1015 Lausanne, Switzerland

M. Long

Department of Physics, University of Birmingham, Edgbaston, Birmingham B15 2TT, United Kingdom

P. Prelovšek

Faculty of Mathematics and Physics, University of Ljubljana, 1000 Ljubljana, Slovenia
J. Stefan Institute, 1001 Ljubljana, Slovenia

Abstract   The Drude weight D, characterizing the reactive part of the conductivity, can be used as a criterion of a metallic or insulating ground state. Here, we will discuss how D remains finite at all temperatures, implying ideal conductivity (ballistic transport) in integrable quantum many body systems commonly used in the description of quasi-one dimensional materials. We will relate this singular behavior to the existence of conservation laws in integrable systems and discuss, in particular, the energy and spin dynamics of the spin 1/2 Heisenberg model.

In a different context, we will show that in a certain limit, the zero temperature reactive Hall constant R_H, is related to the density dependence of the Drude weight. This novel formulation implies a simple picture for the change of sign of charge carriers in the vicinity of a Mott-Hubbard transition.

1.   THE DRUDE WEIGHT

It is by now well known that in strongly correlated systems the zero temperature (T=0) reactive part of the conductivity can be used as a criterion of a metallic or insulating ground state. In particular, following the work of W. Kohn [1], the imaginary part of the conductivity, $\sigma''(\omega \to 0) = 2D/\omega$, characterized by D (now called the "Drude weight" or charge stifness), can be related to the ground state energy density E_0 dependence on an applied fictitious flux

J. Bonča et al. (eds.), Open Problems in Strongly Correlated Electron Systems, 273–282.

ϕ as:

$$D = \frac{1}{2L}\frac{\partial^2 E_0}{\partial\phi^2}\big|_{\phi\to 0}. \tag{1}$$

D is positive for a metallic, ideal conducting state (freely accelerating system), while it is zero in an insulating state (or a normal metal). D also appears in the real part of the conductivity as a δ−function at zero frequency:

$$\sigma'(\omega) = 2\pi D\delta(\omega) + \sigma'_{reg}(\omega > 0). \tag{2}$$

For the simple example of spinless fermions on a one dimensional lattice of L sites, hopping with amplitude $-t$, the eigenvalues in the presence of a flux ϕ become, $\epsilon_k = -2t\cos(k+\phi)$ giving:

$$\begin{aligned}
E_0 &= \sum_{|k|<k_F}(-2t)\cos(k+\phi), \quad k = 2\pi \text{ integer}/L \\
D &= \frac{1}{2L}\frac{\partial^2 E_0}{\partial\phi^2}\big|_{\phi\to 0} = \frac{1}{2L}[\langle 0| - T|0\rangle = \frac{t}{\pi}\sin k_F \tag{3}
\end{aligned}$$

For later reference notice that $D \simeq tn$ for $n \to 0$ while $D \simeq t(1-n)$ for $n \to 1$ where n is the fermion density ($n = k_F/\pi$). A schematic representation of D for a metallic state and with interaction leading to a Mott insulating state at half-filling is given in Fig. (1).

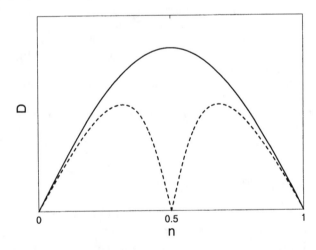

Figure 1   Schematic behavior of the Drude weight as a function of density for a free (full line) and interacting (dotted line) spinless fermion model, with an insulating state at half-filling.

2. INTEGRABLE SYSTEMS

2.1 PROPOSITION

It has recently being proposed that integrable quantum many body systems show dissipationless transport at finite temperatures. This idea was motivated by analytical and numerical studies on a toy model [2] and the fermionic version of the Heisenberg model [3]. It is analogous to the well known effect of transport by solitons in classical nonlinear integrable systems. The key to this idea comes from a study of the Drude weight at finite temperatures. In the "common sense" scenario, the conductivity of a uniform metallic system (no impurity scattering) but with no conserved current, is characterized by a finite Drude weight at zero temperature. The system is an ideal conductor (freely accelerating) and the transport behavior is characterized as ballistic. Upon raising the temperature, the δ−function Drude peak broadens to a function $\sigma'(\omega)$, a Lorentzian within the Drude model, of width $1/\tau$ where τ is a characteristic scattering time. The weight D of the δ−function peak at zero frequency goes to zero exponentially fast with system size while the d.c. conductivity is given by $\sigma_{dc} = \sigma'_{reg}(\omega \to 0)$.

In contrast to the above scenario, in integrable systems, the broadening of the δ−function peak at finite temperatures does not take place, so the system remains an ideal conductor. Only the value of D changes with temperature, going to zero as $\beta = 1/k_B T$ at high temperatures for tight binding models.

The systems for which this singular behavior is observed are often used in the description of quasi-one dimensional electronic or magnetic materials, as the spin 1/2 Heisenberg spin chain, the Hubbard model or the supersymmetric t-J model. They are exactly solvable by the Bethe ansatz (BA) method. Now we will establish and argue on the origin of this behavior by relating the Drude weight to the long time decay of current-current correlation functions.

2.2 *D* AND THE MAZUR INEQUALITY

First, it is easy to generalize relation (1) to finite temperatures,

$$D = \frac{1}{2L} \sum_n w_n \frac{\partial^2 E_n}{\partial \phi^2}\big|_{\phi \to 0}, \quad w_n = \frac{e^{-\beta E_n}}{Z}, \tag{4}$$

as a thermal average of curvatures of many body levels to a flux ϕ. Now, considering the flux dependence of the free energy $F = -k_B T \ln Z$, we obtain:

$$\frac{\partial F}{\partial \phi}\big|_{\phi \to 0} = \langle j \rangle$$

$$\frac{\partial^2 F}{\partial \phi^2}\big|_{\phi \to 0} = 2LD - \beta \sum_n w_n \left(\frac{\partial E_n}{\partial \phi}\right)^2. \tag{5}$$

Noting that the left hand side of the second equation, being the susceptibility for persistent currents, goes to zero in the thermodynamic limit at any non-zero temperature and using that $\partial E_n/\partial\phi|_{\phi\to 0} = \langle j \rangle$, we find:

$$D = \frac{\beta}{2L}\sum_n w_n\langle j \rangle^2 = \frac{\beta}{2L}\langle j(t)j \rangle_{t\to\infty} = \frac{\beta}{2L}C_{jj}. \qquad (6)$$

Thus D is related to the long time asymptotic value of the current-current correlations. Now we can use an inequality proposed by Mazur [4] to obtain a bound for C_{jj} by linking it to the local conservation laws Q_n ($[H,Q_n]=0$) characterizing integrable systems [5]:

$$\lim_{T\to\infty}\frac{1}{T}\int_0^T <A(t)A> dt \geq \sum_n \frac{<AQ_n>^2}{<Q_n^2>} \qquad (7)$$

As the left hand side is basically $\langle A(t)A \rangle_{t\to\infty}$, we can establish a bound for D,

$$D = \frac{\beta}{2L}C_{jj} \geq \frac{\beta}{2L}\sum_n \frac{<jQ_n>^2}{<Q_n^2>}. \qquad (8)$$

This simple argument allows us to understand why integrable systems might show singular transport properties. Of course, if the current in question does not have an "overlap" with any of the conservation laws we are considering then we cannot obtain a useful bound. Furthermore, this inequality implies that a system with even one conservation law coupled to the current would show singular transport without the system being necessarily integrable. We should mention that defining the quantum analogue of a classical "integrable system" is a current issue in the field of quantum chaos. Here, we have in mind BA systems as the ones we mentioned above when refering to integrable models. In this respect, we should note that in such systems the existence of one conservation law in intimately linked to the existence of a macroscopic number of them.

Now we will discuss an application of these ideas to the Heisenberg model, of particular interest considering the existence of excellent quasi-one dimensional magnetic materials candidates for the observation of the ballistic transport we are discussing.

2.3   THE SPIN 1/2 HEISENBERG MODEL

It is described by the Hamiltonian:

$$H = \sum_{l=1}^{L}(J_x S_l^x S_{l+1}^x + J_y S_l^y S_{l+1}^y + J_z S_l^z S_{l+1}^z) \qquad (9)$$

(by a Jordan-Wigner transformation the Heisenberg model is equivalent to a model ("$t-V$ model") of spinless fermions interacting with a nearest-neighbor interaction).

This model is characterized by a macroscopic number of conservation laws. It is worth noticing that the first nontrivial conservation law Q_3 is related to a physical quantity, it coincides with the energy current j^E:

$$Q_3 = j^E = \sum_{l=1}^{L} J_x J_y (S_{l-1}^x S_l^z S_{l+1}^y - S_{l-1}^y S_l^z S_{l+1}^x) + (x, y, z). \tag{10}$$

This is particularly interesting and of actual experimental relevance as it implies ideal thermal conductivity (energy currents do not decay at all, $< j^E(t) j^E > =$ const.) [6].

Regarding the spin current correlations, for $J_x = J_y = 1, J_z = \Delta$, we find that:

$$D = \frac{\beta}{2L} C_{j^z j^z} \geq \frac{\beta}{2L} \frac{< j^z Q_3 >^2}{< Q_3^2 >}, \tag{11}$$

where j^z =spin current= $\sum_{l=1}^{L}(S_l^y S_{l+1}^x - S_l^x S_{l+1}^y)$, implying ideal spin conductivity and the absence of spin diffusion. We can analytically calculate the right hand size of this inequality in the $\beta \to 0$ limit,

$$D \geq \frac{\beta}{2} \frac{8\Delta^2 m^2 (1/4 - m^2)}{1 + 8\Delta^2 (1/4 + m^2)}, \qquad m = < S_l^z > . \tag{12}$$

In this result we remark that we obtain a positive bound only for a finite magnetization m e.g. when the system is in a magnetic field. For $m = 0$ this inequality does not provide a useful bound. However, recent developments in the BA technique allow the analytical calculation of $D(T)$. In particular, the Hubbard model [7] and the nonlinear σ - model [8] have been studied. Regarding the Heisenberg spin 1/2 chain model [9], the exact behavior of $D(T)$ is shown in Fig. (2), indicating that even at zero magnetization, $D(T)$ is positive in the easy axis ($\Delta < 1$) case. The $T = 0$ value D_0 is recovered with a characteristic power law

$$D(T) = D_0 - \text{const.} T^\alpha, \quad \alpha = 2/(\nu - 1), \quad \Delta = \cos(\pi/\nu). \tag{13}$$

2.4   PERSPECTIVES

The main issue raised by these ideas is the robustness of ideal conducting behavior to perturbations breaking the integrability of a system. This issue is

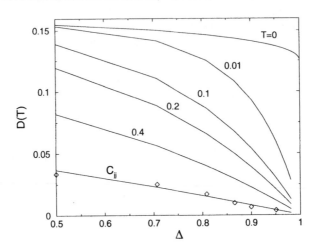

Figure 2 $D(\Delta)$ at various temperatures. The continuous line is the high temperature proportionality constant $C_{jj} = D/\beta$. The symbols indicate exact diagonalization results.

also the most relevant regarding experiments. Studies on classical one dimensional systems indicate that integrable models show nondissipative behavior with surprising robustness to perturbations, leading for instance in the use of solitons in the propagation of signals in optical fibers. On the other hand, some nonintegrable systems show low frequency anomalies, leading to diverging transport coefficients (e.g. thermal conductivity) while others show normal transport. Considering that studies in chaos (e.g. the kicked rotor model) indicate that quantum systems show more "ballistic" behavior than their classical counterparts, we can argue that quantum many body integrable systems might also be relatively immune to perturbations [10].

These issues are expected to be the subject of future theoretical studies: (i) the full spectral dependence of the conductivity should be evaluated analytically for integrable systems (note a recent studies emphasizing the role of conservation laws in a Luttinger liquid approach to one dimensional transport [11]), (ii) numerical simulation studies playing an important role in analyzing the dynamic properties of correlated systems and their robustness to perturbations, (iii) studies in quantum chaos extended to the dynamics in quantum many body systems [10].

On the experimental side, quasi-one dimensional magnetic materials, as the Sr_2CuO_3 compound, are probably at the moment the best candidates for exploring singular transport behavior, via NMR [12] or thermal conductivity experiments [13]. The "zero mode" in the optical absorption of Bechgaard salts, indicative of nearly ballistic charge transport should also be noted, although it is difficult to argue that electronic systems are good representations of an integrable model as the Hubbard model.

3. REACTIVE HALL RESPONSE

3.1 MOTIVATION

The doping of a Mott-Hubbard insulator (due to correlations), poses the problem of the charge carriers sign as probed, for instance, in a Hall experiment. Within a semiclassical approach, for a metallic system described by a single band picture, the Hall constant $R_H \simeq -1/n$, is electron-like at low densities changing to $R_H \simeq +1/\delta$, hole-like (positive) near full-band ($\delta = 2 - n$, n=density). In contrast, for a Mott-Hubbard insulator, it is expected that the Hall constant $R_H \simeq -1/n$ is electron-like at low densities, changing to $R_H \simeq +1/\delta$, hole-like (positive) near half-filling ($\delta = 1 - n$, n=density), the turning point depending on the interaction.

The standard approach for the study of the Hall constant is within the Kubo formalism,

$$R_H(\omega \to 0) = -\frac{1}{B} \frac{\sigma_{yx}(\omega)}{\sigma_{xx}^0(\omega)\sigma_{yy}^0(\omega)}. \tag{14}$$

To evaluate this expression, the low frequency dissipative part of the conductivities should be evaluated at zero (for the diagonal) and nonzero (for the off-diagonal) magnetic field. This is not an easy task for strongly correlated systems by either numerical or analytical methods.

It would be interesting therefore to have a simple description, analogous to the Kohn criterion, of the charge carriers sign as a function of doping. The approach we will discuss is to consider the Hall constant at zero temperature, thus analyze the reactive parts of the conductivities [14]. Such a formulation, appropriate for numerical simulations on ladder systems was presented in [15]. In the following, we will show how this Hall constant can be simply related to the density dependence of the Drude weight [16].

3.2 D AND R_H

For simplicity we consider a two dimensional system with a magnetic field along the $z-$axis and spatial modulation along the $y-$axis of wavevector q, $B^z = B \cos(qy)$. Further, a uniform electric field $E^x(t)$ of frequency ω is applied along the $x-$axis and one of wavevector q, $E^y(t)$ along the $y-$axis. The generated currents of the same frequency are $J^x(t)$ and $J_q^y(t)$ respectively. From linear response theory we find:

$$\begin{aligned} \langle J^x \rangle &= \sigma_{j^x j^x} E^x(t) + \sigma_{j^x j_q^y} E^y(t) \\ \langle J_q^y \rangle &= \sigma_{j_q^y j^x} E^x(t) + \sigma_{j_q^y j_q^y} E^y(t) . \end{aligned} \tag{15}$$

Now, in contrast to the usual derivation of the Hall constant expression, we will keep the $q-$dependence explicit by converting the current-current to current-density correlations using the continuity equation:

$$\langle J^x \rangle = \sigma_{j^x j^x} E^x(t) + \frac{1}{q} \chi_{j^x n_q} E^y(t)$$

$$\langle J_q^y \rangle = -\frac{1}{q} \chi_{n_q j^x} E^x(t) + (\frac{\omega}{q})^2 \chi_{n_q n_q} \frac{i}{\omega} E^y(t). \qquad (16)$$

χ's are susceptibilities and n_q is a density of modulation q.

At T=0, the response is non-dissipative so we will study the reactive (out-of phase) induced currents. Furthermore, at this point we will consider the "screening" (or slow) response in the y−direction, by taking the (q, ω) limits in the order $\omega \to 0$ first and $q \to 0$ last. In contrast, along the x−direction we take the response in the usual "transport" (or fast) regime, with the limits in the opposite order. In this case the conductivities (susceptibilities) can be written as derivatives of the ground state energy ϵ^0 with respect to a fictitious flux ϕ^x along the x−direction and a potential μ_q of modulation q along the y−direction.

$$\langle J^x \rangle_0 = \frac{\epsilon^0_{\phi^x \phi^x}}{\omega} (iE^x(t)) + (\frac{-1}{q}) \epsilon^0_{\phi^x \mu_q} E^y(t)$$

$$\langle J_q^y \rangle_0 = \frac{1}{q} \epsilon^0_{\mu_q \phi^x} E^x(t) - \frac{\omega}{q^2} \epsilon^0_{\mu_q \mu_q} (iE^y(t)). \qquad (17)$$

Finally, setting $\langle J_q^y \rangle_0 = 0$ we determine the "reactive" Hall constant:

$$R_H \equiv -\frac{1}{B} \frac{E^y}{\langle J^x \rangle_0} = \frac{q}{B} \frac{\epsilon^0_{\mu_q \phi^x}}{\epsilon^0_{\phi^x \phi^x} \epsilon^0_{\mu_q \mu_q} + \epsilon^0_{\mu_q \phi^x} \epsilon^0_{\phi^x \mu_q}}. \qquad (18)$$

Considering that $D = \frac{1}{2} \epsilon^0_{\phi^x \phi^x}$ is the Drude weight, $\kappa_q = \epsilon^0_{\mu_q \mu_q} = \partial n_q / \partial \mu_q$ is the compressibility corresponding to the density modulation n_q and taking the $q \to 0$ limit, we obtain a particularly simple expression for R_H:

$$R_H = -\frac{1}{D} \frac{\partial D}{\partial n}. \qquad (19)$$

This expression is appealing as it gives a direct, intuitive understanding for the change of sign of charge carriers in the vicinity of a metal-insulator transition. First, at low densities, $D \propto n$ giving $R_H \simeq -1/n$; close to a Mott insulator $D \propto \delta = 1 - n$, implying $R_H \simeq +1/\delta$. Furthermore, we obtain a change of sign in the vicinity of a Mott transition at a density which depends on the interaction strength and is given by the position of the maximum of D. Second, for independent electrons, where D is proportional to the kinetic energy, by taking the limit $t' \to 0$ and calculating D as a sum of D's for individual $x-$chains, we obtain from (19):

$$D = \frac{2t}{\pi}\sin(\frac{\pi n}{2}), \quad R_H = -\frac{\pi}{2}\frac{1}{\tan(\frac{\pi n}{2})}, \tag{20}$$

an expression used in analysis of the Hall constant of quasi-one dimensional compounds [17].

A further simple application of (19) is the case of a quasi-one dimensional correlated system where by assuming nearly decoupled chains we can calculate $D(n)$ for each chain analytically using the BA method [18, 19], D given by the sum over all chains. For chains described by the Hubbard model, the evaluated D and resulting typical R_H is shown in Fig. (3).

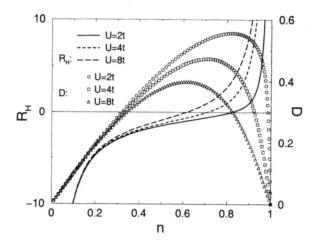

Figure 3 R_H for the Hubbard model from expression (19) for $t' \to 0$.

3.3   DISCUSSION

Several issues are raised by the above formulation; first, its relation to the widely used semiclassical formulation of the Hall constant. Preliminary calculations indicate that for independent electrons described by an arbitrary anisotropic single band model, expression (19) is qualitatively (and even quantitatively) in agreement with the semiclassical results.

Second, the relation to the finite temperature (dissipative) Hall constant evaluated from the Kubo formula is unclear. So far it has not been possible to bring (14) in the limit $T \to 0, \omega \to 0$ to a simple transparent form as the formulation (19).

Regarding applications, the formulation (19) can be used for the calculation of R_H for other correlated systems, although the evaluation of D for two or three dimensional systems is not an easy matter. It can also be applied to systems described within a single (or multi) band picture. Studies of actual interest are the evaluation and comparison with experiment of R_H for quasi-one dimensional compounds and the one-dimensional transport proposed for the striped phases in the high T_c superconductors.

Acknowledgments

Part of this work was done during visits of (P.P.) and (M.L.) at IRRMA as academic guests of EPFL. X.Z. and F.N. acknowledge support by the Swiss National Foundation grant No. 20-49486.96, the EPFL, the Univ. of Fribourg and the Univ. of Neuchâtel.

References

[1] W. Kohn, Phys. Rev. **133**, A171 (1964).

[2] H. Castella, X. Zotos, P. Prelovšek, Phys. Rev. Lett. **74**, 972 (1995).

[3] X. Zotos and P. Prelovšek, Phys. Rev. B**53**, 983 (1996).

[4] P. Mazur, Physica **43**, 533 (1969).

[5] X. Zotos, F. Naef and P. Prelovšek, Phys. Rev. B**55** 11029 (1997).

[6] F. Naef and X. Zotos, J. Phys. C. **10**, L183 (1998); F. Naef, Ph. D. thesis no.2127, EPF-Lausanne (2000).

[7] S. Fujimoto and N. Kawakami, J. Phys. A. **31**, 465 (1998).

[8] S. Fujimoto, J. Phys. Soc. Jpn., **68**, 2810 (1999).

[9] X. Zotos, Phys. Rev. Lett. **82**, 1764 (1999).

[10] T. Prosen, Phys. Rev. Lett. **80**, 1808 (1998).

[11] A. Rosch and N. Andrei, cond-mat/0002306.

[12] M. Takigawa, N. Motoyama, H. Eisaki and S. Uchida, Phys. Rev. Lett. **76**, 4612 (1996).

[13] A.V. Sologubenko, E. Felder, K. Giannò, H.R. Ott, A. Vietkine and A. Revcolenschi, preprint (2000).

[14] P. Prelovšek, Phys. Rev. B **55**, 9219 (1997).

[15] P. Prelovšek, M. Long, T. Markež and X. Zotos, Phys. Rev. Lett. **83**, 2785 (1999).

[16] X. Zotos, F. Naef, M. Long and P. Prelovšek, Phys. Rev. Lett. (2000).

[17] J.R. Cooper *et al.*, J. Phys. (Paris) **38**, 1097 (1977); K. Maki and A. Virosztek, Phys. Rev. B**41**, 557 (1990).

[18] F.D.M. Haldane, Phys. Lett. **81A**, 153 (1981).

[19] N. Kawakami and S-K. Yang, Phys. Rev. B**44**, 7844 (1991).

INHOMOGENEOUS LUTTINGER LIQUIDS: POWER-LAWS AND ENERGY SCALES

V. Meden[1,2], W. Metzner[2], U. Schollwöck[3], and K. Schönhammer[1]

[1] *Inst. f. Theoretische Physik, Universität Göttingen, Bunsenstr. 9, D-37073 Göttingen, Germany*

[2] *Inst. f. Theoretische Physik C, RWTH Aachen, D-52056 Aachen, Germany*

[3] *Sektion Physik, Universität München, Theresienstr. 37, D-80333 München, Germany*

Abstract   We present a study of the one-particle spectral properties for a variety of models of Luttinger liquids with open boundaries. First we show that the Hamiltonian for an interaction which is long range in real space can be written as a quadratic from in bosons (bosonization) and calculate the spectral weight. For weak interactions the boundary exponent of the power-law suppression of the weight close to the chemical potential is dominated by a term linear in the interaction. This motivates us to investigate the spectral properties within the Hartree-Fock approximation. It gives power-law behavior and qualitative agreement with the exact spectral function. For the lattice model of spinless fermions and the Hubbard model we present numerically exact results obtained by using the density-matrix renormalization-group algorithm. Again many aspects of the behavior of the spectral function close to the boundary can be understood within the Hartree-Fock approximation. For the Hubbard model with weak interaction U the spectral weight is enhanced in a large energy range around the chemical potential. Following a crossover at exponentially (in $1/U$) small energies a power-law suppression, as predicted by bosonization, sets in. This shows that for small U bosonization only holds on exponentially small energy scales.

1.   LUTTINGER LIQUIDS: BULK PROPERTIES

Theoretically it is well established that fermions with repulsive interaction which are restricted to move in one spatial dimension do not obey Fermi liquid theory [1]. A detailed picture of the bulk physical properties of such systems emerges if one takes into account the results obtained over the last 50 years analyzing different models and using a variety of analytical and numerical techniques [1]. Already 1950 Tomonaga showed that the low-energy excitations of a model of 1D electrons which only includes electron-electron scattering processes with small momentum transfer, i.e. of a model with a two-particle interaction which is long range in real space, can be described by collective, bosonic degrees of freedom [2], a technique which later became

J. Bonča et al. (eds.), Open Problems in Strongly Correlated Electron Systems, 283–292.

known as bosonization. A mathematically more elaborate discussion and a first calculation of a correlation function for a similar model was given by Luttinger [3] and Mattis and Lieb [4]. Later various correlation functions for the Tomonaga-Luttinger (TL) model have been calculated, many of them showing characteristic power-law behavior. A significant step forward towards a unifying picture covering the low-energy physics of a large class of models was the observation that electron-electron backscattering processes [5] with large momentum transfer of order $2k_F$ - neglected in the TL model - are irrelevant in a renormalization group (RG) sense as long as the bare interaction is repulsive [6]. Here k_F denotes the Fermi momentum. These and further insights led Haldane [7] to conjecture that the low-energy physics of a large class of models with repulsive interaction - the Luttinger liquid (LL) universality class - can be understood by mapping them onto the TL model.

LL theory thus gives a prescription of how to calculate thermodynamic quantities at low temperatures and the exponents of the asymptotic power-law behavior of correlation functions. For a given model with spin rotational invariant interaction one only has to determine the LL parameter K_ρ and the velocities v_ρ and v_σ of charge and spin excitations. The physics at asymptotically large space-time distances, i.e. in the low-energy sector, then follows by comparison with thermodynamic quantities and correlation functions calculated within the TL model. In this sense the low-energy physics of LL's is expected to be universal even though K_ρ, v_ρ, and v_σ in general depend on the details of the model considered, e.g. the strength and range of the interaction, the filling factor, and the one-particle dispersion [7, 1].

For the one-particle spectral function $\rho(\omega)$ entering the description of angular integrated photoemission and energies asymptotically close to the chemical potential μ, LL theory predicts power-law suppression $\rho(\omega) \sim |\omega|^\alpha$ of the bulk spectral weight with an exponent $\alpha = (K_\rho + K_\rho^{-1} - 2)/(2z)$, where $z = 1$ for spinless fermions and $z = 2$ for spin 1/2-fermions [1]. For weak interactions U, K_ρ generically goes like $K_\rho = 1 - cU/v_F + \mathcal{O}([U/v_F]^2)$ with a dimensionless positive constant c and the Fermi velocity v_F. For α this leads to $\alpha \sim (U/v_F)^2$. Thus in second order perturbation theory a sign of the nonanalytic power-law behavior shows up as a term $\sim (U/v_F)^2 \ln |\omega|$ in the one-particle self-energy Σ. A first order calculation of Σ only leads to a finite renormalization of v_F and a shift of μ and does not reveal any of the characteristic LL features.

What is not provided by LL theory is the scale Δ on which the power-law behavior for a given microscopic (lattice) model and a specific correlation function is expected to hold. This is so since the TL model does not contain any energy scale which can directly be related to a characteristic energy of the microscopic model considered. Also from direct numerical calculations of correlation functions for such models not much is known about these scales. A quantum Monte-Carlo calculation of $\rho(\omega)$ for the 1D Hubbard model indicates that Δ is small [8]. This implies that chains of many lattice sites are required to observe the suppression in numerical calculations.

One way to experimentally verify the predicted LL behavior is to probe the one-particle properties using high resolution photoemission spectroscopy [9]. Up to dipole matrix elements angular integrated spectra provide a direct measurement of $\rho(\omega)$.

2. LUTTINGER LIQUIDS WITH BOUNDARIES

In recent years the theoretical expectation that LL's with periodic boundary conditions (PBC) including impurities scale to chains with open ends [10, 11] led to several studies of interacting models with hard walls, usually called "open" (or "fixed") boundary conditions (OBC) [12, 13, 14, 15, 16, 17]. In this paper we mainly want to address three questions related to the the local spectral function $\rho(x, \omega)$ of LL's with OBC:

- Does the same kind of "universality" hold as for PBC?
- Does the Hartree-Fock (HF) approximation provide useful insight?
- What can we learn about the scale on which power-law behavior holds?

In a continuum model of length L and with OBC the noninteracting eigenstates are given by $\varphi_n(x) = \sqrt{2/L} \sin(k_n x)$ where $k_n = n\pi/L$, $n \in \mathbf{N}$. Here the quantum numbers k_n do not have the meaning of momenta and the total momentum is not conserved. To express the interacting part of the Hamiltonian in terms of the operators $a_{n,s}^{(\dagger)}$ which annihilate (create) electrons of spin species s in the one-particle states $|\varphi_n\rangle$ one has to calculate the matrix element of the two-body interaction within these states. A general interaction with potential $V(x - x')$ leads to eight different scattering vertices [17]. They depend on different combinations (differences and sums) of the four external quantum numbers and cannot simply be parameterized by the "momentum transfer" as for PBC. Only two of the vertices can be written as a quadratic form in bosonic density operators similar to the small momentum processes in case of PBC [2, 17]. The charge and spin bosons $b_{n,\nu} = \rho_{n,\nu}/\sqrt{n}$ for $\nu = \rho, \sigma$ are given by the charge and spin densities $\rho_{m,\rho} = (\rho_{m,\uparrow} + \rho_{m,\downarrow})/\sqrt{2}$ respectively $\rho_{m,\sigma} = (\rho_{m,\uparrow} - \rho_{m,\downarrow})/\sqrt{2}$, where $(s = \uparrow, \downarrow)$ $\rho_{m,s} = \sum_n a_{n,s}^{\dagger} a_{n+m,s}$. For OBC there exists no RG calculation which shows that the six processes which cannot be written quadratically in the bosons are irrelevant, as it is the case for PBC's [6]. In some of the previous publications [12, 13] it was nonetheless tacitly assumed that these processes can be neglected. In Ref. [17] we discussed the subtleties of bosonization for a continuum model with OBC. There it is shown that for a two-particle interaction which is of long range in real space and restricting the Fock space to a subspace with no holes deep in the Fermi see and no particles in one-particle states with energies much higher than the Fermi energy (low-energy subspace) the above six vertices indeed vanish. This is the idea originally considered by Tomonaga [2] for PBC and in the following this model will be called TL model with OBC. Neglecting constants and particle number contributions and linearizing the one-particle dispersion around k_F, the Hamiltonian can be written as

$$\hat{H} = \sum_{n=1}^{\infty} n \left[v_F \frac{\pi}{L} \left(b_{n,\rho}^\dagger b_{n,\rho} + b_{n,\sigma}^\dagger b_{n,\sigma} \right) + \frac{z}{4L} \tilde{V}(q_n) \left(b_{n,\rho}^\dagger + b_{n,\rho} \right)^2 \right], \quad (1)$$

where $\tilde{V}(q_n)$ is the Fourier transform of $V(x - x')$. To obtain the spinless model from Eq. (1) one has to drop the terms containing $b_{n,\sigma}^{(\dagger)}$ and set $z = 1$. The above problem of independent, selfcoupled harmonic oscillators can easily be diagonalized leading to

$$\hat{H} = \sum_{n=1}^{\infty} \left[\omega_n \alpha_{n,\rho}^\dagger \alpha_{n,\rho} + v_F \frac{\pi}{L} n b_{n,\sigma}^\dagger b_{n,\sigma} \right], \quad (2)$$

with $\omega_n = v_F k_n \sqrt{1 + z\tilde{V}(k_n)/(\pi v_F)}$ and bosonic operators $\alpha_{n,\rho}$ given by a linear combination of $b_{n,\rho}^\dagger$ and $b_{n,\rho}$. After adding one-particle states with quantum numbers k_n, $n < 1$ to the Hilbert space [18] the fermionic field operator $\psi_s(x) = \sqrt{\frac{2}{L}} \sum_{n=1}^{\infty} \sin(k_n x) a_{n,s}$ can be expressed in terms of the $\alpha_{n,\rho}^{(\dagger)}$ [7] and

$$\rho^<(x, \omega) = \int_{-\infty}^{\infty} \frac{dt}{2\pi} e^{i\omega t} \left\langle \psi_s^\dagger(x, 0) \psi_s(x, t) \right\rangle \quad (3)$$

can be calculated exactly in close analogy to the case of PBC [1]. Here $\langle \ldots \rangle$ denotes the ground state expectation value. For x close to the boundary this leads to

$$\rho^<(x, \omega) \sim |\omega|^{\alpha_B} \Theta(-\omega), \quad \alpha_B = (K_\rho^{-1} - 1)/z \quad (4)$$

with $K_\rho = \left\{ 1 + [z\tilde{V}(0)]/[\pi v_F] \right\}^{-1/2}$. For repulsive interaction ($K_\rho < 1$) the boundary exponent α_B is larger than the bulk exponent α. This fact has been used in the interpretation of experimental spectra [9].

Without explicitly demonstrating that the TL model is the effective low-energy (fixed point) model for all models of LL's with OBC the result of Eq. (4) has been assumed to hold for all LL's [12, 13]. This implies that the boundary exponent α_B can be expressed in terms of the bulk LL parameter K_ρ. The generalization was confirmed in Ref. [14] where methods of boundary conformal field theory were used to calculate α_B for Bethe ansatz solvable models and in Refs. [15] and [17] where we presented numerically exact results for the spectral function of microscopic lattice models (see also Secs. 4. and 5.).

For small $\tilde{V}(0)$ it follows

$$\alpha_B = \frac{\tilde{V}(0)}{2\pi v_F} + \mathcal{O}\left(\left[\frac{\tilde{V}(0)}{\pi v_F}\right]^2\right). \tag{5}$$

This has to be contrasted to the small $\tilde{V}(0)$ behavior of the bulk exponent α which is quadratic in the interaction. Thus signs of the nonanalytic behavior of $\rho(x, \omega)$ can already be obtained using the HF self-energy, which will be analyzed in the next section.

3.   THE HARTREE-FOCK APPROXIMATION

For the continuum model with n_F fermions considered above the (non-self-consistent) HF self-energy is given by

$$\left[\Sigma_s^{HF}\right]_{n,n'} = \delta_{n,n'} \left\{\delta\mu - \frac{1}{2L} \sum_{n_1=1}^{n_F} \left[\tilde{V}(k_n - k_{n_1}) + \tilde{V}(k_n + k_{n_1})\right]\right\}$$

$$+ \frac{1}{2L}\left\{z\tilde{V}(k_n + k_{n'}) - \tilde{V}\left(\frac{k_n - k_{n'}}{2}\right)\right\} f\left(\frac{n+n'}{2}\right)$$

$$- \frac{1}{2L}\left\{z\tilde{V}(k_n - k_{n'}) - \tilde{V}\left(\frac{k_n + k_{n'}}{2}\right)\right\} f\left(\frac{|n-n'|}{2}\right), \tag{6}$$

with

$$f(m) = \begin{cases} 1 & \text{for } m \in \{1, 2, \ldots, n_F\} \\ 0 & \text{otherwise} \end{cases} \tag{7}$$

and $\delta\mu = z\tilde{V}(0)n_F/L$. The self-energy has a nontrivial matrix structure in the quantum numbers n and n' due to the broken translational invariance. From the self-energy the retarded Green function follows by a matrix inversion $[G(\omega)]_{n,n'} = \left[\{\omega - \xi(k_n) + i0\}\mathbf{1} - \Sigma\right]_{n,n'}^{-1}$, with the unity matrix $\mathbf{1}$ and $\xi(k) = \varepsilon(k) - \mu$, where $\varepsilon(k)$ is the one-particle dispersion. The local spectral function is then given by

$$\rho(x, \omega) = -\frac{1}{\pi} \text{Im} \sum_{n,n'=1}^{\infty} \varphi_n^*(x)\varphi_{n'}(x) [G(\omega)]_{n,n'}. \tag{8}$$

To gain a first insight in the behavior of the spectral weight close to μ the matrix can be inverted perturbatively to lowest order in $\tilde{V}$

$$\frac{\rho^{HF}(x, \omega)}{\rho^0(x, \omega)} = \left\{1 + \left[\frac{\tilde{V}(0)}{2\pi v_F} - z\frac{\tilde{V}(2k_F)}{2\pi v_F}\right] \ln\left|\frac{\omega}{v_F k_F}\right| + \mathcal{O}\left(\tilde{V}^2\right)\right\}. \tag{9}$$

The logarithmic divergence in ρ^{HF} is not due to a singular frequency behavior of Σ^{HF}, but emerges in the perturbative approach to the matrix inversion. This has to be contrasted to PBC where the first indication of a break down of perturbation theory can be found in second order. The second order self-energy $\Sigma^{(2)}$ displays a logarithmic divergence $\tilde{V}^2 \ln |\omega|$ leading to the same kind of divergence in the spectral function [6]. The diagram responsible for the second order divergence for PBC does also lead to a logarithmic divergence $\left[\Sigma_s^{(2)}(\omega)\right]_{n,n'} \sim \delta_{n,n'} \tilde{V}^2 \ln |\omega|$ for OBC. Thus logarithmic terms found in perturbation theory for $\rho(x, \omega)$ in the case of OBC can have two different origins. The term $\sim U \ln |\omega|$ is what we already expected to find from the bosonization result Eq. (4).

To study how the leading logarithmic divergence of ρ^{HF} Eq. (9) is modified by higher orders in $\tilde{V}$, one can numerically invert $\left[G^{\mathrm{HF}}(\omega)\right]^{-1}$ for large but finite systems. In Ref. [17] we study the TL model with $\tilde{V}(2k_F) = 0$ [19] following this line. It is shown that the HF spectral function displays power-law behavior, which reveals that the leading logarithmic divergence found in Eq. (9) can be resummed to give a power-law. We find $\alpha_B^{\mathrm{HF}} = \tilde{V}(0)/(2\pi v_{\mathrm{HF}})$ and thus α_B^{HF} and α_B do agree up to leading order in $\tilde{V}(0)/(2\pi v_F)$. Quantitative agreement between exact results, obtained from bosonization and HF can be reached for $\tilde{V}(0)/(2\pi v_F) \ll 1$. This surprising result implies that many aspects of the behavior of the spectral function close to the boundary, including the power-law suppression associated with LL behavior, can be understood within the Hartree-Fock approximation.

4.   LATTICE MODEL OF SPINLESS FERMIONS

To investigate whether or not the bosonization result Eq. (4) also holds for a model with an interaction which is short range in real space we consider the lattice model of spinless fermions with N lattice sites, lattice constant $a = 1$, hopping matrix element $t = 1$, nearest neighbor interaction U, and OBC

$$\hat{H} = -\sum_{j=1}^{N-1} \left(c_j^\dagger c_{j+1} + c_{j+1}^\dagger c_j\right) + U \sum_{j=1}^{N-1} n_j n_{j+1}. \tag{10}$$

$c_j^{(\dagger)}$ denotes the creation (annihilation) operator at site j and $n_j = c_j^\dagger c_j$.

We have performed a density-matrix renormalization-group (DMRG) study [20] for chains of up to $N = 512$ sites calculating matrix elements $w_0(n_F; U) = |\langle E_0^{n_F-1}|c_1|E_0^{n_F}\rangle|^2$, i.e. the spectral weight at the chemical potential and the boundary site [15, 17]. $|E_0^{n_F}\rangle$ denotes the exact n_F-particle ground state. It is expected that provided $\rho(x, \omega)$ shows power-law behavior as a function of ω the same holds for the ratio $w \equiv w_0(x, n_F; \tilde{V})/w_0(x, n_F; 0)$ as a function of $1/N$ both with the same exponent. Smaller systems are sufficient to find power-law behavior in w compared to ρ. Results for w at two different filling factors $n_f = n_F/N$ and U are shown in Fig. 1. $K_\rho(U, n_f)$ for these parameters can be determined [17]. For large N the numerical data nicely follow the solid lines, which are proportional to power-laws with exponent $\alpha_B = K_\rho^{-1} - 1$. We can thus conclude that the spectral weight close to the boundary and the chemical potential shows a suppression with a power-law and exponent α_B as predicted by bosonization.

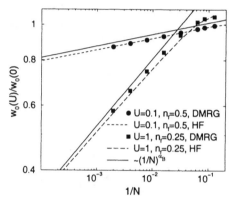

Figure 1 Spectral weight at μ and lattice site 1 for the lattice model of spinless fermions. The symbols show the DMRG data and the dashed lines the corresponding HF results. For comparison the solid lines are power-laws with exponent α_B. The parameters are given in the legend. The numerical error of the DMRG data is smaller than the size of the symbols.

The HF self-energy for spinless fermions is best studied in the site representation. Due to the OBC the site occupancies $\langle n_j \rangle_0$ and the renormalization of the hopping amplitudes $\langle c_{j+1}^\dagger c_j \rangle_0$ depend on j and in order to obtain the site diagonal Green function one numerically has to invert a tridiagonal matrix. Fig. 1 presents HF data for w. The HF approximation shows power law behavior with exponent $\alpha_B^{\mathrm{HF}} = \tilde{V}_{\mathrm{eff}}/(2\pi v_{\mathrm{HF}})$ which has the same form as for the TL model if one replaces $\tilde{V}(0)$ by the effective interaction $\tilde{V}_{\mathrm{eff}} = \tilde{V}(0) - \tilde{V}(2k_F) = 2U[1 - \cos(2k_F)]$. Again α_B and α_B^{HF} do agree up to order $\tilde{V}_{\mathrm{eff}}/(2\pi v_F)$.

5. HUBBARD MODEL

Next we study $\rho(x, \omega)$ for the 1D Hubbard model with OBC ($t = 1$, $a = 1$)

$$\hat{H} = -\sum_s \sum_{j=1}^{N-1} \left(c_{j,s}^\dagger c_{j+1,s} + c_{j+1,s}^\dagger c_{j,s} \right) + U \sum_{j=1}^{N} n_{j,\uparrow} n_{j,\downarrow}. \tag{11}$$

As for the lattice model of spinless fermions we have calculated the spectral weight at $j = 1$ and μ as a function of N using the DMRG algorithm [15, 17]. We were able to obtain results for up to $N = 256$ sites. Fig. 2 shows data for

w, quarter filling $n_f = 2n_F/N = 0.5$, and different U. Instead of decreasing, as predicted by bosonization and in contrast to our findings for the other two models considered, w increases for small and moderate values of U. For moderate $U \approx 2$ a crossover to a suppression occurs at system sizes reachable within DMRG. For smaller U, only the increase can be observed. We expect that the crossover sets in at much larger chain length. Only the $U = 4$ data display a clear suppression of the spectral weight for all the system sizes available and thus only for large U a comparison to the power-law with exponent $\alpha_B = (K_\rho^{-1} - 1)/2$ predicted by bosonization seems meaningful. As discussed in Ref. [17] for $U = 8$ and $U = 16$ the final suppression of weight is consistent with the prediction of bosonization.

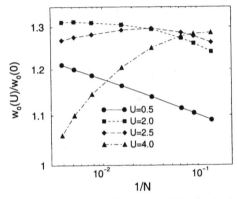

Figure 2   DMRG results for the spectral weight at μ and lattice site 1 for the Hubbard model at quarter filling. The interaction U is given in the legend. The numerical error of the DMRG data is smaller than the size of the symbols.

The surprising new finding is the increase of weight for small and moderate U and the subsequent crossover and suppression. We expect that these features can also be found in the small ω behavior of $\rho(x, \omega)$. This could have been expected already from the lowest order result Eq. (9) for the spectral function. For a k-independent interaction U and $z = 2$ the leading logarithmic correction to ρ is given by $\rho(x, \omega) \sim 1 - U(2\pi v_F) \ln |\omega/(v_F k_F)| + \ldots$. The prefactor of the logarithmic correction has thus the opposite sign as in the case of the TL model. As long as $[U/(2\pi v_F)] \ln (N) \ll 1$ this indicates a logarithmic increase of the weight. The crossover scale can be estimated from this expression to be $\sim \exp [-2\pi v_F/U]$. From the data it is clear that also the exact crossover scale $1/N_c$ strongly depends on U. Going from $U = 2.5$ to $U = 2$ it roughly decreases by one order of magnitude.

Within the Hartree approximation (the Fock term vanishes) for the Hubbard model the hopping is not modified by the interaction and the problem which remains to be solved for an electron of spin species s is the dynamics in an external potential generated by $\langle n_{j,-s} \rangle_0$. The Hartree potential shows

Friedel oscillations, i.e. oscillates as $\cos(2k_F j)$ and slowly decays as $1/j$. As discussed in Ref. [17] the leading small ω dependence of $\rho^H(x, \omega)$ for this one-particle scattering problem can be calculated solving the Lippmann-Schwinger equation. One obtains

$$\frac{\rho_1^H(\omega)}{\rho_1^0(\omega)} = \frac{\left|\frac{2\omega}{v_F k_F}\right|^{-\alpha_B^H}}{1 + \left[\frac{U}{8v_F}\right]^2 \left[\left|\frac{2\omega}{v_F k_F}\right|^{-\alpha_B^H} + \mathrm{sign}(\omega)\right]^2}, \tag{12}$$

where $\alpha_B^H = U/(2\pi v_F)$. This function shows the crossover behavior discussed above with a crossover scale

$$\frac{\Delta^H}{v_F k_F} = \exp\left\{-\frac{\pi v_F}{U} \ln \frac{1 + [U/(8v_F)]^2}{[U/(8v_F)]^2}\right\} \tag{13}$$

and a final power-law suppression with exponent α_B^H. Up to logarithmic corrections Δ is exponentially small in $-1/U$. The exponentially slow scaling of coupling constants [6] not taken into account in the Hartree calculation will change the leading small U behavior of α_B to $\alpha_B = U/(4\pi v_F) + \mathcal{O}\left([U/2\pi v_F]^2\right)$ and will also modify the prefactor and the logarithmic corrections of the crossover scale, but for small U, Δ will remain exponentially small in $-1/U$ [17].

6.   SUMMARY

The results presented here can be summarized by answering the questions raised at the beginning of Sec. 2. :

- For the three models studied and the local spectral function "universality" holds similar to the case of PBC!

- Many aspects of the behavior of the spectral function close to the boundary, including the power-law suppression associated with LL behavior, can be understood within the Hartree-Fock approximation!

- For the Hubbard model with OBC and small U the scale on which the power-law suppression predicted by bosonization can be observed is exponentially (in $-1/U$) small! Thus bosonization only holds on exponentially small energy scales!

Acknowledgments

This work has been supported by the SFB 341 of the Deutsche Forschungsgemeinschaft (V.M. and W.M.). Part of the calculations were carried out on a T3E of the Forschungszentrum Jülich.

References

[1] For a review see: J. Voit, Rep. Prog. Phys. **58**, 977 (1995).

[2] S. Tomonaga, Prog. Theor. Phys. **5**, 544 (1950).

[3] J.M. Luttinger, J. Math. Phys. **4**, 1154 (1963).

[4] D.C. Mattis and E.H. Lieb, J. Math. Phys. **6**, 304 (1965).

[5] Here we do not consider umklapp scattering processes which only occur in lattice models and even there are irrelevant as long as the interaction is sufficient weak or the filling factor sufficient incommensurable (see e.g. Ref. [1]).

[6] J. Sólyom, Adv. Phys. **28**, 201 (1979).

[7] F.D.M. Haldane, J. Phys. C**14**, 2585 (1981).

[8] R. Preuss *et al.*, Phys. Rev. Lett. **73**, 732 (1994).

[9] For a recent review of the experimental situation see: M. Grioni and J. Voit in *Electron spectroscopies applied to low-dimensional materials,* ed. by H. Stanberg and H. Hughes (1999).

[10] D.C. Mattis, J. Math. Phys. **15**, 609 (1974).

[11] C.L. Kane and M.P.A. Fisher, Phys. Rev. Lett. **68**, 1220 (1992).

[12] M. Fabrizio and A. Gogolin, Phys. Rev. B **51**, 17827 (1995).

[13] S. Eggert *et al.*, Phys. Rev. Lett. **76**, 1505 (1996).

[14] Y. Wang *et al.*, Phys. Rev. B **54**, 8491 (1996).

[15] K. Schönhammer *et al.*, Phys. Rev. B **61**, 4393 (2000).

[16] J. Voit *et al.*, Phys. Rev. B **61**, 7930 (2000).

[17] V. Meden *et al.*, cond-mat/0002215 and Eur. Phys. J. B (2000), in press.

[18] The additional states are assumed to be filled in the ground state and thus do not modify the low-energy physics of the model.

[19] This implies that for repulsive interactions ($\tilde{V}(0) > 0$) the prefactor of the logarithm in Eq. (9) is positive and the perturbative expression indicates a suppression of the weight.

[20] *Density-Matrix Renormalization,* ed. by I. Peschel *et al.* (Springer, Berlin, 1999) and references therein.

NODAL LIQUIDS AND DUALITY

Nick E. Mavromatos and Sarben Sarkar

Department of Physics, King's College London, Strand, London WC2R 2LS, United Kingdom.

Abstract

Using a SU(2) × U(1) gauge theory for a t–J model around a node of the Fermi surface, we discuss patterns of dynamical symmetry breaking, which may lead to a pseudogap phase and to the appearance of narrow one-dimensional spatial structures, induced by the presence of holes. A possible connection with stripe phases is briefly discussed by passing to an appropriate dual theory. We discuss confinement properties of spinons and holons and derive the spectrum of the (gapful) physical excitations which are composites of holons corresponding to 'mesons', and composites of spinons corresponding to 'baryons'.

Strong electron correlations are regarded generally as an important ingredient of a theory for high temperature superconductivity. There has been a great deal of ingenuity in theoretical approaches and this has been concentrated on the most appropriate way of incorporating constraints in terms of 'particle' occupancy at sites of the underlying lattice. Symmetry, both global and local and both in the Wigner–Weyl mode or the Nambu–Goldstone mode have been shown to be essential ingredients. However the exact symmety group and the representation and nature of any symmetry breaking forms are the questions which need to be settled.

In this note we will put forward a candidate for this symmetry [1]:

$$G \simeq \mathrm{SU}(2) \times \mathrm{U}(1), \tag{1}$$

where G is a local symmetry. The SU(2) symmetry is present in a Nambu–Goldstone form and only a compact $\mathrm{U}_\tau(1)$ associated with the τ_3 generator of SU(2) remains unbroken. In the physical (2+1)-dimensional model [1] the statistical gauge symmetry U(1) appears in a gauge fixed form, and in fact is associated with exotic statistics of the relevant excitations. This basic symmetry breaking which is achieved dynamically [1] leads to a gauge-invariant mass gap for charge degrees of freedom (the pseudogap) and also to 'one dimensional' structures which we suggest are related to the insulating stripe phases which are of current interest in underdoped cuprates. In addition there is a global

J. Bonča et al. (eds.), Open Problems in Strongly Correlated Electron Systems, 293–302.

SU(2) symmetry related to the spin which in the gauge theory approach will be associated with a 'custodial' SU(2) global symmetry, as we shall discuss at the end.

Our aim is, starting from a reasonable microscopic Hamiltonian, H, to form a continuum field theory for the low energy excitations of the cuprates. Such a model belongs to a class of t–J models with additional interactions [1, 2]. At half-filling H reduces to the Heisenberg Hamiltonian for which adding a spin-up particle is the same as removing a spin-down particle owing to the constraint of single particle occupancy. This in turn leads to a 'hidden' SU(2) gauge symmetry of the Heisenberg model. In the presence of holes this no longer holds, and the theory has a U(1) gauge symmetry. Usually there is a 'discontinuity' in the formulation of the problem at half-filling and away from half-filling. In the former there is a local SU(2) symmetry while in the latter the SU(2) formulation is totally replaced with a U(1) local symmetry from the outset. In our case we have the symmetry group G which reduces to SU(2) at half-filling while away from half-filling, owing to strong gauge fluctuations associated with the U(1) factor in G, the SU(2) is dynamically broken down to $U_\tau(1)$, i.e. the system decides about the local symmetry given a maximal G-symmetry.

In the two-dimensional Heisenberg model Affleck *et al* [3] noted that

$$\underline{S}_i = \mathrm{Tr}\left(\chi_i^\dagger \chi_i \,\underline{\sigma}^T\right) \tag{2}$$

where, in terms of c_1 and c_2 (the annihilation operators for up and down electrons),

$$\chi = \begin{pmatrix} c_1 & c_2 \\ c_2^\dagger & -c_1^\dagger \end{pmatrix} \tag{3}$$

and so there is a local SU(2) symmetry $\chi_i \longrightarrow h_i\chi_i$, $h_i \in$ SU(2). In all the above the index i labels the lattice sites and $\underline{\sigma}$ is the vector of Pauli-matrices.

Away from half-filling we make the *ansatz* [1]

$$\chi = \begin{pmatrix} \psi_1 & \psi_2 \\ \psi_2^\dagger & -\psi_1^\dagger \end{pmatrix} \begin{pmatrix} z_1 & -\bar{z}_2 \\ z_2 & \bar{z}_1 \end{pmatrix} \tag{4}$$

where ψ_α, z_α with $\alpha \in \{1, 2\}$ are fermions and bosons respectively. The ψ_α describe charge degrees of freedom and the z_α describe the spin degrees of freedom. The index α is related to the underlying bipartite (antiferromagnetic) lattice structure. At a site the original formulation had two degrees of freedom: the spin-up hole and the spin-down hole. The above spin-charge separation *ansatz* should have constraints and symmetries which reduce the number of physical degrees of freedom to two. One constraint is that c_α should obey canonical commutation relations. On assuming canonical commutation/anti-commutation relations for z_α and ψ_α we find that the canonical anticommutation relations for the c operators are satisfied on physical states provided the following constraints at each site i hold [1]:

$$\psi_{1,i}\,\psi_{2,i} = 0 = \psi_{2,i}^\dagger\,\psi_{1,i}^\dagger\,,\quad \sum_{\beta=1,2}\left(\bar{z}_{i,\beta}\,z_{i,\beta} + \psi_{\beta,i}^\dagger\,\psi_{\beta,i}\right) = 1. \quad (5)$$

These relations are single occupancy constraints and imply that our formulation is suitable in the limit of strong correlations. It will be convenient to work in the functional integral framework where the ψ are Grassmann variables. If we count the degrees of freedom in χ, $\deg\chi$, we have $\deg\chi = 4 + 4$, where 4 counts the Grassmann degrees of freedom and 4 is the number of real degrees of freedom in $z_1, z_2, \bar{z}_1$ and $\bar{z}_2$.

Just as in the Heisenberg case there is an SU(2) symmetry

$$\Psi_i \longrightarrow \Psi_i\,h_i \qquad \text{where} \qquad \Psi = \begin{pmatrix} \psi_1 & \psi_2 \\ \psi_2^\dagger & -\psi_1^\dagger \end{pmatrix}$$

$$Z_i \longrightarrow h_i^\dagger\,Z_i \qquad \text{where} \qquad Z = \begin{pmatrix} z_1 & -\bar{z}_2 \\ z_2 & \bar{z}_1 \end{pmatrix} \quad (6)$$

where $h_i \in$ SU(2). However, there is also a dynamical U(1) gauge symmetry acting on the Ψ fields, which is due to phase frustration from holes moving in a spin background. Some arguments to justify this from a microscopic point of view have been given in reference [4]. Consequently, this symmetry is associated with exotic statistics of the pertinent excitations [1], which is an exclusive feature of the planar spatial geometry. The existence of these gauge symmetries together with the constraints reduce the effective number of degrees of freedom to the physical degrees of freedom of the system. The *ansatz* gives us a maximal symmetry G. The dynamics will be specified by a Hamiltonian which will determine any spontaneous symmetry breaking that occurs.

As a generic model we will consider a generalized t–J model which is an effective single-bond model derived from a more realistic 5-bond model [2]. This is in terms of the parameters $\{t, t', t'', J, V\}$ where t, t', and t'' are nearest-neighbour, next nearest-neighbour and third nearest-neighbour hoppings respectively, J is the Heisenberg antiferromagnetic interaction and V is a static attractive nearest-neighbour interaction. It turns out that, under certain circumstances, the presence of V allows the theory in the continuum limit to show dynamical supersymmetry between spinon and holon degrees of freedom as has been discussed in ref. [2].

The Heisenberg term can be written in terms of a Hubbard–Stratonovich field, Δ_{ij}, as

$$-\frac{J}{8}\sum_{\langle ij\rangle}\mathrm{Tr}\left(\chi_i\,\chi_j^\dagger\,\chi_j\,\chi_i^\dagger\right)=\sum_{\langle ij\rangle}\mathrm{Tr}\left[\frac{8}{J}\Delta_{ij}^\dagger\,\Delta_{ji}+\left(\chi_i^\dagger\,\Delta_{ij}\,\chi_j+h.c.\right)\right].$$

(7)

The hopping part of the Hamiltonian can be written as

$$-\sum_{\langle ij\rangle}t_{ij}\,c_{\alpha,i}^\dagger\,c_{\alpha,j}=-\sum_{\langle ij\rangle}t_{ij}\left(\chi_{i,\alpha\gamma}^\dagger\,\chi_{j,\gamma\alpha}+\chi_{i,\alpha\gamma}^\dagger\,(\sigma_3)_{\gamma\beta}\,\chi_{j,\beta\alpha}\right)=$$

$$=-\sum_{\langle ij\rangle}t_{ij}\left(\overline{Z}_{i,\beta\kappa}\,\Psi_{i,\kappa\alpha}^\dagger\,\Psi_{j,\alpha\gamma}\,Z_{j,\gamma\beta}+\overline{Z}_{i,\beta\kappa}\,\Psi_{i,\kappa\alpha}^\dagger\,(\sigma_3)_{\alpha\lambda}\Psi_{j,\lambda\gamma}\,Z_{j,\gamma\beta}\right).$$

(8)

The global SU(2) spin symmetry, mentioned above, is explicitly given by $Z_i\to Z_i h$ (and equivalently $\chi_i\to\chi_i h$) where $h\in$ SU(2) is a group element.

In the Hartree–Fock approximation we obtain the Hamiltonian

$$H_{\mathrm{HF}}=\sum_{\langle ij\rangle}\mathrm{Tr}\left[\frac{8}{J}\Delta^\dagger{}_{ij}\,\Delta_{ji}+(-t_{ij}(1+\sigma_3)+\Delta_{ij})\,\Psi_j^\dagger\langle Z_j\,\overline{Z}_i\rangle\Psi_i\right]+$$

$$+\sum_{\langle ij\rangle}\mathrm{Tr}\left[\overline{Z}_i\langle\Psi_i^\dagger\,(-t_{ij}(1+\sigma_3)+\Delta_{ij})\,\Psi_j\rangle Z_j+h.c.\right]. \quad (9)$$

Using the G-symmetry (1) of the *ansatz* we can write in the presence of gauge fixing

$$\langle Z_j\overline{Z}_i\rangle\equiv|A_1|\,\mathcal{R}_{ij}\,U_{ij}\;,\;\;\langle\Psi_i^\dagger\,(-t_{ij}(1+\sigma_3)+\Delta_{ij})\,\Psi_j\rangle\equiv|A_2|\,\mathcal{R}_{ij}\,U_{ij}.$$

(10)

where $\mathcal{R}\in$ SU(2) and $U\in$ U(1) are group elements. The fact that apparently gauge non-invariant correlators are non-zero on the lattice is standard in gauge theories, and does not violate Elitzur's theorem [5], precisely due to the above-mentioned gauge-fixing procedure, which is done prior to any computation. The amplitudes $|A_1|$ and $|A_2|$ are considered frozen which is a standard assumption in the gauge theory approach to strongly correlated electron systems [3].

By standard arguments the low energy lattice action for the fermion part becomes

$$S_{\mathrm{F}}=\frac{1}{2}\kappa'\sum_{i,\mu}\left(\overline{\hat{\Psi}}_i\,\gamma_\mu\,\mathcal{R}_{i,\mu}\,U_{i,\mu}\,\hat{\Psi}_{i+\mu}+\overline{\hat{\Psi}}_{i+\mu}\,\gamma_\mu\,\mathcal{R}_{i,\mu}\,U_{i,\mu}\,\hat{\Psi}_i\right) \quad (11)$$

where $\hat{\Psi}_{i,\alpha}$ is a Nambu–Dirac two-component spinor for each 'colour' $\alpha\in\{1,2\}$. The $\mathcal{R}$ and U matrices act on the colour matrices whereas the 2×2 γ

matrices act on the two components of $\hat{\Psi}_{i,\alpha}$. The components of $\hat{\Psi}$ are linearly related to those of Ψ Moreover, we have kinetic (i.e. Maxwell) terms for the gauge link variables in the form of plaquette terms in the lattice action:

$$S_G = \sum_p [\beta_2(1 - \text{Tr}\mathcal{R}_p) + \beta_1(1 - \text{Tr}U_p)] \tag{12}$$

where p denotes plaquettes, the β_i are inverse couplings, $\beta_2 \equiv \beta_{\text{SU}(2)} \propto 1/g^2$, $\beta_1 \equiv \beta_{\text{U}(1)}$, and $\mathcal{R}_p$, U_p are a product of the link variables over the plaquette p. It is important to notice for our purposes that, as a result of the gauge-fixed form of the U(1) factor in G in the spinon sector, there is no plaquette term for the U(1) field, which implies $\beta_1 = 0$ and hence we are in the strong coupling limit [1]. The coupling β_2 is large [2].

We will find it useful to give the explicit representation of the global G-symmetry whose gauging produces S_G. On writing

$$\tilde{\Psi} = \begin{pmatrix} \hat{\Psi}_1 \\ \hat{\Psi}_2 \end{pmatrix} \tag{13}$$

the generators of G are given by

$$\left\{ 1_4, \begin{pmatrix} 0 & 1_2 \\ 1_2 & 0 \end{pmatrix}, i \begin{pmatrix} 0 & 1_2 \\ -1_2 & 0 \end{pmatrix}, \begin{pmatrix} 1_2 & 0 \\ 0 & -1_2 \end{pmatrix} \right\}. \tag{14}$$

Let us consider some relevant multiplets of this group at any site j. The bilinears

$$\phi_1 = -i(\overline{\hat{\Psi}}_1 \hat{\Psi}_2 - \overline{\hat{\Psi}}_2 \hat{\Psi}_1), \quad \phi_2 = \overline{\hat{\Psi}}_1 \hat{\Psi}_2 + \overline{\hat{\Psi}}_2 \hat{\Psi}_1, \quad \phi_3 = \overline{\hat{\Psi}}_1 \hat{\Psi}_1 - \overline{\hat{\Psi}}_2 \hat{\Psi}_2 \tag{15}$$

form an adjoint representation of SU(2) [1]. There is also a vector adjoint representation

$$(\mathcal{A}_\mu)_1 = i(\overline{\hat{\Psi}}_1 \tilde{\sigma}_\mu \hat{\Psi}_2 - \overline{\hat{\Psi}}_2 \tilde{\sigma}_\mu \hat{\Psi}_1), \quad (\mathcal{A}_\mu)_2 = \overline{\hat{\Psi}}_1 \tilde{\sigma}_\mu \hat{\Psi}_2 + \overline{\hat{\Psi}}_2 \tilde{\sigma}_\mu \hat{\Psi}_1,$$
$$(\mathcal{A}_\mu)_3 = \overline{\hat{\Psi}}_1 \tilde{\sigma}_\mu \hat{\Psi}_1 - \overline{\hat{\Psi}}_2 \tilde{\sigma}_\mu \hat{\Psi}_2 \tag{16}$$

and two singlets

$$\mathcal{S}_4 = \overline{\hat{\Psi}}_1 \hat{\Psi}_1 + \overline{\hat{\Psi}}_2 \hat{\Psi}_2, \quad (\mathcal{S}_\mu)_4 = \overline{\hat{\Psi}}_1 \tilde{\sigma}_\mu \hat{\Psi}_1 + \overline{\hat{\Psi}}_2 \tilde{\sigma}_\mu \hat{\Psi}_2 \tag{17}$$

where $\tilde{\sigma}_0 = -i\sigma_3$, $\tilde{\sigma}_1 = \sigma_1$ and $\tilde{\sigma}_2 = \sigma_2$.

Now we will consider spontaneous symmetry breaking in G due to the interactions. Since we are at the strong coupling limit of U(1), where $\beta_1 = 0$, the U(1) gauge field can be integrated out exactly to give [1]

$$\int \mathcal{DR}\, \mathcal{D}\overline{\hat{\Psi}}\, \mathcal{D}\hat{\Psi} \, \exp(-S_{\text{eff}})$$

where

$$S_{\text{eff}} = \beta_2 \sum_p (1 - \text{Tr}\mathcal{R}_p) + \sum_{i,\mu} \ln I_0(\sqrt{y_{i\mu}}), \tag{18}$$

$$y_{i\mu} = -\kappa^2 \text{Tr}\left(M^{(i)}\left(-\gamma_\mu\right) \mathcal{R}_{i\mu}\, M^{(i+\mu)}\, \gamma_\mu\, \mathcal{R}_{i\mu}^\dagger \right) \tag{19}$$

and

$$M^{(i)} = \sum_{a=1}^{3} \phi_a(i)\sigma_a + \mathcal{S}_4(i)\mathbf{1} + i\left((\mathcal{S}_\mu)_4(i)\gamma^\mu + \sum_{a=1}^{3}(\mathcal{A}_\mu)_a(i)\gamma^\mu\sigma_a \right). \tag{20}$$

Owing to the Grassmann content of M,

$$-\ln I_0^{\text{tr}}(2\sqrt{y_{i\mu}}) = -y_{i\mu} + \frac{1}{4}y_{i\mu}^2 - \frac{1}{9}y_{i\mu}^3 + \frac{11}{192}y_{i\mu}^4, \tag{21}$$

which is exact. This type of analysis leads to

$$\langle M^{(i)} \rangle = u\sigma_3 \tag{22}$$

and

$$\ln I_0^{\text{tr}}(2\sqrt{y_{i\mu}}) \sim M_B^2\left((B_{i\mu}^1)^2 + (B_{i\mu}^2)^2 \right) + \text{interaction terms} \tag{23}$$

with $M_B^2 = \kappa^2 u^2$, and so

$$\langle \phi_3 \rangle \neq 0. \tag{24}$$

(Here $\mathcal{R}_{i\mu} = \cos(|\underline{B}_{i\mu}|) + i\underline{\sigma}\cdot\underline{B}_{i\mu}\sin(|\underline{B}_{i\mu}|)/|\underline{B}_{i\mu}|$). Notice that since $u \propto \beta_1^{-1}$, in the strong coupling limit $u \to \infty$. Hence, a parity invariant mass for the fermions is generated dynamically by the strongly-coupled U(1) interactions in G [1], and thus the theory is gapped to charge excitations. But two of the SU(2) gauge bosons also acquire masses, and hence the SU(2) group is broken down to $U_\tau(1)$.

To make analytical progress towards an understanding of the confinement properties of spinon and holon degrees of freedom, it is convenient to go to a continuum limit for the low-energy excitations of the theory. By linearizing about a node of the Fermi surface on the underlying theory of fermions we can consider a relativistic field theory of the form

$$\mathcal{L} = -\frac{1}{4}F_{\mu\nu}^a F^{a\,\mu\nu} + \frac{1}{2}(D_\mu^{ab}\phi^b)^2 + \mu^2\phi^2 - \lambda(\phi^2)^2 \tag{25}$$

where $D_\mu^{ab}\phi^b = \partial_\mu\phi^a - ig\epsilon^{abc}B_\mu^b\phi^c$ and $\mu^2 > 0$, since we have symmetry breaking (actually, in (2+1)-dimensions the symmetry broken phase is connected analytically to the phase where the symmetry is unbroken [6]). The above form should not be considered as quantitative, but it captures correctly the qualitative features of our approach, and it is sufficient for our purposes here. Terms involving $\mathcal{A}$ and $\mathcal{S}$ are not important for our analysis.

At the perturbative level $B_{3\mu}$ is massless, and in fact the theory is superconducting [1]. However this is not true when non-perturbative effects are taken into account, such as monopoles, which are instantons in the (2+1)-dimensional theory [7]. The monopole is a Euclidean configuration which behaves asymptotically as

$$\hat{\phi}^a = \hat{r}^a, \qquad \tilde{F}_\mu^a(x) = \frac{1}{g}\frac{\hat{r}^a\hat{r}_\mu}{r^2} \tag{26}$$

where the caret indicates a unit vector and the tilde indicates the dual field tensor.

At the perturbative level the current J^μ associated with $U_\tau(1)$ is conserved and in general is given by

$$J^\mu = \frac{1}{g}\epsilon^{\mu\nu\lambda}\partial_\nu(\tilde{F}_\lambda^a\,\hat{\phi}^a). \tag{27}$$

In the context of the model (25), it has been shown [6] that there exists a local gauge invariant field $V(x)$ such that

$$J^\mu = -\frac{i}{4\pi}\epsilon^{\mu\nu\lambda}\partial_\mu(V^*\partial_\lambda V - c.c.). \tag{28}$$

The field V interpolates between defect structures and is a disorder variable. In particular we can identify

$$V^*\partial_\lambda V - c.c. = \frac{4\pi i}{g}\tilde{F}_\lambda^a\,\hat{\phi}^a. \tag{29}$$

At the perturbative level the current $\tilde{F}_\mu$ is given by

$$\tilde{F}_\mu = \tilde{F}_\mu\hat{\phi}^a - \frac{1}{g}\epsilon^{\mu\nu\lambda}\epsilon^{abc}\hat{\phi}^a\,(D_\nu\hat{\phi})^b\,(D_\lambda\hat{\phi})^c \tag{30}$$

and is conserved. However in the presence of monopoles

$$\partial^\mu\tilde{F}_\mu = \frac{4\pi}{g}\delta^{(3)}(x) \tag{31}$$

and conservation is lost; the magnetic flux

$$\Phi = \int d^3x\,\tilde{F}_0(x) \tag{32}$$

does not generate symmetries of the system, i.e. $U_\alpha = \exp(i\alpha\Phi)$ is not a symmetry of the physical Hilbert space for arbitrary α. Since only configurations with an integer number of monopoles have finite energy the discrete flux transformations

$$U_k = \exp(\frac{ig}{2}k\Phi) \tag{33}$$

where k is an integer are still symmetries [6]. Since the fundamental flux is in units of $2\pi/g$ the only independent operators on the Hilbert space are $U_0 = 1$ and U_1:

$$U_0 V(x) U_0^{-1} = V(x) \qquad U_1 V(x) U_1^{-1} = -V(x) \tag{34}$$

and so the theory retains a $\mathbb{Z}_2$ symmetry. The effective Lagrangian for V is [6]

$$\mathcal{L}^{\text{dual}} = \partial_\mu V^* \, \partial^\mu V - \lambda (V^* V - \mu^2)^2 - \frac{m^2}{4}(V^2 + V^{*2}) + h(\epsilon^{\mu\nu\lambda} \partial_\nu V^* \, \partial_\lambda V)^2. \tag{35}$$

The parameters appearing in $\mathcal{L}^{\text{dual}}$ can be calculated perturbatively as

$$\mu^2 = \frac{g^2}{8\pi^2} \,, \quad \lambda = \frac{2\pi^2 M_\phi^2}{e^2} \,, \quad h \propto \frac{M_B}{g^4 M_\phi^2} \,,$$

$$m = m_{\text{ph}} \propto e^{-S_0/2} \propto M_B^2 \exp(-M_B/g^2) \,, \tag{36}$$

where S_0 is the instanton action. At low energies the derivative term can be ignored. The mass M_ϕ is that associated with the ϕ fields from the Higgs mechanism; in our strong-coupling U(1) situation $M_\phi \gg m_{\text{ph}}$ [1]. Notice that the presence of a small 'photon' mass m_{ph} implies a pseudogap phase for the statistical model. Because of the $\mathbb{Z}_2$ symmetry, when external (adjoint) charges are introduced into the theory, narrow string-like structures are produced [6]. It is tempting to conjecture that such structures are related to stripe phases in the cuprates. The width of the stripes in our model is of order [6] $1/m_{\text{ph}}$, which is finite in underdoped situations. The mass $m_{\text{ph}} \to 0$ as superconductivity is approached, and the stripes will become very wide and eventually occupy the entire space ('absence of stripes'). Because in equation (29) there is the unit vector ϕ^a, which is related to a bilinear in the holon fields, the string-like object requires holons. Away from the strip region the Wilson loop shows an area law which is consistent with antiferromagnetic order. This is related to the fact that, as we shall discuss below, the spinon excitations are in the fundamental representation of the SU(2) group, and as such they are directly associated with the Wilson loop which gives the energy of a state with two heavy external charges in the fundamental representation. A detailed correspondence requires further research. It should be stressed, though, that the above dual description provides a natural and clear picture for the confinement of both holons and spinons. For details we refer the reader to the discussion in ref. [6], which parallels our case here.

We would now like to discuss the effects and (confinement) properties of spinons, z_α, $\alpha \in \{1,2\}$. From the construction of the ansatz (4) it becomes clear that the spinon sector in our theory consists of a complex scalar doublet, in the fundamental representation of the SU(2) gauge group. In the terminology of ref. [6], this situation corresponds to including scalar constituent 'quarks' z^A, where $A \in \{1,2\}$ is an SU(2) fundamental representation index. The presence of such constituents leads to the presence of 'baryons' in our picture. In other words, in our model holon composites correspond to 'mesons', whilst spinon composites correspond to 'baryons'. These are the physical excitations of our spin-charge separated non-Abelian gauge model for the planar doped antiferromagnet. In the dual picture, the presence of 'baryons' is described [6] by the introduction of an *additional* field W, which should be such that : (i) the elementary defect (soliton) of the field V should carry 'baryon' number 1/2 (in the SU(2) theory) in order to represent the "constituent (fundamental) quark". For this it is necessary that in the core of the defect (where V vanishes) the field W has a non-zero value, whilst outside the core $W \to \langle W \rangle = 0$, since 'baryon' number symmetry cannot be broken spontaneously. (ii) The interaction potential should favour configurations in which, for V close to its vacuum expectation value, W is small, whereas for small V, W is non zero. This can be easily implemented by either imposing the σ-model constraint $V^*V + W^*W = 1/8\pi^2\beta_2$, or adding the interaction term $\lambda(V^*V + W^*W - 1/8\pi^2\beta_2)^2$ in the Lagrangian. For convenience we choose the σ-model constraint. The long-distance properties, we are interested in, are indistinguishable between the two cases [6]. (iii) The 'baryon' should be attached to the topological soliton, so three-derivative terms are necessary in the dual effective Lagrangian, which on symmetry grounds has the form [6]:

$$\mathcal{L}^{\text{dual/baryon}} = \partial_\mu V^* \, \partial^\mu V + \partial_\mu W^* \, \partial^\mu W + h(V + V^*) +$$

$$+ \frac{1}{4W^*W}(W^*\partial_\mu W - W\partial_\mu W^*)^2 - \qquad (37)$$

$$- \frac{1}{8\pi(W^*W)(V^*V)}\epsilon^{\mu\nu\lambda}(W^*\partial_\mu W - W\partial_\mu W^*)\,\partial_\nu(V^*\partial_\lambda V - V\partial_\lambda V^*)$$

We now remark that it is an exclusive feature of the SU(2) case that the vacuum in the one-soliton sector is degenerate [6], corresponding to solitons with 'baryon' number $1/2$ and $-1/2$. This degeneracy corresponds to an additional global SU(2) 'custodial' symmetry [6], under which the Z-matrix in the ansatz (4) is transformed as $Z \to ZU$. The 'baryon' number is therefore associated with one of the generators of the custodial SU(2). The fermion matrix is neutral under the custodial symmetry. In the condensed-matter framework this symmetry may be associated with the global SU(2) spin group of the electron constituents mentioned earlier. Evidently, 'particles' in this dual picture come in pairs with opposite 'baryon' number. In the dual picture the confinement of spinons is evident [6], as can be demonstrated by means of the area law behaviour of the Wilson loop. This completes our qualitative discussion on the symmetry-breaking and confinement properties of our (zero- (or low-) temperature) effective spin-charge separated theory. More work is clearly needed for a quantitative analysis.

Acknowledgments

We thank A. Kovner and A. Campbell–Smith for discussions. This work is supported in part by the Leverhulme Trust and P.P.A.R.C. (U.K.).

References

[1] K. Farakos and N.E. Mavromatos, Phys. Rev. B **57**, 3017 (1998); Mod. Phys. Lett. A **13** 1019 (1998).

[2] N.E. Mavromatos and Sarben Sarkar, cond-mat/9912323, Physical Review B (2000), in press.

[3] I. Affleck, *et al*, Phys. Rev. **B38** (1988), 745.

[4] N. Dorey and N.E. Mavromatos, Phys. Lett. B **250** 107 (1990).

   Z.Y. Weng *et al*, Phys. Rev. **B 55** 3894 (1997).

[5] S. Elitzur, Phys. Rev. D **12** 3978.

[6] A. Kovner and B. Rosenstein, Int. J. Mod. Phys. A **8** 5575 (1993); JHEP **9809** 003 (1998).

[7] A.M. Polyakov, Phys. Lett. B **59**, 82 (1975); Nucl. Phys. B **120**, 429 (1977).

SPIN–CHARGE SEPARATION IN THE Sr$_2$CuO$_3$ AND SrCuO$_2$ CHAIN MATERIALS

Karlo Penc[1] and Walter Stephan[2]

[1] Research Institute for Solid State Physics and Optics, H-1525 Budapest, P.O.B. 49, Hungary

[2] Department of Physics, Bishop's University, Lennoxville, Québec, Canada J1M 1Z7

Abstract   The photoemission and electron-energy loss spectra (EELS) of the one-dimensional Mott insulators Sr$_2$CuO$_3$ and SrCuO$_2$ are discussed using microscopic models in the strong coupling limit. We show that the EELS data can be understood within an extended effective one–band Hubbard model, and that both the spin–charge separation which occurs in 1D as well as excitonic effects are essential. Furthermore we compare the spectral functions obtained from the Hubbard (one-band) and Emery (two-band) model, and show that some features of the photoemission spectra can be best understood starting from a two-band model.

1.   INTRODUCTION

Quasi–1D materials based on cuprate compounds, of which Sr$_2$CuO$_3$ and SrCuO$_2$ are the best examples, have become new candidates for ideal model systems which allow the study of basic physical concepts in one–dimension. Magnetic susceptibility measurements have shown that they can be regarded as an almost ideal realization of the 1D spin-$\frac{1}{2}$ antiferromagnetic Heisenberg model [1], which describes the magnetic excitations of a Mott insulator. Information on the electronic structure and the dynamics of the charge carriers is highly desirable, especially against the background of spin–charge separation expected in 1D [2, 3]. Angle resolved photoemission spectroscopy (ARPES) data [4, 5, 6] has been interpreted in terms of holon and spinon bands, with bandwidths determined by the hopping term t and the exchange constant J, respectively. In Sec. 2. we analyze the photoemission spectra using a two-band model.

Another extremely important quantity reflecting the electronic structure of a solid is the dielectric function, which may be measured via optical spectroscopy for the special case of zero momentum transfer. Electron energy–loss spectroscopy (EELS) on the other hand, offers the possibility to study the momentum dependence of the electronic excitations, i.e. the dynamical dielectric

J. Bonča et al. (eds.), Open Problems in Strongly Correlated Electron Systems, 303–309.

response. The EELS spectra of Sr_2Cu_3 were measured by R. Neudert *et al.* in Ref. [7]. In Sec. 3. we present the interpretation of the spectra based on the Hubbard model extended with nearest neighbor repulsion.

2.   PHOTOEMISSION SPECTRA

In most cases, measured photoemission spectra have been compared with theoretical ones obtained from an effective one band Hubbard model. There are several difficulties inherent to this procedure: i) the actual material is a charge transfer insulator, while the one-band model is a Mott-Hubbard insulator. Therefore one is lead to question how much of the spectra can be attributed to generic features where the details of the model are not important; ii) For the CuO_2 plane, the t-J model is derived to describe the dynamics of complex objects - the Zhang-Rice singlets [8]. In the CuO_3 chains the O ions are not all identical, and the original picture of Zhang and Rice has to be modified.

To answer these questions, we will consider the 1D model involving the Cu ions and the oxygens between them, the so called two-band Emery model [9], as the simplest extension of the Hubbard model:

$$\mathcal{H} = -t \sum_{i,\delta,\sigma} (d_{i,\sigma}^\dagger p_{i+\delta,\sigma} + \text{H.c.}) + \frac{\Delta}{2} \sum_i \left(n_{i+1/2}^p - n_i^d \right) + U \sum_i n_{i,\uparrow}^d n_{i,\downarrow}^d,$$

$$(1)$$

where $d_{i,\sigma}^\dagger$ and $p_{i+\delta,\sigma}^\dagger$ denote the hole creation operators on copper d and oxygen p orbitals at sites i and $i + \delta$, respectively. The Cu-Cu distance is taken to be unity and $\delta = \pm 1/2$. The phase factors in the hybridization coming from the symmetry of the Cu and O orbitals are absorbed in the definition of the d and p operators. In the potential part we include the on-site energy difference $\Delta = \varepsilon_p - \varepsilon_d$ and the on-site Coulomb repulsion U of the Cu $3d$-orbitals. We choose $U > \Delta$ in order to have a charge transfer insulator with one hole per unit cell [10]. In Ref. [11] we have shown that it is possible to do controlled calculations both analytically and numerically in the strong coupling limit $(U, \Delta \gg t)$.

In the atomic limit of the Emery model we can distinguish two features in the photoemission part: the hole can go onto either the Cu or O site, leading to weights at $\omega = \Delta/2 - U$ and $-\Delta/2$, respectively. We are primarily interested in the spectral function of the oxygen hole, defined by

$$B(k,\omega) = \sum_{f,\sigma} |\langle f|p_{k,\sigma}^\dagger|\text{GS}\rangle|^2 \delta(\omega + E_f - E_{\text{GS}}),$$

$$(2)$$

where the dynamics of the O hole in the final states $|f\rangle$ is described by

$$\mathcal{T} = (t_S + t_T) \sum_{i,\delta,\delta',\sigma,\tau} (2\delta_{\sigma,\tau} - 1)\, \tilde{p}^\dagger_{i+\delta,\sigma} \tilde{d}^\dagger_{i,\bar{\sigma}} \tilde{d}_{i,\tau} \tilde{p}_{i+\delta',\bar{\tau}} + t_T \sum_{i,\sigma,\delta} \tilde{p}^\dagger_{i+\delta,\sigma} \tilde{p}_{i-\delta,\sigma},$$

(3)

with the effective hopping amplitudes $t_S = t^2/(U - \Delta)$ and $t_T = t^2/\Delta$ (see also [12, 13]). By using the Heisenberg model ground state as $|GS\rangle$ the main effect of the (4th order) AF interaction [14] between Cu spins is accounted for.

The O hole can form a singlet or triplet with a neighboring Cu spin. We denote by $|S_L\rangle$ ($|S_R\rangle$) states where the O hole forms a singlet with the Cu spin on its right (left). In general, the resulting spectrum is complicated and the singlets and triplets mix with one another, except for the particular case of t_S finite and $t_T = 0$ [13]. In this case the combination $|S_L\rangle - |S_R\rangle$ is the one-dimensional counterpart of the Zhang-Rice singlet, in terms of which the lowest energy excitations may be described by a one-band model. This singlet moves through the lattice like the empty site in the large-U Hubbard model, leaving the spin sequence unchanged. This extends the known factorization of the wave function of a state with momentum k into charge and spin parts [3] to apply to the two-band model, and allows the calculation of the spectral functions [11]

$$B_S(k,\omega) = \frac{4t_S(1 + \cos k)}{\Delta + 2\omega} B_{\text{HM}}(k,\omega).$$

(4)

In Eq. (4) the prefactor comes from the internal structure of the $|S_L\rangle - |S_R\rangle$ singlets, and $B_{\text{HM}}(k,\omega)$ is the spectral function of the large-U Hubbard model [15] assuming the same dispersion. This singlet part of the spectral function is centered at $\omega = -\Delta/2 + 4t_S$, with bandwidth $4t_S$. Additionally, at $\omega = -\Delta/2$ we find nondispersing solutions, made mostly of 'triplets'.

The influence of finite t_T is shown in Fig. (1). The lower 'singlet' band increases its width, while the overall shape of $B_S(k,\omega)$ does not change significantly. The 'triplet' band extends from $-\Delta/2$ to $-\Delta/2 - 4t_T$ and a sharp dispersion dominates the spectrum. Only a slight weight transfer from the 'singlet' to the 'triplet' band can be observed, as we increase t_T from 0 to t_S.

Comparing with the t-J model for small J (Fig. (1a)), we can see that although (also for $t_T = t_S$) the 'singlet' feature is similar to the t-J model result, there are detailed differences in the distribution of weight, similar to those in Eq. (4), as well as in the dispersion of the upper edge of the 'singlet' continuum. We therefore see that even in parameter regimes where the one-band t-J description accurately predicts low-energy excitation energies, the two-band model may have different properties as far as other physical observables is concerned, exemplified here by the momentum and frequency dependence of the spectral weights. The effect of finite J is to give dispersion to the now dispersionless lower 'spinon' edge in both Emery and t-J model.

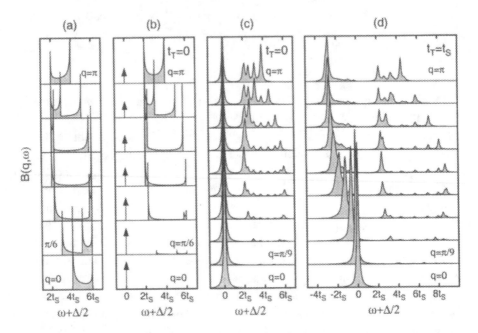

Figure 1   The analytical result for photoemission spectra of (a) $t - J$ model in the $J/t \to 0$ limit, and (b) $t_T = 0$ effective model [Eq. (4)], compared to (c) a Lánczos diagonalization of 18 site effective model for $t_T = 0$ and (d) $t_T = t_S$. The δ functions are plotted as Lorentzians of width 0.1. The $q = k + \pi$ is the momentum when the relative phases of d and p orbitals are properly included and it should be used when we compare with the experiments.

Let us now compare our results with the experiments. For both $SrCuO_2$ and Sr_2CuO_3 the low energy region shows features found in the t-J model, i.e. the holon and spinon bands dispersing with $t \approx 0.5 - 0.6eV$ and $J \approx 0.15 - 0.2eV$, respectively. However, an additional interesting feature is the weight reduction as the zone center ($q = 0$) is approached. In Refs. [5, 6] this is attributed to the different cross sections of Cu and O orbitals, while in our theory it arises quite naturally from the internal structure of the low energy singlets.

3.   EELS SPECTRA

In the right panel of Fig. (2) we show the loss function of Sr_2CuO_3 for selected momenta q, with $q = 0.8 \,\text{Å}^{-1}$ being the zone boundary [7]. In the following we will focus on the loss function in the range of 1.6 eV to $\sim$4 eV since here the spectral features are expected to be exclusively due to transitions within the CuO_3 chain and model calculations are believed to be of relevance. Peaks in the loss function in the low energy range discussed here arise from collective excitations (plasmons) related to interband transitions. In the spectrum for $q = 0.1 \,\text{Å}^{-1}$, the first possible interband transitions across the gap form a broad continuum-like absorption feature in the loss function around 2.4 eV. This broad absorption continuum narrows with increasing momentum transfer, evolving into a single rather sharp peak centered at 2.8 eV for $q = 0.5 \text{Å}^{-1}$ with a width of only ~ 0.5 eV. At the same time, the integrated intensity stays almost constant indicating a transfer of spectral weight from the continuum to the sharp peak. The remainder of the continuum is visible as a shoulder at around 3.3 eV.

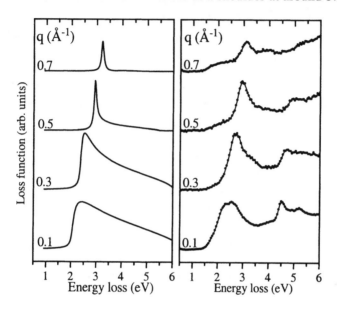

Figure 2   Loss function of Sr₂CuO₃ (right panel) measured with the momentum transfer $\vec{q}$ parallel to the chain direction. The left panel shows the calculated loss function $\mathcal{N}(q, \omega)$ plotted with an energy resolution of 0.115 eV, and scaled to the experimental peak heights.

Since the one-band model was successful in describing the main features of the photoemission spectra, it is natural to study the density response function $\mathcal{N}_0(k, \omega)$ of the Hubbard model, extended to include a nearest neighbor Coulomb repulsion term of strength V. The results of this analysis were presented in Ref. [16], where the imaginary part of the density response function was calculated using a combination of numerical and analytic methods. We have shown that in the $U \gg t, V$ limit the $\mathcal{N}_0(k, \omega)$ may be calculated within an effective $t-J$ like model, where exactly one hole and one "double occupancy" (doublon) are explicitly included in the states excited across the Mott–Hubbard gap, and that to very good approximation the spin degrees of freedom decouple from the problem. We are then left with an effective (spinless) particle–hole model, with a nearest neighbor attraction V between the doublon and holon, with opposite signs of hopping matrix element for the two carriers, and with the band centers separated by U. In the small V limit one then expects an optical gap of $U-4t$, followed by a continuum of interband transitions up to an energy of $U + 4t$. Going to higher momentum transfers the range of possible interband transitions decreases, leading to an excitation energy U at the zone boundary. The inclusion of finite nearest-neighbor repulsion V leads to the possibility of the formation of an excitonic state. For $V < 2t$ the exciton lies within the continuum in the optical limit ($q \to 0$), and is therefore not a well defined excitation there, but will appear at the zone boundary at energy $U - V$, accounting for almost all of the spectral weight. The narrowing of the low-energy feature in the EELS data with increasing q can be explained within this scheme.

In comparing with the experimental measurements it is necessary to worry about the effects of the longer range part of the Coulomb interaction, since these materials are insulators and the longitudinal response is appropriate. This was done in Ref. [7] at an RPA-level of approximation by using the response function calculated for the short–range interaction model as the "Lindhard function" within the RPA. Some results are shown in Fig. (2) (left panel) where the parameters $t = 0.55 \, \text{eV}$, $U = 4.2 \, \text{eV}$, $V = 1.3 \, \text{eV}$ were used, which led to the best description of the experimental data. These values are consistent with the parameters obtained from photoemission experiments. Note that the momentum dependence of the lineshape provides a strong constraint on the parameters in addition to that given by the dispersion. Of course, the exact lineshape observed at low q is not reproduced by our theory, but considering the simplifications inherent in our model the agreement with the measured loss function (right panel) is reasonable.

Acknowledgments

We would like to thank Prof. P. Fulde for his kind hospitality in Max-Planck-Institute für Physik komplexer Systeme in Dresden, where the present work started. This work was part-funded by Hungarian OTKA D32689, AKP98-66, and Bolyai 118/99.

References

[1] T. Ami et al., Phys. Rev. B **51**, 5994 (1995). N. Motoyama, H. Eisaki, S. Uchida, Phys. Rev. Lett. **76**, 3212 (1996).

[2] J. Sólyom, Adv. Phys. **28**, 201 (1979). J. Voit, Rep. Prog. Phys. **58**, 977 (1995).

[3] F. Woynarovich, J. Phys. C: Solid State Phys. **15**, 97 (1982); M. Ogata and H. Shiba, Phys. Rev. B **41**, 2326 (1990).

[4] C. Kim et al., Phys. Rev. Lett. **77**, 4054 (1996).

[5] C. Kim et al., Phys. Rev. B**56**, 15589 (1997).

[6] N. Nagasako et al., J. Phys. Soc. Jpn. **66**, 1756 (1997); H. Fujisawa et al., Solid State Commun. **106**, 543 (1998); Phys. Rev. B**59**, 7358 (1999).

[7] R. Neudert *et al.*, Phys. Rev. Lett. **81**, 657 (1998).

[8] F.C. Zhang and T.M. Rice, Phys. Rev. B **37**, 3759 (1988).

[9] V.J. Emery, Phys. Rev. Lett. **58**, 2794 (1987).

[10] J. Zaanen, G.A. Sawatzky and J.W. Allen, Phys. Rev. Lett. **55**, 418 (1985).

[11] K. Penc and W. Stephan, unpublished.

[12] J. Zaanen and A.M. Oleś, Phys. Rev. B **37**, 9423 (1988); P. Prelovšek, Phys. Lett. A **126**, 287 (1988).

[13] V.J. Emery and G. Reiter, Phys. Rev. B **38**, 11938 (1988); F.C. Zhang, Phys. Rev. B**39**, 7375 (1989).

[14] J.E. Hirsch, Phys. Rev. Lett. **59**, 228 (1987).

[15] S. Sorella and A. Parola, J. Phys. Condens. Matter **4**, 3589 (1992).

[16] W.Stephan and K. Penc, Phys. Rev. B **54**, 17269 (1996).

FRUSTRATED QUANTUM ISING MODEL AND CHARGED KINKS

M. V. Mostovoy, D. I. Khomskii, J. Knoester

Materials Science Center, University of Groningen, Nijenborgh 4, 9747 Groningen,

The Netherlands

N.V. Prokof'ev

Physics and Astronomy, Hasbrouck Laboratory, University of Massachusetts, Amherst,

MA 01003, USA

Abstract   We study the frustrated two-dimensional Ising model in a transverse field, which describes the charge ordering in the quarter-filled ladder material NaV_2O_5. The Monte Carlo simulation shows a cascade of transitions, at which the wave vector of the periodic modulation of the longitudinal magnetization has a discontinuity. We discuss this peculiar behavior by mapping the original model on the system of charged kinks.

1.   INTRODUCTION

In this contribution we discuss the anisotropic Ising model on the triangular lattice in a transverse field:

$$H = \sum_{\langle ij \rangle} J_{ij}\sigma_i^z\sigma_j^z - h\sum_i \sigma_i^x, \qquad . \qquad (1)$$

where $\langle ij \rangle$ denotes a pair of neighboring sites of the triangular lattice. We assume that in one direction the exchange is antiferromagnetic (AFM) $J_1 = J_A > 0$, while in other directions it is ferromagnetic, $J_2 = J_3 = -J_F < 0$ (see Fig. 1). Furthermore, we consider the case $J_F \leq J_A$ (otherwise the ground state of the model is ferromagnetic and not frustrated). We note, that by a canonical transformation the exchange can be made antiferromagnetic in all three directions.

This model describes the charge ordering in α'-NaV_2O_5. In this material vanadium ions form quarter-filled two-leg ladders organized into layers (see Fig. 2). The ladder rungs are predominantly occupied by a single electron.

J. Bonča et al. (eds.), Open Problems in Strongly Correlated Electron Systems, 311–316.

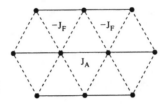

Figure 1   The triangular lattice represented as a an array of antiferromagnetic chains coupled ferromagnetically.

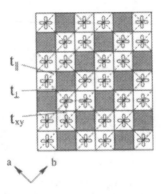

Figure 2   The crystal structure of the V-O plane in NaV_2O_5: V ions, located at the centers of plaquettes form two-leg ladders (dashed lines).

The two different positions of an electron on a rung are described as the spin up and spin down states, $\sigma^z = \pm 1$. Then the transverse field h in Eq.(1) corresponds to the hopping amplitude $t_\perp$ along rungs, while the Ising interaction between neighboring spins describes the Coulomb repulsion between electrons on neighboring rungs [1].

At $T_c = 34K$ sodium vanadate undergoes a phase transition, which first was believed to be the spin-Peierls transition. Later, however, detailed experimental studies showed that the transition is mainly due to electronic charge ordering in vanadium ladders (see, *e.g.*, Ref. [1], and references therein). The transition temperature is, however, very small compared to typical values of the hopping amplitude and Coulomb repulsion, and we argue that this may result from the frustration in the interrung Coulomb interaction between electrons.

2.   MEAN FIELD AND MONTE CARLO RESULTS

At zero transverse field the model becomes a classical Ising model, which is exactly solvable [2, 3]. For $J_F \leq J_A$ the system is disordered at any temperature. For nonzero h, we first use the mean field approximation to obtain the phase diagram shown in Fig. 3. At large transverse field and high temperature the system is disordered. It is also disordered at small h and

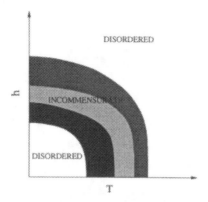

Figure 3   The schematic mean field phase diagram of the model.

T. In the latter case, the AFM correlations within chains are strong, while correlations between spins in different chains are absent due to the frustration. The two disordered domains are separated by a region of an incommensurate state. In this state the wave vector of the periodic modulation of $\langle \sigma_n^z \rangle$ is smaller than π, which gives some gain in the interchain interaction energy. In a finite system the wave vector of the ordered state, instead of varying smoothly with h and T, jumps from one commensurate value to another.

We also performed the continuous time Monte Carlo simulation for the spin lattice of $N_y = 20$ AFM chains of length $N_x = 96$. In particular, we calculated the susceptibilities, χ_n, to the periodic magnetic field in z-direction with the wave vector $q_n = \pi - \frac{2\pi n}{N_x}$. The h-dependence of the susceptibilities χ_n, for $n = 1, 2, \ldots, 6$ and $T = 0.25$, $J_A = 1$, and $J_F = 1/2$ is shown in Fig. 4. We see the cascade of transitions: At low h the susceptibility χ_0 (the staggered susceptibility) is large. At $h \approx 0.22$ this state is replaced by the state with large χ_1. In general, the index n of the largest susceptibility at a given value of the transverse field grows with h.

3.   KINK CRYSTAL

The key to understanding of the Monte Carlo results is provided by a transformation from the Ising spins to kinks in the staggered magnetization of spins in the AFM chains. Such a kink corresponds to two parallel spins on a chain bond. In an isolated chain the creation energy of a static kink is $2J_A$. It can be, however, reduced by interchain interactions. We note, that the total interaction energy between two neighboring chains is only nonzero when there are kinks in these chains. The interchain interaction energy can be written as a sum over the interactions between pairs of kinks from neighboring chains. If we assign charges to kinks, $+1$ and -1, that alternate along chains, then the energy of the

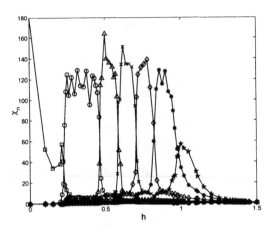

Figure 4   The plot of h-dependence of the susceptibilities χ_n to the periodic modulation of the staggered magnetization obtained by Monte Carlo simulations for $n = 0$ (squares), $n = 1$ (circles), $n = 2$ (triangles), $n = 3$, (crosses), $n = 4$ (diamonds), $n = 5$ (asterisks), and $n = 6$ (pentagrams).

interchain interactions has the form:

$$H_{int} = 2J_F \sum_{ij} q_i^e \, q_j^o \, \text{sign}(x_i - x_j). \tag{2}$$

where q_i^e and q_j^o are the charges of kinks in two neighboring even and odd chains with coordinates in the chain direction, respectively, x_i and x_j. Note, that there are two inequivalent types of the AFM chains in the triangular lattice, shifted with respect to each other in the chain direction, which we call even and odd chains. From Eq.(2) we see, that the minimal kink creation energy is lowered due to interchain interactions from $2J_A$ down to $2(J_A - J_F)$.

The interaction of spins with the transverse field results in the hopping of the kinks along the AFM chains and it also creates and annihilates pairs of kinks. This leads to a further lowering of the minimal kink creation energy.

At low temperatures the configurations, in which all kinks have the lowest possible energy have the highest weight. In this configurations the sequence, in which the positively and negatively charged kinks occur in neighboring chains, is fixed (see Fig. 5). However, to explain the large values of the susceptibilities found in Monte Carlo simulation, we have to understand first how the interaction Eq.(2) that depends not on distances between kinks, but only on their relative order, can lead to a formation of a kink crystal.

To that end we find the one-chain distribution function for the state with N kinks in each chain, all in energetically favorable positions. At zero transverse field, *i.e.*, for the classical Ising model, this distribution function is obtained by performing the summation in the partition function of the model over the kink coordinates in all chains except one. The summation only goes over the

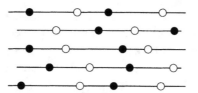

Figure 5   The configuration of kinks minimizing the interchain energy. Here, the black(white) circles denote the positively(negatively) charged kinks.

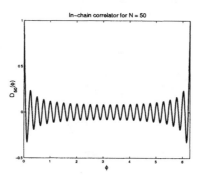

Figure 6   The plot of the alternating part of the in-chain spin correlator, $D_N(\phi)$, for $N = 50$.

kink positions that keep the interchain energy at a minimal possible value. The result is:

$$P_N(\phi_1, \phi_2, \dots, \phi_N) = \prod_{i<j} |z_i - z_j|^2, \qquad (3)$$

where the angle ϕ_j is related to the position of kink n_j in a chain of length N_x by $\phi_j = \frac{2\pi n_j}{N_x}$, and $z_n = e^{i\phi_n}$.

We see that, even though the interaction between kinks in two neighboring chains is very "loose", the integration over kink positions in many chains results in the long-range logarithmic repulsion between kinks,

$$E(\phi_1, \phi_2, \dots \phi_N) = -T \ln P_N = -2T \sum_{i<j} \ln |z_j - z_i|,$$

which favors the equidistant ordering of kinks (the so-called "order-from-disorder"). The latter can be seen from the in-chain spin-spin correlation function: $\langle \sigma_n \sigma_m \rangle \equiv (-)^{n-m} D\left(2\pi \frac{(n-m)}{N}\right)$, which we calculated using the distribution Eq.(3). The function $D_N(\phi)$ for $N = 50$ is plotted in Fig. (6). It clearly shows the long-range spin correlations resulting from the periodic ordering of kinks.

We can now explain the peculiar transverse field dependence of the spin susceptibilities, obtained in our Monte Carlo simulation (see Fig. 4). At low temperature temperature $T = 0.25$, the states with all kinks in energetically favorable positions give the main contribution to the partition function. At $h < 0.22$, the state with no kinks is dominant, which is why the staggered magnetization χ_0 is large. At $0.22 < h < 0.46$, there is one pair of kinks in each chain, which gives large χ_1, the state with high χ_2 is the state two pairs of kinks per chain, etc.

We now show that the number of kinks per chain indeed increases with h. For small h and T, the density of kinks is small and their motion can be treated semiclassically. Then the density of kinks, minimizing the free energy, is:

$$n = \frac{N}{N_x} = \frac{1}{\pi} \int_{-\pi}^{\pi} \frac{dk}{2\pi} e^{-\beta E_k}, \tag{4}$$

where E_k is the kink dispersion. For small h:

$$
\begin{aligned}
E_k &= 2(J_A - J_F) + h \cos k \tag{5} \\
&+ \frac{h^2}{J_A} \left[\sin^2 k + \frac{J_F}{(4J_A + 2J_F)} \right] + O(h^3).
\end{aligned}
$$

When h increases, the gap in the kink excitation spectrum, Δ, becomes smaller, which explains why the kink density increases with the transverse field. For finite N_x, the kink density n changes in steps of $\frac{2}{N_x}$, resulting in the cascade of the first order transitions found in the Monte Carlo simulation.

4.   SUMMARY

In conclusion, we studied the frustrated quantum Ising model using the mean field approximation and Monte Carlo simulations. In both cases we find a cascade of transitions between the states with a periodic modulation of the spin magnetization. We show that these states are crystals made of charged kinks.

Acknowledgments

This work is supported by the "Stichting voor Fundamenteel Onderzoek der Materie (FOM)', the National Science Foundation (DMR-0071767) and in part under Grant No. PHY94-07194.

References

[1] M.V. Mostovoy and D.I. Khomskii, Solid State Comm. **113**, 159 (2000).

[2] G.H. Wannier, Phys. Rev. **79**, 357 (1950).

[3] R.M.F. Houtappel, Physica **16**, 425 (1950).

ERGODIC PROPERTIES OF QUANTUM SPIN CHAINS: KICKED TRANSVERSE ISING MODEL

Tomaž Prosen

Physics department, Faculty of Mathematics and Physics, University of Ljubljana, SI-1000 Ljubljana, Slovenia

Abstract   The problem of calculation of time-correlation functions and relaxation phenomena in interacting quantum many-body systems or quantized fields is put into the framework of *ergodic theory* and *dynamical systems*. In this spirit we discuss spectral analysis of the *adjoint* propagator in a suitable Hilbert space of quantum observables in Heisenberg picture as an alternative approach to characterize many-body dynamics. We illustrate our approach by working out the time-autocorrelation functions at infinite temperature of a large class of observables for a spin $s = 1/2$ Ising chain in a periodically kicking transverse magnetic field, and show that they decay to their asymptotic values as $\propto t^{-3/2}$.

General understanding of time-correlation functions and relaxation phenomena of interacting quantum many-body systems is a challenging problem with only very partial existing solutions. Situation is much better if one considers the motion of a one-body or few-body systems where the classical ergodic theory (see e.g. [1]) describes a variety of different types of motion, ranging from quasi-periodic integrable motion characterized by a discrete spectrum of the (Liouville) evolution operator, through KAM quasi-integrability, to *ergodicity*, *mixing* and *chaos* characterized by a continuous evolution spectrum. *Mixing*, which is equivalent of saying that correlation functions of an arbitrary pair of observables (A,B) decay in time, $\lim_{t\to\infty}(\langle A(t)B \rangle - \langle A \rangle \langle B \rangle) = 0$, is the necessary dynamical property needed to justify the relaxation to equilibrium (micro/grand/canonical) state and the laws of statistical mechanics, such as the fluctuation-dissipation theorem and transport laws. Mixing in quantum many-body system has been investigated by numerical experiment only quite recently [2]. In another numerical experiment [3], a suggestive possibility of *dynamical phase transition* from non-ergodic/non-mixing to ergodic/mixing dynamics in a non-integrable periodically kicked chain of interacting spinless fermions (*kicked t-V model*) has been proposed. General absence of mixing in *completely integrable* quantum many-body systems (such as those solvable by Bethe ansatz), being the consequence of an infinite sequence of conservation

J. Bonča et al. (eds.), Open Problems in Strongly Correlated Electron Systems, 317–322.

laws, is directly responsible for finite temperature anomalous transport properties of such systems [4, 5]. Quantum mixing property is defined (here) relative
with respect to a certain (Hilbert) (sub)space $\mathfrak{S}$ of observables and is most con-
veniently discussed in Heisenberg picture where the equations of motion for
quantum observable $A(t)$ are addressed. Our approach may be interpreted also
as a kind of generalization of Lee's *recurrence-relations method* of constructing
continued-fraction representation of time-correlation functions [6].

After a short outline of the main ideas we will illustrate the method by
describing a recent calculation [8] of the time-correlation functions for Ising
$s = 1/2$ spin chain periodically kicked with transverse magnetic field.
Quantum dynamics in Heisenberg picture is governed by the Heisenberg equa-
tion $(d/dt)A(t) = i(\operatorname{ad} H(t))A(t)$, where $(\operatorname{ad} H)A := [H, A] = HA - AH$.
Our aim is to include description of dynamics of periodically time-dependent
(e.g. kicked) quantum systems so we will adopt Floquet formalism and integrate equations of motion over one period of time τ (for autonomous systems,
τ may be arbitrary, e.g. infinitesimally small), defining the so called *quantum
Floquet map* or one-step propagator $U := \mathcal{T} \exp(-i \int_0^\tau dt H(t))$ which induces
the *adjoint Floquet map* U^{ad} acting on the (sub)space $\mathfrak{S}$ of observables

$$A(t + \tau) = U^{\text{ad}} A(t), \quad U^{\text{ad}} A := U^\dagger A U.$$

We should require the subspace $\mathfrak{S}$ to be invariat under dynamics, namely
$A \in \mathfrak{S} \Rightarrow U^{\text{ad}} A \in \mathfrak{S}$. Linear space $\mathfrak{S}$ is turned into the *Hilbert space* by
introducing an innner product of two *extensive* observables in thermodynamic
limit (size $L \to \infty$) as

$$(A|B) = \langle A^\dagger B \rangle, \quad \langle A \rangle := \lim_{L \to \infty} \frac{1}{L} \frac{\operatorname{tr} A}{\operatorname{tr} 1} \tag{1}$$

which has a meaning of a canonical average at infinite temperature. [1]

We note that the adjoint Floquet map U^{ad} is a *unitary operator* in $\mathfrak{S}$ w.r.t.
inner product (1), or equivalently, that $\operatorname{ad} H$ is *self-adjoint*. Thus a pair
$(U^{\text{ad}}, \mathfrak{S})$ defines a *unitary evolution* to which we apply standard terminology
and results of *spectral theory of dynamical systems* (see e.g. part III of ref.
[1]). To any observable $A \in \mathfrak{S}$ we assign *spectral measure* $d\mu_A(\vartheta)$, where
the *spectral parameter* lies on a unit circle $\theta \in S^1 \equiv [0, 2\pi)$, which is directly
related to the time-autocorrelation function of $A(t)$

$$\langle A^\dagger A(m\tau) \rangle = (A | [U^{\text{ad}}]^m A) = \int d\mu_A(\vartheta) e^{im\vartheta}. \tag{2}$$

So, ergodic properties of the unitary evolution are dual to its spectral properties:

---

[1] This is the simplest choice which makes sense also for time-dependent systems. For autonomous systems
one may work with more general finite-temperature canonical inner product (see e.g. [6]).

- Quantum dynamics is *ergodic* in $\mathfrak{S}$ iff 1 is (at most) *simple (non-degenerate) eigenvalue* of U^{ad} (with eigenvector 1 if $1 \in \mathfrak{S}$). Then $\lim_{T \to \infty} \frac{1}{T} \int_0^T dt \{(A(t)|B) - (A|1)(1|B)\} = 0$.

- Quantum dynamics is *mixing* in $\mathfrak{S}$ if $\lim_{t \to \infty}(A(t)|B) = (A|1)(1|B)$ for arbitrary $A, B \in \mathfrak{S}$. Then U^{ad} has an *absolutely continuous spectrum*, except possibly for a simple eigenvalue 1.

- If 1 is a *multiply degenerate* eigenvalue of U^{ad} then the corresponding eigenvectors Q_n are the *conservation laws* of $\mathfrak{S}$, $U^{\text{ad}} Q_n = Q_n$

Let Q_n be a complete set of *orthonormalized* $(Q_n|Q_m) = \delta_{nm}$ eigenvectors with eigenvalue 1. The averaged time-autocorrelation function is the *spectral weight at point* 1

$$D_A := \lim_{T \to \infty} \frac{1}{2T} \int_{-T}^T dt(A(t)|A) = \sum_n |(Q_n|A)|^2. \qquad (3)$$

For example, if $A = J$ is the particle/spin current D_J is proportional to the charge/spin stiffness at infinite temperature and $D_J > 0$ indicates ballistic transport. Even when the set of known conservation laws is incomplete, the above relation gives rigorous lower bound on D_A [7, 5].

Next we propose a simple but nontrivial application of the above picture. We consider a Hilbert space $\mathfrak{S}$ of observables over infinite spin 1/2 chains described by Pauli operators $\sigma_j^s, s \in \{x, y, z\}, j \in \mathbb{Z}$, which is spanned by two infinite sequences of selfadjoint observables U_n and $V_n, n \in \mathbb{Z}$, namely

$$U_n = \sum_{j=-\infty}^{\infty} \begin{cases} \sigma_j^x (\sigma_j^z)_{n-1} \sigma_{j+n}^x, & n \geq 1, \\ -\sigma_j^z, & n = 0, \\ \sigma_j^y (\sigma_j^z)_{-n-1} \sigma_{j-n}^y, & n \leq -1, \end{cases}$$

$$V_n = \frac{1}{\sqrt{2}} \sum_{j=-\infty}^{\infty} \sigma_j^x (\sigma_j^z)_{n-1} \sigma_{j+n}^y + \sigma_j^y (\sigma_j^z)_{n-1} \sigma_{j-n}^x, \quad n > 0,$$

and $V_{-n} = -V_n$, where $(\sigma_j^z)_k := \prod_{l=1}^k \sigma_{j+l}^z$ for $k \geq 1$, $(\sigma_j^z)_0 := 1$, forming a Lie algebra:

$$\begin{aligned} [U_m, U_n] &= \sqrt{2}iV_{m-n}, \\ [V_m, V_n] &= 0, \\ [U_m, V_n] &= \sqrt{8}i(U_{m+n} - U_{m-n}). \end{aligned}$$

Operators $U_m, m \in \mathbb{Z}$ and $V_n, n \in \mathbb{N}$ form an orthonormal basis of Hilbert space $\mathfrak{S}$ w.r.t. metric (1). Let us consider Heisenberg dynamics generated by some hamiltonian $H(t) \in \mathfrak{S}$. Using the following representation of an arbitrary time-evolving observable

$$A(t) = \psi_0(t)U_0 + \sum_{n=1}^{\infty} \vec{\psi}_n(t) \cdot (U_n, U_{-n}, V_n) = \underline{\psi}(t) \cdot \underline{E},$$

in terms of a scalar $\psi_0(t)$ and a sequence of vector coefficients $\vec{\psi}_n(t)$, or in the ordered basis $\underline{E} = (U_0, U_1, U_{-1}, V_1, U_2, U_{-2}, V_2, \dots)$, the evolution of $A(t)$ is governed by the Schrödinger equation for the 'wave-function' $\underline{\psi}$

$$
\begin{aligned}
(d/dt)\underline{\psi}(t) &= -i\underline{H}^{\,\text{ad}}(t)\underline{\psi}(t), &\text{or}&& \underline{\psi}((m+1)\tau) &= \underline{U}^{\,\text{ad}}\,\underline{\psi}(m\tau), \\
(H^{\,\text{ad}})_{jk} &= -(E_j|[H, E_k]), &\text{or}&& (U^{\,\text{ad}})_{jk} &= (E_j|U^{\,\text{ad}}\,E_k).
\end{aligned}
$$

The unitary spectral problem (or its autonomous hermitean version)

$$\underline{U}^{\,\text{ad}}\,\underline{\psi}(\vartheta) = e^{i\vartheta}\underline{\psi}(\vartheta) \tag{4}$$

for a (generalized) eigenfunction $\underline{\psi}(\vartheta)$ can be interpreted as *one-particle tight-binding scattering problem on a semi-infinite lattice.*

We will now illustrate the above by considering the unitary motion in $\mathfrak{S}$ generated by Ising chain with coupling J which is periodically kicked by transverse magnetic field h (see [8] for details)

$$
\begin{aligned}
H_{\text{Ki}}(t) &= \sum_{j=-\infty}^{\infty} \left(J\sigma_j^x \sigma_{j+1}^x + \tau\delta_\tau(t)h\sigma_j^z\right) = JU_1 - \tau\delta_\tau(t)hU_0, \\
U_{\text{Ki}} &= \exp(-i\tfrac{\alpha}{2}U_1)\exp(i\tfrac{\beta}{2}U_0),
\end{aligned}
\tag{5}
$$

where $\alpha := 2\tau J$, $\beta := 2\tau h$ and $\delta_\tau(t) := \sum_{m=-\infty}^{\infty} \delta(t - m\tau)$ is a periodic Dirac delta. In the limit $\tau \to 0$, $\tau\delta_\tau(t) \to 1$, we obtain the static transverse Ising chain whose dynamics and time-correlation functions have been studied in the literature [9] and can be reproduced by the present method [8]. The spectral problem (4) for (5) is formally rewritten as one-particle tight-binding model (TBM) with next-nearest neighbour hopping

$$A_\beta \left(B_\alpha \vec{\psi}_{n+2} - C_\alpha \vec{\psi}_{n+1} + F_\alpha \vec{\psi}_n + C_\alpha^T \vec{\psi}_{n-1} + B_\alpha^T \vec{\psi}_{n-2}\right) = e^{i\vartheta}\vec{\psi}_n, \tag{6}$$

$$
A_\beta = \begin{pmatrix} c_\beta^2 & s_\beta^2 & -\sqrt{2}c_\beta s_\beta \\ s_\beta^2 & c_\beta^2 & \sqrt{2}c_\beta s_\beta \\ \sqrt{2}c_\beta s_\beta & -\sqrt{2}c_\beta s_\beta & c_{2\beta} \end{pmatrix}, \quad B_\alpha = s_\alpha^2 \begin{pmatrix} 0 & 0 & 0 \\ 1 & 0 & 0 \\ 0 & 0 & 0 \end{pmatrix}
$$

$$
C_\alpha = \sqrt{2}c_\alpha s_\alpha \begin{pmatrix} 0 & 0 & 0 \\ 0 & 0 & 1 \\ 1 & 0 & 0 \end{pmatrix}, \quad F_\alpha = \begin{pmatrix} c_\alpha^2 & 0 & 0 \\ 0 & c_\alpha^2 & 0 \\ 0 & 0 & c_{2\alpha} \end{pmatrix},
$$

where we write $c_x := \cos x$, $s_x := \sin x$, $n \geq 2$. The above scattering problem can be solved by 'the method of images', namely it can be verified that eq.(6) holds on the entire lattice $n \in \mathbb{Z}$ and includes the boundary conditions if we introduce the reflection symmetry $n \to -n$ with the images

$$\vec{\psi}_{-n} := K \vec{\psi}_n, \quad K = \begin{pmatrix} 0 & 1 & 0 \\ 1 & 0 & 0 \\ 0 & 0 & -1 \end{pmatrix}, \quad \vec{\psi}_0 := \begin{pmatrix} 0 \\ \psi_0 \\ 0 \end{pmatrix}$$

so the solutions of (6) are simple Bloch waves $\vec{\psi}_n = \vec{u}(\varphi)e^{i\varphi n}$. Straightforward calculation [8] yields two continous spectral bands

$$\vartheta_{1,2}(\varphi) = \pm 2 \arccos(\cos \alpha \cos \beta + \sin \alpha \sin \beta \cos \varphi) \quad (\text{mod } 2\pi),$$

and one degenerate spectral band $\vartheta_3(\varphi) \equiv 0$, which can be used to deduce an infinite sequence of conservation laws of the kicked Ising chain [10]

$$Q_k = s_\alpha c_\beta (U_{k+1} + U_{-k+1}) - c_\alpha s_\beta (U_k + U_{-k}) + \frac{1}{\sqrt{2}} s_\alpha s_\beta (V_{k+1} + V_{-k+1}).$$

implying *non-ergodicity* of the model. Two continuous spectral bands $\vartheta_{1,2}(\varphi)$ generate, for any observable $A \in \mathfrak{S}$, an *absolutely continuous* spectral measure having a square-root dependence $d\mu_A/d\vartheta \propto |\vartheta - \vartheta_c|^{1/2}$ near the band edges $\vartheta_c = \pm 2|\alpha - \beta|, \pm 2|\alpha + \beta|$. After performing the Fourier transformation (2) this results in an asymptotic $t^{-3/2}$ relaxation of time-correlation functions, namely for any $A \in \mathfrak{S}$, $\exists C_A > 0$, $|\langle A^\dagger A(\tau m) \rangle - D_A| < C_A m^{3/2}$. For example, the result for z−magnetization $M = -U_0$ reads

$$\langle M(m\tau)M \rangle \approx D_M + \frac{1}{4} \left| \frac{s_\alpha}{\pi s_\beta^3} \right|^{\frac{1}{2}} \left\{ \frac{\sin(2|\alpha - \beta|m - \frac{\pi}{4})}{|s_{\alpha-\beta}|^{1/2}} - \frac{\sin(2|\alpha + \beta|m + \frac{\pi}{4})}{|s_{\alpha+\beta}|^{1/2}} \right\} m^{-\frac{3}{2}},$$

The weight of the point spectrum D_A (stiffness) can be calculated either from the conserved charges (3) or from the sum rule

$(A|A) = D_A + \int_{0+}^{2\pi^-} d\vartheta (d\mu_A(\vartheta)/d\vartheta)$.

One can generalize all the formalism described above to describe dynamics of spatially modulated spin chains [11], e.g. Ising or Heisenberg XY models in spatially quasi-periodic transverse field [12]. In such cases one can map infinite temperature quantum dynamics of spin chains in Heisenberg picture to one-particle TBM on a 2d lattice or Schrödinger equation on 2d torus. The wavenumber of the field modulation maps to an effective Planck's constant of the one-particle image problem, and when its value is decreased the dynamics may become truly (quantum) mixing and even chaotic in 'the classical limit' [11]. For example, it has been shown [11] that the dynamics of XX chain in (periodically kicking) spatially modulated quasi-periodic transverse field is equivalent to quantum (kicked) Harper model [13]. However, we should stress that the analytic treatment of Heisenberg quantum dynamics of the spin chains that were discussed here is possible only due to bilinearity of the hamiltonian when expressed in Jordan-Wigner fermionic operators. In more general (non-integrable) models the structure of Hilbert space $\mathfrak{S}$ would be much more complex (see e.g. [3] where certain atempts have been made to describe Heisenberg dynamics in a kicked t-V model or kicked XX-Z spin chain).

Acknowledgments

The author is grateful to Indubala Satija for illuminating discussions and Ministry of Science and Technology of Slovenia for financial support.

References

[1] I. P. Cornfeld, S. V. Fomin and Ya. G. Sinai, *Ergodic Theory* (Springer, New York 1982).

[2] G. Jona-Lasinio and C. Presilla, Phys. Rev. Lett. **77** (1996) 4322; P. Castiglione, G. Jona-Lasinio and C. Presilla, J. Phys. A **29** (1996) 6169.

[3] T. Prosen, Phys. Rev. Lett. **80** (1998) 1808; Phys. Rev. **E60** (1999) 3939; J. Phys. A **31** (1998) L645.

[4] H. Castella, X. Zotos and P. Prelovšek, Phys. Rev. Lett. **74** (1995) 972; X. Zotos and P. Prelovšek, Phys. Rev. **B53** (1996) 983.

[5] X. Zotos, F. Naef and P. Prelovšek, Phys. Rev. **B55** (1997) 11029.

[6] M. H. Lee, Phys. Rev. Lett. **49** (1982) 1072; Phys. Rev. **B26** (1982) 2547; H. Mori, Prog. Theor. Phys. **34** (1965) 399.

[7] P. Mazur, Physica **43** (1969) 533; M. Suzuki, Physica **51** (1971) 277.

[8] T. Prosen, *to appear in* Prog. Theor. Phys. Suppl. (2000).

[9] Th. Niemeijer, Physica **36** (1967) 377; H. W. Capel and J. H. H. Perk, Physica **87A** (1977) 211.

[10] T. Prosen, J. Phys. A **31** (1998) L397.

[11] T. Prosen, Phys. Rev. **E60** (1999) 1658.

[12] Quasi-particle spectral properties of such systems have been studied extensively by Satija, e.g. in I. Satija, Phys. Rev. **B49** (1994) 3391.

[13] see e.g. Artuso *et al*, Int. J. Mod. Phys. **B8** (1994) 207.

VII

MOTT-HUBBARD TRANSITION, INFINITE DIMENSION

STRONGLY CORRELATED ELECTRONS: A DYNAMICAL MEAN FIELD PERSPECTIVE

Gabriel Kotliar

Center for Materials Theory and Physics Department
Rutgers University, Piscataway NJ 0854, USA

Abstract   We discuss the role of the Mott transition in the strong correlation problem and describe new insights gained using dynamical mean field techniques (DMFT).

1.   STRONGLY CORRELATED MATERIALS

The last two decades has witnessed a revival in the study of strongly correlated electron systems. A large variety of transition metal compounds, rare earth and actinide based materials have been synthesized. Strong correlation effects are also seen in organic metals, and carbon based compounds such as Bucky balls and carbon nanotubes. These systems display a wide range of physical properties such as high temperature superconductivity, heavy fermion behavior, and colossal magnetoresistance to name a few [1].

Strongly correlation effects are the result of competing interactions. They often produce at low temperatures several thermodynamic phases which are very close in free energy, resulting in complex phase diagrams.

As a result of these competing tendencies, strongly correlated electron systems are very sensitive to small changes in external parameters, i.e. pressure, temperature, composition, stress. This view is supported by a large body of experimental data as well as numerous controlled studies of various models of strongly correlated electron systems [1].

At the heart of the strong correlation problem is the competition between localization and delocalization, i.e. between the kinetic energy and the electron electron interactions. When the overlap of the electrons among themselves is large, a wave like description of the electron is natural and sufficient. Fermi liquid theory explains why, in a wide range of energies, systems such as alkali and noble metals behave as weakly interacting fermions, i.e. they have a Fermi surface, linear specific heat and a constant magnetic susceptibility and charge compressibility. The one electron spectra form quasi-particles and quasi-hole

J. Bonča et al. (eds.), Open Problems in Strongly Correlated Electron Systems, 325–336.

bands. The transport properties, are well described by Boltzmann theory applied to long lived quasi-particles, an approach that makes sense as long as $k_f l \gg 1$. Density functional theory in the local approximation, is able to predict most physical properties with remarkable accuracy.

When the electrons are very far apart, a real space description becomes valid. A solid is viewed as a regular array of atoms where each atom binds an integer number of electrons. These atoms carry spin and orbital quantum numbers, giving rise to a natural spin and orbital degeneracy. Transport occurs via activation with the creation of vacancies and doubly occupied sites. Atomic physics calculations together with perturbation theory around the atomic limit allows us to derive accurate spin -orbital Hamiltonians. The one electron spectrum of Mott insulators is composed of atomic excitations which are broaden to form bands which have no single particle character, known as Hubbard bands. In the large majority of known compounds the spin and orbital degrees of freedom generally order at low temperatures breaking spin rotation and spatial symmetries. However, when quantum fluctuations are strong enough to prevent the ordering, possible new forms of quantum mechanical ground states emerge.

These two limits, well separated atoms, and well overlapping bands, are by now well understood and form the basis of the "standard model" of solid state physics. One of the frontiers in strongly correlated electron physics problem, is the description of the electronic structure of solids, away from these well understood limits. The challenge is to develop new concepts and new computational methods, capable of describing situations where both itineracy and localization are simultaneously important. The "standard model " of solids breaks down in this situation and strongly correlated electron systems have anomalous properties such as resistivities which far exceed the Ioffe Regel Mott limit $\rho_{Mott}^{-1} \approx (e^2/h)k_f$, non Drude like optical conductivities, and spectral functions which are not well described by band theory [1]. To treat this problem one needs a technique which is able to treat Kohn Sham bands and Hubbard bands on the same footing, and which is able interpolate continuously between the atomic and the band limit. Dynamical Mean Field Theory (DMFT) is the simplest approach satisfying this requirement and will be used to describe some of recent advances in our understanding of the Mott transition problem [2].

2.   DYNAMICAL MEAN FIELD THEORY

The dynamical mean field theory Hubbard model Hamiltonian, Eq. (1), plays the role of the "Ising model " of strongly correlated electrons. It is the simplest model describing the competition between localization and itineracy.

$$H = - \sum_{<ij>,\sigma} (t_{ij} + \mu\delta_{ij})(c_{i\sigma}^+ c_{j\sigma} + c_{j\sigma}^+ c_{i\sigma}) + U \sum_i n_{i\uparrow} n_{i\downarrow}. \qquad (1)$$

The essential parameters in the model are δ doping, or chemical potential μ, T temperature, U/t magnetic frustration ($t_{ij} = t$ n.n. , t' t" n.n.n.). We will denote the set of relevant parameters that can be varied as α. On a Bethe lattice with large coordination with $t_{ij} = (\frac{1}{\sqrt{(t)}})^{|i-j|}$ [3] the local properties can be computed from an Anderson impurity model [4],

$$
\begin{aligned}
H_{AIM} &= \sum_{K\sigma} \epsilon_K C_{k\sigma}^+ C_{k\sigma} + \sum_{K\sigma} V_k(C_{k\sigma}^+ f_\sigma + f_\sigma^+ C_{k\sigma}) + \epsilon_f f_\sigma^+ f_\sigma^+ + \\
&+ U f_\uparrow^+ f_\uparrow f_\downarrow^+ f_\downarrow,
\end{aligned} \tag{2}
$$

with an hybridization function $\Delta(i\omega_n) = \sum_{K\sigma} \frac{V_K{}^2}{(i\omega_n - \epsilon_k)}$ obeying the self consistency condition:

$$
t^2 G_{imp}(i\omega_n)[\Delta, \alpha] = \Delta(i\omega_n). \tag{3}
$$

In the limit of large lattice coordination, all the physical quantities can be expressed in terms of the local quantities and the hopping matrix elements. This is convenient since it allows the evaluation of all the transport properties. A great deal of work has gone into solving and understanding the dynamical mean field equations (3) [2]. Furthermore the method is easily extended to treat situations with broken spin and spatial symmetries. An extension to treat realistic band structure and orbital degeneracy has recently been carried out in [25], [20], [23]. In the following sections we will describe some of the insights that have emerged and some lessons that we learned about strongly correlated electron systems.

3. INSIGHTS FROM DMFT

DMFT phase diagrams, Frustration, Complexity and Universality. The low temperature phase diagram of even simple Hamiltonians treated within DMFT, has several distinct phases, and is fairly complex. Even the simplest, bare bones Hamiltonian (one band Hubbard model with partial frustration) described in the previous section, has at least a metallic antiferromagnetic phase and a paramagnetic insulating phase in addition to a paramagnetic metal phase and the antiferromagnetic insulating phase. The phase diagram of Ref. [8] shown in Fig. 1 bears certain qualitative similarity of this phase diagram vis a vis the phase diagram of Vanadium Oxide. This observation lead Rozenberg et al. [8] to suggest several optical experiments which confirmed some qualitative predictions of DMFT.

It is important to emphasize, however, that the main lesson that should be drawn from the qualitative similarity, in the low temperature region, between the DMFT phase diagram of one of the simplest model of correlated electron systems and that of some real oxides, is the ability of the DMFT method to capture multiplicity of possible ordered states. The detailed nature of these phases, and the character of the transitions between them, depend on many details of the Hamiltonian describing the specific crystal structure and chemistry of the compound. To approach this problem, realistic versions of DMFT have been constructed and are being developed [25] [20].

The strong dependence of the low temperature phases and of the low temperature physical properties of each material on its crystal structure and chemical composition should be contrasted with the remarkable degree of universality that is predicted to occur at higher temperatures. All that is required to produce the high temperature features of the DMFT phase diagram, is a large degree of magnetic frustration to suppress the long range order and to allow for a localized phase with a large entropy entropy content. In systems without magnetic frustration, the onset of magnetism or other forms of order preempts us from accessing this strongly correlated regime. The origin of the magnetic frustration is crucial for understanding the low temperature part of the phase diagram, with its myriad of ordered phases, but is rather irrelevant in the high temperature regime, where thermal fluctuations average over all the various different configurations, leading to a more universal description which is captured by a relatively local approach such as DMFT in its single site or its clusters versions. In systems such as the titanates and in the Vanadium oxides, the origin of frustration arises from the orbital degeneracy which is unique to those materials. In the Nickel selenide sulfide mixtures, the crystal structure is such that a sizeable ring exchange term competes with the nearest neighbor superexchange interaction resulting in a reduced Neel temperature. Still, these systems display very similar phenomena around the Mott transition endpoint.

The contrast between the highly universal behavior at high temperature and the dependence of low temperature properties on additional parameters in the Hamiltonian, was discussed in Ref. [29] in connection with the comparison of the physical properties of the Vanadium Oxide and the Nickel Selenide Sulfide mixtures. The phase diagram of the two dimensional organic compound κ BEDTTF [27], where the frustration originates in its underlying chiral triangular lattice of dimers, strengthen the validity of this point of view. Indeed many of the high temperature physical properties of this have been accounted for by the DMFT studies of McKenzie and Merino [14].

To summarize, since magnetic frustration and competition of kinetic and interaction energy is all that is required for obtaining the high temperature part of the "canonical " phase diagram of a correlated electron system where the transition between the localized and extended regime as a function of $\frac{U}{t}$ takes place via a first order transition [7] [8], this is faithfully reproduced by the simplest model containing these ingredients treated within DMFT.

The phase diagram (Fig. 1) displays two crossover lines. The dotted line is a coherence incoherence crossover (i.e. the continuation of the U_{c2} line where

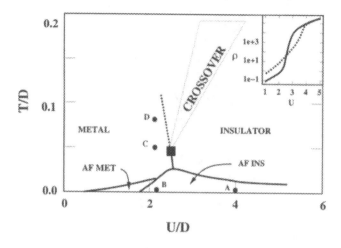

Figure 1 Schematic phase diagram of partially frustrated Hubbard model from Ref. [8], the inset illustrates the behaviour of the resistivity above but near the Mott endpoint

metallicity is lost). The shaded area is a continuation of the U_{c1} line, where the temperature becomes comparable with the gap. Both were observed in the V_2O_3 and $NiSeS$ system [9] , [28]. Further justification for this point of view, and a refined description of the localization delocalization transition around the Mott transition endpoint was achieved by the development of a Landau like description [21], [22].

Coherent and Incoherent Spectra. The mapping of the Hubbard model onto an Anderson impurity model described in previous section resulted in an important insight: that the one electron spectral function of the Hubbard model in the strongly correlated metallic region contains both atomic features (i.e. Hubbard bands) and quasiparticle features, in its spectra [4]. Further investigations revealed [5] that as the transition at zero temperature is approached there is a dramatic transfer of spectral weight from the low lying quasiparticles to the Hubbard bands, which results in a Mott transition point where the quasiparticle mass diverges, but a discontinuous gap opens in the quasi-particle spectra. These calculations were in agreement with the pioneering work of Fujimori et. al. [6] who arrived essentially to the same picture on the basis of his experimental data.

Anomalous resistivities. Fig. 2 describes the anomalous resistivities near these crossover regions. Notice the anomalously large metallic resistivity which is typical of many oxides [10]. While the curves in this figure far exceed the Ioffe Regel limit (using estimates of k_f from T=0 calculations) there is no violation of any physical principle. At low temperatures, a k space based Fermi

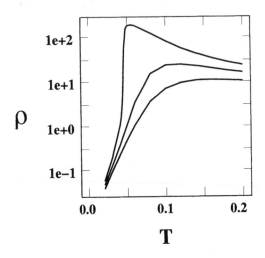

Figure 2 $\rho_{dc}(T)$ around the coherence incoherence crossover near the finite temperature Mott endpoint. $U/D = 2.1, 2.3, 2.5$ (bottom to top), obtained with the IPT method from Ref. [8].

liquid theory description works but in this regime the resistivity is low (below the Ioffe Regel limit). Above certain temperature the resistivity exceeds the Ioffe Regel limit but then quasiparticle description becomes inadequate. There is no breakdown or singularities in our formalism, the spectral functions remain smooth, (above the Mott transition endpoint), only the physical picture changes. At high temperatures we have an incoherent regime to which the Ioffe Regel criteria does not apply, because there are no long lived excitations with well defined crystal momentum in the spectra. The electron is strongly scattered off orbitals and spin fluctuations, and is better described in real space. In this regime, there is no simple description in terms of k space elementary excitations, but one can construct a simple description and perform quantitative calculations if one adopts the spectral function as a basic object in terms of which one formulates the theory.

Only the anomalously large magnitude of the resisitivity (which follows from a Greens function which has branch cuts rather than well defined poles), is universal as can be seen by comparison of the detailed temperature dependence at half filling (as in Fig. 2) and away from half filling as in Fig. 3 and 4. The temperature dependence of the transport in the high temperature incoherent regime, depends on whether the system is at integer filling or doped, as can be shown numerically [12] and analytically [11] in the example of the doped Mott insulator. The low temperature and the high temperature anomalously large resistivities also occur in strongly coupled electron phonon systems, as discussed in Ref. [13].

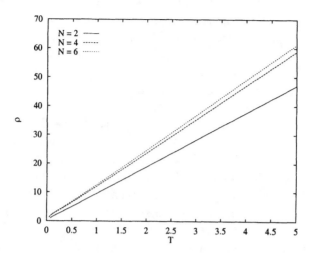

Figure 3 $\rho_{dc}(T)$ in units of $z\hbar a/e^2$, vs T (in units of D) for different values of orbital degeneracy N for a fixed doping $\delta = .1$ obtained with the NCA method which is valid at high temperatures, from the work of Palsson [11].

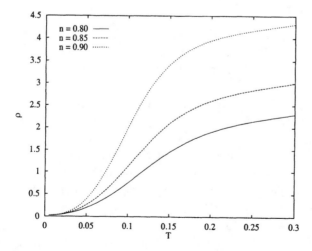

Figure 4 $\rho_{dc}(T)$ vs T in units of $z\hbar a/e^2$, vs T (in units of D) obtained with the IPT method, for different dopings at $\frac{U}{D} = 2.8$ from Ref. [11].

Anomalous Transfer of Spectral Weight.   Another manifestation of the same physics is the anomalous transfer of spectral weight which is observed in the one electron and in the optical spectra of correlated systems as parameters such as doping or pressure are varied. This surprising aspect of strong correlation physics was noted and emphasized by many authors [16]. Transfer of spectral weight can also take place as a function of temperature. For example the "kinetic energy "which appears in the lo energy optical sum rule can have sizeable temperature dependence, an effect that was discovered experimentally by [17] and explained theoretically by DMFT calculations [18]

Once more thinking about this problem in terms of well defined quasiparticles is not useful. It is more fruitful to formulate the problem   in terms of spectral functions describing on the same footing coherent and incoherent excitations. The relative weights of these components in the spectra evolves smoothly with temperature and leads to sizeable variations in the integrated optical intensity. This fundamental role of the spectral function, becomes even more prominent, in the Landau theory approach to the Mott transition where we allow the Green's function to fluctuate away from its physical saddle point value, in order to explore different non perturbative states which may not be accessible in perturbation theory in the interaction strength.

4.   STRONGLY CORRELATED ELECTRON SYSTEMS, RENORMALIZATION GROUP FLOWS AND OUTLOOK

To understand better the DMFT description of the high temperature regime, and its connection with the underlying critical points of self consistent impurity models, it is illuminating to make a comparison with the theory of quantum critical points.

In this theory [15] a phase transition at zero temperature at a critical value of a control parameter x_c, cause anomalies in a finite region of the temperature control parameter phase diagram (see Fig. 5). This region is the quantum critical regime. In the language of the renormalization group (RG), in this region, the renormalization group trajectories are very close to the fixed point describing the critical point. If the initial conditions depart from the critical surface, the renormaliation group trajectories at sufficiently low temperatures are driven towards two different fixed points describing the two stable phases.

In a strongly correlated regime, it is fruitful to think of the phase diagram of a model Hamiltonian, as one varies several independent parameters, so the parameter space in this case is multidimensional. We describe this schematically, in Fig. 5 by two axis x and y . Examples of coordinates in the relevant space are parameters which control the degree of magnetic frustration, or parameters

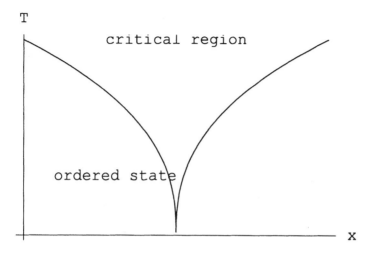

Figure 5 Schematic view of a phase diagram of a system undergoing at x_c a quantum phase transition as a function of a control parameter x. The anomalous region is the quantum critical regime [15].

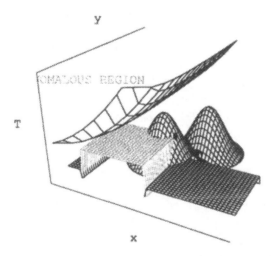

Figure 6 Schematic view of the DMFT phase diagram of a strongly correlated electron system. x and y are multidimensional control parameters, there are many underlying low temperature phases separated by first and second order phase transitions. The anomalous region is the finite temperature manifestation of the localization delocalization crossover rather than the result of proximity to one particular instability.

which control the relative stability of two competing phases. The phase diagram of a correlated system in this parameter space is more complex and is described by several competing phases and first and second order phase transitions separating them as shown schematically in Fig. 6. Inside this complex phase diagram, but probably in a physically inaccessible region, are transition lines reminiscent of the Mott transition in fully frustrated models. The shaded region at high temperature, is not directly controlled by the vecinity to a single unstable fixed point. There are two many instabilities which are nearby, and at the temperature scale considered, they are all two close in energy to be able to nudge the RG trajectory towards a specific critical trajectory (wich has a relatively small basin of attraction). DMFT is able to describe this intermediate behavior at high temperatures by means of a weakly k dependent but frequency dependent Weiss field. This results in the anomalies described in the previous subsections. Of course, at lower temperatures some form of ordering is established and more conventional behavior is restored. A somewhat analogous, but much more extreme situation (macroscopic rather than finite number of competing states), occurs in glassy systems. In the context of glassy DMFT is known to describe successfully many experimentally observed phenomena [26].

One can also think of the intermediate asymptotic regime of the disordered phase at finite temperatures in a functional integral representation. If there are several competing instabilities one is forced to introduced several Hubbard Stratonovich fields to describe them. Theories of individual phase transitions would have a phase diagram such as the one in Fig. 5a, with a Hubbard stratonovic fields condensing at a special ordering wavevector. However as long as all the instabilities are competing on the same footing, the Hubbard Stratonovich fields have low energy but no specific wavevector. In this case a local picture, with a q independent Hubbard Stratonovich master field, reproduces the correct physics at high temperatures.

It is useful to think of the program of performing realistic electronic structure calculations for correlated materials in the light of the previous discussion. The Hamiltonian describing the electrons at short distances is known and easily written down. This is the formal starting point of all electron first principles calculations. However, to describe the physics at a lower energy scale Λ, one would like to eliminate the degrees of freedom which have energies much larger than that scale, and derive an effective Hamiltonian which is more transparent and contains only the relevant or active degrees of freedom. The effective Hamiltonian at that scale, is the model Hamiltonian which is usually written down by the solid state physicist on physical grounds. As Λ is reduced a renormalization group flow in the space of all Hamiltonians is defined. Different initial conditions at short distances describe different materials, different pressures, lattice spacings concentration etc..

If one starts with conditions that correspond to weakly correlated systems (e. g. atomic numbers involving s or p electrons, high densities, etc.). The RG flows are relatively simple and converge at low energies to reach simple fixed points describing band metals or insulators.

On the other hand when we start from more correlated situations (e.g. open shells, containing relatively localized d or f electrons, lower densities), the RG trajectories are diverging from one another, reflecting the diversity of phases nearby. This situation calls for quantitative methods for realistic modelling of the material in question. One of the most serious difficulties in carrying out the Wilson RG program described above, is the continuous change in the *form* of the effective Hamiltonian from scale to scale. A typical example is the formation of a heavy fermion liquid state at a coherence energy scale. At high energies the effective Hamiltonian contains atomic configurations and conduction electrons, at low frequencies only heavy quasiparticles are the relevant degrees of freedom. In spite of these difficulties, an RG analysis taking account some quantum chemistry in the initial conditions has been carried out *in the local approximation* aided by developments in DMFT, see e.g. Ref. [19].

While following the RG flows down to very low temperatures and predicting physical properties in the most strongly correlated situations may prove to be very difficult, we optimistically hope that cluster DMFT, with small sizes will be accurate in a wide range to interesting situations (not too close to phase transitions, not too low temperatures). This view is supported by the recent success of S. Savrasov [20] in describing within realistic DMFT some of the most puzzling properties of δ Pu, a strongly correlated electron system.

Acknowledgments

This research was supported by the NSF, Division of Materials Research DMR 95-29138

References

[1] For a recent review see M. Imada, A. Fujimori, and Y. Tokura, *Rev. Mod. Phys.* **70**, (1998) 1039.

[2] For review and references to literature see A. Georges, G. Kotliar, W. Krauth, and M. Rozenberg, *RMP. Reviews of Modern Physics* **68**, (1996) 13.

[3] W. Metzner and D. Vollhardt, *Phys. Rev. Lett.* **62**, (1989) 324.

[4] A. Georges and G. Kotliar, *Phys. Rev. B* **15**, (1992) 6479.

[5] X.Y. Zhang, M. Rozenberg and G. Kotliar, *Phys. Rev. Lett.* **70**, (1993) 1666.

[6] A. Fujimori et al. *Phys. Rev. Lett.* **69**, (1992), 1796.

[7] A. Georges and W. Krauth, *Phys. Rev. B* **48**, (1993) 7167.

[8] M. Rozenberg, G. Kotliar and X.Y. Zhang, *Phys Rev. B* **49**, (1994) 10181. M. Rozenberg et al. *Phys. Rev. Lett.* **75**, (1995) 105.

[9] Kuwamoto, Honig and Appell, *Phys. Rev. B.*, **22**, 2626 (1980).

[10] V. J. Emery and S. Kivelson, *Phys. Rev. Lett.*, **74**, 3253 (1995).

[11] G. Palsson and G. Kotliar, *Phys. Rev. Lett.* **80**, 4775 (1988).

[12] M. Jarrell and T. Pruschke, *Phys. Rev. B* **49**, 1458 (1994).

[13] A.J. Millis, Hu. Jun, S. Das Sarma, *Phys Rev Lett*, **82**, (1999), 2354.

[14] R. H. McKenzie Comments Cond. Mat. Phys. 18, 309 (1988); J. Merino and R. McKenzie cond-mat/9909041

[15] J. A. Hertz, *Phys. Rev. B* **14**, pp. 1165-1184 (1976);
A.J. Millis, *Phys. Rev. B* **48**, pp. 7183-7196 (1993).

[16] for an early discussion see H. Eskes, M.B.J. Meinders, G.A. Sawatzky, *Phys Rev Lett*, **67** (1991) 1035.

[17] Schlesinger, Z. et. al. *Phys. Rev. Lett.* **71**, (1993) 1748

[18] M. Rozenberg, G. Kotliar, and H. Kajueter, *Phys. Rev. B* **54**, (1996) 8452

[19] G. Kotliar and Q. Si, *Physica Scripta*, **49**, 165-171 (1993).

[20] S. Savrasov et. al., in preparation

[21] G. Kotliar, *European Journal of Physics B*, **11**, (1999) 27.

[22] G. Kotliar, E. Lange, and M . Rozenberg, *Phys. Rev. Lett.* **84**, 5180-5183 (2000).

[23] M. Katsenelson and A. Lichtenstein, *Phys. Rev. B* **61**, 8906-8912 (2000).

[24] N. Furukawa and M. Imada *J Phys. Soc. Jpn* **61**, 331 (1992), **62** 2557 (1993).

[25] V. Anisimov et al., *J. Phys. Cond. Matt.* **9** 7359 (1997)

[26] For a recent review see Bouchaud, Cugliandolo Mezard, and Kurchan, *Out of Equilibrium Dynamics in Spin Glasses and other Glassy Systems*, A. P. Young Editor. Singapore, World Scientific.(1997)

[27] E. Lefebvre et al., cond-mat/0004455 .

[28] Miyasaka and Takagi, *unpublished*

[29] R. Chitra and G. Kotliar, *Phys. Rev. Lett* **83**, 2386 (1999);
G. Kotliar, *Physica B*, **259-261** (1999) 711.

d-WAVE PAIRING IN THE STRONG-COUPLING 2D HUBBARD MODEL

Th. Pruschke[1], Th. Maier[1], J. Keller[1], M. Jarrell[2]

[1]*Institut für Theoretische Physik, Universität Regensburg, 93040 Regensburg*

[2]*Department of Physics, University of Cincinnati, Cincinnati, OH 45221-0011*

Abstract   The superconducting instabilities of the doped repulsive 2D Hubbard model are studied in the intermediate to strong coupling regime with help of the Dynamical Cluster Approximation (DCA). To solve the effective cluster problem we employ an extended Non Crossing Approximation (NCA), which allows for a transition to the broken symmetry state. At sufficiently low temperatures we find stable d-wave solutions with off-diagonal long range order. The maximal $T_c \approx 150K$ occurs for a doping $\delta \approx 20\%$ and the doping dependence of the transition temperatures agrees well with the generic high-T_c phase diagram.

1.   INTRODUCTION

Even more than a decade after the initial discovery of the discovery of high-T_c superconductors we are still far from a complete understanding of the rich physics observed in these materials [1]. Angle resolved photoemission experiments on doped materials show a *d*-wave anisotropy of the gap in the superconducting state [2]. In underdoped materials even in the normal state this pseudogap persists [2, 3], which is believed to cause the unusual non-Fermi-liquid behavior in the normal state. This emphasizes the importance of achieving a better understanding of the superconducting phase, i.e. the physical origin of the pairing mechanism, the nature of the pairing state and the character of low energy excitations.

Relativley early, Anderson suggested, that simple models like the Hubbard model or the closely related t-J model should be sufficient to capture the essential physics of the high-T_c cuprates [5]. However, despite years of intensive studies, the precise properties of these models are still largely unknown, except for one or infinite dimensions, and especially the physics in $D = 2$ remains a controversial issue.

Finite size QMC calculations for the doped 2D Hubbard model for Coulomb repulsion U less than or equal to the bandwidth W, support the idea of a spin

J. Bonča et al. (eds.), Open Problems in Strongly Correlated Electron Systems, 337–345.

fluctuation driven interaction mediating d-wave superconductivity [4]. But the fermion sign problem limits these calculations to temperatures too high to observe a possible Kosterlitz-Thouless transition for the 2D system [4] and also poses severe restrictions on the values of U and doping. Another problem encountered in QMC calculations is their finite size character, which makes statements for the thermodynamic limit dependent on a scaling ansatz.

Results within approximate many particle methods like the Fluctuation Exchange Approximation (FLEX) [6, 7] are in agreement with QMC results, i.e. they show evidence for a superconducting state with d-wave order parameter at moderate doping for sufficiently low temperatures [6, 7]. But the FLEX method as an approximation based on a perturbation expansion in the two-particle interaction U breaks down in the strong coupling regime $U > W$, where W is the bare bandwidth.

However, experimentally the cuprates, or more precisely their parent compounds with a half-filled d band, are Mott-Hubbard insulators. This already gives a lower bound for the value of U; viz, in order to be a Mott-Hubbard insulator at half filling, it is necessary that $U > W$ for the one-band Hubbard model. Thus, the application of the above mentioned results to the cuprates is at least problematic and a theory that allows to do calculations in the strong coupling regime highly desirable.

Calculations within the Dynamical Mean Field Approximation (DMFA) [8] can be performed in the strong coupling regime and in the thermodynamic limit. But the lack of non-local correlations inhibits a transition to a state with a non-local (like e.g. d-wave) order parameter. The recently developed Dynamical Cluster Approximation (DCA) [9, 10, 11] is a fully causal approach which systematically incorporates non-local corrections to the DMFA by mapping the lattice problem onto an embedded periodic cluster of size N_c. For $N_c = 1$ the DCA is equivalent to the DMFA and by increasing the cluster size N_c the dynamic correlation length can be gradually increased while the DCA solution remains in the thermodynamic limit.

Using a Nambu-Gorkov representation of the DCA we observe a transition to a superconducting phase in doped systems at sufficiently low temperatures. This occurs in the intermediate to strong coupling regime $U > W$ and the corresponding order parameter has d-wave symmetry.

2.   METHOD

A detailed discussion of the DCA formalism was given in previous publications [9, 10, 11] where it was shown to systematically restore momentum conservation at internal diagrammatic vertices which is relinquished by the DMFA. However, the DCA also has a simple physical interpretation based on the observation that the self energy is only weakly momentum dependent for systems where the dynamical intersite correlations have only short spatial range. The corresponding self-energy is a functional of the interaction U and the Green function propagators and may be calculated on a coarse grid of N_c selected **K**-points only. Knowledge of the momentum dependence on a finer

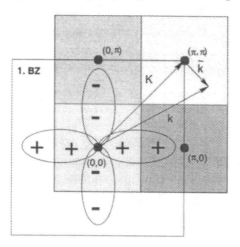

Figure 1   Choice of the $N_c = 4$ cluster **K**-points (filled circles), corresponding coarse graining cells (shown by different fill patterns) and a sketch of the d-wave symmetry of the order parameter.

grid may be discarded to reduce the complexity of the problem. To this end the first Brillouin zone is divided into N_c cells around the cluster momenta **K** (see Fig. 1). The Green functions used to form the self-energy $\Sigma(\mathbf{K}, \omega)$ are coarse grained, or averaged over the momenta $\mathbf{K} + \tilde{\mathbf{k}}$ surrounding the cluster momentum points **K** (cf. Fig.1).

Thus, the *coarse grained* Green function is

$$\hat{\bar{G}}(\mathbf{K}, \omega) = \frac{N_c}{N} \sum_{\tilde{\mathbf{k}}} \hat{G}(\mathbf{K} + \tilde{\mathbf{k}}, \omega) \quad , \tag{1}$$

where the sum runs over all vectors $\mathbf{k} = \mathbf{K} + \tilde{\mathbf{k}}$ within a cell around the cluster momentum **K**. Note that the choice of the coarse grained Green function has two well defined limits: For $N_c = 1$ the sum over $\tilde{\mathbf{k}}$ runs over the entire Brillouin zone, $\hat{\bar{G}}$ is the local Green function, thus the DMFA algorithm is recovered. For $N_c = \infty$ the $\tilde{\mathbf{k}}$-summation vanishes and the DCA becomes equivalent to the exact solution. The dressed lattice Green function takes the form

$$\hat{G}(\mathbf{k}, \omega) = \left(\omega \tau_o - \epsilon_\mathbf{k} \tau_3 - \hat{\Sigma}(\mathbf{K}, \omega) \right)^{-1} \quad , \tag{2}$$

with the self-energy $\hat{\Sigma}(\mathbf{k}, \omega)$ approximated by the cluster self-energy $\hat{\Sigma}(\mathbf{K}, \omega)$. To allow for a possible transition to the superconducting state we utilized the Nambu-Gorkov matrix representation [13] in (2) where the self-energy matrix $\hat{\Sigma}$ is most generally written as an expansion $\hat{\Sigma} = \sum_i \Sigma_i \tau_i$ in terms of the Pauli matrices τ_i. The diagonal components of $\hat{\Sigma}$ represent quasiparticle renormalizations, whereas the off diagonal parts are nonzero in the superconducting state only.

Since the self-energy $\hat{\Sigma}(\mathbf{K}, \omega)$ does not depend on the integration variable $\tilde{\mathbf{k}}$, we can write

$$\hat{G}(\mathbf{K}, \omega) = (\omega \tau_o - \bar{\epsilon}_{\mathbf{K}} \tau_3 - \hat{\Sigma}(\mathbf{K}, \omega) - \hat{\Gamma}(\mathbf{K}, \omega))^{-1}, \qquad (3)$$

where $\bar{\epsilon}_{\mathbf{K}} = N_c/N \sum_{\tilde{\mathbf{k}}} \epsilon_{\mathbf{K}+\tilde{\mathbf{k}}}$. This has the form of the Green function of a cluster model with periodic boundary conditions coupled to a dynamic host described by $\hat{\Gamma}(\mathbf{K}, \omega)$. Here we employ a perturbation theory about $hat\Gamma(\mathbf{K}, \omega) = 0$ to calculate the cluster Green function and self-energy respectively. A detailed discussion of this algorithm and the approximation used (NCA) is given in Ref. [10] for the paramagnetic state. For the superconducting state the theory has to be extended in order to account for the hybridization to the anomalous host, which couples cluster states with different particle numbers.

The self-consistent iteration is initialized by calculating the coarse grained average $\hat{\bar{G}}(\mathbf{K})$ (Eq. 1) and with Eq. 3 the host function $\hat{\Gamma}(\mathbf{K})$, which is used as input for the NCA. The NCA result for the cluster self-energy $\hat{\Sigma}(\mathbf{K})$ is then used to calculate a new estimate for the coarse grained average $\bar{G}(\mathbf{K})$ (Eq. 1). The procedure continues until the self-energy converges to the desired accuracy.

3.   RESULTS

Our aim is to study the one-particle properties of the doped 2D Hubbard Model

$$H = \sum_{ij,\sigma} t_{ij} c_{i\sigma}^{\dagger} c_{j\sigma} + U \sum_i n_{i\uparrow} n_{i\downarrow} , \qquad (4)$$

where $c_i^{\dagger}$ (c_i) creates (destroys) an electron at site i with spin σ and U is the on-site Coulomb repulsion. For the Fourier transform of the hopping integral t_{ij} we use

$$\epsilon_{\mathbf{k}} = \epsilon_o - \mu - 2t(\cos k_x + \cos k_y) - 4t' \cos k_x \cos k_y, \qquad (5)$$

accounting for both, nearest neighbor hopping t and next nearest neighbor hopping t'. We set $t = 0.25\text{eV}$ and $U = 3\text{eV}$, well above the bandwidth $W = 8t = 2\text{eV}$. For this choice of parameters the system is a Mott-Hubbard insulator at half filling as required for a proper description of the high-T_c cuprates.

To allow for symmetry breaking we start the iteration procedure with finite off diagonal parts of the self-energy matrix $\hat{\Sigma}$. As mentioned earlier, one expects the order parameter of a possible superconducting phase to have d-wave symmetry. Therefore we work with a 2x2-cluster ($N_c = 4$), the smallest cluster size incorporating nearest neighbor correlations. For the set of cluster points we choose $\mathbf{K}_{\alpha l} = l\pi$, where $l = 0, 1$ and $\alpha = x$ or y. Fig. 1 illustrates this choice of $\mathbf{K}$-points along with a sketch of the d-wave order parameter and the coarse graining cells. Obviously, for symmetry reasons, in the case of d-wave superconductivity, we expect the coarse grained anomalous Green function to vanish at the zone center and the point (π, π). Whereas the anomalous parts at the points $(0, \pi)$ and $(\pi, 0)$ should be finite with opposite signs. Fig. 2 shows a typical result for the local density of states (DOS) in the

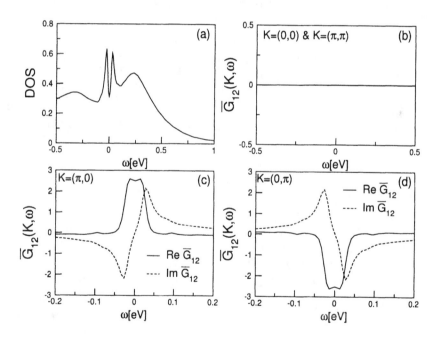

Figure 2 (a) The local density of states (DOS) near the Fermi energy and the anomalous coarse grained Green functions at the cluster points (b) $\mathbf{K} = (0,0)$ and $\mathbf{K} = (\pi, \pi)$, (c) $\mathbf{K} = (\pi, 0)$ and (d) $\mathbf{K} = (0, \pi)$ in the superconducting state. The nearest neighbor hopping integral $t = 0.25$eV, next nearest neighbor hopping integral $t' = 0$, bandwidth $W = 2$eV, the on-site Coulomb repulsion $U = 3$eV, temperature $T = 137$K and the doping $\delta = 0.19$. The anomalous parts of the Green function (b)-(d) are consistent with a d-wave order parameter.

superconducting state along with the anomalous coarse grained Green function $\bar{G}_{12}(\mathbf{K}, \omega) = N_c/N \sum_{\bar{k}} \langle\langle c_{\mathbf{K}+\bar{k}\uparrow}; c_{-(\mathbf{K}+\bar{k})\downarrow}\rangle\rangle_\omega$ at the cluster $\mathbf{K}$-points for $t' = 0$, temperature $T = 137\text{K}$ and doping $\delta = 0.19$. The anomalous coarse grained Green function vanishes at the cluster points $(0, 0)$ and (π, π) but is finite at the points $(\pi, 0)$ and $(0, \pi)$, consistent with a d-wave order parameter. Note that this result is independent of the initialization of the self-energy, i.e. an additional initial s-wave contribution vanishes in the course of the iteration. Thus a possible s-wave contribution to the order parameter can be ruled out.

The finite pair amplitude is also reflected in the local density of states (DOS) depicted in Fig. 2a, where we show the lower sub-band of the full spectrum near the Fermi energy. It displays the superconducting state pseudogap at zero frequency as expected for a d-wave order parameter.

Fig. 3 shows the DOS near the Fermi energy for the same parameters as in Fig. 2, fixed temperature $T = 137\text{K}$, but for various dopings. Obviously, the size of the pseudo gap in the superconducting state, measured as the peak to peak distance, as well as the density of states at the Fermi energy do not depend strongly upon doping. However the drop in the density of states from the gap edge to the $\omega = 0$ value first increases, reaches a maximum at about 19% doping, then decreases again. This behavior originates in the doping dependence of the anomalous Green function. In the inset we plot the coarse grained anomalous equal time Green function $\bar{G}_{12}(\mathbf{K}, \tau = 0) = N_c/N \sum_{\bar{k}} \langle c_{\mathbf{K}+\bar{k}\uparrow} c_{-(\mathbf{K}+\bar{k})\downarrow}\rangle$ for $\mathbf{K} = (\pi, 0)$. This number as a measure of the superconducting gap shows exactly the same behavior as the density of states.

The anomalous components $\bar{G}_{12}(\mathbf{K}, \omega)$ and hence the superconducting state pseudogap in the DOS become smaller with increasing temperature and eventually vanish at a critical temperature T_c depending on the set of parameters. Assembling these values for T_c as function of doping leads to the phase diagram shown in Fig. 4. As a function of doping, $T_c(\delta)$ has a maximum $T_c^{max} \approx 150\text{K}$ at $\delta \approx 19\%$ and strongly decreases with decreasing or increasing δ. The qualitative behavior of $T_c(\delta)$ in the calculated $T - \delta$ region agrees well with the generic phase diagram of the high-T_c cuprates. Unfortunately, due to the breakdown of the NCA at very low temperatures we are not able to extend the phase diagram beyond the region shown in Fig. 4. This means in particular that we cannot predict reliable values for $\delta_c(T = 0)$, beyond which superconductivity vanishes.

Another question concerns magnetic phases, especially antiferromagnetism. There is indeed a region of antiferromagnetism for a 2×2 cluster with Néel temperatures above T_c. However, T_N drops quickly with doping, and we do not find any long-range order other than superconductivity close to optimal doping. For a $2D$ system we do not expect to find antiferromagnetism at finite T, while superconductivity can prevail as Kosterlitz-Thouless transition. Indeed, preliminary results for larger clusters show that T_N decreases with

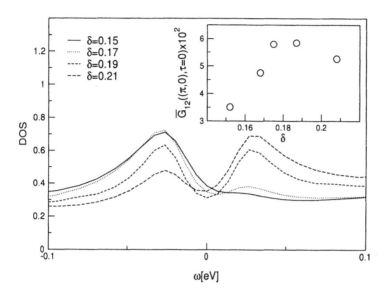

Figure 3   Density of states in a narrow region at the Fermi energy for the same parameters as in Fig. 2 but for various dopings. The gap size and the density of states at $\omega = 0$ are independent of doping.   Inset: Equal time coarse grained anomalous Green function $\bar{G}_{12}(\mathbf{K}, \tau = 0)$ at $\mathbf{K} = (\pi, 0)$.

cluster size, in accordance with the theorem of Mermin and Wagner, while T_c actually weakly increases.

The inset of Fig. 4 shows the transition temperature dependence $T_c(t', \delta = $ const.) on the next nearest neighbor hopping amplitude t' for fixed doping $\delta = 0.18$. As compared to $t' = 0$ T_c strongly decreases with growing negative t' but increases for $t' > 0$. At first this result seems counterintuitive as decreasing t' would tend to increase the density of states at the Fermi surface and so increase T_c, and a $t' < 0$ would seem to provide a mixing between next-near-neighbor sites that promotes d-wave symmetry. However a similar result was found by White *et al.*. They argue that due to short-ranged antiferromagnetic correlations, the relative phase between the states with N and $N + 2$ holes already has d-wave symmetry; therefore, $t' > 0$ provides a mixing compatible with d-wave pairing, while $t' < 0$ works against it.

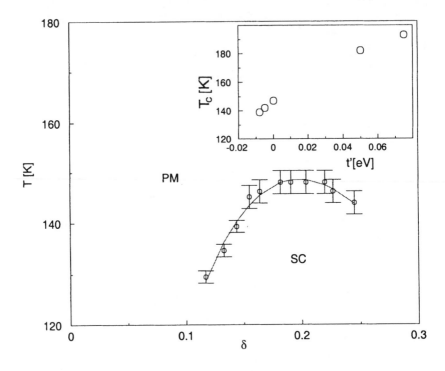

Figure 4   Temperature-doping phase diagram for the 2D Hubbard model via DCA for a $N_c = 4$ cluster. The nearest neighbor hopping $t = 0.25\text{eV}$, next nearest neighbor hopping $t' = 0$ and the Coulomb repulsion $U = 3\text{eV}$. The error bars result from the finite resolution in temperature. Inset: Transition temperature $T_c(t')$ for fixed doping $\delta = 0.18$ as a function of the next nearest neighbor hopping amplitude t'.

4.   SUMMARY

We have used the recently developed DCA to study the long open question of whether the 2D Hubbard model shows instabilities towards a superconducting state in the intermediate to strong coupling regime. We find conclusive evidence that at moderate doping a transition to a state with off diagonal long range order occurs and that the corresponding order parameter has pure d-wave symmetry. The corresponding temperature-doping phase diagram agrees qualitatively with the generic high-T_c phase diagram.

Acknowledgments

It is a pleasure to acknowledge useful discussions with P.G.J. van Dongen, M. Hettler and H.R. Krishnamurthy. This work was supported by NSF grants DMR-9704021, DMR-9357199 and the Graduiertenkolleg "Komplexität in Festkörpern". Computer support was provided by the Ohio Supercomputer Center and the Leibnitz-Rechenzentrum, Munich.

References

[1] For a review, see M.B. Maple, cond-mat/980202.

[2] H. Ding *et al.,* Nature **382**, 51 (1996).

[3] F. Ronning *et al.,* Science (1998).

[4] For a review, see D.J. Scalapino, cond-mat/9908287.

[5] P.W. Anderson, **The Theory of Superconductivity in the High-T_c Cuprates**, Princeton University Press , Princeton, NJ (1997)

[6] N.E. Bickers, D.J. Scalapino, and S.R. White, Phys. Rev. Lett., **62**, 961 (1989).

[7] T. Moriya, Y. Takahashi, and K. Ueda, J. Phys. Soc. Japan, **59**, 2905, (1990).

[8] W. Metzner and D. Vollhardt, Phys. Rev. Lett. **62**, 324 (1989).

[9] M.H. Hettler *et al.*, Phys. Rev. B **58**, 7475 (1998). M.H. Hettler *et al.,* preprint cond-mat/9903273.

[10] Th. Maier *et al.*, Eur. Phys. J. B **13**, 613 (2000).

[11] C. Huscroft et al., cond-mat/9910226.

[12] D.F. Elliot and K.R. Rao, *Fast Transforms: Algorithms, Analyses, Applications* (Academic Press, New York, 1982).

[13] J.R. Schrieffer, **Theory of Superconductivity**, Addison Wesley, Reading, MA (1993).

[14] H. Keiter, J.C. Kimball, Intern. J. Magnetism, **1**, 233 (1971).

STUDIES OF THE MOTT–HUBBARD TRANSITION IN ONE AND INFINITE DIMENSIONS

R.M. Noack, C. Aebischer, and D. Baeriswyl
Institut de Physique Théorique, Université de Fribourg, CH-1700 Fribourg, Switzerland

F. Gebhard
Fachbereich Physik, Philipps-Universität Marburg, D-35032 Marburg, Germany

Abstract   We discuss the Mott–Hubbard transition, the metal–insulator transition as a function of interaction strength U, for the half–filled Hubbard model in two numerically tractable cases: on a one–dimensional chain with nearest–neighbor hopping t', and in the limit of infinite dimensions. In the one–dimensional model, we calculate the electric susceptibility using the Density Matrix Renormalization Group and show that the transition is infinite order, irrespective of t'. In infinite dimensions, we use the Random Dispersion Approximation to calculate the behavior of the quasiparticle weight and the single–particle gap for the transition from the paramagnetic metal to the paramagnetic insulator. Within the accuracy of our calculations, we find no evidence of the discontinuous behavior found in other approaches to the infinite–dimensional limit.

1.   INTRODUCTION

As emphasized in the pioneering work of Mott [1], a strongly correlated system such as the Hubbard model can undergo a metal–insulator transition at half–filling which comes about due to the competition between kinetic and local potential energy and is characterized by the formation of local moments. This Mott–Hubbard transition is associated with a zero–temperature critical point and is expected to occur when the Hubbard U is of the order of the bandwidth [2]. The zero–temperature critical behavior is hard to investigate in detail for two reasons: first, in many realistic models on regular lattices, anti-ferromagnetic ordering can occur at finite temperature and at U–values below the Mott–Hubbard transition. Second, methods sufficiently well–controlled to give rigorous results in two and three dimensions are not available; for example, the fermion sign problem crops up in quantum Monte Carlo calculations on lattices sufficiently frustrated to suppress antiferromagnetic ordering. Our aim

J. Bonča et al. (eds.), Open Problems in Strongly Correlated Electron Systems, 347–359.

here is to study the finite–U transition in two cases in which well–controlled calculations can be performed: a one–dimensional chain with next–nearest neighbor hopping (t–t'–U chain) and the limit of infinite dimensions ($d = \infty$). In the t–t'–U chain, Section 2., we will use the Density Matrix Renormalization Group (DMRG) to calculate the electric susceptibility and the correlation length, appropriately defined for open boundary conditions. We will characterize the transition in both the $t' = 0$ case, for which there is a Bethe Ansatz solution, and in the case of sufficiently large t', for which there is a transition between two magnetically non–ordered phases at finite interaction strength. In the $d = \infty$ limit, section 3., we will use the Random Dispersion Approximation (RDA) to directly calculate the quasiparticle weight and single–particle gap as a function of interaction strength in the ground state. We will discuss our results for the behavior in the vicinity of the metal–insulator transition and compare them to the results obtained via other approaches to the $d = \infty$ limit. In section 4. we will summarize and conclude.

2.   THE t–t' HUBBARD CHAIN

In this section, we study the $U - t - t'$ model,

$$\hat{H} = - \sum_{i\sigma} \left(t\, c_{i\sigma}^{\dagger} c_{i+1\sigma} + t'\, c_{i\sigma}^{\dagger} c_{i+2\sigma} + \text{h.c.} \right) + U \sum_{i} n_{i\uparrow} n_{i\downarrow}, \qquad (1)$$

the one–dimensional Hubbard model with an additional next–nearest–neighbor hopping t'. Here, in the usual notation, $c_{i\sigma}^{\dagger}$ creates an electron of spin σ on site i and $n_{i\sigma} = c_{i\sigma}^{\dagger} c_{i\sigma}$. We consider a lattice of L sites with open boundary conditions, and we measure all energies in units of $t = 1$. At half–filling, this model undergoes a Mott metal–insulator transition as a function of U. For $t' = 0$, it is known through Bethe Ansatz calculations that the transition occurs at critical interaction strength $U_c = 0$ and that the insulating phase has gapless critical spin correlations. For $t' < 0.5$ the behavior of the system should not qualitatively change, since the noninteracting dispersion

$$\varepsilon(k) = -2t \cos k - 2t' \cos 2k \qquad (2)$$

still leads to two Fermi points with unchanged Fermi velocities and thus the same behavior in weak–coupling. (Note that the sign of t' is irrelevant at half–filling due to particle–hole symmetry.) For $t' > 0.5$, there are four Fermi points, and the weak–coupling analysis via the renormalization group and bosonization becomes more complicated. The general prediction is that there is a *metallic* phase with one low–lying charge mode and a spin gap [3]; this is the one–dimensional analog of a superconductor. In strong coupling, one expects the system to behave as the frustrated Heisenberg chain (with couplings J and J'), which is an insulator and has a spin gap when $J'/J \sim (t'/t)^2 > 0.241$ [4]. Therefore, the Mott metal–insulator transition should occur at finite U_c. This phase diagram has been confirmed numerically using the DMRG [5, 6].

Here we investigate the critical behavior near the Mott transition in more detail in both regimes of t' using the DMRG. In order to do this, we would like to calculate a quantity which can provide accurate information about the critical behavior. Usually, the susceptibility associated with the order parameter of the transition would be such a quantity; however the order parameter for a Mott transition is difficult to define. One can nevertheless define a susceptibility in terms of the response to an applied electric field that couples to the system via

$$\hat{H}_{\text{ext}} = -E\hat{X} = -E\sum_i x_i \hat{n}_i , \qquad (3)$$

where $\hat{X}$ is the dipole operator (we have taken the charge $q = 1$), x_i is the x–coordinate of the i-th site and $\hat{n}_i$ measures the occupation of this site. Note that this coupling is only well–defined for open boundary conditions, which we have already applied to Hamiltonian (1). The response to an applied electric field, however, has recently been generalized to periodic boundary conditions [7]. The electric field induces a polarization

$$P = \frac{1}{L}\langle X \rangle = -\frac{1}{L}\frac{\partial E_0}{\partial E} , \qquad (4)$$

where E_0 is the ground state energy and the average is taken with respect to the full Hamiltonian $\hat{H} + \hat{H}_{\text{ext}}$. The zero–field electric susceptibility is then defined as

$$\chi = \frac{\partial P}{\partial E}\bigg|_{E=0} = -\frac{1}{L}\frac{\partial^2 E_0}{\partial E^2}\bigg|_{E=0} . \qquad (5)$$

The electric susceptibility is a quantity characterizing the insulator; as we shall see in the following, it diverges when a continuous metal–insulator transition is approached from the insulating side, and is infinite in a metal [8].

In order to calculate the electric susceptibility, we perform DMRG calculations using the finite–system method on lattices of up to $L = 100$ sites, retaining up to 2400 states for the system block. This allows us to keep the sum of the discarded density–matrix eigenvalues to below 10^{-8}. We apply a small electric field so that the system is in a linear response regime (typically $EL = 0.001$) and measure

$$\chi = \frac{P}{E} = \frac{1}{LE}\sum_i x_i \langle \hat{n}_i \rangle . \qquad (6)$$

Tests at $U = 0$ (the most difficult case to to treat numerically) show that we can reproduce analytic results to within less than one percent.

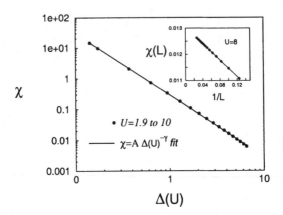

Figure 1   Electric susceptibility as a function of the gap in a log–log scale for $t' = 0$. The straight line represents a power-law fit. Each point results from a finite-size scaling, as shown in the inset for $U = 8$.

A further test of the method is the $t' = 0$ case: exactly known results from the Bethe Ansatz can provide insight into the behavior expected as well as providing a numerical check. In particular, it is known that the charge gap opens exponentially with U,

$$\Delta(U) \simeq \frac{8}{\pi}\sqrt{U}e^{-2\pi/U}, \tag{7}$$

for small U. A general closed–form expression is also available [9]. Fig. 1 compares this closed–form expression for the charge gap with the numerically calculated electric susceptibility. In order to do this, the finite–size susceptibility must first be extrapolated to $L \to \infty$, as shown in the inset to obtain χ_∞. As can be seen, we obtain a very good fit (to within 0.5%) to the form $\chi \sim 1/\Delta^2$ for U approximately between 2 and 10. For smaller U, the correlation length, which also can be obtained from the Bethe Ansatz [10], becomes too large to perform reliable finite–size scaling. Assuming that the relation $\chi \sim 1/\Delta^2$ remains valid for $U < 2$, we find that at $t' = 0$ the electric susceptibility diverges exponentially as

$$\chi \sim \frac{1}{U}\exp(4\pi/U) \tag{8}$$

for $U \to 0$. For $U > 10$, it can be shown using a strong–coupling analysis that $\chi \sim U^3 \sim 1/\Delta^3$.

For $t' > 0.5$, we must perform a similar finite–size scaling to obtain χ_∞, but expect to find a finite U_c. Fig. 2 shows the finite–size scaling of χ for $t' = 0.7$. There are two characteristically different behaviors: at small U, the system is metallic, and the susceptibility diverges with system size. A fit to a power law

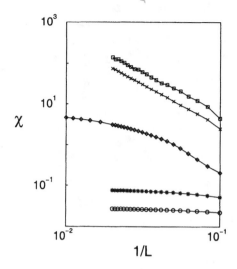

Figure 2   Electric susceptibility, χ, as a function of $1/L$ for $t' = 0.7$ and $U = 1$ (squares), $U = 2.5$ (crosses), $U = 4$ (diamonds), $U = 5.5$ (stars), and $U = 7$ (circles).

in L yields an exponent very close to 2 (within 5%) for the small U values. For $U = 0$, it can be shown analytically that $\chi \sim L^2$ for large L for all values of t'. We conjecture that such a L^2 divergence of χ is *generic* for a one–dimensional perfect metal. For larger U, χ extrapolates to a finite value as $L \to \infty$. While this is clear for the two larger U–values in Fig. 2, care must be taken near the transition because the system appears metallic up to a length scale on the order of the correlation length which diverges at the transition. Such a crossover from metallic to insulating behavior is evident in the $U = 4$ curve, for which we have taken lattice sizes of up to $L = 100$ to show that χ scales to a finite value, i.e. that the system is insulating.

In the insulating regime, we perform finite–size scaling for large L using a linear fit and extrapolating to $1/L = 0$. The result, χ_∞, is shown in Fig. 3 for $t' = 0$, 0.7, and 0.8 as a function of U. For $t' = 0$, our results confirm that the transition takes place at $U_c = 0$. For $t' = 0.7$ and 0.8, it is clear that χ diverges at progressively larger U. However, one must perform careful fitting in order to accurately determine U_c and the form of the divergence at $t' > 0$, as an analytical result for the charge gap exists only at $t' = 0$. For $t' = 0.7$, we have calculated χ at many U–values near the transition and have fitted to both a power law, $\chi \sim (U - U_c)^\gamma$ and the exponential form

$$\chi_\infty = A \exp\left[\frac{B}{(U - U_c)^\sigma}\right] \qquad (9)$$

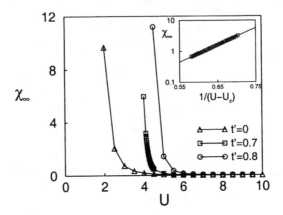

Figure 3 Electric susceptibility $\chi_\infty(U, t')$ of the infinite–size system for $t' = 0, 0.7, 0.8$, as a function of U; the lines are guides to the eye. Inset: $\chi_\infty(U, t' = 0.7)$ for $U = 4.1$ to 4.4 (squares) as a function of $1/(U - U_c)$, on a semilog scale. The line is a fit to an exponential form.

Here we have not included power–law prefactors which would represent logarithmic corrections because their form is unknown. We find that the fit to the power law form yields $U_c \simeq 3.4$, a point at which careful finite–size scaling of χ yields a finite value of χ_∞. Therefore, this U_c is clearly too large. The exponential fit yields $\sigma \simeq 1.049$, $B \simeq 12.45$ and $U_c \simeq 2.67$, a more reasonable value of U_c. Note that the values for σ and B are again very close to the ones obtained for $t' = 0$. The inset of Fig. 3 shows a semilog plot of χ_∞ versus $1/(U - U_c)$ as well as the fit itself, illustrating its good quality. We therefore find that the exponential form, Eq. (9), expected in an infinite–order transition characterizes the transition at *all* t', irrespective of whether a spin gap exists or whether U_c is finite or zero.

In one dimension, one can also define a correlation length associated with the insulating phase [8],

$$\xi = \frac{1}{L}\langle\Psi_0|\hat{X}^2|\Psi_0\rangle \qquad (10)$$

where $\hat{X}$ is the dipole operator defined previously and $|\Psi_0\rangle$ is the ground state of the system. This quantity measures the fluctuations of the polarization, and is also related to a weighted spatial average of the density–density correlation function. We have checked that this correlation length is proportional to that extracted directly by fitting the density–density correlation function to an exponential.

The correlation length defined in this manner can be used to test scaling near the quantum critical point. In particular, hyperscaling, the assumption

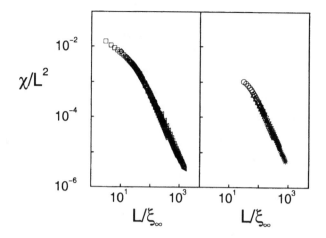

Figure 4   Scaling plots of $\chi(L, U, t')/L^2$ versus $L/\xi_\infty(U, t')$ in a log–log scale: $t' = 0$ (left), $t' = 0.8$ (right). Different symbols correspond to different values of U.

that there is one relevant length scale, ξ_∞, as $L \to \infty$, leads to the finite–size scaling form for the electric susceptibility [8],

$$\chi = L^{2+z-d}C\Phi(L/\xi_\infty) , \qquad (11)$$

where C is a non–universal constant that depends on microscopic details, Φ is a universal function [11], z is the dynamic critical exponent and here $d = 1$. The hyperscaling assumption implies that Φ tends to a finite value as $L/\xi_\infty \to 0$. This is the region in which the system appears metallic and in which χ tends to scale like L^2. Thus $z = 1$ is the only consistent value in Eq. (11), in agreement with exact results for $t' = 0$ [10]. In the opposite limit, $L/\xi_\infty \to \infty$, the system behaves as an insulator for all sizes and χ tends to a finite value χ_∞. The scaling form (11) with $z = d = 1$ thus implies $\lim_{x\to\infty} \Phi(x) = A/x^2$ and $\chi_\infty \sim \xi_\infty^2$.

In order to confirm the scaling form Eq.(11) for our model, we plot the DMRG results for χ/L^2 as a function of L/ξ_∞ in Fig. 4. The quantity ξ_∞ is obtained by calculating ξ on finite systems using Eq. (10) and then performing a finite–size extrapolation similar to that used to obtain χ_∞. Notice that *all* L and U points for a particular t' collapse onto the same curve, confirming hyperscaling. Therefore, ξ_∞ behaves as the correlation length.

3.   THE RANDOM DISPERSION APPROXIMATION

The Mott transition as a function of interaction strength in the half–filled Hubbard model has been quite extensively studied within the Dynamical Mean Field Theory (DMFT), which is exact in the limit of infinite dimensions. The standard approach to the DMFT involves taking advantage of the local nature of the self–energy to map the infinite–dimensional Hubbard model to a local impurity self–consistent approximation (LISA). While the LISA can be solved using a number of methods, the general picture for the metal–insulator transition is provided by iterated second–order perturbation theory (IPT). Within IPT, one finds a region of coexistence of a metallic solution and an insulating solution for $U_{c,1} \le U \le U_{c,2}$. Within this region, the metallic solution can be argued to be more stable at zero temperature, leading to a discontinuous behavior of the insulating gap at $U_{c,2}$. At finite temperature, this coexistence leads to a line of first–order transitions ending in a second–order tricritical point. For a review of these methods and results, see Ref. [12].

The Random Dispersion Approximation is an alternate route to the limit of infinite dimensions. It takes as starting point the properties of the kinetic energy operator

$$\hat{T} = \sum_{\mathbf{k}\sigma} \varepsilon(\mathbf{k}) \, \hat{n}_{\mathbf{k}\sigma} \tag{12}$$

in the limit of infinite dimensions. In particular, the kinetic energy operator is characterized by the noninteracting density of states

$$D_0(\varepsilon) = \frac{1}{L} \sum_{\mathbf{k}} \delta(\varepsilon - \varepsilon(\mathbf{k})) \tag{13}$$

and by the joint density of states

$$D_{\mathbf{q}}(\varepsilon_1, \varepsilon_2) = \frac{1}{L} \sum_{\mathbf{k}} \delta(\varepsilon_1 - \varepsilon(\mathbf{k}))\delta(\varepsilon_2 - \varepsilon(\mathbf{k} + \mathbf{q})) \ . \tag{14}$$

The joint density of states can be interpreted as the momentum–averaged probability density that a momentum transfer $\mathbf{q}$ generates a transfer of kinetic energy from ε_1 to ε_2. In infinite dimensions, the joint density of states factorizes, $D_{\mathbf{q}}(\epsilon_1, \epsilon_2) = D_0(\epsilon_1)D_0(\epsilon_2)$ for almost all $\mathbf{q}$. Higher correlation functions factorize accordingly. This generic feature of momentum space in infinite dimensions is use to *define* the Random Dispersion Approximation (RDA). In the RDA, we treat a finite system in which the dispersion $\varepsilon_{\mathbf{k}}$ is replaced by the random dispersion $\varepsilon_{\mathbf{k}}^{\text{RDA}}$ with the same density of states, $D_0^{\text{RDA}}(\varepsilon) = D_0(\varepsilon)$, but a completely uncorrelated joint density of states, $D_{\mathbf{q}\neq 0}^{\text{RDA}}(\varepsilon_1, \varepsilon_2) = D_0(\varepsilon_1)D_0(\varepsilon_2)$. In order to satisfy these conditions, we randomly assign an energy $\varepsilon_{\mathbf{k}}^{\text{RDA}}$ chosen from the probability distribution $D_0(\varepsilon)$ to each momentum $\mathbf{k}$. In this version of the RDA all forms of magnetic order are suppressed ("fully frustrated hopping" [12]) so that magnetic order does not conceal the Mott-Hubbard transition.

It can be shown that the RDA leads to the same collapse of diagrams as the more conventional limit of taking the number of nearest neighbors to infinity [2].

The proof is based on the observation that the nonlocal part of the noninteracting Green function behaves like $1/\sqrt{L}$, i.e. it will vanish in the thermodynamic limit, while the local part is of order unity and, for a given $D_0(\varepsilon)$, is identical to that found in the conventional limit. This leads to a purely local self–energy and the same skeleton expansion as obtained in the infinite–dimensional limit of the Hubbard model.

In order to numerically calculate the properties of the Hubbard model within the RDA, we treat an ensemble of formally one–dimensional systems, each with Hamiltonian

$$\hat{H}^{\mathcal{Q}} = \sum_{k,\sigma} \varepsilon_k^{\mathcal{Q}} \hat{c}_{k,\sigma}^{+} \hat{c}_{k,\sigma} + U \sum_l \hat{n}_{l,\uparrow} \hat{n}_{l,\downarrow} \, . \tag{15}$$

The realizations $\varepsilon_k^{\mathcal{Q}}$ of the dispersion are chosen so that the noninteracting density of states $D_0(\varepsilon)$ is properly reproduced, and k is taken so that the boundary conditions are antiperiodic for an even number of sites L and periodic boundary conditions for odd L. For more details on how we choose the set of configurations over which to average, see Ref. [13]. We calculate the ground–state properties for each realization $\mathcal{Q}$ at lattice size L by performing an exact diagonalization in the momentum basis using the Davidson algorithm [14].

Thus we can calculate physical quantities such as the one-particle gap

$$\Delta^{\mathcal{Q}} = E_0^{\mathcal{Q}}(L+1) + E_0^{\mathcal{Q}}(L-1) - 2E_0^{\mathcal{Q}}(L) \, , \tag{16}$$

and the momentum distribution

$$n_k^{\mathcal{Q}} = \frac{1}{2} \sum_{\sigma} \langle \Psi_0^{\mathcal{Q}} | \hat{c}_{k,\sigma}^{+} \hat{c}_{k,\sigma} | \Psi_0^{\mathcal{Q}} \rangle \, , \tag{17}$$

for a particular U, L and configuration $\mathcal{Q}$. Here $|\Psi_0^{\mathcal{Q}}\rangle$ is the normalized, paramagnetic ground-state wave function at half band-filling, and $E_0^{\mathcal{Q}}(N)$ is the N-particle ground-state energy.

To obtain the mean value of a physical quantity, we average over R_L selected configurations $\mathcal{Q}$, e.g., $\Delta(U;L) = (1/R_L)\sum_{\mathcal{Q}} \Delta^{\mathcal{Q}}(U;L)$. The error is then given by the standard deviation divided by $\sqrt{R_L}$ due to the random sampling [13]. We then must extrapolate the result to infinite system sizes, e.g., $\Delta(U) = \lim_{L\to\infty} \Delta(U;L)$ in order to obtain results equivalent to the $d = \infty$ limit. In the following, we present results for a semielliptic density of states, $D(\varepsilon) = 4/(\pi W)[1 - (2\varepsilon/W)^2]^{1/2}$ for $|\varepsilon| < W/2$, and measure energies in units of $t = W/(2\pi)$.

Fig. 5(a) shows the one-particle gap $\Delta(U;L)$ as a function of $1/L$ for various values of U. The lines are separate $\mathcal{O}(1/L^2)$ extrapolations for odd and even lattice sites. The error bars on the symbols, calculated as described above, are small enough so that finite-size corrections are the dominant effect. The extrapolated gap, $\Delta(U)$, is shown in Fig. 6(a).

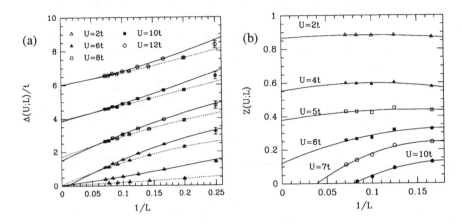

Figure 5 (a) One-particle gap $\Delta(U; L)$ plotted as a function of $1/L$ for various U values. Separate extrapolations are done using a second-order polynomial in $1/L$ for even L (solid lines) and odd L (dotted lines). (b) The quasiparticle weight $Z(U; L)$ plotted versus $1/L$ for various U values, also fitted with a second-order polynomial in $1/L$.

In order to determine the quasiparticle weight $Z(U)$, we perform a fit to the averaged momentum distribution function for a particular U, L using a Fermi–liquid form $n(\varepsilon) = [1 + Z(U; L)]/2 + \alpha(\varepsilon/W) \ln (\varepsilon/W) + \beta(\varepsilon/W)$, valid for $\varepsilon < 0$ [13]. We can obtain $n(-\varepsilon) = 1 - n(\varepsilon)$ due to particle-hole symmetry. We use the even lattice sizes only because the k–points are better placed relative to the Fermi level than those in odd–sized lattices. In Fig. 5(b), we display $Z(U; L)$ obtained from this fit. Then, as is shown in the figure, we perform a $(1/L) \to 0$ extrapolation by fitting to a second order polynomial in $1/L$. The extrapolated value, $Z(U)$, is displayed in Fig. 6.

As can be seen in Fig. 6(a), both the gap, $\Delta(U)$, and the quasiparticle weight, $Z(U)$, go to zero continuously at the same $U_c \approx W$, to within the errors, which we estimate to be approximately $\pm 0.1W$. The gap $\Delta(U)$ is zero to within the error of the finite-size extrapolation (approximately the symbol size) for $U < U_c \approx W$, and grows approximately linearly, i.e. $\Delta(U) \approx U - W$ for $U > U_c$. The fact that we obtain the same value of U_c for both $Z(U)$ and $\Delta(U)$ is a consistency check for the method: since we are treating a well–defined system in its ground state, we expect the system to be either metallic or insulating at a particular U. A discontinuous transition would appear as a jump in either $Z(U)$, or, as expected from the discontinous scenario found in IPT, in $\Delta(U)$ rather than a region in which both quantities are finite.

Comparison of our results with a number of other methods, all of which are based on solving the single–impurity problem self–consistently (the LISA), yield both qualitative and quantitative discrepancies. Quantitatively, the RDA

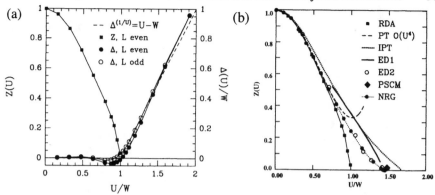

Figure 6   (a) The quasiparticle weight $Z(U)$ and the one-particle gap, $\Delta(U)$, extrapolated to the thermodynamic limit separately for even and odd system sizes. (b) A comparison of $Z(U)$ for various methos, as described in the text.

yields a $U_c^{\mathrm{RDA}} \approx W$ that is substantially lower than the critical $U_{c,2} \approx 1.4 - 1.5W$ found with other methods, with fairly good consistency between the methods. Even the lower bound on the coexistence region, $U_{c,1}$, tends to be somewhat higher than U_c^{RDA}. This is illustrated in a comparison of $Z(U)$, shown in Fig. 6(b). Here PT is standard fourth order perturbation theory on the original Hubbard model. IPT is second–order perturbation theory [12], ED1 [15] and ED2 [16] are exact diagonalization studies, PSCM is a projective self–consistent method [17], and NRG is a numerical renormalization group study [18], all within the LISA. In addition, two finite–temperature quantum Monte Carlo studies [19, 20] both yield extrapolated zero–temperature critical points at $U_c \approx 1.4 - 1.5W$, although there has been some disagreement as to the presence or absence of a coexistence region.

Qualitatively, we also find no evidence of discontinuous behavior due to the coexistence found in most of the LISA–based calculations. However, the large quantitative differences in the calculations ($Z \approx 0.3$ for NRG and ED2 at $U = W$ versus $Z = 0$ for RDA) near the critical point would have to be understood before a comparison of the qualitative behavior would be meaningful.

As far as the origin of these discrepancies are concerned, one possibility that has been suggested is that the relatively small lattice sizes available to the RDA calculations do not allow sufficient resolution, i.e. that the finite–size effects would become larger at larger lattice sizes. This possibility can only be addressed by treating substantially larger lattices, larger than can be treated with exact diagonalization. Another possibility is that there is some intrinsic difference between the RDA and the LISA approaches, especially near the transition. In the LISA, the mapping of the Hubbard model onto an effective single–site model is based on perturbation theory in U and $1/U$; it is conceivable that the corresponding radii of convergence are not big enough to cover the critical regime.

4.   SUMMARY

We have discussed the Mott transition as a function of interaction strength at zero temperature in two examples of the half–filled Hubbard model: the t–t'–U chain and the infinite–dimensional model with a semi–elliptical density of states. In both models, the critical interaction strength U_c is of the order of the bandwidth. In the t–t'–U chain, U_c becomes nonzero for $t' > 0.5t$, where the spin degrees of freedom develop a gap. By calculating the electric susceptibility using the Density Matrix Renormalization Group, we have been able to show that the critical behavior is nevertheless the same as in the $t' = 0$ case: the susceptibility diverges exponentially when the transition is approached from the insulating side. This implies that the effective model for the critical behavior of the charge degrees of freedom must be the same in both cases.

In the limit of infinite dimensions, we use the Random Dispersion Approximation, an approach that takes advantage of the decorrelation of momenta, to calculate the behavior of the quasiparticle weight and the one–particle gap. We find continuous behavior of both quantities and a transition at $U_c = (1.0 \pm 0.1)W$, where W is the bandwidth. Both the value of U_c and the continuous behavior of the gap are in contradiction with the majority of calculations based on the self–consistent solution of a single–inpurity problem. The origin of this discrepancy is not yet clear.

Acknowledgments

R.M.N., C.A., and D.B. thank the Swiss National Foundation for financial support under Grant No. 20–53800.98 , and thank F.F. Assaad, S. Daul, E. Jeckelmann, G.I. Japaridze, S. Sachdev, C.A. Stafford and X. Zotos for valuable discussions. In addition, F.G. and R.M.N. acknowledge helpful discussions with R. Bulla, D. Logan, P. Nozières, D. Vollhardt, P.G.J. van Dongen, S.R. White, and X. Zotos. Parts of the numerical computations were done at the Swiss Center for Scientific Computing.

References

[1] N.F. Mott, *Metal-Insulator Transitions*, 2nd ed. (Taylor and Francis, London, 1990).

[2] F. Gebhard, *The Mott Metal-Insulator Transition*, Springer Tracts in Modern Physics Vol. 137 (Springer, Berlin, 1997).

[3] M. Fabrizio, Phys. Rev. B **54**, 10054 (1996).

[4] S. Eggert, Phys. Rev. B **54**, 9612 (1996); K. Okamoto and K. Nomura, Phys. Lett. A **169**, 422 (1992).

[5] R. Arita, K. Kuroki, H. Aoki, and M. Fabrizio, Phys. Rev. B **57**, 10 324 (1998).

[6] S. Daul and R. M. Noack, Phys. Rev. B **61**, 1646 (2000).

[7] R. Resta, Phys. Rev. Lett. **80**, 1800 (1998); R. Resta and S. Sorella, Phys. Rev. Lett. **82**, 370 (1999).

[8] C. Aebischer, D. Baeriswyl, and R. M. Noack, cond-mat/0006354.

[9] A. A. Ovchinnikov, Sov. Phys. JETP **30**, 1160 (1970).

[10] C. A. Stafford and A. J. Millis, Phys. Rev. B **48**, 1409 (1993).

[11] V. Privman and M. E. Fisher, Phys. Rev. B **30**, 322 (1984).

[12] A. Georges, G. Kotliar, W. Krauth, and M. J. Rozenberg, Rev. Mod. Phys. **68**, 13 (1996).

[13] R. M. Noack and F. Gebhard, Phys. Rev. Lett. **82**, 1915 (1999).

[14] E. R. Davidson, Comp. in Physics **7**, 519 (1993).

[15] M. J. Rozenberg, G. Moeller, and G. Kotliar, Mod. Phys. Lett. B **8**, 535 (1994).

[16] M. Caffarel and W. Krauth, Phys. Rev. Lett. **72**, 1545 (1994).

[17] G. Moeller, Q. Si, G. Kotliar, M. J. Rozenberg, and D. S. Fisher, Phys. Rev. Lett. **74**, 2082 (1995).

[18] R. Bulla, Phys. Rev. Lett. **83**, 136 (1999).

[19] J. Schlipf *et al.*, Phys. Rev. Lett. **82**, 4890 (1999).

[20] M. Rozenberg, R. Chitra, and G. Kotliar, Phys. Rev. Lett. **83**, 3498 (1999).

QUANTUM CRITICAL BEHAVIOR OF CORRELATED ELECTRONS: RESONANT STATES

V. Janiš

Institute of Physics, Academy of Sciences of the Czech Republic,
Na Slovance 2, CZ-18221 Praha, Czech Republic

Abstract   Coulomb repulsion drives correlated electrons toward a critical point where bound electron-hole pairs condense. However, before the critical point is reached quantum fluctuations dominate the critical region and can lead to formation of resonant pair states. We use a two-particle parquet approach to describe quantum criticality. We show with a simplified version of the parquet equations that due to a two-particle self-consistence of the nonlinear parquet equations, a new phase with anomalous vertex functions may arise. The new solution describes a phase with resonant pair states where the effective interaction between the quasiparticles becomes complex. Although the two-particle vertex functions display symmetry breaking with order parameters, the low-energy one-particle spectrum of the phase with resonant states remains Fermi-liquid like.

1.   INTRODUCTION

Low-energy charge excitations in metals are in most situations well described by renormalized Bloch waves within the Fermi-liquid theory. Motion of elementary excitations, quasi-particles, is in Fermi liquid only weakly correlated and the electrons act as individual objects. A picture of almost free electrons with weak dynamical fluctuations is appropriate. However, a number of low-temperature phenomena in transition and heavy-fermion metals and in particular in underdoped cuprates and manganites display experimentally relevant deviations from the Fermi-liquid behavior. In these systems correlation-driven fluctuations are no longer negligible and the effects of the Coulomb repulsion cannot be reduced to a mere renormalization of the Fermi-gas parameters. The electrons are much more strongly forced to cooperate to establish a thermodynamic equilibrium than in the Fermi-liquid phase. The most prominent examples of non-Fermi-liquid states are Luttinger liquid, nearly antiferromagnetic Fermi liquids, nearly charge-ordered phases (stripes), and quantum critical behavior. In particular the quantum critical behavior characterized by a macroscopic quantum coherence has attracted great attention among condensed

J. Bonča et al. (eds.), Open Problems in Strongly Correlated Electron Systems, 361–370.

matter theorists and experimentalists in recent years [1]. The reason for the increased interest in quantum criticality lies in the opportunity to understand on a microscopic level an observed specific non-Fermi-liquid behavior in the metallic states beyond the standard quasi-particle scheme.

Quantum critical points are linked to zero-temperature phase transitions where only correlation-induced fluctuations are present. However, quantum critical and cooperative phenomena with long-range quantum coherence go beyond the zero-temperature magnetic and superconducting phase transitions. Quantum criticality in itinerant models can generally be characterized by proximity of singularities in correlation or vertex functions and include collective response such as metamagnetism, metal-insulator transition, Kondo effect, and quantum transport in strongly interacting (high-temperature) superconductors and in disordered electron systems near band edges. Quantum critical regions can be very large and can extend to rather high temperatures and relatively small interaction strengths.

Quantum critical behavior in correlated electrons is caused by formation and condensation of quasiparticle bound pairs. The pairing mechanism can either include an electron and a hole or two electrons for magnetic and superconducting transitions, respectively. Except for the weak-coupling BCS pairing theory we do not have many appropriate tools how to investigate correlation-induced quantum fluctuations in quantum criticality. Most of the approaches are numerical techniques such as Monte-Carlo simulations and numerical and density-matrix renormalization groups. Numerical solutions deliver quantitative results and global trends, but do not say much about the microscopic origin of quantum pairing. Among analytic means, a mapping onto a Landau-Ginzburg-Wilson functional seems to be most suitable for employing the renormalization-group reasoning and scaling also to quantum critical phenomena [2]. In this scheme, however, we loose contact with the original fermionic degrees of freedom and cannot trace down how the Fermi-liquid phase gets unstable. To understand the microscopic origin of the non-Fermi-liquid behavior we have to use direct analytic methods and solutions for models of interacting electrons.

We use the generic Hubbard model at half filling and zero temperature and investigate its behavior in the transition region between weak and strong-coupling regimes where the Fermi-liquid solution is expected to break down due to formation of electron-hole pairs that can finally lead to a metal-insulator transition. We use the parquet approach with two singlet electron-hole scattering channels as a most suitable analytic tool for investigating quantum criticality of the metal-insulator transition. The leading-order potentially divergent contributions to the effective interaction are comprised in this approximation. We explicitly demonstrate that quantum critical behavior has a much more complex structure than the classical one and may lead to new quantum effects with symmetry breaking terms in two-particle vertex functions.

2.   CLASSICAL VS. QUANTUM CRITICALITY

Classical phase transitions are best described within the Landau-Ginzburg-Wilson functional where only relevant degrees of freedom are kept so that the low-energy and universal critical behavior is reproduced

$$H[\Psi] = \frac{1}{2}\sum_q (m(T) + q^2)|\Psi(q)|^2$$

$$+ \frac{U}{4N^2}\sum_{q_i} \Psi(q_1)\Psi(q_2)\Psi(q_3)^*\Psi(q_1 + q_2 - q_3)^*. \qquad (1)$$

Here m is an effective mass of the excitations and Ψ the fluctuating ordering field. The critical point is defined $m(T_c) = 0$.

We can construct a standard many-body perturbation theory with Feynman diagrams to evaluate the free energy for Hamiltonian (0). Since the fluctuating field Ψ is the ordering field the propagator in such an expansion is the appropriate susceptibility that can be represented in the low-energy limit as

$$\chi(\mathbf{q}) = \frac{1}{m(T)^2}\frac{C}{1 + D(T)^2 q^2/m(T)^2}, \qquad (2)$$

where C and $D(T)$ are Curie and diffusion constants. We see that classical critical points are characterized by a diverging two-point propagator, or bond in the diagrammatic language. This function defines a diverging scale, the correlation length $\xi = m^{-1}$. Above the upper critical dimension this is the only diverging scale and a mean-field theory with a saddle-point solution $\Psi = \langle\Psi\rangle$ correctly predicts the nonanalyticity of the critical point. Below the upper critical dimension another divergent scale in the diffusion constant emerges. We have $D^2 \sim \xi^\eta$, where η is an anomalous dimension. Then also the effective interaction defined with a four-point vertex function $U_{eff} = \Gamma(0, 0; 0)$ diverges. Perturbation theory around the mean-field saddle point gets singular and we have to use renormalization-group ideas to capture the nonanalyticity of the critical point correctly [3]. Nevertheless all divergences can be scaled out with the single diverging scale, the correlation length. The effective interaction in the new length scale ($\xi = 1$) turns finite and we can use a loop and small-parameter expansion to determine the universal critical behavior.

Description of a quantum critical behavior of correlated fermions is more involved. First, we do not have a proper mean-field (saddle-point) theory for correlation-induced transitions. Second, we cannot work with only ordering (bosonic) fields if we are interested in the way the Fermi-liquid solution breaks down. It is essential to keep track of the one-electron (fermionic) propagator determining the Fermi-liquid behavior. Its low-energy asymptotics is

$$G(\mathbf{k}, \omega) = \frac{a}{(1 - \Sigma')\omega - v_F(|\mathbf{k}| - k_F) + i\delta\,\text{sign}(|\mathbf{k}| - k_F)}. \qquad (3)$$

This two-point function in the diagrammatic expansion for models of interacting electrons does not diverge at quantum critical points. Quantum critical points are characterized by a divergence in either the effective mass $|\Sigma'| \to \infty$ or the Fermi velocity $v_F \to \infty$. In the former case frequency fluctuations dominate while in the latter momentum ones. In both cases, however, first a two-particle (four-point) vertex function diverges. It means that the effective interaction between quasiparticles always diverges at quantum critical points. Unlike the classical case we do not have an appropriate diverging correlation length with which we could scale out the divergence in the effective interaction and use a small-parameter (loop) or weak-coupling expansion. We are faced with a singular perturbation theory with divergences in two-particle vertex functions. Only approximate schemes self-consistent at the two-particle level can properly deal with the singularities in the vertex functions. A first step should be a construction of a mean-field theory describing at least qualitatively correctly the transition from the weak-coupling, Fermi-liquid solution to the strong-coupling, local-moment state.

3.   PARQUET APPROACH

When approaching quantum criticality from the Fermi-liquid side we need one-electron propagators to control the Fermi-liquid behavior and simultaneously two-particle vertex functions to treat the singularities properly. We hence cannot use the Landau-Ginzburg-Wilson approach of [2]. A systematic way beyond the weak-coupling perturbation theory is a skeleton expansion with simultaneous renormalization of the particle mass (self-energy) and charge (vertex-functions). Such a systematic renormalization scheme was introduced and discussed in [4] and is called parquet approach. The lowest-order approximation within this scheme is the parquet approximation being the simplest approximation with a two-particle self-consistence, i.e., vertex functions are determined self-consistently [5].

3.1   PARQUET EQUATIONS

The idea behind the parquet-type approximations is to utilize the ambiguity in the definition of a two-particle irreducibility and a respective Bethe-Salpeter equation for the full vertex function. Unlike the one-particle case we have three ways how to define a two-particle irreducibility, i.e, the way of cutting two one-particle lines without disconnecting the diagram. They are called electron-hole (eh), electron-electron (ee), and interaction (U) channels of multiple two-particle scatterings [6]. Each definition leads to a Bethe-Salpeter equation defining the full two-particle vertex from a two-particle irreducible function. The Bethe-Salpeter equations in the three topologically different channels can be represented as

$$\Gamma(k, k'; q) = \Lambda^\alpha(k, k'; q) - [\Lambda^\alpha GG \odot \Gamma](k, k'; q). \tag{4}$$

where the symbol $\odot$ stands for a multiplication scheme pertinent for the particular channel, i.e., the way the two-particle variables are convoluted [6]. Graphically the parquet equations are represented in Fig. 1. Although we have three different irreducible vertex functions Λ^α the full vertex Γ must remain channel independent. Actually, the various Bethe-Salpeter equations are connected by symmetry transformations.

The set of Bethe-Salpeter equations (4) is turned to parquet equations if we exclude the full vertex Γ from them. We introduce a completely irreducible two-particle vertex I, being irreducible in all the channels simultaneously, with the help of which we can represent the irreducible functions as

$$\Lambda^\alpha = I + \sum_{\alpha' \neq \alpha} \left[\Gamma - \Lambda^{\alpha'} \right]. \tag{5}$$

Reducible function $\mathcal{K}^\alpha = \Gamma - \Lambda^\alpha$ in the α channel is irreducible in the others. Knowing the completely irreducible vertex I we determine all irreducible functions Λ^α as well as the full vertex Γ. In parquet equations, the completely irreducible vertex is chosen to be the bare interaction. Otherwise it can be approximated with Feynman skeleton diagrams [6].

Parquet equations contain renormalized one-particle propagators with renormalized self-energies. To close the system of equations we have to determine the self-energy from the vertex function. It is the Schwinger-Dyson equation that serves this purpose. It reads

$$\begin{aligned}
\Sigma_\sigma(k) &= \frac{U}{\beta N} \sum_{k'} G_{-\sigma}(k') - \frac{U}{\beta^2 N^2} \sum_{k'q} \Gamma_{\sigma-\sigma}(k, k'; q) \\
&\quad \times G_\sigma(k+q) G_{-\sigma}(k'+q) G_{-\sigma}(k').
\end{aligned} \tag{6}$$

It is actually the only really dynamical equation of motion. It is a projection of the functional Schrödinger equation onto Fock space. The last equation to be added is the Dyson equation connecting the self-energy with the one-electron propagator: $G_\sigma(k)^{-1} = G_\sigma^{(0)}(k)^{-1} - \Sigma_\sigma(k)$.

3.2 SIMPLIFIED PARQUET EQUATIONS

Parquet equations (4) with $I = U$, the bare interaction, is the simplest theory with a two-particle self-consistence, i.e., the effective scattering potentials, irreducible vertex functions Λ^α, are renormalized by the presence of other scattering effects and are determined form self-consistent nonlinear equations. A solution from one Bethe-Salpeter equation enters as input into the other

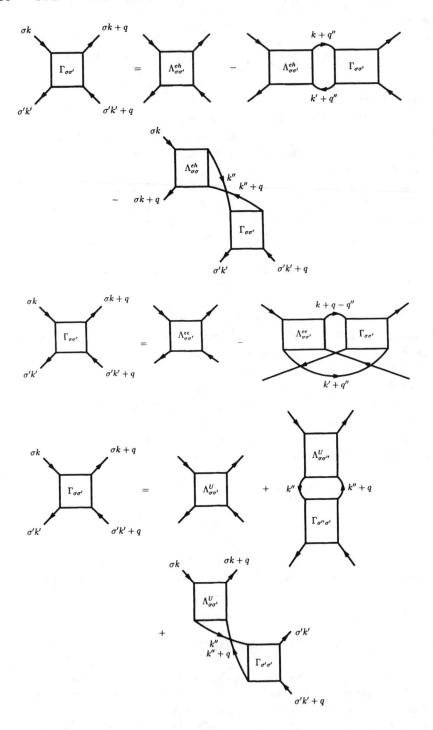

Figure 1 Parquet diagrams with electron-hole, electron-electron, and interaction channels, respectively. The double-primed indices are summed over. We used a four-momentum representation $k = (\mathbf{k}, i\omega_n)$ for fermionic and $q = (\mathbf{q}, i\nu_m)$ for bosonic variables

parquet equations. The parquet equations are hence difficult to solve. However, only nonperturbative solutions of the parquet equations make sense, otherwise we were unable to leave the weak-coupling, Fermi-liquid regime. Our aim now is to find a simplified version of the parquet equations that can be diagonalized analytically and solved nonperturbatively.

A repulsive interaction drives the electron system toward a metal-insulator transition. Electrons and holes tend to form singlet bound pairs. It can be demonstrated on a non-self-consistent version of the parquet equations (fluctuation-exchange (FLEX) approximation) where the irreducible functions Λ^α are replaced by the bare interaction. At half filling of the Hubbard model we observe a pole in the electron-hole and interaction channels at a critical interaction strength U_c. Infinite sums of singlet electron-hole scatterings in the two channels get divergent and must be kept in the critical region of the metal-insulator transition. We can, however, neglect as less important all the triplet and electron-electron irreducible functions in the critical region of the metal-insulator transition and end up with a so-called dipole approximation [7].

The dipole approximation consists of two coupled integral equations that cannot be solved or diagonalized exactly. Since only the bosonic transfer momenta matter in quantum criticality we neglect the incoming fermionic momentum in the vertex function and put it zero in all intermediate states. It means that only the fermionic states near the Fermi energy dominate the pair formation. Beware that such an approximation is allowed only at zero temperature where the bosonic and fermionic Matsubara frequencies are continuous and identical.

It is astonishing that after these two simplifications we are able to diagonalize the parquet equations analytically using a Fourier space-time transformation. The resulting equations for the Hubbard model read

$$\overline{\Lambda^U}_L(x,q) = \frac{U\delta(x) + \overline{\langle \Lambda^{eh} G_\uparrow G_\downarrow \rangle}_L(x,q)\left[U - \overline{\Lambda^{eh}}_L(x,q)\right]}{1 + \overline{\langle \Lambda^{eh} G_\uparrow G_\downarrow \rangle}_L(x,q)}, \qquad (7)$$

$$\overline{\Lambda^{eh}}_R(q,x) = \frac{U\delta(x) - \prod_\sigma \overline{\langle \Lambda^U G_\sigma G_\sigma \rangle}_R(q,x)\left[U - \overline{\Lambda^U}_R(q,x)\right]}{1 - \prod_\sigma \overline{\langle \Lambda^U G_\sigma G_\sigma \rangle}_R(q,x)}. \qquad (8)$$

The bar with a subscript denotes a Fourier transformation from a four-momentum to a space-time representation $x = (\mathbf{x}, \tau)$ in the left, right variable, respectively. The angular brackets denote a convolution $\langle \Lambda^{eh} G_\sigma G_{\sigma'} \rangle(q_L, q_R) = (N\beta)^{-1} \sum_{q'} \Lambda^{eh}(q', q_R) G_\sigma(q') G_{\sigma'}(q_R + q')$ for the electron-hole channel. The variables q_L and q_R interchange their roles in the interaction channel. Note that equations (7,8) have an exact diagonal form of the full solution to the dipole approximation enabling a control of potential poles of the right-hand

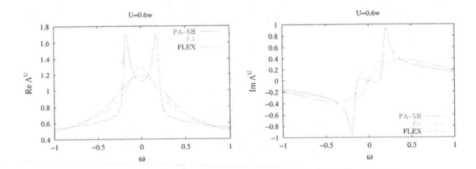

Figure 2   Two particle (irreducible) vertex functions for the FLEX, simplified parquet and simplified parquet equations with symmetry breaking.

sides. The only difference in the exact solution is that the diagonalizing transformation is not the Fourier but a generalized transform. Equations (7,8) must be completed with the Schwinger-Dyson equation (5) with $\Gamma = \Lambda^U + \Lambda^{eh} - U$.

4.   RESONANT STATES: ANOMALOUS TWO-PARTICLE FUNCTIONS

Equations (7,8) can in principle be solved numerically on the temperature axis. But we can get a qualitative insight into the behavior of the solution if we separate the coherent (δ-function) contributions from the rest. Then the two-particle functions retain only one bosonic variable and we can easily decouple the solutions from the two active channels:

$$\Lambda^{eh}(q) = \frac{U}{1 - U^2 \left\langle \frac{G_\uparrow G_\uparrow}{1 + \langle \Lambda^{eh} G_\uparrow G_\downarrow \rangle} \right\rangle (q) \left\langle \frac{G_\downarrow G_\downarrow}{1 + \langle \Lambda^{eh} G_\uparrow G_\downarrow \rangle} \right\rangle (q)}, \qquad (9)$$

$$\Lambda^U(q) = \frac{U}{1 + U \left\langle \frac{G_\uparrow G_\downarrow}{1 - \langle \Lambda^U G_\uparrow G_\uparrow \rangle \langle \Lambda^U G_\downarrow G_\downarrow \rangle} \right\rangle (q)}. \qquad (10)$$

The angular brackets have the same convolutive meaning as in (7,8). Note that numerical complexity of (9,10) is comparable with single-channel or FLEX approximations. Unlike the FLEX-type approximations, (9,10) are nonlinear equations for the vertex functions for given one-electron propagators. Multiple solutions to these equations can exist.

We resort to a mean-field, $d = \infty$, situation with local functions where only frequencies are dynamical variables. This corresponds to the situation where the Mott-Hubbard metal-insulator transition has been intensively studied [8]. We solved equations (9,10) for the model of the $d = \infty$ Bethe lattice with half-bandwidth $w = 1$. The Mott-Hubbard critical point lies at $U \approx$

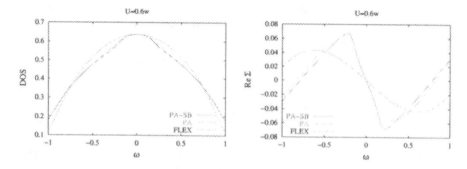

Figure 3 One particle functions for the FLEX, simplified parquet and simplified parquet equations with symmetry breaking.

$3w$. We found that at $U \approx 0.5w$, i. e., deep in the weak-coupling regime, the solution of the parquet equations goes through a bifurcation point. At this point two solutions meet and split the value of the effective interaction $U_{eff} = \Lambda^{\alpha}(0)$ into the complex plane. We then have two complex-conjugate solutions for the irreducible vertex functions Λ^{α}. We obtain a new phase with "anomalous" vertex functions breaking the symmetry of the weak-coupling solution $\Lambda^{\alpha}(-i\omega) = \Lambda^{\alpha}(i\omega)^{*}$. The order parameter can be chosen $\text{Im}\Lambda^{\alpha}(0)$. This order parameter shifts the denominator of the parquet equations at the Fermi energy to the complex plane and enables a by-pass of the pole that cannot be reached due to nonintegrability of the divergence in the vertex functions. The bifurcation point is the only possibility to continue the parquet solution to the strong-coupling regime, since only integrable divergences can exist in an exact theory. This property is correctly reproduced by the parquet approach.

Beyond the bifurcation point the new solution and a solution without symmetry breaking coexist, but the former has a lower energy. Fig. 2 shows the vertex function from the interaction channel calculated in the simplified parquet approximation with and without symmetry breaking and in FLEX. The effect of the symmetry breaking is apparent: shift of the quasiparticle weight away from the Fermi level.

One-particle quantities are calculated from the Schwinger-Dyson equation (5) with the full vertex being a sum of the two complex-conjugate solutions. The low-energy behavior even in the symmetry-broken phase remains Fermi-liquid like, Fig. 3.

The new solution to the parquet equations has a straightforward physical interpretation. A complex effective interaction means that the test particles from IN and OUT asymptotic spaces can be captured (absorbed) to form a temporary bound electron-hole pair. Since the bound pairs do not exist for ever the electron-hole pair is broken after some macroscopically long time and the particles are emitted back to their asymptotic spaces. Such a situation corresponds to a phase with pair resonant states. This is exactly what one expects to happen when the metal-insulator transition is approached.

5.   SUMMARY

We have shown that quantum critical behavior is very complex and cannot be described by a mere analogy with classical phase transitions. First, we cannot get rid of nonuniversal fermionic degrees of freedom and one-particle propagators. Second, we have to take into account all potentially divergent two-particle functions. Quantum critical points cause divergences in two two-particle irreducibility channels simultaneously and only parquet-type approximations with a two-particle self-consistence can comprise the leading singular terms. A mean-field version of the parquet approximation leads to a new phase with resonant pair states and anomalous vertex functions. The new solution shows deviations from the Fermi-liquid behavior in correlation functions and susceptibilities but not in the low-energy limit of the one-electron propagator.

Acknowledgments

The work was supported in part by grant No. 202/98/1290 of the Grant Agency of the Czech Republic. I thank K. Netočný for his help with the programming of the parquet equations.

References

[1]   J. Phys.: Condens. Matter **8** (1996) 9675; special issue on non-Fermi-liquid behavior in metals, edited by P. Coleman, B. Maple, and A. Millis, Proceedings of the M2S-HTSC VI Conference, Houston 2000, to appear in Physica C; see also this volume.

[2]   J. A. Hertz, Phys. Rev. B**14** (1976) 1165 and A. J. Millis, Phys. Rev. B**48** (1993) 7183.

[3]   J. J. Binney, N. J. Dowrick,, A. J. Fisher, and M. E. J. Newman, *The Theory of Critical Phenomena*, Clarendon Press (Oxford 1992).

[4]   C. De Dominicis, J. Math. Phys. **3**, 983 (1962), *ibid* **4**, 255 (1963), and C. De Dominicis and P. C. Martin, J. Math. Phys. **5**, 14, 31 (1964).

[5]   N. E. Bickers, D. J. Scalapino, and S. R. White, Phys. Rev. Lett. **62**, 961 (1989) and N. E. Bickers and D. J. Scalapino, Ann. Phys. (NY) **193**, 206 (1989).

[6]   V. Janiš, Phys. Rev. B**60**, 11345 (1999).

[7]   V. Janiš, J. Phys. C.: Condens. Matter **10**, 2915 (1998).

[8]   A. Georges, G. Kotliar, W. Krauth, and M. Rozenberg, Rev. Mod. Phys. **68**, 13 (1996) and G. Kotliar, this volume.

THEORY OF VALENCE TRANSITIONS
IN YTTERBIUM-BASED COMPOUNDS

V. Zlatić[1,3] and J. K. Freericks[2,3]

[1] Institute of Physics, 10000 Zagreb, Croatia

[2] Department of Physics, Georgetown University, Washington, DC 20057, USA

[3] Isaac Newton Institute, Cambridge CB3 0EH, UK

Abstract   The anomalous behavior of YbInCu$_4$ and similar compounds is modeled by the exact solution of the spin one-half Falicov-Kimball model in infinite dimensions. The valence-fluctuating transition is related to a metal-insulator transition caused by the Falicov-Kimball interaction, and triggered by the change in the f-occupancy.

1.   INTRODUCTION

The intermetallic compounds of the YbInCu$_4$ family exhibit an isostructural transition from high-temperature state with trivalent Yb ions in the $4f^{13}$ configuration to the low-temperature mixed-valent state with Yb ions fluctuating between $4f^{13}$ and $4f^{14}$ configurations [1]. The transition is particularly abrupt in high-quality stoichiometric YbInCu$_4$ samples [2] with a transition temperature equal to $T_v = 42$ K at ambient pressure; the susceptibility and the resistivity drop at T_v by more than one order of magnitude in cooling, while the volume expansion is small, $\Delta V/V \simeq 0.05$. The valence change inferred from $\Delta V/V$ by using the usual ionic radii of Yb^{3+} and Yb^{2+} is about $\Delta n_f \simeq 0.1$, which is consistent with the valence measurements by the L_{III}-edge absorption [1, 4]. The critical temperature depends strongly on external pressure, magnetic field, and alloying [5, 6]. A recent review of the experimental data is given in Ref. [7] and here we just recall the main points which motivate our choice of model.

The integer-valent phase ($T \geq T_v$) is characterized by a Curie-Weiss susceptibility [1, 6] with very small Curie-Weiss temperature $\Theta \ll T_v$. The Curie constant corresponds to the free moment of one magnetic f-hole in a $J = 7/2$ spin-orbit state with $\mu_{eff} = 4.53\mu_B$. The electrical resistance is large and has a small positive slope; it remains almost unchanged in magnetic fields up to 30 T [5]. In some systems, like Yb$_{1-x}$Y$_x$InCu$_4$, the magnetoresistance is slightly

J. Bonča et al. (eds.), Open Problems in Strongly Correlated Electron Systems, 371–380.

negative, while in YbInCu$_4$ (or YbIn$_{1-x}$Ag$_x$Cu$_4$ for $x = 0.15$) it is slightly positive [5]. The Hall constant is large and negative, indicating a small number of carriers [4, 8]. The thermoelectric power has a rather small slope which one finds in a semiconductor with a nearly symmetric density of states [9]. Recent data on the optical conductivity of YbInCu$_4$ [10] shows the absence of a Drude peak at high temperatures and a pronounced maximum of the optical spectral weight at about 1 eV. The high-temperature ESR data for Gd^{3+} embedded in YbInCu$_4$ resemble those found in integer-valence semi-metallic or insulator hosts [11]. Thus, the high-temperature phase indicates the presence of a well defined local moment but gives no signature of the Kondo effect. The overall behavior of the high-temperature phase is closer to that of a semi-metal or paramagnetic small-gap semiconductor than to a Kondo metal.

The mixed-valent phase $(T < T_v)$ behaves like a Pauli paramagnet with moderately enhanced susceptibility and specific heat coefficient [6]. The electrical resistance and the Hall constant are one order of magnitude smaller than in the high-temperature phase [4, 8]. The thermoelectric power [9] has a very large slope typical of a valence fluctuator with large asymmetry in the density of states. The susceptibility, the resistivity and the Hall constant do not show any temperature dependence below T_v, which is also typical of valence fluctuators. The optical conductivity shows a major change with respect to the high-temperature shape. The peak around 1 eV is reduced, the Drude peak becomes fully developed, and an additional structure in the mid-infrared range appears quite suddenly below T_v [10]. A large density of states at the chemical potential μ is indicated by the ESR data as well [12]. Thus, the transition at T_v seems to be from a paramagnetic semimetal to a valence fluctuator.

In contrast to usual valence-fluctuators, which are quite insensitive to the magnetic field, the YbInCu$_4$ family of compounds also exhibit metamagnetic transitions when $T < T_v$. The Yb moment is fully restored at a critical field $H_c(T)$, with a Zeeman energy $\mu_B H_c$ comparable to the thermal energy $k_B T_v$. The metamagnetic transition defined by the magnetoresistance or the magnetization data [7] gives an H-T phase boundary $H_c(T) = H_c^0 \sqrt{1 - (T/T_v)^2}$. The zero-temperature field H_c^0 is related to T_v as $k_B T_v / \mu_B H_c^0 = 1.8$ [7].

To account for these features we need a model in which the non-magnetic, valence-fluctuating, metallic ground state can be destabilized by increasing temperature or magnetic field. Above the transition, we need a paramagnetic semiconductor with an average f-occupancy that is not changed much with respect to the ground state. The correct model for this system is a periodic Anderson model supplemented with a large Falicov-Kimball (FK) interaction term. The temperature or field induced transition suggests that one should place the narrow f-level just above the chemical potential μ. The hybridization keeps the f-count finite below the transition, while large f-f correlations allow only the fluctuations between zero- and one-hole (magnetic) configurations. The low-temperature phase is close to the valence fluctuating fixed point and shows no Kondo effect. However, because of the Falicov-Kimball term, there is a critical f-occupation at which there is a transition into the high-temperature state with a large gap in the d- and f-excitation spectrum. The n_f is driven to criticality either by temperature or magnetic field. In the high-temperature phase the hybridization can be neglected because the f-level width is already large due to thermal fluctuations, and quantum fluctuations are irrelevant. Unfortunately, the above model would be difficult to solve in a controlled way, and here we consider a simplified model in which the hybridization is neglected at all temperatures. This leads to a spin-degenerate Falicov-Kimball model which explains the collapse of the non-magnetic metallic phase at T_v or H_c, and gives a good qualitative description of the high-temperature paramagnetic phase. However, the deficiency of the simplified model is that it yields a negligible f-count in the metallic phase and predicts a large change in the Yb valence at T_v or H_c. It is clear that we can not obtain the valence fluctuating ground state and maintain the average f-occupancy below the transition without hybridization-induced quantum fluctuations. In what follows, we describe the model, explain the method of solution, and present results for static and dynamic correlation functions.

2.   CALCULATIONS

The Hamiltonian of the Falicov-Kimball model [13] consists of two types of electrons: conduction electrons (created or destroyed at site i by $d_{i\sigma}^\dagger$ or $d_{i\sigma}$) and localized electrons (created or destroyed at site i by $f_{i\sigma}^\dagger$ or $f_{i\sigma}$). The conduction electrons can hop between nearest-neighbor sites on the D-dimensional lattice, with a hopping matrix $-t_{ij} = -t^*/2\sqrt{D}$; we choose a scaling of the hopping matrix that yields a nontrivial limit in infinite-dimensions [14]. The f-electrons have a site energy E_f, and a chemical potential μ is employed to conserve the total number of electrons $n_{d\uparrow} + n_{d\downarrow} + n_{f\uparrow} + n_{f\downarrow} = n_{tot}$. The Coulomb repulsion U_{ff} between two f-electrons is infinite and there is a Coulomb interaction U between the d- and f-electrons that occupy the same lattice site. An external magnetic field h couples to localized electrons with a Landé g-factor. The resulting Hamiltonian is [15, 16]

$$H = \sum_{ij,\sigma}(-t_{ij} - \mu\delta_{ij})d_{i\sigma}^\dagger d_{j\sigma} + \sum_{i,\sigma}(E_f - \mu)f_{i\sigma}^\dagger f_{i\sigma}$$
$$+ U\sum_{i,\sigma\sigma'}d_{i\sigma}^\dagger d_{i\sigma}f_{i\sigma'}^\dagger f_{i\sigma'} + U_{ff}\sum_{i,\sigma}f_{i\uparrow}^\dagger f_{i\uparrow}f_{i\downarrow}^\dagger f_{i\downarrow}$$
$$- \mu_B h\sum_{i,\sigma}\sigma(2d_{i\sigma}^\dagger d_{i\sigma} + gf_{i\sigma}^\dagger f_{i\sigma}). \qquad (1)$$

The model can be solved in the infinite-dimensional limit by using the methods of Brandt-Mielsch [15]. We consider the hypercubic lattice with Gaussian density of states $\rho(\epsilon) = \exp[-\epsilon^2/t^{*2}]/\sqrt{\pi}t^*$, and take t^* as the unit of energy ($t^* = 1$). Our calculations are restricted to the homogeneous phase.

The local conduction-electron Green's function satisfies Dyson's equation

$$G^\sigma(z) = \int \frac{\rho(\epsilon)}{z + \mu - \Sigma^\sigma(z) - \epsilon} d\epsilon, \tag{2}$$

where z is a complex variable and Σ^σ is the local self energy which does not depend on momentum [14]. In infinite dimensions, Σ^σ is defined by a sum of skeleton diagrams, which depend on the local d-propagator G^σ but not on t_{ij}. The exact self-energy functional for the FK model is obtained by calculating the thermodynamic Green's function [17] of an atomic system coupled to an external time-dependent field $\lambda^\sigma(\tau)$

$$G^\sigma_{\text{atom}}(\tau) = -\frac{1}{Z} \text{Tr}_{df} \left\langle T_\tau e^{-\beta H_{\text{atom}}} d_\sigma(\tau') d_\sigma^\dagger(\tau) S(\lambda) \right\rangle, \tag{3}$$

where the S-matrix for the λ-field is

$$S(\lambda) = T_\tau e^{-\int_0^\beta d\tau \int_0^\beta d\tau' \lambda(\tau,\tau') d_\sigma^\dagger(\tau) d_\sigma(\tau')}, \tag{4}$$

and H_{atom} is obtained from the Hamiltonian (1) by removing the hopping and keeping just a single lattice site. The exact solution for $G^\sigma_n[\{\lambda_m\}]$ at Matsubara frequency $i\omega_n = i\pi T(2n+1)$ is given by,

$$G^\sigma_n = \frac{w_0}{[G^\sigma_{0n}]^{-1}} + \frac{w_1}{[G^\sigma_{0n}]^{-1} - U}, \tag{5}$$

where w_0 and w_1 are the f-occupation numbers ($w_1 = 1 - w_0$, $w_0 = Z_0/Z$) and [15]

$$Z_0(\lambda, \mu) = 2e^{\beta\mu/2} \prod_n \frac{1}{(i\omega_n)G^\uparrow_{0n}} 2e^{\beta\mu/2} \prod_n \frac{1}{(i\omega_n)G^\downarrow_{0n}}, \tag{6}$$

with

$$Z(\lambda, \mu) = Z_0(\lambda, \mu) + 2e^{-\beta(E_f - \mu)} Z_0(\lambda, \mu - U). \tag{7}$$

The bare Green's function satisfies

$$G^\sigma_{0n} = \frac{1}{i\omega_n + \mu - \lambda^\sigma_n}, \tag{8}$$

with λ_n the Fourier transform of the external time-dependent field.

The self-energy functional $\Sigma^\sigma_n[G^\sigma_n]$ can now be obtained [15] by using the Dyson equation for the atomic propagator,

$$\Sigma^\sigma_n = [G^\sigma_{0n}]^{-1} - [G^\sigma_n]^{-1}, \tag{9}$$

and eliminating $G_{0n}^\sigma[\{\lambda_m\}]$ from Eqs. (5) and (9). The mapping onto the lattice is achieved by adjusting G_{0n}^σ in such a way that $G_n^\sigma[\{\lambda_m\}]$ satisfies the lattice Dyson equation (2).

The numerical implementation of the above procedure is as follows: We start with an initial guess for the self energy Σ^σ and calculate the local propagator in (2). Using (9) we calculate the bare atomic propagator G_{0n}^σ and find $\mathcal{Z}_0$ and $\mathcal{Z}$. Next we obtain w_0, w_1 and find G_n^σ from (5). Using G_{0n}^σ and G_n^σ, we compute the atomic self energy and iterate to the fixed point.

The iterations on the imaginary axis give static properties, like n_f, the f-magnetization $m_f(h,t)$, and the static spin and charge susceptibilities. Having found the f-electron filling w_1 at each temperature, we iterate Eqs. (2) to (9) on the real axis and obtain the retarded dynamical properties, like the spectral function, the resistivity, the magnetoresistance, and the optical conductivity. At the fixed point, the spectral properties of the atom perturbed by λ-field coincide with the local spectral properties of the lattice.

3.   RESULTS AND DISCUSSION

We studied the model for a total electron filling of 1.5 and for several values of E_f and U. The main results can be summarized in the following way.

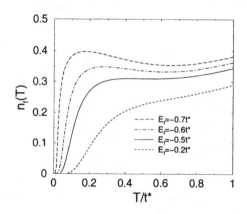

Figure 1   Number of the f-holes plotted versus T/t^* for $U/t^* = 4$. The E_f/t^* increases from top to bottom, and is given by -0.7, -0.6, -0.5, and -0.2, respectively.

The occupancy of the f-holes at high temperatures is large and there is a huge magnetic degeneracy. The f-holes are energetically unfavorable but are maintained because of their large magnetic entropy. In Fig.(1) we show n_f as a function of temperature, plotted for $U = 4t^*$, and E_f/t^* from -0.2 to -0.7. Below a certain temperature, which depends on U and E_f, there is a rapid transition to the low-temperature phase. The transition becomes sharper and is pushed to lower temperatures as E_f decreases at constant U. However, we restrict ourselves to continuous crossovers here, since the region with first-order transitions leads to numerical instabilities.

The uniform f-spin susceptibility is obtained by calculating the spin-spin correlation function [15, 16] and is given by $\chi(T) = Cn_f(T)/T$, where $C = g_L^2\mu_B^2 J(J+1)/3k_B$ is the Curie constant. The $\chi(T)/C$ is shown in

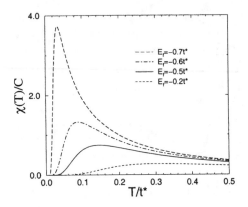

Figure 2 Uniform static magnetic susceptibility of the f-holes plotted versus T/t^* for $U/t^* = 4$. The values of E_f/t^* are the same as in Fig(1). The corresponding values of T_v/t^* are estimated from the maximum of $\chi(T)$, and are given by 0.03, 0.08, 0.15, 0.35, respectively. The T_v increases from top to bottom.

Fig. (2) for $U/t^* = 4$ and for E_f as quoted in Fig.(1). The T_v is obtained from the maximum of the $\chi(T)/C$ and the values corresponding to various parameters used in this paper are quoted in the caption of Fig. (2). The high-temperature susceptibility follows an approximate Curie-Weiss law, but the Curie-Weiss parameters depend on the fitting interval.

The interacting density of states $\rho_d(\omega)$ for the conduction electrons is shown in Fig.(3) for $U/t^* = 4$ and $E_f/t^* = -0.5$, and for several temperatures. (The energy is measured with respect to μ.) The high-temperature DOS has a gap of the order of U, and the chemical potential is located within the gap. Below the transition n_f is small, the correlation effects are reduced, and $\rho_d(\omega)$ assumes a nearly non-interacting shape, with large $\rho_d(\mu)$ and halfwidth $W \simeq t^*$.

The transport properties of the high-T phase are dominated by the presence of the gap, which leads to a small dc conductivity with a weak temperature dependence. The transport properties of the paramagnetic phase are unrelated to the spin-disorder Kondo scattering (there is no spin-spin scattering in the FK model). Below the transition the conductivity increases and assumes large metallic values.

The intraband optical conductivity $\sigma(\omega)$ is plotted in Fig.(4) as a function of frequency, for several temperatures. Above T_v, we observe a reduced Drude peak around $\omega = 0$ and a pronounced high-frequency peak around $\omega \simeq U$.

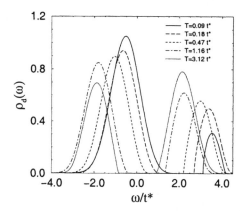

Figure 3 Interacting density of states plotted versus ω/t^* for $U/t^* = 4$, $E_f/t^* = -0.5$ ($T_v/t^* = 0.14$), and for various temperatures, as indicated in the figure.

The shape of $\sigma(\omega)$ changes completely across T_v. Below T_v the Drude peak is fully developed and there is no high-energy (intraband) structure. However, if the renormalized f-level is close to μ, the interband d-f transition could lead to an additional mid-infrared peak. The ratio of the high-frequency peak in Fig(4) and the corresponding value of $T_v = 0.15t^*$, is $U/T_v = 26$. For the same value of U and $E_f = -0.7t^*$ ($T_v = 0.03t^*$) we obtain $U/T_v = 130$, while for $E_f = -0.75t^*$ ($T_v = 0.02t^*$) we find $U/T_v \simeq 200$ (not shown). If we

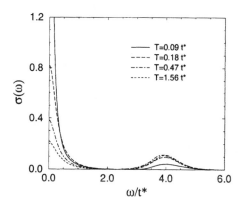

Figure 4 Optical conductivity plotted versus ω/t^* for various temperatures. The U, E_f, and T_v, are the same as in Fig.(3).

estimate the f-d correlation in YbInCu$_4$ from the 8000 cm^{-1} peak in the optical conductivity data [10], we obtain the experimental value $U \simeq 1$ eV. Together with $T_v = 42$ K [7] this gives the ratio $U/T_v \simeq 200$. If we take $U/t^* = 4$ and adjust E_f/t^* so as to bring the theoretical value of T_v in agreement with the the thermodynamic and transport data on YbInCu$_4$, we get a high-frequency peak in $\sigma(\omega)$ at about $8000\ cm^{-1}$, $6000\ cm^{-1}$, and $1500\ cm^{-1}$, for $E_f = -0.75t^*$, $E_f = -0.7t^*$, and $E_f = -0.5t^*$, respectively.

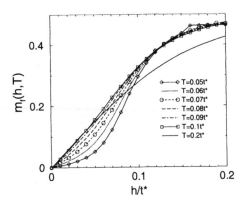

Figure 5   The f-electron magnetization m_f is plotted as a function of h/t^* for various temperatures. The U, E_f, and T_v, are the same as in Fig.(3).

The f-electron magnetization $m_f(h)$ is plotted in Fig.(5) versus reduced magnetic field h/t^*, for several temperatures. Above the characteristic temperature $T_v^* \simeq T_v/2$, the $m_f(h)$ curves exhibit typical local moment behavior. Below T_v^* we find a metamagnetic transition at a critical field H_c; the $m_f(h)$ is negligibly small below H_c and the local moment is fully restored above H_c. Taking the inflection point of the $m_f(h)$ curves, calculated for several values

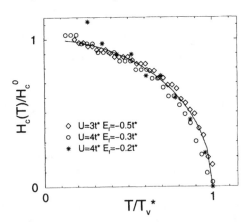

Figure 6   Normalized critical field is plotted as a function of reduced temperature T/T_v^* for several values of E_f/t^* and U/t^*. The full line represents $\sqrt{1 - (T/T_v^*)^2}$ and $T_v^* = T_v/2$.

of U and E_f, as an estimate of $H_c(T)$ we obtain the phase boundary which is shown in Fig.(6), together with the expression $H_c(T)/H_c^0 = \sqrt{1 - (T/T_v^*)^2}$.

Note, the T_v^* values in Fig.(6) differ by more than an order of magnitude, while the ratio $k_B T_v^* / \mu_B H_c^0$ is only weakly parameter dependent.

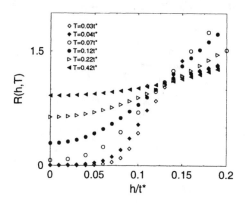

Figure 7 Field-dependent resistivity plotted versus h/t^*. The different symbols correspond to different temperatures, as indicated in the figure. The U and E_f are the same as in Fig.(3).

The metamagnetic transition is also seen in the field-dependent electrical resistance $R(h, T)$ which is plotted in Fig.(7) as a function of h/t^*, for several temperatures. A substantial change in the $R(h, T)$ across T_v^* or H_c is clearly seen.

4.   SUMMARY

From the preceding discussion it is clear that Falicov-Kimball model captures the main features of the experimental data for $YbInCu_4$ and similar compounds. The temperature- and field-induced anomalies are related to a metal-insulator transition, which is caused by large FK interaction and triggered by the temperature- or the field-induced change in the f-occupancy. At high temperatures, we find a large gap in $\rho_d(\omega)$; we expect a similar gap in the f-electron spectrum as well. At low temperatures, both gaps are closed, and the renormalized f-level renormalizes down to the chemical potential.

Our calculations describe doped Yb systems with broad transitions but appear to be less successful for those compounds which show a first-order transition. The numerical curves can be made sharper (by adjusting the parameters) but they only become discontinuous in a narrow parameter range. The main difficulty with the FK model is that it predicts a substantial change in the f-occupancy across the transition and associates the loss of moment with the loss of f-holes. But in the real materials the loss of moment seems to be due to the valence fluctuations, rather than to the reduction of n_f. The description of the valence fluctuating ground state would require the hybridization and is beyond the scope of this work. The actual situation pertaining to Yb ions in the mixed-valence state might be quite complicated, since one would have to consider an extremely asymmetric limit of the Anderson model, in which the ground state is not Kondo-like, there is no Kondo resonance, and there is no single universal energy scale which is relevant at all temperatures [18].

We speculate that the periodic Anderson model with a large FK term will exhibit the same behavior as the FK model at high temperatures. Indeed, if the conduction band and the f-level are gapped, and the width of the f-level is large, then the effect of the hybridization can be accounted for by renormalizing the parameters of the FK model. On the other hand, if the low-temperature state of the full model is close to the valence-fluctuating fixed point with the conduction band and hybridized f-level close to the Fermi level, then the likely effect of the FK correlation is to renormalize the parameters of the Anderson model.

Acknowledgments

We acknowledge discussions with Z. Fisk, B. Lüthi, M. Miljak, M. Očko, and J. Sarrao. This research was supported by the National Science Foundation under grant DMR-9973225.

References

[1] I. Felner and I. Novik., Phys. Rev. B **33**, 617 (1986).

[2] J.L. Sarrao et al., Physics B, **223&224**, 366 (1996).

[3] J. M. Lawrence et al., Phys. Rev. B,**55**, 14 467 (1997).

[4] A. L. Cornelius et al., Phys. Rev. B,**56**, 7993 (1997).

[5] C. D. Immer et al., Phys. Rev. B,**56**, 71 (1997).

[6] J.L. Sarrao et al., Phys. Rev. B,**58**, 409 (1998)

[7] J.L. Sarrao, Physica B, **259&261**, 129 (1999)

[8] E. Figueroa et al., Solid State Commun. **106**, 347 (1998)

[9] M. Očko, J. Sarrao, Z. Fisk, unpublished.

[10] S. R. Garner et al., preprint (2000).

[11] T. S. Altshuler et al., Z. Phys. B**99**, 57 (1995).

[12] C. Rettori et al., Phys. Rev. B,**55**, 1016 (1997).

[13] L. M. Falicov and J. C. Kimball, Phys. Rev. Lett. **22**, 997 (1969).

[14] W. Metzner and D. Vollhardt, Phys. Rev. Lett. **62**, 324 (1989).

[15] U. Brandt and C. Mielsch, Z.Phys. B **75**, 365 (1989); U. Brandt and M. P. Urbanek, ibid. **89**, 297 (1992).

[16] J. Freericks and V. Zlatić, Phys. Rev. B **58**, 322 (1998).

[17] L.P. Kadanoff and G. Baym, *Quantum Statistical Physics* (W. A. Benjamin, Menlo Park, CA), 1962

[18] H. B. Krishnamurti et al., Phys. Rev. B **21**, 1044 (1980).

STRONG ELECTRONIC CORRELATIONS AND LOW ENERGY SCALES

R. Bulla[1] and Th. Pruschke[2]

[1] *Theoretische Physik III, Elektronische Korrelationen und Magnetismus, Universität Augsburg, 86135 Augsburg, Germany*

[2] *Institut für Theoretische Physik, Universität Regensburg, 93040 Regensburg, Germany*

Abstract   One of the many fascinating aspects of strongly correlated electron systems is the appearance of low energy scales: interesting physical phenomena are observed on an energy (and temperature) scale much lower than e.g. local Coulomb repulsion, hybridization, etc. Well known examples are heavy fermion compounds and systems in the vicinity of a metal-insulator transition. These phenomena can be described using simple (but nevertheless strongly correlated) electronic models. For these models, the great difference in energy scales in general represents a difficult task for the theoretical investigation. The numerical renormalization group method turns out to be a very efficient tool to deal with these problems. This method was developed by Wilson for the investigation of the Kondo model and has been recently (successfully) applied to lattice models (such as Hubbard model and periodic Anderson model) within the dynamical mean field theory.

1.   INTRODUCTION

The strong correlations we are dealing with in this paper are those between electrons in d- or f-orbitals in a solid, where the Coulomb interaction between electrons on the same orbital can take values of typically a few eV (corresponding to a few 10 000 K). On the other hand, these strong correlations can be responsible for physical phenomena on a much lower energy scale, say a few meV (corresponding to a few 10 K), and even further below. These phenomena are, e.g. the Kondo effect, heavy fermion behaviour, and under certain conditions, the vicinity to a Mott metal-insulator transition. An understanding of these phenomena can be achieved by investigating simple models for electronic correlations such as the Kondo model, the single impurity Anderson model, the periodic Anderson model, and the Hubbard model.

The presence of such a variety of energy scales, however, imposes a great technical problem for the theoretical investigation. A method especially designed to treat with both strong correlations and low energy scales is Wilson's

J. Bonča et al. (eds.), Open Problems in Strongly Correlated Electron Systems, 381–386

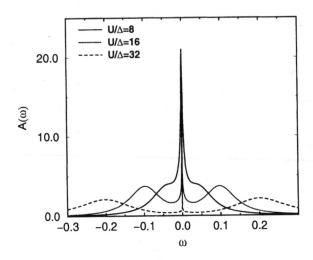

Figure 1   Spectral functions for the single impurity Anderson model for various values of U/Δ (see [4]).

numerical renormalization group method (NRG) [1, 2]. As discussed in this paper, the NRG can indeed provide some understanding of the above mentioned phenomena and the related models. The results will also help to understand the underlying principle for *all* these phenomena where low energy scales appear: the competition between the formation of local moments and the screening of these moments.

2.   RESULTS

Typical NRG results for the spectral function of the single impurity Anderson model [3] are shown in Fig. 1. The charge fluctuation peaks at $\omega \approx \pm U/2$ (U is the Coulomb repulsion at the impurity site) correspond to the unscreened magnetic moment of the impurity. The screening of this magnetic moment at lower energies is manifest in the appearance of a quasiparticle peak at $\omega = 0$. The spectra clearly show the narrowing of the resonance on increasing the ratio U/Δ – corresponding to the exponential dependence of the low energy scale T_K (the Kondo temperature) on U/Δ (Δ is the coupling between impurity and conduction electrons). In the language of the renormalization group, the crossover from the local moment fixed point to that of a screened local moment (the strong coupling or Fermi liquid fixed point) occurs at lower and lower energy scales.

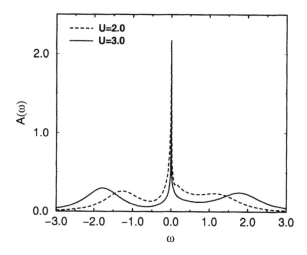

Figure 2   Spectral functions for the periodic Anderson model for $V = 0.45$, $\epsilon_f = -U/2$, $T = 0$ and different values of U (see also [8]).

In the standard single impurity Anderson model, screening is possible for *any* value of U/Δ, except for $U/\Delta = \infty$. There exists, consequently, a critical value $U_c = \infty$ separating the local moment and Fermi liquid phases, with the low energy scale $T_K \to 0$ for $U \to U_c$.

Results for the f-spectral function of the periodic Anderson model in the paramagnetic metallic phase are shown in Fig. 2 (the calculations are within the dynamical mean field theory [5, 6]). A conduction band filling of $n_c \approx 0.6$ has been used in this calculation to avoid the hybridization gap at the Fermi level which is present in the half-filled (particle-hole symmetric) case [7, 8].

Apart from the asymmetry, the overall features are very similar to those of the single-impurity version of the model (for a detailed comparison, see [8]). The spectrum again consists of charge-fluctuation peaks at $\omega \approx \pm U/2$ (corresponding to local magnetic moments on the f-sites) and a quasiparticle resonance at $\omega = 0$ (corresponding to a Fermi liquid ground state with screened magnetic moments). The low energy scale T^* (the width of the quasiparticle resonance) corresponds to the temperature where intersite coherence sets in.

In the case studied here, the low energy scale vanishes for $U \to U_c = \infty$, leaving behind unscreened local moments on the f-orbitals. Recently, it has been found that in the periodic Anderson model with nearest neighbour hybridization, the critical U_c for $T = 0$ is finite [9]. In this case, the transition for $U \to U_c$ shows very similar features as the Mott transition in the Hubbard model — with the difference that the system does not undergo a metal-insulator

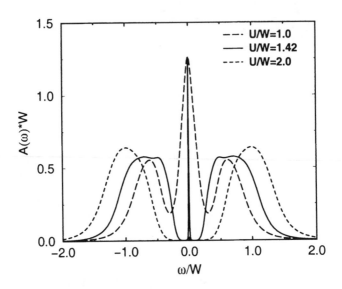

Figure 3   Spectral functions for the Hubbard model on a Bethe lattice for various values of U. A narrow quasiparticle peak develops at the Fermi level which vanishes at the critical $U_c \approx 1.47W$ (see also [11]).

transition as the conduction electrons still show a finite density of states at the Fermi level.

Let us now discuss the NRG results for the Hubbard model [10] at $T = 0$. The spectral function $A(\omega)$ for a Bethe lattice is shown in Fig. 3 for $U = 0.8U_c$, $U = 0.99U_c$ and $U = 1.1U_c$ ($U_c \approx 1.47W$, W: bandwidth) In the metallic phase (for large enough values of U) the spectral function shows the typical three-peak structure with upper and lower Hubbard bands centered at $\pm U/2$ and a quasiparticle peak at the Fermi level. The quasiparticle peak vanishes at $U_c \approx 1.47W$, i.e., the low energy scale vanishes for $U \to U_c$.

3.   CONCLUSIONS

We have discussed the appearance of low energy scales within the context of three different physical phenomena and the NRG results for the corresponding microscopic models: the screening of a local magnetic moment in the single impurity Anderson model, the formation of the heavy fermion liquid in the periodic Anderson model, and the vicinity to a Mott transition in the Hubbard model. The similarities between these phenomena and models are the following:

- The physics of a single impurity and that of a system close to the Mott transition have some striking similarities concerning the structure of the local spectral function; the basic difference appears to be that the quasi-particle peak vanishes only at $U_c \to \infty$ in the impurity model while the U_c is finite for the Hubbard model.

- The relation between single impurity and periodic Anderson models is again obvious from the spectra. The energy scale T^* can be further reduced in the periodic model as compared to T_K and it might even vanish at a finite critical U_c (for nearest neighbour hybridization).

- Systems with a periodic array of impurity (f-) orbitals might undergo a Mott transition at a finite critical U_c — very similar to the Mott transition in the Hubbard model. On the other hand, systems close to the Mott transition are expected to behave as heavy fermion systems.

What is the underlying principle for the appearance of low energy scales in all these cases? It is always the competition between the formation of local moments (leading to an insulating ground state) and the screening of these moments (leading to a Fermi liquid or singlet ground state). The critical U_c might be finite or infinite, depending on the details of the model, but in any case the low energy scale vanishes for $U \to U_c$.

The question, of course, arises, why such a vanishing low energy scale is *not* observed in systems such as V_2O_3 or VO_2. To actually see a vanishing energy scale, the transition from a paramagnetic metal to a paramagnetic insulator has to occur at arbitrary low temperatures. In V_2O_3, however, this transition line ends at ≈ 190 K and the system evolves into an antiferromagnetic ground state.

What is required therefore, is a mechanism to suppress the magnetic ordering and to stabilize the paramagnetic phases. This is probably the case in LiV_2O_4, the first heavy fermion system among the transition metal oxides [12]. Here, magnetic ordering is suppressed due to the spinell structure. Furthermore, the system is believed to be close to a metal-insulator transition. The considerations in this paper show that a variety of models can in principle generate the observed low energy scale – provided that the system is very close to a state with well defined local moments. The microscopic model could be a periodic Anderson model as proposed by Anisimov et al. [13] or a generalized Hubbard model.

These models can be investigated with the NRG method, in particular in the very low temperature regime where the coherence sets in for LiV_2O_4. Furthermore, a whole set of experimentally accessible properties can be calculated with the NRG. These calculations will provide a deeper understanding of systems with strong correlations and low energy scales.

Acknowledgments

The authors would like to thank D.E. Logan, A.C. Hewson, and D. Vollhardt for stimulating discussions. Part of this work was supported by the Deutsche Forschungsgemeinschaft, through the Sonderforschungsbereich 484.

References

[1] K.G. Wilson, Rev. Mod. Phys. **47**, 773 (1975).

[2] H.R. Krishna-murthy, J.W. Wilkins, and K.G. Wilson, Phys. Rev. B **21**, 1003 & 1044 (1980).

[3] A.C. Hewson, *The Kondo Problem to Heavy Fermions* (Cambridge Univ. Press, Cambridge 1993).

[4] R. Bulla, A.C. Hewson, and Th. Pruschke, J. Phys.: Condens. Matter **10**, 8365 (1998).

[5] A. Georges, G. Kotliar, W. Krauth, and M.J. Rozenberg, Rev. Mod. Phys. **68**, 13 (1996).

[6] W. Metzner and D. Vollhardt, Phys. Rev. Lett. **62**, 324 (1989); for an introduction, see D. Vollhardt, Int. J. Mod. Phys. B **3**, 2189 (1989).

[7] M. Jarrell, Phys. Rev. B **51**, 7429 (1995); A.N. Tahvildar-Zadeh, M. Jarrell, and J.K. Freericks, Phys. Rev. B **55**, 3332 (1997).

[8] Th. Pruschke, R. Bulla, and M. Jarrell, preprint cond-mat/0001357.

[9] K. Held and R. Bulla, preprint (2000).

[10] J. Hubbard, Proc. R. Soc. London A **276**, 238 (1963); M.C. Gutzwiller, Phys. Rev. Lett. **10**, 59 (1963); J. Kanamori, Prog. Theor. Phys. **30**, 275 (1963).

[11] R. Bulla, Phys. Rev. Lett. **83**, 136 (1999).

[12] S. Kondo et al., Phys. Rev. Lett. **78**, 3729 (1997); A. Krimmel, A. Loidl, M. Klemm, S. Horn, and H. Schober, Phys. Rev. Lett. **82**, 2919 (1999); C.M. Varma, Phys. Rev. B **60**, 6973 (1999).

[13] V.I. Anisimov, M.A. Korotin, M. Zölfl, Th. Pruschke, K. Le Hur, and T.M. Rice, Phys. Rev. Lett. **83**, 364 (1999).

NON-MAGNETIC MOTT INSULATOR PHASE
AND ANOMALOUS CONDUCTING STATES
IN BARIUM VANADIUM TRISULPHIDE

P. Fazekas[1], H. Berger[2], L. Forró[2], R. Gaál[2], I. Kézsmárki[2,3], G. Mihály[2,3], M. Miljak[4], K. Penc[1], and F. Zámborszky[3]

[1]*Research Institute for Solid State Physics and Optics, P.O.B. 49, H-1525 Budapest, Hungary*

[2] *Institut de Géniue Atomique, Ecole Polytechnique Federale de Lausanne, CH-1015 Lausanne, Switzerland*

[3]*Department of Physics of the Technical University of Budapest, H-1111 Budapest, Hungary*

[4] *Institute of Physics of the University, P.O.B. 304, Zagreb, Croatia*

Abstract   Resistivity and susceptibility measurements on single crystal samples were used to construct the phase diagram of the 3D Mott system $BaVS_3$ in the pressure–temperature (p–T) plane. The metal to non-magnetic Mott insulator transition is gradually suppressed until it vanishes at the critical pressure $p_{cr} \approx 20$ kbar. The high-pressure metallic phase is the first $3d$-electron based non-fermi-liquid with a low-T resistivity obeying $\rho - \rho_0 \propto T^{1.25}$ over at least one and a half decade of T. For $p < p_{cr}$, the metal–insulator transition is preceded by a wide precursor range of critically increasing resistivity. We discuss the nature of the various regimes, and point out similarities and dissimilarities to other strongly correlated d-electron systems.

1.   INTRODUCTION

Transition metal compounds show a great variety of correlation effects ranging from the well understood two-sublattice antiferromagnetism of some simple Mott insulators to the recently discovered triplet superconductivity of Sr_2RuO_4. In systems with integer band filling, the basic phenomenon is the Mott transition which sets in if the interaction strength exceeds a critical value of the order of the bandwidth [1]. It follows that the Mott transition should become suppressed by the application of a sufficiently large pressure. The expected canonical form of the phase diagram in the pressure–temperature (p–T) plane (Fig. 1, left) is quite similar[1] to that observed for the V_2O_3 system [2].

---

[1]Here we disregard the existence of the low-T SDW phase in $V_{2-y}O_3$.

J. Bonča et al. (eds.), Open Problems in Strongly Correlated Electron Systems, 387–392.

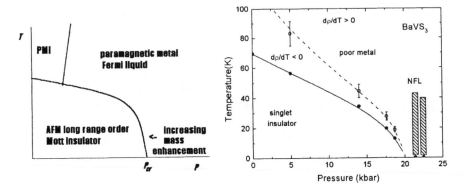

Figure 1 Left: the canonical form of the phase diagram of Mott systems is quite similar to that of V_2O_3. *Right*: a simplified phase diagram of BaVS$_3$ based on resistivity measurements [5]. Notable features: the wide $d\rho/dT < 0$ resistivity precursor region which survives under pressure as long as there is an insulating phase; and the high-pressure non-fermi-liquid (NFL).

This schematic "Mott phase diagram" should be compared to the one obtained from our resistivity and susceptibility measurements on BaVS$_3$ [4, 5], which we show on the right of Fig. 1. The overall shape of the metal–insulator phase boundary (MIPB) is very similar, and even the numerical value of $p_{cr} \approx 20$kbar is almost the same as for V_2O_3. You might say: well, BaVS$_3$ is just another vanadium compound, However, there are also important differences: the nature of the states on either side of the MIPB is quite different from what one finds for V_2O_3. For V_2O_3 it is well known that the Mott insulator is an ordered antiferromagnet; orbital order is still controversial. As for the metallic side, it had been customarily held that pressure induces a decently metallic state with the character of a mass-enhanced Fermi liquid with a concomitant $\rho - \rho_0 \approx AT^2$. Actually, a closer look at this system also revealed some disquieting features: perhaps $\delta\rho \propto T^2$ holds only in certain crystallographic directions while at least in one direction $\delta\rho \propto T^{1.5}$ (a hint at non-fermi-liquid behaviour?) may fit better [3]. On the other hand, there is nothing ambiguous about the statement that all states outside the MIPB of BaVS$_3$ are bad metals. As we are going to see, BaVS$_3$ may well be the best characterized material in an important subgroup of Mott systems, namely those 3D systems which are non-magnetic on the insulating side of the MIPB, and non-fermi-liquids on the metallic side. The resistivity precursor is yet another ill-understood feature which is shared by a number of curious compounds whose overall physical behaviour is in other respects quite dissimilar.

Like many other well-known transition metal compounds, BaVS$_3$ has been known since the sixties, but all previous studies were confined to polycrystalline samples [6, 7]. Some salient features had been known from early on: the existence of a metal-insulator transition (MIT) at $T_{MI} = 69$K; the sharp susceptibility peak; the lack of any magnetic or structural ordering which might accompany the MIT; and the pronounced resistivity minimum at $T \approx 125K$ (qualitatively the same features are seen in Fig. 2 which shows the results of our single crystal measurements). Purely structurally, $BaVS_3$ has well-defined V chains[2] [8], and it was an obvious inference that the substance would be quasi-1D also electrically; this had, in fact, been assumed to be the case. Quasi-one-dimensionality would have offered an apparently easy way to interpret the curious features of $BaVS_3$. In particular, the sharp rise of the resistivity with decreasing T below 125K, which looks so much like a precursor to the onset of the insulating state, might have been explained away as a regime of 1D gaping fluctuations (a soft gap), followed by a lock-in to a truly gaped state at T_{MI}.

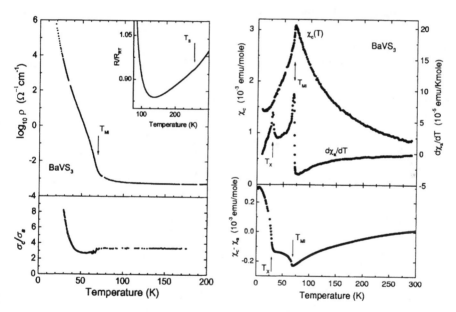

Figure 2 Results obtained on single-crystal $BaVS_3$ samples [4]. *Left*: the resistivity on logarithmic scale; lower part: the conductivity anisotropy; inset: the resistivity minimum is better seen on a linear scale. The $d\rho/dT < 0$ part is referred to as the resistivity precursor. *Right*: the c-axis susceptibility χ_c (top); two phase transitions are revealed in $d\chi_a/dT$ (middle); and in the susceptibility anisotropy (bottom).

2.   EXPERIMENTAL RESULTS AND DISCUSSION

The preparation of the first single crystal samples led to a radical revision of our views on $BaVS_3$. The results of our resistivity and susceptibility mea-

---

[2]We call the chain direction c; the ab planes are triangular lattices of V atoms.

surements (under atmospheric pressure) are shown in Fig. 2. Consider first the conductivity anisotropy which remains $\sigma_c/\sigma_a \approx 3$ while σ itself changes by four orders of magnitude. This shows that the electrical isotropy of $BaVS_3$ is a robust feature.

Lack of space prevents us from discussing the susceptibility data. It should suffice to point it out that the two peaks of $d\chi_a/dT$ demonstrate that there are two phase transitions: one at $T_{MI} = 69K$, and one at $T_X = 30K$. The current understanding is that entropy gain sustains a valence bond liquid state at $T_X < T < T_{MI}$, and magnetic long range order sets in at 30K [4]. It is an important question whether T_X is suppressed under pressure at least at the rate of T_{MI}. Our preliminary magnetoresistivity data lead us to believe that it is. It is a particularly intriguing possibility that the physics near the $T = 0$ quantum phase transition may be dominated by the nearness of a *non-magnetic* Mott insulating phase: it holds the promise of having the chance to study a novel kind of quantum critical behaviour [10].

The Mott transition is gradually suppressed under increasing hydrostatic pressure, and at $p > p_{cr} \approx 20kbar$, we see only metallic behaviour, with $d\rho/dT > 0$ for all T (Fig. 3, left).

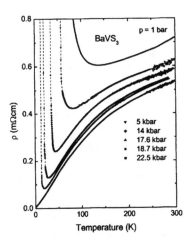
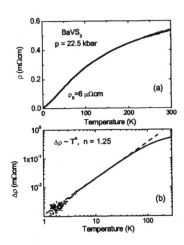

Figure 3   The resistivity under pressure (after [5]). *Left:* $\rho(T)$ for a series of pressures. The insulating phase vanishes at $p_{cr} \approx 20kbar$. *Right:* The low-T resistivity at $p = 22.5kbar$ (top) is well fitted with the fractional power law $\rho - \rho_0 \propto T^{1.25}$ (bottom).

As shown by the $p = 22.5kbar$ data (Fig. 3, right), the high-pressure metal does not follow the $\Delta\rho = \rho(T) - \rho_0 \approx AT^2$ law expected for a fermi liquid. Instead, the fractional power law fit $\Delta\rho \propto T^{1.25}$ was successful over at least one and a half decade of T (1K < T < 40K). It has become accepted to consider such behaviour as a hallmark of the non-fermi-liquid (NFL) state. In particular, $\Delta\rho \propto T^n$ with $n < 1.5$ is associated with good quality samples of nearly antiferromagnetic 3D systems [11]. In this context, $n = 1.5$ would be the 'dirty limit', expected for disorder-dominated samples.

$\Delta\rho \propto T^n$ with $n = 1.2 - 1.3$ was observed on an increasing number of f-electron systems [12], but it has remained a rarity with d-electron systems. We mention, however, CaRuO$_3$ which follows $\Delta\rho \propto T^{1.5}$ at low T [13], and the most recent finding of $\Delta\rho \propto T^{1.4}$ near the MIT of Ca$_{2-x}$Sr$_x$RuO$_4$ [14]. Let us note that though NFL behaviour might have been expected primarily at $p \approx p_{cr}$ at finite T [10], we find it instead in a broad range of $p > p_{cr}$. An analogous observation was made in Ref. [12].

A puzzling feature of the observed resistivity is the existence of a $d\rho/dT < 0$ regime within the nominally metallic phase which we consider as a precursor to the MIT. Though extrinsic effects (e.g. Kondo impurities) can give rise to a resistivity minimum, we rather think that in BaVS$_3$ it is a collective effect, since it is suppressed by pressure like the MIT (see Fig. 3 (left), and Fig. 1 (right)). The resistivity precursor is not a universal feature of the Mott phase diagram (Fig. 1, left), but something similar is seen in a number of systems of considerable physical interest: magnetite [15], underdoped high-T_C cuprates [16], some ruthenates [17], and certain bilayer molybdenum oxides [18].

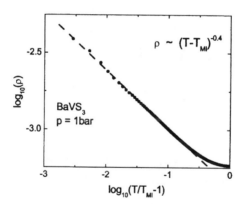

Figure 4   The apparent divergence of the precursor ($d\rho/dT < 0$) part of the resistivity at ambient pressure is well fitted with the power law $\rho \propto (T - T_{MI})^{-0.4}$ (after [5]).

Returning to BaVS$_3$, the apparent divergence of the precursor resistivity tempted us to try a power law fit[3] which was indeed successful (Fig. 4). We have to admit that we know of no physical mechanism which should give us $\rho \propto (T - T_{MI})^{-0.4}$ over at least two decades of the reduced temperature $(T - T_{MI})/T_{MI}$, but we believe nonetheless that the empirical result is significant.

---

[3]Lacking a clear motivation, we did not try more complicated fits which would bear out the obvious fact that $\rho(T_{MI})$ is finite.

We still have to suggest a reason why the $T_X < T < T_{MI}$ Mott insulator phase is non-magnetic. The fact that $BaVS_3$ is built up of triangular planes of V ions, does not suffice: for a pure Heisenberg model, the triangular lattice is not sufficiently frustrated to rule out Néel-type long range order. Though one may invoke magnetic anisotropy (whose existence is evident in Fig. 2), we rather prefer to think in terms of dynamical frustration caused by orbital fluctuations of d-electrons [4, 19]. The details of this scenario will be published elsewhere.

Acknowledgments

P.F. thanks Peter Prelovšek and Tony Ramšak for their kind hospitality and for support during the meeting. At various times, the authors were supported by the grants OTKA T025505 and D32689, FKFP 0355 and B10, AKP 98-66 and Bolyai 118/99 in Budapest, and by the Swiss National Foundation for Scientific Research in Lausanne.

References

[1] P. Fazekas, *Lecture Notes on Electron Correlation and Magnetism*, World Scientific, Singapore (1999).

[2] W. Bao *et al.*, Phys. Rev. B **58**, 12727 (1998).

[3] S. Klimm *et al.*: J. Magn. Magn. Mater. (in press); S. Horn, private comm.

[4] G. Mihály *et al.*, Phys. Rev. B **61**, R7831 (2000).

[5] L. Forró *et al.*, cond-mat/0001371 (2000); submitted for publication.

[6] M. Nakamura *et al.*, Phys. Rev. B **49**, 16191 (1994).

[7] T. Graf *et al.*, Phys. Rev. B **51**, 2037 (1995).

[8] F. Sayetat *et al.*, J. Phys. C **15**, 1627 (1982).

[9] M. Shiga and H. Nakamura, RIKEN Review No. **27**: (April 2000).

[10] S. Sachdev, Science **288**, 475 (2000).

[11] A. Rosch, Phys. Rev. Lett. **82**, 4280 (1999).

[12] G. Knebel *et al.*, Phys. Rev. B**59**, 12390 (1999).

[13] L. Klein *et al.*: Phys. Rev. B**60**, 1448 (1999).

[14] S. Nakastuji and Y. Maeno, Phys. Rev. Lett. **84**, 2666 (2000).

[15] S.K. Park, T. Ishikawa, and Y. Tokura, Phys. Rev. B**58**, 3717 (1998).

[16] P. Prelovšek *et al.*, Phys. Rev. Lett. **81**, 3745 (1998).

[17] G. Cao *et al.*, Phys. Rev. B**56**, R2916 (1997).

[18] T. Katsufuji *et al.*, Phys. Rev. Lett. **84**, 1998 (2000).

[19] A.M. Oleś *et al.*: Phys. Rev. B **61**, 6257 (2000).

VIII

SHORT CONTRIBUTIONS

LOW-ENERGY EXCITATIONS
IN THE ANISOTROPIC SPIN-ORBITAL MODEL

J. Bała[1], A. M. Oleś[1], and G. A. Sawatzky[2]

[1]*Institute of Physics, Jagellonian University, Reymonta 4, PL-30059 Kraków, Poland*
[2]*Laboratory of Applied and Solid State Physics, Materials Science Center,*
University of Groningen, Nijenborgh 4, 9747 AG Groningen, The Netherlands

Abstract   The elementary excitations of the ferromagnetic spin-orbital model which represents the low energy excitations for strongly correlated electrons described by the Hubbard Hamiltonian with a twofold-degeneracy are studied. For a completely degenerate system with the Ising-like orbital interactions the spin-and-orbital bound states can have much lower energy than the other excitations of the model and might thus be responsible for the instability of the ground state.

The discovery of colossal magnetoresistance in the manganites [1] renewed the interest in strongly correlated electron systems which involve orbitally degenerate [2] d and f states. Here we investigate an idealized ferromagnetic (FM) spin-orbital model with the anisotropy of superexchange interactions between orbital pseudospins, derived from the Hubbard Hamiltonian with a twofold degeneracy and hoppings between orbitals of the same type,

$$H = -J_s \sum_{\langle ij \rangle} \mathbf{S}_i \mathbf{S}_j - \sum_{\langle ij \rangle} [J_t + 4J_{st}\mathbf{S}_i\mathbf{S}_j] \left[\frac{1}{2}\alpha \left(T_i^+T_j^- + T_i^-T_j^+ \right) + T_i^zT_j^z \right].$$

(1)

The first term describes the exchange interaction between spins $S = \frac{1}{2}$, and fulfills the SU(2) symmetry. The orbital degrees of freedom for twofold degenerate d orbitals are described by an effective pseudospin operators $\mathbf{T}_i$ at site i, and interact by an anisotropic interaction $\propto J_t$. The orbital pseudospins couple to the spins by the term $\propto J_{st}$. The isotropic case ($\alpha = 1$) was considered by van den Brink *et al.* [3].

The z-th component is conserved of both spin and pseudospin for the Hamiltonian (1), and thus rigorous results can be obtained. Using the equation of motion method we have calculated the exact energies of single spin $[\omega_s(\mathbf{k})]$ and orbital $[\omega_t(\mathbf{k})]$ excitations, as well as of the composite spin-and-orbital excitations, $|\mathbf{k}, \mathbf{q}\rangle = S^-_{\frac{\mathbf{k}}{2}-\mathbf{q}} T^-_{\frac{\mathbf{k}}{2}+\mathbf{q}}|0\rangle$, in the FM spin and pseudospin (FF) state

J. Bonča et al. (eds.), Open Problems in Strongly Correlated Electron Systems, 395–397

with **k** and **q** being the total and relative momenta of the combined spin-and-pseudospin excitation, respectively. For the Fourier transform of the equation of motion one finds:

$$\left[\omega - \omega_s\left(\frac{\mathbf{k}}{2} - \mathbf{q}\right) - \omega_t\left(\frac{\mathbf{k}}{2} + \mathbf{q}\right)\right]|\mathbf{k}, \mathbf{q}\rangle =$$

$$- J^+ \sum_{\mathbf{a}, \mathbf{q}'}\left[\cos(\frac{\mathbf{ka}}{2}) - \cos(\mathbf{qa})\right]\left[\cos(\frac{\mathbf{ka}}{2}) - \cos(\mathbf{q}'\mathbf{a})\right]|\mathbf{k}, \mathbf{q}'\rangle$$

$$- J^- \sum_{\mathbf{a}, \mathbf{q}'}\left[\sin(\frac{\mathbf{ka}}{2}) - \sin(\mathbf{qa})\right]\left[\sin(\frac{\mathbf{ka}}{2}) - \sin(\mathbf{q}'\mathbf{a})\right]|\mathbf{k}, \mathbf{q}'\rangle, \quad (2)$$

with $J^{\pm} = J_{st}(1\pm\alpha)/N$. Here, we are seeking solution for $0 < \omega < \omega_c(\mathbf{k})$ below the boundary of the continuum $\omega_c(\mathbf{k}) = \min_{\mathbf{q}}\left[\omega_s(\frac{\mathbf{k}}{2} - \mathbf{q}) + \omega_t(\frac{\mathbf{k}}{2} + \mathbf{q})\right]$.

The simplest solution of the above equation of motion is obtained in the one-dimensional (1D) model. As shown in Fig. 1 for $\alpha = 0.1$ two bound states are present in the whole Brillouin zone (BZ) with small dispersion and the lower one can have a much lower energy than other excitations except for $k \simeq 0$ due to the spin-wave (Goldstone) mode. [see Fig. 1(b)]. Such a behavior of bound states is quite different from that found in the Heisenberg model [4], where the anisotropy would also lead to the dispersionless bound states but such states do not cause any instability of the FM state.

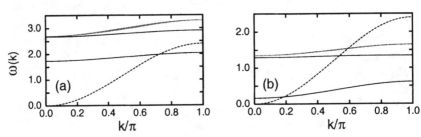

Figure 1   Dispersions of the spin-wave (dashed lines), orbital-wave (dotted lines) excitations, and spin-and-orbital bound states (solid lines) of a 1D model with: $J_s = 0.2J_{st}$, $\alpha = 0.1$, and different orbital interactions: (a) $J_t = 2.0J_{st}$ and (b) $J_t = 0.5J_{st}$. The boundary of spin-and-orbital continuum is almost degenerate with the orbital excitations.

Next, we concentrate on the case $\alpha = 0$ and present the phase diagrams (Fig. 2) illustrating parameter regions where different number of bound states is present in the whole BZ. In both 1D and 2D cases one can find bound states for all momenta only when $J_s < J_{st}$. With increasing α the extent of different regions shrinks and they disappear for $\alpha \to 1$ as there are no bound states at $k = 0$ for $\alpha = 1$. The FF state is stable only above the solid line and the most interesting situation arises when one is close to the threshold of the FF state with small values of α where a dispersionless bound state with very low energy [see Fig. 1(b)] can lead to the instability of the FF state.

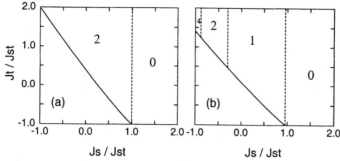

Figure 2   Phase diagram of a 1D (a) and 2D (b) model for $\alpha = 0.0$. FF state is stable above the solid line, and is characterized by 4, 2, 1, or 0 spin-and-orbital bound states, respectively, present at all momenta.

The collective modes with such low energies, although gapped, should contribute to the thermodynamic properties of the system and make it possible to observe them in susceptibility or specific heat measurements. Moreover, the structure of the mixed excitation $|\mathbf{k}, \mathbf{q}\rangle$ guarantees that it carries both spin and orbital component in equal proportion at all momenta $\mathbf{k}$ and such a bound state could be seen in the neutron spectroscopy.

We hope that similar behavior of the model can be found in the case of e_g orbitals with predominantly AF interactions. However, in that limit the quantum fluctuations may alter the ground state, e.g. leading to spin and orbital dimerized states [5], characterized by a spin gap.

Acknowledgments

We acknowledge the financial support by the Committee of Scientific Research (KBN) of Poland, Project No. 2 P03B 175 14.

References

[1] R. von Helmolt *et al.*, Phys. Rev. Lett. **71**, 2331 (1993); S. Jin *et al.*, Science **264**, 413 (1994); P. Schiffer *et al.*, Phys. Rev. Lett. **75**, 3336 (1995).

[2] For a review see: M. Imada, A. Fujimori, and Y. Tokura, Rev. Mod. Phys. **70**, 1039 (1998).

[3] J. van den Brink *et al.*, Phys. Rev. B **58**, 10 276 (1998).

[4] M. Wortis, Phys. Rev. **132**, 85 (1963).

[5] L. F. Feiner, A. M. Oleś, and J. Zaanen, Phys. Rev. Lett. **78**, 2799 (1997).

DYNAMICAL MEAN-FIELD SELFCONSISTENCY RELATION FOR MULTIBAND SYSTEMS

G. Czycholl

Institut für Theoretische Physik, Universität Bremen,
Postfach 330 440, D-28 334 Bremen, Germany

Abstract   Within the dynamical mean field theory (DMFT) the Hubbard model is mapped on a single impurity Anderson model (SIAM), for which the hybridization function has to be determined selfconsistently. But in DMFT treatments of the periodic Anderson model (PAM) unphysical results may be obtained when mapping it on an effective SIAM. Therefore, in a proper DMFT the PAM must be mapped on a two-level effective impurity model so that the DMFT selfconsistency relation becomes a $2 * 2$ matrix equation.

Within the dynamical mean-field theory (DMFT) [1] a correlated electron lattice model is mapped on an effective single-impurity problem with a coupling to an effective bath to be determined selfconsistently. This mapping becomes exact in the limit of large dimension, $d \to \infty$ [2].

The Hubbard model (HM)

$$H_{HM} = \sum_{\vec{k}\sigma} \varepsilon_{\vec{k}} c_{\vec{k}\sigma}^{\dagger} c_{\vec{k}\sigma} + U \sum_{\vec{R}} c_{\vec{R}\uparrow}^{\dagger} c_{\vec{R}\uparrow} c_{\vec{R}\downarrow}^{\dagger} c_{\vec{R}\downarrow} \qquad (1)$$

can be mapped on an effective single-impurity Anderson model (SIAM). The DMFT selfconsistency condition requires that the site-diagonal matrix element of the (retarded) Green function of the large-d HM is equal to the f-Green function of the effective SIAM, i.e.

$$G_{HM}(z) = \frac{1}{N} \sum_{\vec{k}} \frac{1}{z - \Sigma(z) - \varepsilon_{\vec{k}}} = G_{SIAM}(z) = \frac{1}{z - \Sigma(z) - \Gamma(z)} \qquad (2)$$

where $\Gamma(z)$ denotes the bath Green function (hybridization function with an effective conduction band). $\Gamma(z)$ fulfills

$$\Gamma(z) = z - \Sigma(z) - G_{HM}^{-1}(z) = z - \Sigma(z) - G_0^{-1}(z - \Sigma(z)) \qquad (3)$$

where $G_0(\tilde{z}) = \int d\varepsilon \rho_0(\varepsilon)/(\tilde{z} - \varepsilon)$ denotes the site-diagonal Green function of the uncorrelated lattice model, ρ_0 being the unperturbed density of states. This

J. Bonča et al. (eds.), Open Problems in Strongly Correlated Electron Systems, 399–401.

$\Gamma(z)$ always has the proper analytic properties of a retarded (effective "bath" or conduction band) Green function; it fulfills, in particular, Im $\Gamma(z) < 0$ for Im $z > 0$. Therefore, no unphysical results can occur in the standard DMFT selfconsistency loop for the HM.

But for the periodic Anderson model (PAM)

$$
\begin{aligned}
H_{PAM} &= \sum_{\vec{k}\sigma} \epsilon_{\vec{k}} c^{\dagger}_{\vec{k}\sigma} c_{\vec{k}\sigma} + \sum_{\vec{R}\sigma} E_f f^{\dagger}_{\vec{R}\sigma} f_{\vec{R}\sigma} \\
&+ V \sum_{\vec{R}\sigma} \left(c^{\dagger}_{\vec{R}\sigma} f_{\vec{R}\sigma} + f^{\dagger}_{\vec{R}\sigma} c_{\vec{R}\sigma} \right) + U \sum_{\vec{R}} f^{\dagger}_{\vec{R}\uparrow} f_{\vec{R}\uparrow} f^{\dagger}_{\vec{R}\downarrow} f_{\vec{R}\downarrow} \quad (4)
\end{aligned}
$$

the site-diagonal Green function (per spin direction) is a $2 * 2$-matrix

$$
\hat{G}_{PAM}(z) = \frac{1}{N} \sum_{\vec{k}} \begin{pmatrix} z - \epsilon_{\vec{k}} & -V \\ -V & z - E_f - \Sigma(z) \end{pmatrix}^{-1} \quad (5)
$$

In all existing DMFT-treatments [3, 4, 5] the PAM, too, is mapped on the SIAM and the equality of the f-electron Green function of the PAM with the f-electron Green function of the effective SIAM is used as the DMFT selfconsistency requirement:

$$
G^{ff}_{PAM}(z) = \frac{1}{N} \sum_{\vec{k}} \frac{1}{z - E_f - \Sigma(z) - \frac{V^2}{z - \epsilon_{\vec{k}}}} = \frac{1.}{z - E_f - \Sigma(z) - \Gamma(z)} \quad (6)
$$

This selfconsistency condition yields

$$
\Gamma(\tilde{z}) = \frac{\tilde{z} V^2 G_0(z - \frac{V^2}{\tilde{z}})}{\tilde{z} + V^2 G_0(z - \frac{V^2}{\tilde{z}})} \quad (7)
$$

with $\tilde{z} = z - E_f - \Sigma(z)$. But this Γ does not necessarily fulfill the correct analytic properties. This is explicitly demonstrated in Fig.1, where for a semielliptic model density of states, i.e. $G_0(z) = 2(z - \sqrt{z^2 - 1})$, fixed $E_f, \Sigma(z)$ and different V^2 the resulting imaginary part Im $\Gamma(\tilde{z})$ is plotted as a function of Re z. Obviously, there are regions of positive imaginary part of $\Gamma(\tilde{z})$ for certain choices of the parameters, i.e. $\Gamma(\tilde{z})$ has not the correct analytic properties for a (retarded) bath Green function. Unphysical behavior within DMFT treatments of the PAM has also been obtained recently by Keiter and Leuders [6] (also using the mapping of the PAM on the SIAM).

A possible explanation for this unphysical behavior is the following: The selfconsistency relation (6) is not the appropriate one, but for a proper DMFT treatment of the PAM it should be mapped on a single impurity model with two levels per spin direction and corresponding couplings to bath Green functions. Then the DMFT selfconsistency condition also is a $2 * 2$-matrix equation:

$$
\hat{G}_{PAM}(z) = \begin{pmatrix} z - \Gamma_c(z) & -\Gamma_{cf}(z) \\ -\Gamma_{cf}(z) & z - E_f - \Sigma(z) - \Gamma_f(z) \end{pmatrix}^{-1} \quad (8)
$$

Im $\Gamma(E + i0)$

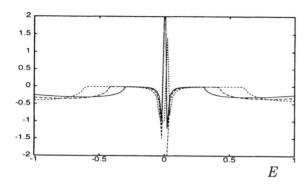

E

Figure 1   Energy dependence of Im $\Gamma(E+i0)$ for the PAM according to (7) for fixed selfenergy imaginary part Im $\Sigma(E + i0) = -0.002$, $E_f = 0$. and different $V^2 = 0.4, 0.6, 0.8$

This means the SIAM is only the appropriate single-impurity model for the DMFT-mapping of the HM, but a more general two-level single-impurity model has to be used for the DMFT-mapping of the PAM. More realistic lattice models with n bands require corresponding n-level single-impurity models with the corresponding number of bath Green functions. This observation is certainly important for recent attempts to combine the DMFT with ab-initio methods, because for real correlated electron systems one usually needs multi-band descriptions (for instance the five correlated d-bands for the transition metals like iron and nickel).

References

[1] Georges A., Kotliar G., Krauth W. and Rozenberg M.J., Rev. Mod. Phys. **68** (1996) 13

[2] Metzner W. and Vollhardt D., Phys. Rev. Lett. **62** (1989) 324

[3] Jarrell M., Akhlaghpour H. and Pruschke T., Phys. Rev. Lett. **70** (1993) 1670

[4] Hülsenbeck G. and Qin Q., Solid State Commun. **90** (1994) 195

[5] Pruschke T., Bulla R. and Jarrell M., preprint (2000)

[6] Keiter H. and Leuders T., Europhys. Lett. **49** (2000) 801

SUPERFLUIDITY IN FERMI-SYSTEMS WITH REPULSION

M.S. Mar'enko, M.Yu. Kagan, D.V. Efremov
P.L. Kapitza Institute for Physical Problems, Moscow 117334, Russia

M.A. Baranov
Russian Research Center "Kurchatov Institute", Moscow 123182, Russia

Abstract   We discuss the superfluid transition in Fermi-system with repulsive interaction. In the model of weakly interacting Fermi gas with short range repulsion the presence of Kohn singularity in the effective interaction leads to the Cooper pairing in the p-wave channel. We show that the superfluid critical temperature in spin-polarized system exhibits a strong nonmonotonic behavior as a function of polarization. We also study the superfluid p-wave phase diagram for a 2D Fermi-liquid. In the weak-coupling limit the absolute minimum corresponds both to planar and axial phases. To lift this degeneracy, we calculate the strong-coupling corrections (*i. e.* next order of T_c/ε_F) to the quartic terms and find the most possible candidates to the phase stabilized in the system.

1.   INTRODUCTION

Unconventional mechanisms of Cooper pairing have recently started to attract a greater attention. This is primarily due to the discovery of high-T_c superconductivity, superconductivity in organic compounds, heavy-fermion and ruthenium based materials. The aim of present paper is to show the possibility of superfluidity in the Fermi-systems with repulsion already in a simplest case of a Fermi-gas with short range repulsive interaction. Of course the real strongly-correlated systems such as HTSC-materials or heavy fermion compounds require more rigorous description. The advantage of our approach is that we are far from structural and magnetic instabilities, so we can study the generic superfluid transition in a system. The presence of small parameter in the theory allows us to use a standard perturbative technique and control the order of the diagrams. The considered form of the interparticle interaction assumes that only s-wave scattering takes place in the system, characterized by the scattering length a. (In the principal order of perturbation theory $p_F a = p_F m g/4\pi$.) The

J. Bonča et al. (eds.), Open Problems in Strongly Correlated Electron Systems, 403–408.

corresponding small dimensionless parameter, the gas parameter λ, is given by $\lambda = 2|a|p_F/\pi$.

In the considered model the bare interparticle interaction can be either attractive or repulsive. In the attractive case the system is unstable towards the traditional BSC-type singlet Cooper pairing with relative orbital momentum of the pair $\ell = 0$ [1]. In the systems with repulsive interaction, the formation of $\ell = 0$ Cooper pairs is clearly impossible and in order to investigate the existence of superfluidity, we need to study the possibility of $l \neq 0$ Cooper pairing. The possible existence of superfluidity in Fermi systems with repulsion was first emphasized by Kohn and Luttinger in 1965. In [2] they examined the contribution of collective effects to the scattering amplitude in a particle-particle channel which leads to an effective attractive interaction between quasiparticles at the Fermi surface via polarization of the Fermionic background. A principal role in the formation of attractive harmonics in the effective interaction and consequently the superfluidity is played by the Kohn's singularity in the effective interaction. In the three-dimensional case, it has the form

$$\tilde{\Gamma}^{\text{sing}}_{\text{eff}}(q) \sim \left[(2p_F)^2 - q^2\right] \ln \left|(2p_F)^2 - q^2\right| + \Gamma_{\text{reg}}(q^2). \qquad (1)$$

It should be noted that the above contribution to the effective interaction decreases over large distances more slowly that the bare interaction $U_0(r - r')$ and consequently corresponds to the main contribution to the scattering amplitude in the limit of large momenta ℓ [2]: $\tilde{\Gamma}^{(l)}_{\text{eff}} \sim (-1)l^{-4}$.

It was shown in [3, 4] that effective attraction occurs in the second order of perturbation theory also for angular momentum $\ell = 1$, which gives the following result for T:

$$T_{c1} \sim \tilde{\varepsilon} \exp\left\{-\frac{5\pi^2}{4(2\ln 2 - 1)(ap_F)^2}\right\} \approx \tilde{\varepsilon} \exp\left\{-\frac{12.9}{\lambda^2}\right\}. \qquad (2)$$

where a is the s-wave scattering length, $\tilde{\varepsilon}$ the energy cut-off of the order of Fermi-energy.

It can be seen from formula (2) that in order to determine the preexponential factor $\tilde{\varepsilon}$ we need to retain terms up to the order of λ^4. (This follows from the fact that since $\tilde{\Gamma}_1$ begins from terms of the order of λ^2, to obtain terms of the order of λ^0 in the exponent, we need to know $\tilde{\Gamma}_1$ up and including to the terms of λ^4).

The calculation of third- and fourth-order diagrams in terms of the gas parameter λ to the effective interaction together with taking into account the retardation effects and renormalization of the singular part of the Green function gives the following result [5]:

$$T_{c1} = \frac{2}{\pi} e^C \varepsilon_F \exp\left\{-(0.077\lambda^2 + 0.33\lambda^3 + 0.26\lambda^4)^{-1}\right\}. \qquad (3)$$

Note that this formula is important for experimental search of fermionic superfluidity at ultralow temperatures in diluted mixtures of ^{3}He in ^{4}He and in fermionic isotope of Li (^{6}Li) in a confined geometry of magnetic traps.

2.   TWO-DIMENSIONAL CASE

We consider now a case of two-dimensional weakly interacting Fermi-gas with repulsion. In the low density limit $r_0 p_F \ll 1$, where r_0 is the range of potential, the perturbative expansion is also legitimate. The corresponding dimensionless small parameter f_0 is related to the s-wave scattering amplitude between two particles on the Fermi surface. For the considered case the gas parameter is given by

$$f_0 = \frac{mU_0/4\pi}{1 + (mU_0/4\pi)\ln(r_0 p_F)^{-2}}, \tag{4}$$

where U_0 is the zeroth Fourier component of the potential. In the low density limit the formula (4) reduces to $f_0 = -1/\ln(r_0 p_F)$. It turns out that the Kohn's singularity is in 2D even more pronounced than in 3D. In second order of perturbation theory the singular part of the effective interaction is given by:

$$\frac{m}{4\pi}\tilde{\Gamma}_{\text{eff}}^{\text{sing}}(q) = -f_0^2 \mathrm{Re}\sqrt{1 - (2p_F/q)^2}. \tag{5}$$

However, the effective vortex in 2D is zero for $q \leq 2p_F$, and does not contribute to the harmonics of effective interaction. The situation changes if we go beyond the second order of perturbation theory. In next, third order of expansion the Kohn singularity exists already for $q \leq 2p_F$:

$$\frac{m}{4\pi}\tilde{\Gamma}_{\text{eff}}^{\text{sing}}(q) \sim -f_0^3 \mathrm{Re}\sqrt{1 - (q/2p_F)^2}, \tag{6}$$

and it leads to an effective attraction in the p-wave channel. Numerical calculations of the third-order diagrams give the following result for the p-wave critical temperature:

$$T_{c1} = \tilde{\varepsilon}\exp\left(-\frac{1}{6.1 f_0^3}\right), \tag{7}$$

where $\tilde{\varepsilon}$ is unknown prefactor of the order of the Fermi-energy. Note that in the complete analogy with 3D case, the partial harmonics with higher values of ℓ have the absolute values much smaller than that for $\ell = 1$. Therefore, the leading instability towards the superfluid pairing in 2D Fermi-gas is again in the p-wave channel.

Note the presence of a rather large numerical coefficient 6.1 in (7). It provides the reasonable order of magnitude for T_{c1} in spite of the fact that it is given by third power of small parameter f_0.

The typical value of scattering amplitude obtained from the experiments on ^{3}He atoms on the surface of ^{4}He is $f_0 \leq 0.3$ for ^{3}He coverages corresponding to the dilute Fermi gas situation. In this regime the critical temperature reads $T_{c1} \sim 10^{-3} K$.

Note that in 2D case the bare p-wave scattering amplitude has an order of magnitude $f_1 \sim (r_0 p_F)^3$. At the same time, the leading term in perturbative expansion for the effective interaction $\tilde{\Gamma}$ in T_{c1} has an order of $(1/\ln(r_0 p_F))^3$ in low density limit. Thus, $f_0^3 \gg f_1, \ldots$, and in formula (7) we can restrict the description to an expansion in terms of f_0 only.

Of course, in real 2D systems only Kosterlitz-Thouless transition is possible. However the calculations of Beasley et al [6] show that in dilute fermi-systems where $T_c \ll \varepsilon_F$ the BCS mean field critical temperature yields a very good estimate from above for a real transition temperature:

$$\frac{T_{BCS} - T_{KT}}{T_{KT}} \sim \frac{T_c}{\varepsilon_F} \ll 1$$

3. T_C ENHANCEMENT IN A SPIN-POLARIZED FERMI-SYSTEM

The critical temperature of a superfluid transition in repulsive Fermi-system can be significantly increased by applying external magnetic field. In polarized system the singlet s-wave pairing is paramagnetically suppressed because of the flip of one spin in the Cooper pair. The effect of polarization in system with the triplet p-wave pairing is not so obvious.

It turns out that an external magnetic field provides the separation of channels in superfluid system. Namely, the Cooper pair is formed by two particles with spin "up" while effective interaction is prepared by two particles with spin "down".

In 3D case the singular part of the effective interaction in second order of perturbation theory has a form $\tilde{\Gamma}_{eff}^{sing}(\theta) \sim (\pi - \theta)^2 \ln(\pi - \theta)$ at $H = 0$ (here θ is the angle between incoming and outgoing momenta). In the presence of a magnetic field it changes to $\tilde{\Gamma}_{eff}^{sing}(\theta) \sim (\theta_c - \theta) \ln(\theta_c - \theta)$, where $\theta_c \neq \pi$. As a result the Kohn's anomaly increases, leading to the enhancement of partial harmonics in the effective interaction.

At the same time there is an opposite effect, namely: with the increase of a polarization the density of states for particles with spin "down" decreases. In the limit of completely polarized system ($\alpha = 1$, where $\alpha = (n_\uparrow - n_\downarrow)/(n_\uparrow + n_\downarrow)$) there are no spin "down" particles ($n_\downarrow = 0$) in the system, so the Cooper pairing is impossible.

The competition of the above two processes leads to a strongly non-monotonic behavior of $T_{c1}(\alpha)$. The critical temperature has a maximum at $\alpha = 0.4$ and goes to zero at $\alpha \to 1$ (see Fig.1a).

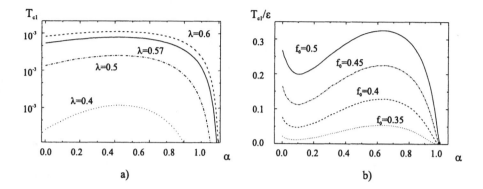

Figure 1 The p-wave superfluid critical temperature T_{c1} as a function of polarization α for different values of gas parameter: a) 3D case; b) 2D case.

In 2D case at $H \neq 0$ the Kohn's anomaly changes from (5) to

$$\frac{m}{4\pi} \tilde{\Gamma}_{\text{eff}}^{\text{sing}}(q_\uparrow) = -f_0^2 \text{Re}\sqrt{1 - (2p_{F\downarrow}/q_\uparrow)^2}, \qquad (8)$$

and becomes effective for superfluid pairing already in a second order of perturbation theory. It can be seen that a competition of contributions to $\tilde{\Gamma}_{\text{eff}}$ from particle-particle and particle-hole channels in third order provides the minimum at the plot $T_{c1}(\alpha)$ at small $\alpha \approx 0.1$. For typical values of $f_0 = 0.3$ the critical temperature in minimum is approximately two times smaller than at $\alpha = 0$. At the same time, the maximum of $T_{c1}(\alpha)$ occurs at $\alpha = 0.6$ and is even broader than in 3D: from $\alpha = 0.3$ to $\alpha = 0.9$. The value of $T_{c1}(\alpha = 0.6)$ is approximately 8 times larger than $T_{c1}(\alpha = 0)$ for $f_0 = 0.3$ (see Fig. 1b).

Note that the effect of a strong enhancement of T_c also takes place in two band situation. In this case the role of spins up is played by the electrons (or holes) of the first band while the role of spins down by electrons of the second band. That is why the proposed mechanism can be also effective in 2D or quasi-2D metallic systems such as superconductive Sr_2RuO_4 where two bands electrons cross the fermi-level.

4.   PHASE DIAGRAM IN TWO DIMENSIONAL CASE

In the weak coupling (BCS) limit for zero magnetic field, two different phases correspond to the minimum of the free energy in 2D case (the axial and the planar phases). To lift this degeneracy one has to consider next order ($\sim T_c/\varepsilon_F$) strong-coupling corrections. Following [7], we have calculated the difference between the free energies of axial and planar phases to this order:

$$\Phi_{axial} - \Phi_{planar} \sim -\frac{N(0)}{T_c^2}\left(\frac{T_c}{\varepsilon_F}\right) \tag{9}$$
$$\times \int \frac{d\varphi}{2\pi}|\sin\varphi|T_a^2(\hat{k}_1,\hat{k}_2;\hat{k}_1,\hat{k}_2),$$

where $N(0)$ is the density of states on the Fermi surface, $T_a(\hat{k}_1,\hat{k}_2;\hat{k}_3,\hat{k}_4)$ is the spin-antisymmetric part of quasiparticle scattering amplitude with momenta k_i on the Fermi surface and φ is the angle between $\hat{k}_1$ and $\hat{k}_2$. Clearly, the energy difference (2) is always negative, therefore in the absence of a magnetic field in the 2D Fermi-gas undergoes a superfluid transition to an axial phase (resembling A-phase of ^{3}He) in contrast with a 3D case where at low densities the isotropic B-phase corresponds to the global minima of a Free-energy [8].

Acknowledgments

We acknowledge fruitful discussions with A.F. Andreev, H.W. Capel, I.A. Fomin, Yu. Kagan, V.I. Marchenko, D. Rainer, and I.M Suslov. This work was supported by INTAS grant 98-963, RFBR grants 00-02-16255, 00-02-26684 and grant of the President of Russian Federation 15-1-2000/22. D.V.E. is also grateful to Landau scholarship for financial support.

References

[1] L.P. Gorkov and T.K. Melik-Barkhudarov, Sov. Phys. JETP **13**, 1018 (1961).

[2] W. Kohn and J.H. Luttinger, Phys. Rev. Lett. **15**, 524 (1965).

[3] D. Fay and A. Layzer, Phys. Rev. Lett. **20**, 187 (1968).

[4] M.Yu. Kagan, and A.V. Chubukov, JETP Letters **47**, 525 (1988).

[5] D.V. Efremov, M.S. Mar'enko, M.A. Baranov, and M.Yu. Kagan, JETP **90**, 861 (2000).

[6] M.R. Beasley and J.E. Mooij, Phys. Rev. Lett. **42**, 1165 (1979).

[7] D. Rainer and J.W. Serene, Phys.Rev B **113**, 4745 (1976).

[8] M.Yu. Kagan, M.A. Baranov, D.V. Efremov, M.S. Mar'enko, and H.W. Capel, JETP Lett. **59**, 268 (1994).

DYNAMIC JAHN-TELLER EFFECT AND DISTORTIONAL DISORDER IN MANGANITES

Louis Felix Feiner[1,2] and Andrzej M. Oleś[3]

[1] *Utrecht University, Princetonplein 5, NL-3584 CC Utrecht, The Netherlands*

[2] *Philips Research Laboratories, Prof. Holstlaan 4, NL-5656 AA Eindhoven, The Netherlands*

[3] *Institute of Physics, Jagellonian University, Reymonta 4, PL-30059 Kraków, Poland*

Abstract   We study the (local) dynamic Jahn-Teller (JT) effect in the context of a spin-and-orbital t-J model for the manganites. It is shown that allowing the correlated e_g electrons to form local vibronic states with variationally optimized distortion amplitude, leads to a "JT-liquid" with local tetragonal distortions and yet macroscopic cubic symmetry. At intermediate doping ($x \simeq 0.25$) a transition from a large-distortion polaronic insulator to a small-distortion metal with correlated electrons takes place, as observed in the ferromagnetic phase in the manganites.

Recent theoretical work on the manganites has reestablished the importance of the e_g-type orbital degeneracy of Mn^{3+}, in particular in the low-doping regime where orbital order occurs in conjunction with a cooperative Jahn-Teller (JT) effect. In the high-doping phases, although orbitally disordered, the JT effect should still make itself felt: since the local JT energy is as large as the e_g hopping, local JT distortions are expected to persist when the cooperative JT order is suppressed. In fact, there is ample experimental evidence for such *local* distortions (e.g. [1]) even though the *global* symmetry is cubic [2].

Here we investigate the role of the (local) dynamic JT effect in the context of a spin-and-orbital t-J model, derived in the strong correlation (large U) limit [3, 4]. This model corresponds to Mn^{4+} ions (d^3 Hund's rule $^4A_2$ state) sitting in a background of Mn^{3+} ions (d^4 Hund's rule ^5E state), and interacting by orbital-dependent superexchange. We are interested in the ferromagnetic phases, where the e_g electron hopping $\propto t$ involves only the orbital and charge degrees of freedom, and may be represented by spinless orbital fermions $c_{\pm}$ referring to complex orbitals $\{|+\rangle, |-\rangle\}$ [4], with double occupancy forbidden.

We supplement this electronic model by the local JT coupling between each e_g electron and the vibrational modes $q_\theta = q\cos\phi$ and $q_\epsilon = q\sin\phi$ of the oxygen octahedron surrounding the Mn^{3+} ion on which the electron resides, described by the standard linear $E \otimes \epsilon$ JT Hamiltonian [5] with coupling strength k. All its eigenstates are doubly degenerate, and the ground doublet

J. Bonča et al. (eds.), Open Problems in Strongly Correlated Electron Systems, 409–411.

$\{|\Psi_+(k)\rangle, |\Psi_-(k)\rangle\}$ is connected continuously (when $k \to 0$) to the electronic doublet $\{|+\rangle, |-\rangle\}$. The associated energy gain is $E_{JT} \simeq \frac{1}{2}k^2\hbar\omega \simeq 350$ meV for the manganites, while $\hbar\omega \simeq 70$ meV, so $k \simeq 3.3$. In the vibronic states $|\Psi_+(k)\rangle$, $|\Psi_-(k)\rangle$ there is a tetragonal distortion with amplitude $\simeq k$ (i.e. $\langle \Psi_+|q_\theta^2 + q_\epsilon^2|\Psi_+\rangle \simeq k^2$), but it is "randomly oriented" ($\langle\Psi_+|q_{\theta(\epsilon)}|\Psi_+\rangle = 0$, because the phase ϕ is undetermined), while in a coherent state $|\hat{n}\rangle = \cos(\psi/2)e^{-i\theta/2}|\Psi_+\rangle + \sin(\psi/2)e^{+i\theta/2}|\Psi_-\rangle$ the distortion is generally finite, $\langle q_\theta\rangle \simeq k\sin\psi\cos\theta$, $\langle q_\epsilon\rangle \simeq k\sin\psi\sin\theta$.

Since E_{JT} is large, there is a strong tendency for the e_g electrons *not* to remain in the purely electronic states $|+\rangle$, $|-\rangle$, but to form vibronic states instead. However, formation of vibronic states with full amplitude k strongly suppresses the hopping by the familiar polaronic overlap [$\propto \exp(-k^2/2)$], and therefore becomes unfavorable at high hole doping. We study these opposing tendencies by assuming that the e_g electrons (or more precisely the ^5E states) form ground-doublet-like vibronic states $|\Psi_+(k_1)\rangle$, $|\Psi_-(k_1)\rangle$ with a distortion amplitude k_1 which may be different from k, and which we treat as a variational parameter. We take as an explicit variational wavefunction

$$|\Psi_+(k_1; q, \phi)\rangle = \sqrt{p(k_1)/\pi} \; e^{-q^2/2} [\cosh(k_1 q)|+\rangle + \sinh(k_1 q)e^{i\phi}|-\rangle], \tag{1}$$

where $p(k_1) = [1 + 2k_1\exp(k_1^2)\mathrm{Erf}(k_1)]^{-1}$, which has been shown to be an excellent approximation for all k when $k_1 = k$ [5]. This yields analytical expressions for the local energy gain $E_{\mathrm{vibr}}(k_1) < E_{JT}(k)$ and the polaronic factor $w(k_1) = |\langle\Psi_+(k_1)|c_+^\dagger|0\rangle|^2 < 1$ reducing the hopping t.

After the fermions $c_\pm$ have been formally replaced by dressed fermions $a_\pm$ referring to the vibronic states $|\Psi_\pm\rangle$, the remaining calculation can be performed in exact analogy with the corresponding electronic problem (see [4]): first the fermions are expressed as $a_\pm^\dagger = B_\pm^\dagger f_\pm^\dagger b$, where the Kotliar-Ruckenstein bosons $B_\pm$ and the pseudofermions $f_\pm$ represent the orbital degrees of freedom and the slave boson b represents the charge, and next the mean-field replacements $B_+ \mapsto \cos(\psi/2)e^{-i\theta/2}$, $B_- \mapsto \sin(\psi/2)e^{+i\theta/2}$, $b^2 \mapsto x$ are made. Thus electronic correlations are accounted for by the KR bosons, giving a further bandwidth renormalization [in addition to that by $w(k_1)$]. In the total energy we also include the contribution from the superexchange.

Similar to the electronic case [4], phases with different vibronic order differ in kinetic energy because of a different density of states. We have considered the following: (i) two vibronically (i.e. orbital- and lattice) ordered phases: Vz and Vx, with states $|\Psi_z\rangle \propto |\Psi_+\rangle + |\Psi_-\rangle$ and $|\Psi_x\rangle \propto |\Psi_+\rangle - |\Psi_-\rangle$, respectively, at all ($Mn^{3+}$) sites; (ii) two "coherently disordered" phases: V+, with $|\Psi_+\rangle$ at all sites (similar to the orbital ordered $|+\rangle$ phase [6]), and V+/-, with $|\Psi_+\rangle$, $|\Psi_-\rangle$ alternating; (iii) the vibronically disordered (VD) phase, with $\psi = \pi/2$ and random θ. The results are shown in Fig. 1. In all phases an

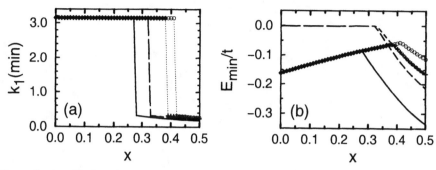

Figure 1   Amplitude of local JT distortion k_1 (a) and total energy E_{min}/t [with t defined as in Ref. [4]] (b) versus hole doping x in various phases (defined in the text): Vz (long dashed line), Vx (short dashed line), V+ (circles), V+/- (crosses), VD (full line).

abrupt crossover occurs from a strongly vibronic state [large k_1 ($\simeq k$)] at small doping to a nearly electronic state ($k_1 \simeq 0$) at large doping, corresponding to an insulator-metal transition. Further, in the metallic regime the VD phase is stable compared to the ordered phases by a large margin, as in the electronic case [4]. This is characteristic for orbital physics in contrast with spin physics and is related to the absence of SU(2) symmetry (pseudospin nonconservation).

We conclude that at finite doping (roughly $x > 0.15$) a zero-temperature "Jahn-Teller liquid" can be stable, which is characterized by the coexistence of local tetragonal (JT) distortion [1] and macroscopic cubic symmetry [2]. At intermediate doping ($x \simeq 0.25$) a transition occurs from an insulating phase with large distortion amplitude to a metallic phase with small (but non-zero) distortions, corresponding to the transition between the two ferromagnetic phases occurring in the manganites.

We acknowledge the support by KBN of Poland, Project 2 P03B 175 14.

References

[1] D. Louca *et al.*, Phys. Rev. B **56**, R8475 (1997).

[2] T.G. Perring *et al.*, Phys. Rev. Lett. **77**, 711 (1996).

[3] L.F. Feiner and A.M. Oleś, Phys. Rev. B **59**, 3295 (1999); A.M. Oleś and L.F. Feiner, Acta Phys. Polon. A **97**, 193 (2000).

[4] A.M. Oleś and L.F. Feiner, these Proceedings (2000).

[5] R. Englman, The Jahn-Teller Effect in Molecules and Crystals, Wiley, London, (1972).

[6] D.I. Khomskii, unpublished, and these Proceedings (2000).

DIAGRAMMATIC THEORY OF THE ANDERSON IMPURITY MODEL WITH FINITE COULOMB INTERACTION

K. Haule[a], S. Kirchner[b], H. Kroha[b], P. Wölfle[b]

[a] J. Stefan Institute, SI-1000 Ljubljana, Slovenia

[b] Institut für Theorie der Kondensierten Materie, Universität Karlsruhe
D-76128 Karlsruhe, Germany

Abstract   We developed a self-consistent conserving pseudo particle approximation for the Anderson impurity model with finite Coulomb interaction, derivable from a Luttinger Ward functional. It contains an infinite series of skeleton diagrams built out of fully renormalized Green's functions. The choice of diagrams is motivated by the Schrieffer Wolff transformation which shows that singly and doubly occupied states should appear in all bare diagrams symmetrically. Our numerical results for T_K are in excellent agreement with the exact values known from the Bethe ansatz solution. The low energy physics of non-Fermi liquid Anderson impurity systems is correctly described while the present approximation fails to describe Fermi liquid systems, since some important coherent spin flip and charge transfer processes are not yet included. It is believed that CTMA (Conserving T-matrix approximation) diagrams will recover also Fermi liquid behavior for Anderson models with finite Coulomb interaction as they do for infinite Coulomb interaction.

Anderson impurity models have been of considerable interest recently for studying the transport in mesoscopic structures [1] as well as for describing strong correlations on the lattice in the limit of large dimensions [2]. In both cases the doubly occupied impurity site, which is characterized by an additional energy cost U, is essential for the correct physical description.

In terms of pseudofermion (f_σ) and slave boson operators (a, b) the Anderson model is defined by the Hamiltonian

$$
\begin{aligned}
H &= \sum_{\vec{k},\sigma} \epsilon_{\vec{k}} c^+_{\vec{k}\sigma} c_{\vec{k}\sigma} + \epsilon_d (\sum_\sigma f^+_\sigma f_\sigma + 2a^+ a) + \\
&+ \sum_{\vec{k},\sigma} V_k (c^+_{\vec{k}\sigma} b^+ f_\sigma + \sigma c^+_{\vec{k}\sigma} f^+_{-\sigma} a + h.c.) + U a^+ a,
\end{aligned}
\tag{1}
$$

413

J. Bonča et al. (eds.), Open Problems in Strongly Correlated Electron Systems, 413–416.

where ϵ_d is the energy of the local d level, U is the Coulomb interaction and $V_{\vec{k}}$ are the hybridization matrix elements. A physical electron in the local level is created with the operator $d_\sigma = b^+ f_\sigma + \sigma f^+_{-\sigma} a$, where b represents the empty level and a the doubly occupied one (in the following we will call a and b the heavy and the light boson, respectively). For the representation to be exact, the local operator constraint $Q \equiv \sum_\sigma f^+_\sigma f_\sigma + b^+ b + a^+ a = 1$ must be fulfilled at all times. This is guaranteed by a suitable projection technique [4].

1.   METHOD

It is well known that the Non-Crossing Approximation (NCA) gives the correct Kondo temperature T_K in the limit of large Coulomb repulsion U while it fails to recover the correct Kondo temperature for finite U. The reason may be seen from the Schriffer-Wolff transformation [3] where the effective coupling constant has two equally important terms for $U \simeq -2\epsilon_d > 0$

$$J = -\frac{V^2}{\epsilon_d} + \frac{V^2}{\epsilon_d + U},$$ (2)

where just one of them survives in the large U limit. In Eq. (2) the first (second) term is generated by exchange of a bare light (heavy) boson. This implies that the light (b) and heavy (a) bosons should appear symmetrically in all bare diagrams. In order to generate the corresponding diagrams of the Luttinger-Ward functional one may start with the NCA diagram for the $U \to \infty$ case (first diagram in Fig. 1a)) and replace the light boson line by a heavy boson, exchanging the conduction electron lines appropriately. The result is the second diagram. This is not sufficient, however, because the diagrams should be symmetric in a, b on the level of bare diagrams. One may show that this exchange generates an infinite series of diagrams, each containing a circle of pseudoparticle lines with one light boson and n heavy bosons (or vice versa) dressed by conduction electron lines spanning at most three pseudoparticle lines (see Fig. 1a)). The corresponding self-energies are obtained by functional differentiation and are shown in Fig. 1b) in terms of vertex functions (the dash- and cross-shaded half-circle areas).

The Green's function of the physical d electrons can be calculated from the exact relation that can be derived with the help of the equation of motion

$$G_{\vec{k}\vec{k}'}(\omega) = \delta_{\vec{k}\vec{k}'} G^0_{\vec{k}\vec{k}}(\omega) + \sum_{\vec{k}\vec{k}'} G^0_{\vec{k}\vec{k}}(\omega) V_{\vec{k}} G_d(\omega) V^*_{\vec{k}'} G^0_{\vec{k}'\vec{k}'}(\omega).$$ (3)

The local conduction electron Green's function can be expressed in terms of the local conduction electron self energy Σ_c (see Fig. 1c)),

$$G_d(\omega) = \frac{1}{V^2} \Sigma_c(\omega).$$ (4)

Figure 1   a)The Luttinger Ward functional, b) the Bethe-Salpeter equations for the T-matrix vertices (dark and light shaded areas), c) the Contributions to the irreducible self energies for pseudo-particles as well as for conduction electrons are shown.

2.   NUMERICAL RESULTS

The Kondo temperature was extracted from the width of the Abrikosov-Suhl peak in the local electron Green's function. The values we get from the present approximation are in excellent agreement with the exact values known from the Bethe ansatz solution, as can be seen in Fig. 2. The simple Non-Crossing Approximation (the first two terms in Φ in Fig. 1a) gives Kondo temperatures several orders of magnitude too low. Including one additional diagram does not increase T_K sufficiently ($UNCA$) (here we disagree with the corresponding claim in Ref. [5]). An *infinite* series of *skeleton* ladder diagrams is needed to get the correct scaling of T_K with the Coulomb energy U and the local electron energy ϵ_d.

3.   SUMMARY

In summary, we have shown that an infinite number of skeleton diagrams is necessary to recover the correct dynamic energy scale (T_K) in the Anderson impurity model with finite Coulomb interaction energy. We found a systematic way to obtain those diagrams and have shown that the approximation is deriv-

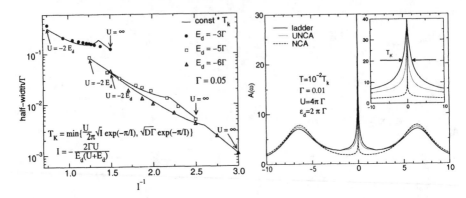

Figure 2   Left: Kondo temperature for various parameters (E_d and U). Solid lines are values of T_k from the Bethe-ansatz solutions. Right: Local electron spectral function calculated with NCA, UNCA and the ladder approximation. The width of the peak is proportional to the Kondo temperature.

able a from Luttinger-Ward functional and therefore conserving. Numerical results show excellent agreement with the exact values of T_K.

This work was funded by the ESF Program on *"Fermi-liquid instabilities in correlated metals"* (FERLIN), and by SFB 195 der Deutschen Forschungsgemeinschaft.

References

[1] For an overview see *Mesoscopic Electron Transport* L. L. Sohn, L. P. Kowenhoven, G. Schön, NATO ASI Series E, Vol. 345 (Kluwer, 1997).

[2] A. Georges, G. Kotliar, W. Krauth, and M. J. Rozenberg, Rev. Mod. Phys. **68**, 13 (1996) .

[3] J. R. Schrieffer in P. A. Wolff, Phys. Rev. **149**, 491 (1966).

[4] J. Kroha, P. Wölfle, and T. A. Costi, Phys. Rev. Lett. **79**, 261 (1997); J. Kroha and P. Wölfle, Acta Phys. Pol. B **29**, (12) 3781 (1998), cond-mat/9811074; J. Kroha and P. Wölfle, Andvances in Solid State Phys. **39**, 271 (1999).

[5] Pruschke et al. Z. Phys. B **74**, 439-449 (1989).

MAGNETOTRANSPORT OF THE CUPRATES
IN THE QUANTUM CRITICAL POINT SCENARIO

R. Hlubina

Department of Solid State Physics, Comenius University, SK-842 48 Bratislava, Slovakia

Abstract In the cuprates, critical scattering of the electrons on the antiferromagnetic spin fluctuations, stripe fluctuations, d-wave pair fluctuations, and singular forward scattering have been invoked as the cause of their anomalous magnetotransport properties. Here we show that within standard transport theory, none of these approaches leads to the observed phenomenology.

At optimal doping, the physical observables in the cuprates exhibit simple power law behaviour as a function of external parameters such as temperature or applied frequency. The absence of any internal energy scale is taken by many as an indication of an underlying quantum critical point (QCP) close to optimal doping [1]. Several QCP (or QCP-like) theories have been suggested for the cuprates, all of which share the assumption that the quantum disordered state of the overdoped cuprates is more or less Fermi liquid-like. Since in the QCP scenario the quantum phase transition is continuous, as the QCP is approached from the overdoped side the fluctuations towards ordering increase, and the Fermi liquid quasiparticles are assumed to scatter predominantly on these fluctuations. The various proposed theories differ in their choice of the relevant quantum order in the cuprates, or equivalently, in the choice of the relevant scattering mechanism: The nearly antiferromagnetic Fermi liquid theory of Pines and collaborators [2] stresses the importance of scattering on antiferromagnetic spin fluctuations, whereas the stripe scenario of the Rome group [3] assumes dominant scattering on charge density wave fluctuations. From the point of view of standard Boltzmann transport theory, in both cases the electrons scatter on a particle-hole collective mode of the electron liquid, the dominant scatterings occuring at a finite momentum transfer $\mathbf{Q}$.

Some time ago, it has been pointed out that scattering on a finite-momentum collective mode affects only (hot) electrons in those parts of the Fermi surface, which are connected by $\mathbf{Q}$. The remaining (cold) electrons do not experience the presence of critical fluctuations and short-circuit the transport at low energy. This introduces an energy scale, a fraction of the spin fluctuation energy J in

J. Bonča et al. (eds.), Open Problems in Strongly Correlated Electron Systems, 417–419.

the nearly antiferromagnetic case, below which the response is Fermi-liquid like even precisely at the QCP [4].

A different QCP-like phenomenology has been proposed by Ioffe and Millis [5]. In their so-called cold spot model they assume that the electrons scatter predominantly on superconducting fluctuations:

$$H = \sum_{k,q} \frac{g_k}{\sqrt{2}} \left[\left(c_{-k+q/2\downarrow} c_{k+q/2\uparrow} - c_{-k+q/2\uparrow} c_{k+q/2\downarrow} \right) \left(a_q^\dagger + b_{-q} \right) + \text{H.C.} \right],$$

where $g_k = (\cos k_x a - \cos k_y a)/2$ is a form-factor respecting the d-wave symmetry of the superconducting state and $a_q^\dagger$ and $b_q^\dagger$ create particle-like and hole-like pair fluctuations, respectively. Assuming that the dynamical critical exponent for the superconducting fluctuations $z = 1$, Ioffe and Millis obtain for the electron scattering rate

$$\frac{\hbar}{\tau_k} = \Gamma g_k^2 + \frac{T^2}{\varepsilon_0}. \tag{1}$$

The scattering rate is thus finite as $T \to 0$ leading to a hot quasiparticle behaviour along the whole Fermi surface, except for the cold spots in the neighborhood of the $(\pm 1, \pm 1)$ directions, where the scattering rate is Fermi liquid-like. For $T \ll T_0 = \sqrt{\varepsilon_0 \Gamma}$ the transport is dominated by the cold spots, whose dimension (and thus also the effective number of charge carriers n) grows as T/T_0. Within the relaxation-time approximation, the conductivity $\sigma = ne^2\tau/m$ and the Hall number $R_H = 1/ne$. Thus the cold-spot model apparently leads to $R_H \propto T^{-1}$ and $\sigma \propto T^{-1}$, in agreement with experiment.

A more detailed analysis shows, however, that scattering on superconducting fluctuations does not lead to the above phenomenology [6]. In fact, an electron with momentum k scatters on a superconducting fluctuation in that it annihilates another electron with momentum $k' \approx -k$. Since the total momentum of the annihilated pair of electrons is close to zero, the electric current changes only little due to such scattering. Scattering on superconducting fluctuations is thus equivalent to forward scattering. Therefore one has to distinguish the single particle lifetime Eq. (1) from the transport lifetime τ_k^{tr}, for which we obtain (within the model of Ref. [5]) $\hbar/\tau_k^{tr} \sim (T^2/\varepsilon_0)g_k^2$. Note that τ_k^{tr} is Fermi liquid-like even for the hot electrons, spoiling the agreement with experiment.

In [6] it was furthermore shown by a direct solution of the Boltzmann equation that quite generally, Fermi liquids with forward scattering are not likely to exhibit the two-lifetime phenomenology of the cuprates, even if the scattering is strongly anisotropic along the Fermi line. This has twofold implications: a) The cold-spot phenomenology does not follow from scattering on pair fluctuations even if modified spectral functions of the fluctuations are assumed. b) The spinon-holon transport phenomenology [7] can not be derived perturbatively by assuming singular forward scattering of the electrons.

From the experimental point of view, the recent ARPES data [8] seems to exclude theories of the charge transport with large-$\mathbf{Q}$ scattering, since in such theories τ_k and τ_k^{tr} should exhibit the same T-dependence. On the other hand, scattering on d-wave pair fluctuations might play the same role which has been very recently assigned to impurity scattering [9]: it produces an angle-dependent (and T-independent) off-set of the single-particle scattering rate, whereas it is not visible in the in-plane transport, being suppressed by the $1 - \cos\theta$ factor. Moreover, since the superconducting fluctuations for T not too close to T_c are likely to have a 2D character, the c-axis resistivity measures the single-particle scattering rate [10] and thus the mechanism for the c-axis charge transport proposed in [5] does apply. However, the in-plane magnetotransport has to be due to a different mechanism and this question is not solved in [9] either.

We conclude that none of the proposed theories of the normal state of the cuprates which invoke singular scattering of the electrons on a collective mode is free of problems when applied to the in-plane magnetotransport. This is the case if the exchanged boson is a spin fluctuation [2], charge fluctuation [3], pair fluctuation [5], or a collective excitation leading to the marginal Fermi-liquid phenomenology [9].

Acknowledgments

This work was part-funded by the Slovak Grant Agency VEGA under No. 1/6178/99.

References

[1] For a review, see S. Sachdev, Science **288**, 475 (2000).

[2] B. Stojković and D. Pines, Phys. Rev. B **55**, 8576 (1997).

[3] C. Castellani, C. Di Castro, and M. Grilli, Phys. Rev. Lett. **75**, 4650 (1995).

[4] R. Hlubina and T. M. Rice, Phys. Rev. B **51** (1995) 9253.

[5] L. B. Ioffe and A. Millis, Phys. Rev. B **58**, 11 631 (1998).

[6] R. Hlubina, unpublished.

[7] P. W. Anderson, *The Theory of Superconductivity in the High-T_c Cuprates* (Princeton Univ. Press, Princeton, 1997).

[8] T. Valla *et al.*, preprint cond-mat/0003407.

[9] E. Abrahams and C. M. Varma, preprint cond-mat/0003315.

[10] R. Hlubina, Phys. Rev. B **60**, 3068 (1999).

2.   EFFECTIVE HOLE-HOLE INTERACTION

We consider a standard three-band model of the CuO_2 planes, which includes explicitly $d_{x^2-y^2}$ orbitals on copper and $p_{x/y}$ (p_σ) orbitals on oxygen [3]; see Eq. (1) of Ref. [3]. The model includes two oxygen phonon modes: breathing and buckling, with frequencies ω_b and ω_a, respectively. The copper on-site energy is modulated by oxygen displacements: along Cu-O bond (u_l, e-ph strength λ_d) and perpendicularly to the plane (z_l, e-ph strength λ_a). Cu-O hopping is modulated by u_l and the coupling is denoted by λ_{dp}.

The details of the cell perturbation method were given elsewhere [3, 4]. Here we investigate the values of V_{hh}, which is the effective interaction between two excess (doped) holes located on nearest-neighbor sites. We study the dependence of V_{hh} on the on-site copper correlation energy U_d. In Fig. 1 we plot V_{hh} as a function of U_d, for four different values of λ_{dp} and with fixed $\lambda_a = 0$. For $\lambda_{dp} = 0$, the interaction is always repulsive and decreases with

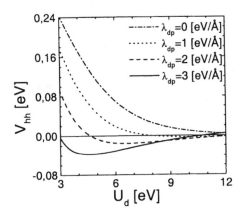

Figure 1 Effective hole-hole interaction between the charge carriers, V_{hh}, as a function of the on-site copper correlation energy U_d and for different values of the electron-phonon coupling λ_{dp}. All the parameters (except U_d) are fixed according to the so-called standard set [5]. For the phonon frequencies we have adopted the values $\omega_b = 0.07$ eV and $\omega_a = 0.04$ eV. All the curves are for $\lambda_d = 3$ eV/Åand $\lambda_a = 0$.

increasing U_d. With increasing λ_{dp}, a possibility of an attractive interaction arises. For the attraction between the charge carriers, the optimum is a moderate value of correlation, though for larger values of the e-ph coupling, this optimum value of correlation gets smaller. The presented results demonstrate that the most favorable conditions for superconductivity are when correlation are neither too weak nor too strong, and this is indeed the case for the cuprates.

We have also studied the effect of the buckling mode on V_{hh}, putting $\lambda_a \neq 0$. In Fig. 2, we plot the dependence of V_{hh} on U_d for $\lambda_a = 1$ eV/Å, with the rest parameters as in Fig. 1. The overall picture is similar, though the e-ph

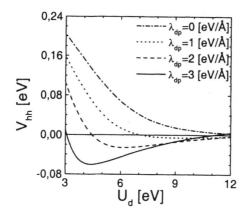

Figure 2   The same as in Fig. 1 but for $\lambda_a = 1$ eV/Å.

coupling, related to the buckling mode, stabilizes the attraction furthermore.

3.   SUMMARY

We have shown that in presence of the breathing phonon mode and corresponding electron-phonon interactions, moderate electron correlation gives the most attractive interaction between the charge carriers in the CuO_2 planes. We have also shown that the buckling mode stabilizes the attraction.

Acknowledgments

This work is partially supported by the Polish Committee for Scientific Research (KBN) through the Project No. 2 P03B 175 14. J.K. also acknowledges a partial support from the Polish-French inter-government scheme "Polonium".

References

[1] P. W. Anderson, Science **235**, 1196 (1987); E. Dagotto, Rev. Mod. Phys. **66**, 763 (1994).

[2] J. Ranninger, Z. Phys. B – Condensed Matter, **84**, 167 (1991) and the references therein; A.S. Alexandrov and N.F. Mott, *High-Temperature Superconductors and other Bose Liquids* (Taylor and Francis, 1994).

[3] P. Piekarz, J. Konior, and J. H. Jefferson, Phys. Rev. B **59**, 14697 (1999).

[4] J.H. Jefferson, Physica B **165-166**, 1013 (1990); J.H. Jefferson, H. Eskes, and L. F. Feiner, Phys. Rev. B **45**, 7959 (1992); L.F. Feiner, J.H. Jefferson, and R. Raimondi, Phys. Rev. B **53**, 8751 (1996).

[5] M. S. Hybertsen, M. Schlüter, and N. E. Christensen, Phys. Rev. B **39**, 9028 (1989).

SUPERCONDUCTIVITY IN DISORDERED Sr₂RuO₄

G. Litak

Department of Mechanics, Technical University of Lublin, 20-618 Lublin, Poland

J.F. Annett, B.L. Györffy

H.H. Wills Physics Laboratory, University of Bristol, Tyndall Ave, Bristol, BS8 1TL, UK

Abstract   We discus the influence of disorder on the critical temperature T_C of a p-wave superconductor. To describe disordered Sr₂RuO₄ we use extended Hubbard model with random site energies treated in the Coherent Potential Approximation.

1.   INTRODUCTION

Recent experimental evidence suggests that the Cooper pairs in superconducting Sr₂RuO₄ are triplets with p-wave internal symmetry, as in the case of superfluid ³He [1, 2, 3, 4]. One of characteristic features of this exotic state is the strong influence of impurities on its superconducting properties. Studies of the electronic structure in Sr₂RuO₄ [2] have identified an extended van Hove singularity close to the Fermi energy E_F. In this note we investigate the interplay between the van Hove singularity and disorder in a model p-wave superconductor.

2.   THE MODEL

We consider a simple extended Hubbard Hamiltonian [4, 5]:

$$H = \sum_{i\sigma}(\varepsilon_i - \mu)\hat{n}_{i\sigma} + \sum_{ij\sigma} t_{ij}c_{i\sigma}^+ c_{j\sigma} + \frac{1}{2}\sum_{ij\sigma\sigma'} U_{ij}\hat{n}_{i\sigma}\hat{n}_{i\sigma'} \qquad (1)$$

where as usual $c_{i\sigma}^+$ and $c_{i\sigma}$ are the Fermion creation and annihilation operators for an electron on site i with spin σ, $\hat{n}_{i\sigma}$ is the number operator and μ is the chemical potential. Disorder is introduced into the problem by allowing the local site energy $\varepsilon_i = \pm\delta/2$ to vary randomly from site to site with equal

425

probability. U_{ij} is the attractive interaction $(i \neq j)$ between nearest sites and t_{ij} is the hopping integral from site j to site i which takes nonzero values between nearest and next nearest sites. In $k-$space: $\varepsilon_k = \sum_j t_{ij} \exp R_{ij}k = -2t(\cos k_x + \cos k_y) - 4t' \cos k_x \cos k_y$, where the hopping parameter $t' =$

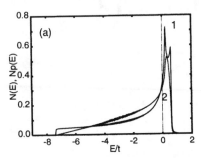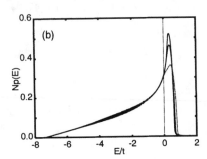

Figure 1   (a) $N(E)$ (1) and $N_p(E)$ (2) in a clean system, (b) $\overline{N}_p(E)$ for disordered system with $\delta/t = 0.0, 0.3, 0.6$ starting from the top line ($\mu = 0$).

$0.45t$ as well as the band filling $n_\gamma/2 = 0.66$ were fitted to the experimental cyclotron masses and corresponding carriers occupations for the γ band of Sr_2RuO_4, the interaction $U_{ij}/t = 0.446$ was chosen to get $T_C = 1.5K$ [1, 3, 4].

3.   DENSITY OF STATES AND T_C

The linearized gap equation for the critical temperature T_C of p-wave pairing from the Hamiltionian (1) reads [4]:

$$1 = \frac{|U|}{\pi} \int_{-\infty}^{\infty} dE \mathrm{Tanh} \frac{E}{2k_B T_C} \mathrm{Im} \frac{\overline{G}_{11}^p(E)}{2E - \mathrm{Tr}\Sigma(E)}, \qquad (2)$$

where $\overline{G}_{11}^p(E)$ is an averaged electron Green function which defines the weighted density of states (DOS) of p-wave electron states $\overline{N}_p(E)$:

$$\overline{N}_p(E) = -\frac{1}{\pi}\overline{G}_{11}^p(E) = -\frac{1}{\pi N}\sum_k \mathrm{Im} \frac{2\sin^2 k_x}{E - \Sigma_{11}(E) - \varepsilon_k + \mu}, \qquad (3)$$

where $\Sigma_{11}(E)$ is a Coherent Potential which describe the electron self energy in the disordered system. In case of a clean system $(\mathrm{Im}\Sigma_{11}(E) = 0)$ we get a conventional gap equation with the DOS $N(E)$ substituted by $N_p(E)$ under the integral (Eq. 2). In Fig 1a we show $N(E)$ and $N_p(E)$ for the clean system. Note that the van Hove singularity in $N(E)$ produces the maximum in $N_p(E)$. The singularity in $N_p(E)$ is smeared by the presence of the term $\sin^2 k_x$ in Eq. 3. This leads to a maximum in T_C with changing n (Fig. 1a). Equations 2 and 3 are influenced by disorder by different effects. Firstly, the peak in $\overline{N}_p(E)$ is smeared (Fig. 1b) leading to small decrease of T_C. The second and more interesting effect arises from Eq. 2, where $\Sigma_{11}(E)$ acts as a pair breaker. Using

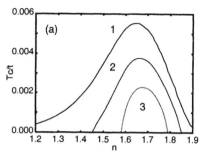 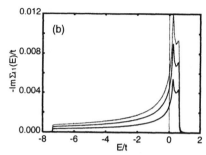

Figure 2   (a) T_C as a function of n for various values of disorder potential ($\delta/t = 0.0, 0.10, 0.13$ for curves 1, 2, 3 respectively, (b) Imaginary part of self energy for various values of disordered potential $\delta/t = 0.10, 0.13, 0.15$ starting form the bottom line ($\mu = 0$).

the arguments of Ref. [5], Eq. 2 can be evaluated to yield:

$$\ln\left(\frac{T_C}{T_{C0}}\right) = \psi\left(\frac{1}{2}\right) - \psi\left(\frac{1}{2} + \frac{|\mathrm{Im}\Sigma_{11}(0)|}{2\pi T_C}\right), \tag{4}$$

where T_{C0} denotes the critical temperature in a clean system. The full influence of disorder on $\overline{N}_p(E)$ is illustrated in Fig. 1b. Note that the position of the maximum value is not affected by small disorder. On the other hand the critical temperature T_C, plotted in Fig. 2a is degradated strongly with disorder. This is due to the pair-breaking term $-\mathrm{Im}\Sigma_{11}(E)$, which is shown in Fig. 2b.

4.   SUMMARY

When the Fermi energy is close to a Van Hove singularity relatively weak disorder can cause very rapid T_C degradation in a p-wave superconductor. The case of Sr₂RuO₄ may be an example of this phenomenon.

Acknowledgments

This work was partially founded by the Committee of Scientific Research (Poland) through the grant KBN 2P03B9018, and the Royal Society (UK).

References

[1] A.P. Mackenzie *et. al.*, Phys Rev. Lett. **76**, 3786 (1998).

[2] D.H. Lu *et. al.*, J. Low Temp. Phys. **105**, 1587 (1996).

[3] D.F. Agtenberg, T.M. Rice, M. Sigrist, Phys. Rev. Lett. **78**, 3374 (1997).

[4] G. Litak, J.F. Annett and B.L. Györffy, Acta Phys. Pol. A **97**,249 (2000).

[5] A.M. Martin *et. al.*, Phys Rev. B **60** 7523 (1999).

FINE ELECTRONIC STRUCTURE AND MAGNETISM OF LaMnO$_3$ AND LaCoO$_3$

R.J. Radwański[1,2], and Z. Ropka[2]

[1]*Inst. of Physics, Pedagogical University, 30-084 Kraków, Poland.*
[2]*Center for Solid State Physics, 31-150 Kraków, Poland.*

Abstract   By analysis of all known electronic and magnetic properties we came out to the conclusion that in LaMnO$_3$ and LaCoO$_3$ exists the fine electronic structure associated with the atomic-like states of the Mn^{3+} and Co^{3+} ions. It turns out that for the description of electronic and magnetic properties of real 3d-ion systems the fine electronic structure, resulting from crystal-field, spin-orbit coupling, local lattice distortions (Jahn-Teller effect) has to be taken into account. In our atomic-like approach the d electrons form the highly-correlated system 3d^n and have the discrete energy spectrum in contrary to the band picture that yields the continuous energy spectrum. Our studies indicate that the orbital moment has to be unquenched in the description of 3d-ion compounds.

LaMnO$_3$ and LaCoO$_3$ belong to the class of compounds known as Mott insulators. LaMnO$_3$-based compounds exhibit the giant magnetoresistance [1, 2]. The uniqueness of LaCoO$_3$ is mostly related with its non-magnetic ground state at low temperatures and the violation of the Curie-Weiss law [3, 4]. Despite of enormous long lasting theoretical efforts the description of such compounds is still under very strong debate. The fundamental controversy "how to treat the d electrons" starts already at the beginning - should they be treated as localized or itinerant. Directly related with this problem is the structure of the states: do they form the continuous energy spectrum like it is in the band picture [3, 4], schematically shown in Fig. (1), or the discrete energy spectrum typical for the localized states.

The aim of this paper is to point out the importance of the single-ion-like electronic states of the Mn^{3+} and Co^{3+} ions in LaMnO$_3$ and LaCoO$_3$. We make use of our general theoretical approach to a solid, from atomic physics to solid-state physics [5], considering the atomic-like states of the Mn^{3+} and Co^{3+} ions. We make use of the fact that the both compounds are isostructural and isoelectronic.

In the insulating LaMnO$_3$ and LaCoO$_3$ compounds the 3d cation (M) is in the trivalent state as is anticipated from the compensated valences La^{3+}Mn^{3+}O$_3^{2-}$

J. Bonča et al. (eds.), Open Problems in Strongly Correlated Electron Systems. 429–432

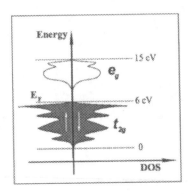

Figure 1   Schematic description of the d states in LaCoO₃ within the band approach - there is the continuum energy spectrum. After Ref. 4.

and $La^{3+}Co^{3+}$. In the perovskite structure the 3d cation is situated in the octahedral surrounding of the oxygen ions - thus it is reasonable to consider the cluster MO_6. The Mn^{3+} and Co^{3+} ion has 4 and 6 d-electrons in the incomplete outer shell. These electrons are treated as forming the highly-correlated electron system 3d n, i.e like in the atom [6]. In a zero-order approximation these electron correlations within the incomplete 3d shell are accounted for by the phenomenological two Hund's rules. They yield for both the $3d^4$ and $3d^6$ electron configuration the same ground term ^5D with S=2 and L=2. Under the action of the dominant cubic crystal field, the ^5D term splits into the orbital triplet $^5T_{2g}$ and the orbital doublet ^5E_g spread over 2-3 eV. In the octahedral oxygen surrounding the subterm ^5E_g is lower for the $3d^4$ configuration, Fig. (2) - the left side, whereas the orbital triplet $^5T_{2g}$ is lower for the $3d^6$ system as shown in the right side of Fig. (2).

The low-energy electronic structure can be calculated from the single-ion-like Hamiltonian considered within the 25-fold LS space and containing the crystal-field and spin-orbit coupling. The cubic CEF interactions are dominant of H_{CF}^{cub} = 2-3 eV whereas others are H_{s-o}=0.1 eV, H_t=0.05 eV, H_z=0.01 eV. The electronic structure for the Mn^{3+} and Co^{3+} ions contains the 25 spin-orbital states spread over 2-3 eV that are largely grouped. From the physical point of view the presence of the states close to the ground state is important. For the Mn^{3+} ion 10 states are confined to 2 meV only - the existence of so many states helps in the development of the magnetic state as is observed for LaMnO₃ with T_N of 140 K despite of the fact that the lowest state is a singlet. Then the spin-dependent inter-site interactions can induce easily the magnetic moment. The existence of the fine electronic structure in the energy window below 10 meV causes anomalous temperature behaviour of many physical properties as we well know from studies of rare-earth compounds [7]. It is visible in particular in the calculated temperature dependence of the paramagnetic susceptibility of

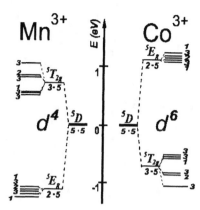

Figure 2   The realistic fine electronic structure produced by the octahedral CEF interactions in the presence of the spin-orbit coupling for the Mn^{3+} (d^4) in LaMnO₃ and for the Co^{3+} ion in LaCoO₃. The trigonal off-cubic distortion causes further splitting leading to the non-magnetic ground state in LaCoO₃.

the Co^{3+} ions [8, 9]. The trigonal distortion breaks down completely the Curie law and forms the non-magnetic state at the atomic scale. The calculated c(T) confirms the low-temperature non-magnetic state and exhibits the pronounced maximum of the susceptibility at 90 K with the Curie-Weiss law at temperatures above 130 K reproducing the experimental data. The trigonal distortion term H_t with of +15.5 meV yields for the LaCoO₃ the spin-like gap of 11 meV (in the presence of B$_4$=+200 K and λ of -630 K). With increasing temperature higher states become populated - as these states are strongly magnetic it manifests in the susceptibility experiment as a temperature induced low-high spin transition.

We would like to point out that the properties of LaMnO₃ and LaCoO₃ are described with the same CEF parameter B$_4$, -200 K and +200K. The change of the sign is related with the change of the Stevens coefficient for the d^4 and d^6 electronic system - one could also discus this change of the sign in terms of the hole-particle symmetry. We have to mention that in the current literature, apart of the band picture, often appears a simplified localized electronic structure like presented in Fig. (3) [10]. Everybody can see that such the structure is really very simplified and only schematic. Our calculations show that the discrete energy spectrum is more complex. Our approach allows the calculations of the eigenfunctions and spin and orbital moments.

In conclusion, the very consistent description of magnetic properties of two different, though isostructural and isoelectronic, compounds LaMnO₃ and LaCoO₃ has been obtained within our approach - from atomic physics to solid state physics. The d electrons form the highly-correlated atomic-like system and exhibit the discrete energy spectrum originating from the Mn^{3+} and Co^{3+} ions, Fig. 2. The diamagnetic state of LaCoO₃ is associated with the non-

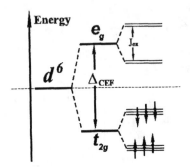

Figure 3   Oversimplified discrete energy spectrum of the Co^{3+} ion in $LaCoO_3$, after ref. 10.

magnetic singlet ground state of the Co^{3+} ion. We would like to point out that our approach should not be considered as the treatment of an isolated ion - we consider the cation in the oxygen octahedron MO_6. The perovskite structure is built up from the corner sharing octahedra MO_6 - thus such the atomic structure occurs at each cation. Experiment confirms that even in the macroscopic crystalline system such the atomic structure is largely preserved. These observations give the base for the developed by us the quantum atomic solid-state theory (QUASST). The present calculations prove the importance of the intra-atomic s-o coupling and the local scale off-cubic lattice distortions for the description of low-temperature properties of compounds containing 3d ions.

References

[1] Y.Tokura *et al.*, Physica C **263**, 544 (1996).

[2] W. Koshibae *et al.*, J.Phys.Soc. Japan **66**, 957 (1997).

[3] M.A.Korotin *et al.*, Phys. Rev. B **54**, 5309 (1996).

[4] P.Ravindran *et al.*, Phys. Rev. B. **60**, 16423 (1999).

[5] R.J.Radwanski and Z.Ropka,The preservation of the individuality of 3d atoms in a solid - http://xxx.lanl.gov/abs/cond-mat/0006092

[6] R.J.Radwanski and Z.Ropka,Relativistic effects in the electronic structure of 3d paramagnetic ions - http://xxx.lanl.gov/abs/cond-mat/9907140

[7] R.J.Radwanski, J.Phys.: Condens. Matter. **8**, 10467 (1996).

[8] R.J.Radwanski, Z.Ropka, Physica B **281-282**, 507 (2000).

[9] R.J.Radwanski and Z.Ropka, Solid State Commun. **112**, 621 (1999). http://xxx.lanl.gov/abs/cond-mat/9906437.

[10] M.R.Ibarra *et al.*, Phys. Rev. B. **57**, R3217 (1998).

ANOMALIES IN THE CONDUCTION EDGE OF QUANTUM WIRES

T. Rejec[1] and A. Ramšak[1,2]

[1] *J. Stefan Institute, SI-1000 Ljubljana, Slovenia*

[2] *Faculty of Mathematics and Physics, University of Ljubljana, SI-1000 Ljubljana, Slovenia*

J.H. Jefferson

DERA, St. Andrews Road, Great Malvern, Worcestershire WR14 3PS, England

Abstract   We study the conductance threshold of clean nearly straight quantum wires in which an electron is bound. We show that such a system exhibits spin-dependent conductance structures on the rising edge to the first conductance plateau, one near $0.25(2e^2/h)$, related to a singlet resonance, and one near $0.75(2e^2/h)$, related to a triplet resonance. As a quantitative example we solve exactly the scattering problem for two-electrons in a wire with circular cross-section and a weak bulge. From the scattering matrix we determine conductance via the Landauer-Büttiker formalism. The conductance anomalies are robust and survive to temperatures of a few degrees. With increasing magnetic field the conductance exhibits a plateau at e^2/h, consistent with recent experiments.

1.   INTRODUCTION

Following the pioneering work in Refs. [1, 2] many groups have now observed conductance steps in various types of quantum wire. These first experiments were performed on gated two-dimensional electron gas (2DEG) structures, while similar behaviour of conductance are shown in "hard-confined" quantum wire structures, produced by cleaved edge over-growth [3], epitaxial growth on ridges [4], heteroepitaxial growth on "v"-groove surfaces [5] and most recently in $GaAs/Al_\delta Ga_{1-\delta}As$ narrow "v"-groove [6] and low-disorder [7] quantum wires.

These experiments strongly support the idea of ballistic conductance in quantum wires and are in surprising agreement with the now standard Landauer-Büttiker formalism [8, 9] neglecting electron interactions [10]. However, there are certain anomalies, some of which are believed to be related to electron-electron interactions and appear to be spin-dependent. In particular, already in

J. Bonča et al. (eds.), Open Problems in Strongly Correlated Electron Systems, 433–439.

early experiments a structure is seen in the rising edge of the conductance curve [1], starting at around $0.7G_0$ with $G_0 = 2e^2/h$ and merging with the first conductance plateau with increasing energy. Under increasing in-plane magnetic field, the structure moves down, eventually merging with a new conductance plateau at e^2/h in very high fields [11, 12]. Theoretically this anomaly has not been adequately explained, despite several scenarios, including spin-polarised sub-bands [13], conductance suppression in a Luttinger liquid with repulsive interaction and disorder [14] or local spin-polarised density-functional theory [15]. Recently we have shown that these conductance anomalies near $0.7G_0$ and $0.25G_0$ are consistent with an electron being weakly bound in wires of circular and rectangular cross-section, giving rise to spin-dependent scattering resonances [16, 17, 18].

In this paper we develop further the single bound-electron picture and give new results for wires of circular cross-section, including magnetic field dependence.

2.   THE MODEL

We consider quantum wires which are almost perfect but for which there is a very weak effective potential, which has at most two bound states. Such an effective potential can arise, for example, from a smooth potential due to remote gates. Alternatively it could arise from a slight buldge in the an otherwise perfect wire. We consider this latter situation for the cases of quantum wires with both circular cross-section [16], appropriate for 'hard-confined' v-groove wires; or rectangular cross-section [17], which approximate 'soft-confined' wires resulting from gated 2DEGs. These are shown schematically in Figure 1. The cross-sections of these wires are sufficiently small that the lowest transverse channel approximation is adequate for the energy and temperature range of interest. The smooth variation in cross-section also guarantees that inter-channel mixing is negligible. Restricting ourselves to this lowest transverse channel, the Schrödinger equation on a finite-difference grid in the z-direction may be written,

$$H = -t \sum_{i\sigma} \left(c_{i+1,\sigma}^\dagger c_{i\sigma} + c_{i\sigma}^\dagger c_{i+1\sigma} \right) + \sum_{i\sigma} \epsilon_{i\sigma} n_{i\sigma} + \tag{1}$$

$$+ \sum_i U_{ii} n_{i\uparrow} n_{i\downarrow} + \frac{1}{2} \sum_{i \neq j} U_{ij} n_i n_j,$$

where $c_{i\sigma}^\dagger$ creates an electron with spin σ at the $z = z_i$ in the lowest transverse channel; $n_i = \sum_\sigma n_{i\sigma}$ with $n_{i\sigma} = c_{i\sigma}^\dagger c_{i\sigma}$; $t = \hbar^2/(2m^*\Delta^2)$, where $\Delta = z_{i+1} - z_i$; $\epsilon_i = \hbar^2/(m^*\Delta^2) + \epsilon(z_i+) + g^*\mu_B \sigma B$, where $\epsilon(z_i)$ is the energy of the lowest transverse channel at z_i and $g^*\mu_B \sigma B$ is the Zeeman energy for a magnetic field B in the z-direction, as in Refs. [16, 17]. U_{ij} is an effective

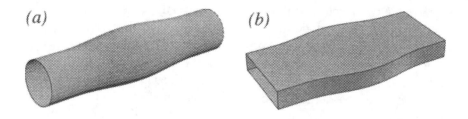

Figure 1 The wire shape is symmetric around the z-axis. The potential is constant $V(x, y, z) = 0$ within the boundary, and $V_0 > 0$ elsewhere. (a) Circular cross-section, defined by, $r_0(z) = \frac{1}{2}a_0(1 + \xi \cos^2 \pi z/a_1)$ for $|z| \leq \frac{1}{2}a_1$ and $x_0(z) \equiv \frac{1}{2}a_0$ otherwise. (b) Rectangular cross-section is defined by $x_0(z) = \frac{1}{2}a_0(1 + \xi \cos^2 \pi z/a_1)$.

screened Coulomb interaction which was obtained by starting with a full 3D Coulomb interaction, integrating over the lowest transverse modes and then adding screening phenomenologically. The dielectric constant is taken as $\varepsilon = 12.5$, appropriate for GaAs. Note that this is a general form, the difference between the two cases shown in Figure 1 being reflected entirely in the energy parameters ϵ and U. We note that this Hamiltonian also has the form for a perfectly straight wire subject to a smooth potential variation, defined by the ϵ.

3.   TWO-ELECTRON APPROXIMATION

Over a range of parameters in which the deviation from a perfect straight wire is small, one, and only one, electron resides in a bound state. This is because the weak effective potential provided by the bulge will always contain at least one bound state and since the binding energy is small, a second electron cannot be bound due to Coulomb repulsion. For larger bulges or deeper potential wells more electrons may be bound but these situations will not be considered here. Note that even if there is more than one bound state (we consider one or two) there can still only be one bound electron due to the Coulomb repulsion. The actual number of electrons in the wire may be changed by varying the Fermi energies in the leads and reservoirs to which the wire is connected. In experiments, this is achieved simply by changing the voltages on one or more gate electrodes. When there is more than one electron in the wire, a current will flow from source to drain contacts and at low temperatures the motion of these electrons is ballistic. The conduction electrons will scatter from the bound electron giving rise to resistance. If we neglect the mutual interaction between conduction electrons, then the transport problem reduces to a two-electron scattering problem described by equation (1). This is a reasonable approximation provided that the mean electron density is not too low, i.e. that the mean electron separation is of order the effective Bohr radius or less.

We solve the two-electron scattering problem exactly subject to the boundary condition that the asymptotic states consist of one bound electron in the ground state and one free electron. The main features of these scattering solutions may be understood by the following simple physical picture. The effective potential due to the single bound electron and the effective potential well gives rise to a symmetric double barrier structure since an incident electron will initially feel the Coulomb repulsion due to the bound electron but will then pass through a weak local minimum, e.g., at the the point of maximum diameter in the wire of circular cross-section. This gives rise to a resonance, the peak of which corresponds to perfect transmission. A more refined analysis shows that this resonance is spin dependent, singlet and triplet resonances occuring at different energies, with the singlet always lowest. In fact this spin-dependent scattering is the quasi one-dimensional analogue of the three-dimensional case, discussed some 70 years ago by Oppenheimer and Mott [19].

From the scattering solutions we compute the conductance using again a Landauer-Büttiker formula which, incorporating the results of spin-dependent scattering, takes the following form,

$$
G = -\frac{2e^2}{h} \int \frac{\partial f(\epsilon - E, T)}{\partial \epsilon} \tag{2}
$$
$$
\times \left[\frac{1}{4}[\mathcal{T}_s(\epsilon - E_B) + \mathcal{T}_t(\epsilon - E_B)] + \frac{1}{2}\mathcal{T}_t(\epsilon + E_B) \right] d\epsilon,
$$

where the subscripts $\mathcal{T}_s$ and $\mathcal{T}_t$ refer to singlet and triplet transmission probabilities respectively, with $E_B = \frac{1}{2}g^*\mu_B B$ and E is the Fermi energy in the leads.

4.   RESULTS

We have performed detailed calculations for both the circular and rectangular wire cross sections shown in Figure 1. Apart from small quantitative differences, the results are very similar and hence, for brevity, we shall only show results for wires with circular cross-section.

In Fig. 2(a) we show plots at zero temperature and magnetic field of $\mathcal{T}_s(E)$ and $\mathcal{T}_t(E)$ for a typical wire with the geometry of Fig. 1(a). The thin dotted line represents the non-interacting result, independent of spin. We see clearly the sharp singlet resonance at low energy followed by the broader triplet at higher energy. In Fig. 2(b) the conductance G in units of $2e^2/h$ is shown, as calculated for various temperatures. The resonances have a strong temperature dependence and, in particular, the sharper singlet resonance is more readily washed out at finite temperatures. However, it should be noted that resonances survive to relatively high temperatures, because the width of the wire, which dictates the energy scale, is small ($a_0 = 10$ nm) [20]. Note that for weak coupling, the energy scale is set by the x-energy of the lowest channel, $\sim a_0^{-2}$

and hence the conductance vs. Ea_0^2 with Ua_0 fixed is roughly independent of a_0 (the scaling would be exact for $V_0 \to \infty$).

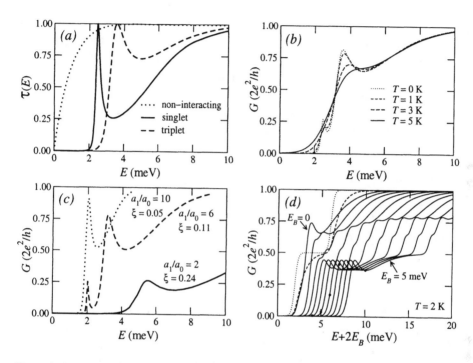

Figure 2 Resonance and conductance curves. (a) Typical transmission probabilities for singlet and triplet resonances in zero magnetic field. Here [and in (b) and (d)] $a_0 = 10$ nm, $a_1 = 60$ nm, $\xi = 0.1$ and $V_0 = 0.4$ eV. (b) Total conductance in zero magnetic field showing temperature dependence of anomalies near 0.25 and 0.7. (c) Dependence of conductance on wire geometry. (d) Magnetic field dependence of conductance showing weak resonance and saturation to e^2/h in a high magnetic field.

In Fig. 2(c) we show the zero temperature and field conductance curves for three different bulge shapes, which become longer and flatter as we move from right to left. For the case with the shortest bulge region (right) we see only a singlet resonance. This is because the effective potential well has only one bound state and hence, even in the absence of Coulomb interaction, would not support a triplet. On the other hand, the other two cases have two one-electron bound states. In the absence of Coulomb interaction, these levels give rise to singlet and triplet bound states. When the Coulomb interaction is switched on, the states develop into two resonance peaks (dotted line). Here the position of both peaks nearly coincides, because the bulge is relatively long and the singlet-triplet splitting is small. For the remaining case (dashed line), both the lowest singlet and triplet develop into resonances when the Coulomb interaction is switched on, though the singlet is only just unbound with the resonance lying close to the conduction edge energy and is thus very sharp.

In Fig. 2(d) conductance for $T = 2$ K is presented for magnetic field increasing from zero in steps with $\Delta E_B = 0.5$ meV and for clarity the curves have been shifted by $2E_B$ to the right with increasing E_B. We present results for $a_0 = 10$ nm, but note that E_B also obeys the above mentioned scaling $E_B a_0^2$ with varying a_0. Magnetic fields which would give substantial effects in e.g. narrow "v"-groove wires [6], would have to be very large, since $E_B = 1$ meV corresponds to large $g^* B \sim 35$ T. However, due to "$E_B a_0^2$" scaling, the corresponding value for a wider wire with $a_0 \sim 50$ nm would be only ~ 1.4 T. Also plotted in Fig. 2(d) for comparison are the corresponding results for the non-interacting electron case (dotted) and the perfectly straight wire (dashed), with $E_B = 2$ meV. In this figure we have indicated with a dot the points $E = E_B$. To the left of these points G simplifies, $G(E, B) = \frac{e^2}{h} \mathcal{T}_t(E + E_B)$, whereas at high energies spin-flip transmission probabilities $t_{\uparrow\downarrow \to \uparrow\downarrow}$ and $t_{\uparrow\downarrow \to \downarrow\uparrow}$, contained within the remaining terms of Eq. (2), are non-zero. These parts of the curves should be treated with caution though they are expected to be more reliable at lower fields.

5.   SUMMARY

In summary, we have shown that quantum wires with weak longitudinal confinement, or open quantum dots, can give rise to spin-dependent, Coulomb blockade resonances when a single electron is bound in the confined region. The emergence of a specific structure at $G(E) \sim \frac{1}{4} G_0$ and $G \sim \frac{3}{4} G_0$ is a consequence of the singlet and triplet nature of the resonances and the probability ratio 1:3 for singlet and triplet scattering and as such is a universal effect. A comprehensive numerical investigation of open quantum dots using a wide range of parameters shows that singlet resonances are always at lower energies than the triplets, in accordance with the corresponding theorem for bound states [21]. With increasing in-plane magnetic field, the resonances shift their position and a plateau $G(E) \sim e^2/h$ emerges. The effect of a magnetic field is observable only in relatively wider quantum wires, due to the intrinsic energy scale $\propto a_0^{-2}$.

Acknowledgments

The authors wish to acknowledge A.V. Khaetskii and C.J. Lambert for helpful comments. This work was part-funded by the U.K. Ministry of Defence and the EU.

References

[1] B.J. van Wees *et al.*, Phys. Rev. Lett. **60**, 848 (1988).

[2] D.A. Wharam *et al.*, J. Phys. C **21**, L209 (1988).

[3] A. Yacoby *et al.*, Phys. Rev. Lett. **77**, 4612 (1996).

[4] P. Ramvall *et al.*, Appl. Phys. Lett. **71**, 918 (1997).

[5] M. Walther, E. Kapon, D.M. Hwang, E. Colas, and L. Nunes, Phys. Rev. B **45**, 6333 (1992); M. Grundmann *et al.*, Semicond. Sci. Tech. **9**, 1939 (1994); R. Rinaldi *et al.*, Phys. Rev. Lett. **73**, 2899 (1994).

[6] D. Kaufman *et al.*, Phys. Rev. B **59**, R10433 (1999).

[7] B.E. Kane *et al.*, Appl. Phys. Lett. **72**, 3506 (1998); D.J. Reilly, cond-mat/0001174.

[8] R. Landauer, IBM J. Res. Dev. **1**, 223 (1957); **32**, 306 (1988).

[9] M. Büttiker, Phys. Rev. Lett. **57**, 1761 (1986).

[10] H. van Houten, C.W.J. Beenakker, and B.J. van Wees, in *Semiconductors and Semmimetals*, edited by R.K. Willardson, A.C. Beer, and E.R. Weber, (Academic Press, 1992).

[11] K.J. Thomas *et al.*, Phys. Rev. Lett. **77**, 135 (1996); Phys. Rev. B **58**, 4846 (1998); **59**, 12252 (1999).

[12] K.J. Thomas *et al.*, Phil. Mag. B **77**, 1213 (1998).

[13] G. Fasol and H. Sakaki, Jpn. J. Appl. Phys. **33**, 879 (1994).

[14] D.L. Maslov, Phys. Rev. B **52**, R14368, 1995.

[15] Chuan-Kui Wang and K.-F. Berggren, Phys. Rev. B **57**, 4552 (1998).

[16] T. Rejec, A. Ramšak, and J.H. Jefferson, cond-mat/9910399.

[17] T. Rejec, A. Ramšak, and J.H. Jefferson, J. Phys.: Condens. Matter **12**, L233 (2000).

[18] V.V. Flambaum and M.Yu. Kuchiev, cond-mat/9910415.

[19] J.R. Oppenheimer, Phys. Rev. **32**, 361 (1928); N.F. Mott, Proc. Roy. Soc. A **126**, 259 (1930).

[20] A. Ramšak, T. Rejec, and J. H. Jefferson, Phys. Rev. B **58**, 4014 (1998).

[21] E. Lieb and D. Mattis, Phys. Rev. **125**, 164 (1962).

UNDERDOPED REGION OF THE 2D t-J MODEL

A. Sherman[1] and M. Schreiber[2]

[1]*Institute of Physics, University of Tartu, Estonia*

[2]*Institut für Physik, Technische Universität, 09107 Chemnitz, Germany*

We apply the spin-wave theory modified for the paramagnetic state [1] to investigate normal-state properties of the 2D t-J model. Self-energy equations for hole and magnon Green's functions are self-consistently solved in the Born approximation for the ranges of hole concentrations $x \lesssim 0.17$ and temperatures $T \lesssim 150$ K. In this region the hole spectrum differs essentially from a conventional metallic spectrum. A distinctive feature of the low-concentration spectrum is the spin-polaron band with the minimum energy (in the hole picture) at $(\pi/2, \pi/2)$, small dispersion along the boundary of the magnetic Brillouin zone, and the extended van Hove singularity near $(0, \pi)$. The shape and narrow width of this band (of the order of the exchange constant $J \sim 0.1$ eV) are determined by the short-range hole-magnon interaction which remains practically unchanged in the transition from long- to short-range order with the correlation length much larger than the lattice spacing. As a result, a part of the spin-polaron band is retained in the paramagnetic state and gives an intensive peak in the spectral function for wave vectors near the boundary of the magnetic Brillouin zone. In addition to this peak a broader less intensive maximum is formed from the incoherent continuum with the transition to paramagnetic state. The dispersion of the maximum resembles the dispersion of the nearest-neighbor band with the hopping constant t. For moderate doping this wider band crosses the Fermi level along the line shown in the inset of Fig. 1a. However, just at the Fermi level crossing the maximum is lost within the foot of a more intensive spin-polaron peak. Its position along the Fermi surface is shown in Fig. 1a for two hole concentrations. In the underdoped case the peak lies slightly above the Fermi level. Near the Fermi surface the position of the photoemission leading edge is determined by the intensive spin-polaron peak and this peculiarity of its dispersion leads to the photoemission pseudogap near $(0, \pi)$ with a finite intensity on the Fermi level provided by the less intensive maximum. The symmetry and size of the obtained pseudogap agree nicely with experiment (Fig. 1b; in our calculations we set $J/t = 0.2$, $t = 0.5$ eV). Besides, in agreement with experiment the pseudogap is closed

J. Bonča et al. (eds.), Open Problems in Strongly Correlated Electron Systems, 441–442.

at optimal doping $x \approx 0.17$. With the further growth of x a sharp transition to the conventional metallic band is observed.

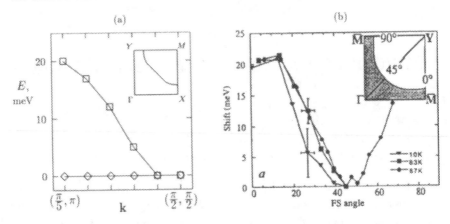

Figure 1   (a) The position of the spin-polaron peak along the Fermi surface (inset) for $x = 0.12$ (box) and $x = 0.17$ (diamond) at $T = 115$ K. (b) The position of the leading edge of the photoemission spectrum measured [2] at $T = 14$ K along the Fermi surface (inset) in Bi2212 with various T_c. The points $\overline{M}$ and Y correspond to $X = (\pi, 0)$, $Y = (0, \pi)$ and $M = (\pi, \pi)$ in part (a).

The obtained Green's functions were used for calculating the magnetic susceptibility, correlation length, spin-lattice relaxation times at the Cu and O sites, and the Cu spin-echo decay time. The calculated T dependences of these quantities are similar to those observed in underdoped $YBa_2Cu_3O_{6+\delta}$. The obtained values have proper order of magnitude. Besides, the calculated dependences of the static spin susceptibility, correlation length, and spin-echo decay time on x also agree with experiment. The static spin susceptibility and spin-lattice relaxation rates decrease with decreasing T. Such behaviour observed at $T > T_c$ in underdoped cuprates is typical for the quantum disordered regime and is connected with a pseudogap in the spectrum of magnetic excitations.

This work [3] was partially supported by the ESF grant No. 4022 and by the WTZ grant (Project EST-003-98) of the BMBF.

References

[1] M. Takahashi, Phys. Rev. B **40**, 2494 (1989).

[2] H. Ding *et al.*, Nature **382**, 51 (1996).

[3] A. Sherman and M. Schreiber, Physica C **303**, 257 (1998); *Studies of High Temperature Superconductors*, Ed. A. V. Narlikar, vol. 27, p. 163, Nova Science, New York (1999); cond-mat/9808087.

LATTICE FERMIONS WITH OPTIMIZED WAVE FUNCTIONS: EXACT RESULTS

J. Spałek[1], A. Rycerz[1], W. Wójcik[2], and R. Podsiadły[1]

[1]*Institute of Physics, Jagiellonian University, Reymonta 4, 30-059 Kraków, Poland*
[2]*Institute of Physics, Technical University, 30-084 Kraków, Poland*

Abstract   We determine renormalized single-particle (Wannier) wave functions for the model of correlated systems diagonalized first in the Fock space. The ring with up to $N = 10$ atoms is taken as an example, for which *all* pair-site interactions and *all* hopping terms are included. The ground state energy, the orbital size, the dimerization energy, and the value of the microscopic parameters, are all calculated as a function of lattice constant.

1. The physics of correlated systems relies on parametrized models: The Hubbard, $t - J$, Kondo, and Anderson (impurity and lattice) models, each containing microscopic parameters expressed in terms of integrals involving the single-particle (Wannier, $w_i(\mathbf{r})$) functions. It is customary to present results as a function of parameters such as Hubbard interaction U, hopping integral t_p, intersite Coulomb interaction K_p, etc., where $p = 1, 2, ...$ is the range of the wave function overlap taken into account. The solution of those models, particularly in low dimension and/or in the strong-correlation limit are of nonperturbative character. Therefore, we have proposed a novel scheme [1, 2] of the wave function readjustment in the correlated state, i.e. *after* the exact diagonalization has been carried out. Such approach leads to the *renormalized wave equation and functions*, a procedure, which closes the second-quantization scheme devised by Fock. The most important feature of the results is the discussion of the system properties as a function of the interatomic distance R, particularly for short rings up to $N = 10$ atoms, which can be regarded as an example of a correlated quantum dot.

2. In Figure 1 (left) we present the ground-state energy versus R (in units of Bohr radius a_0) for a ring of up to $N = 10$ atoms, each contributing with one adjustable $1s$ orbital, for a model containing *all* overlap and hopping integrals, and *all* pair-site interactions. Apart from the case of H_2 molecule ($N = 2$),

J. Bonča et al. (eds.), Open Problems in Strongly Correlated Electron Systems. 443–445.

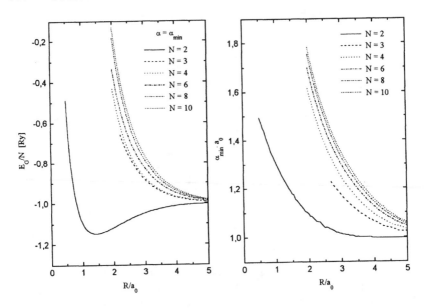

Figure 1   Ground-state energy (left) and the inverse renormalized orbital size, both vs. R

the solution including the ionic repulsions makes such rings unstable. Hence a trapping potential (not added for clarity) e.g. from the substrate, must be added to stabilize them. The right part is the adjusted atomic orbit size $a \equiv \alpha^{-1}$ (in units of a_0). The orbit shrinks remarkably even though all t_ps are taken into account.

In Figure 2 we display the contribution to E_G from dimerization. The optimized values of the shorter lattice constant R_1 are about $10 \div 15\%$ smaller than the average distance R. The zero-point-motion contribution in the case of hydrogen ring is of the same magnitude, but of the opposite sign.

We have also determined the values of the microscopic parameters; listed in Table 1 versus R/a_0. We have calculated the first three integrals t_p and interactions K_p, as well as U, exchange integral J, and the first two correlated-hopping parameters V_p (all quantities are in units of Rydbergs or mRy). The parameter $K_{ij} \approx 2/R_{ij}$ has a longer-range tail.

In summary, by determining exact one-band renormalized Wannier functions we have provided the second step in solving the general one-band model of correlated electrons. The approach was applied to short linear rings. The method can be extended to more realistic cluster calculations involving p- and d-type wave functions.

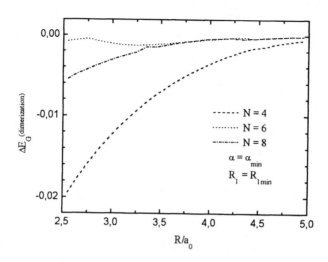

Figure 2 The ground-state energy lowering due to the dimerization (with the shorter lattice constant R_1 minimized), vs. average spacing R.

Table 1 Microscopic parameters (in Ry) versus R for neighbors $p = 1,2,3$.

R/a_0	t_1	$10^3 t_2$	$10^3 t_3$	U	K_1	K_2	K_3	$10^3 J_1$	$10^3 V_1$	$10^3 V_2$
2.0	−0.59	89.6	−98.3	2.30	1.08	0.68	0.45	9.5	−18.1	33.6
2.5	−0.33	45.5	−45.0	1.95	0.84	0.45	0.33	7.4	−17.5	19.6
3.0	−0.20	24.4	−21.9	1.72	0.69	0.39	0.26	5.6	−16.1	12.0
4.0	−0.08	7.4	−5.3	1.45	0.51	0.27	0.18	2.9	−12.9	4.5
5.0	−0.04	4.2	−2.7	1.33	0.40	0.21	0.14	1.3	−9.6	1.6

Acknowledgments

The paper was supported by KBN of Poland, grant No. 2P03P 092 18.

References

[1] J. Spałek, R. Podsiadły, W. Wójcik, and A. Rycerz, Phys. Rev. B **61**, June 15 (2000).

[2] A. Rycerz and J. Spałek, submitted for publication.

ENTROPY SATURATION AND THE BRINKMANN-RICE TRANSITION IN A RANDOM-TILING MODEL

D. K. Sunko

Department of Physics, Faculty of Science, Bijenička 32, HR-10000 Zagreb, Croatia.

Abstract   The parameter regime in which a Brinkmann-Rice (BR) transition appears near half-filling is investigated for a model of of one kind of electrons traversing a plane randomly tiled with CuO_4 molecules, simulating the copper-oxide planes of high-T_c superconductors. As the hole doping is increased, the BR transition evolves continuously into a state characterized by Kauzmann-like plateaus in the entropy vs. temperature curves. Despite clear analogies with the glass transition, these are equilibrium properties of the model. This is because the spin interactions, responsible for ordering in real space, are not included.

Introduction.   One difficulty in describing the metal-insulator transition, caused by strong electron repulsion, is that there is no obvious classical solution in the limit of infinite repulsion. This is in contrast to 'strong-coupling' problems with attractive interactions, which are difficult when the coupling is competitive with the hopping scale, but simplify when it becomes much greater.

Like some other approaches, saddle-point slave-boson [1] and dynamical mean-field theory [2], the random-tiling (RT) model [3] tries to get around this problem by simply postulating a 'heavy mode'. The RT heavy mode is essentially the one proposed by Gutzwiller: electrons of one spin see those of the other as static. This is implemented as a Falicov-Kimball limit of the three-band Emery model, in which only up-electrons move, by projected hopping. There are no spin interactions in the model. The parameters are the hopping overlap t and copper-oxygen splitting Δ_{pd}.

Brinkmann-Rice regime.   Here I concentrate on the limit $t \ll \Delta_{pd}$, which was identified in Ref. [3] as having a crossover of Brinkmann-Rice [4] type. The novelty is in what happens when one moves away from half-filling. In Fig. (1), I show the entropy of the mobile spins. There appear saturation plateaus, reminiscent of Kauzmann plateaus [5] in vitreous liquids. One of the curves even exhibits a 'Kauzmann paradox', extrapolation from the high-temperature

447

part yielding a negative entropy at low temperatures. Entropy saturation due to 'freezing' of kinetic motion was also observed in ^{3}He, and even modelled by the Hubbard model [6].

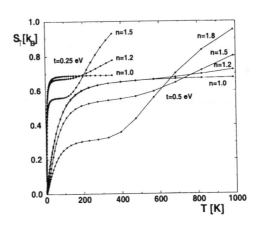

Figure 1   Entropy of mobile spins, for $t = 0.25$ eV (left three curves), and $t = 0.5$ eV (right four curves). Curves are marked by the corresponding concentration. Saturation plateaus are clearly visible. Here $\Delta_{pd} = 3$ eV, $U = 10$ eV, and $n = 2n_\uparrow = 2n_\downarrow$.

The value of the entropy at the plateaus corresponds to complete static disorder of the tiles, so it is the maximum one-band entropy, the rise beyond being due to the oxygen degrees of freedom. These play a role analogous to classical translation in the case of ^{3}He. Once the rise occurs, the oxygen gas rapidly becomes Maxwellian (the chemical potential moves out of the effective band). The plateaus signal 'pseudogaps' in the *system* density of states. The single-particle model on which the problem is mapped reflects this by a very narrow effective band for the parameters in question, so the Maxwellian regime corresponds to the temperature becoming comparable to this effective band-width. The input parameters, as shown, are still on the usual 'electron' scale, indicating strong model renormalization when $t/\Delta_{pd} < 1/5$. The effective *one-particle* density of states shows no pseudogaps.

For realistically large values of the hopping overlap, *e.g.* $t = 1$ eV for $\Delta_{pd} = 3$ eV, the entropy has a fairly smooth 'metallic' rise at finite fillings, and there is no Brinkmann-Rice regime near half-filling, either.

Discussion.   The Brinkmann-Rice 'transition' in the RT model is the end-point of a continuous range of glass-like crossovers, tuned by doping. The model interpolation, between quantum order at low temperatures and classical disorder at high ones, is not smooth in this parameter regime, despite the input t/Δ_{pd} being fairly large, though smaller than suggested by experiment.

The expression 'glass-like' is used only conditionally, to describe the shape of the entropy curves. In the absence of spin interactions, there is no spatially ordered state of lower free energy, even at half-filling. In fact, the RT Mott-Hubbard transition occurs slightly below half-filling, immediately producing a less ordered state [7]. This tendency of charge correlations to disorder the system, presumably counteracting spin correlations on the hole-doped side, is one of the more interesting aspects of the RT model.

Experimentally, a pseudogap can be noticed in the temperature dependence of the entropy of LSCO [8]. For underdoped systems, the curve is first flat, then rises, so that extrapolation from the high-temperature part indicates a disappearance of states at the Fermi level. However, the parameter regime needed to obtain the observed *values* of the entropy in the RT model is the above-mentioned realistic one, $t/\Delta_{pd} \sim 1/3$, where the entropy curves are smooth, not like the ones shown here. This discrepancy of qualitative and quantitative fits is probably due to the overly simplistic Falicov-Kimball limit.

Acknowledgements. Conversations with S. Barišić and E. Tutiš, and one with T.M. Rice, are gratefully acknowledged. This work was supported by the Croatian Government under Project 119 204.

References

[1] G. Kotliar, P.A. Lee and N. Read, Physica **C153–155** (1988) 538.

[2] A. Georges, G. Kotliar, W. Krauth and M.J. Rozenberg, Rev. Mod. Phys. **68** (1996) 13.

[3] D.K. Sunko and S. Barišić, Europhys. Lett. **36** (1996) 607. *Erratum:* **37** (1997) 313.

[4] W.F. Brinkmann and T.M. Rice, Phys. Rev. **B2** (1970) 4302.

[5] W. Kauzmann, Chem. Rev. **43** (1948) 219.

[6] K. Seiler, C. Gros, T.M. Rice, K. Ueda, and D. Vollhardt, J. Low Temp. Phys. **64** (1986) 195–221.

[7] D.K. Sunko, Fizika A (Zagreb) **8** (1999) 311–318, cond-mat/0005170

[8] J.W. Loram, K.A. Mirza, J.R. Cooper, N. Athanassopolou, W.Y. Liang, *Proceedings of the 10th Anniversary HTS Workshop on Physics, Materials and Applications,* B. Batlogg, C.W. Chu, W.K. Chu, D.U. Gubser, K.A. Muller, eds., 1996, pp. 341–4. See also J.R. Cooper, this meeting.

DIAMAGNETIC PROPERTIES OF DOPED ANTIFERROMAGNETS

D. Veberič[1], P. Prelovšek[1,2], and H.G. Evertz[3]

[1] *J. Stefan Institute, SI-1000 Ljubljana, Slovenia*

[2] *Faculty of Mathematics and Physics, University of Ljubljana, SI-1000 Ljubljana, Slovenia*

[3] *Institute for Theoretical Physics, Technical University Graz, AT-8010 Graz, Austria*

Abstract   Finite-temperature diamagnetic properties of doped antiferromagnets as modeled by the two-dimensional t-J model were investigated by numerical studies
of small model systems. Two numerical methods were used: the worldline
quantum Monte Carlo method with a loop cluster algorithm (QMC) and the
finite-temperature Lanczos method (FTLM), yielding consistent results. The
diamagnetic susceptibility introduced by coupling of the magnetic field to the orbital current reveals an anomalous temperature dependence, changing character
from diamagnetic to paramagnetic at intermediate temperatures.

The dc orbital susceptibility of the system in the external magnetic field is

$$\chi_d = -\mu_0 \frac{\partial^2 F}{\partial B^2} = -\frac{\chi_0}{\beta}\left[\frac{1}{Z}\frac{\partial^2 Z}{\partial \alpha^2} - \left(\frac{1}{Z}\frac{\partial Z}{\partial \alpha}\right)^2\right], \qquad (1)$$

where $\chi_0 = \mu_0 e^2 a^4/\hbar^2$ and $\alpha = eBa^2/\hbar$. In the previous studies [1] it was
realized that results are quite sensitive to finite-size effects, so we also used the
QMC method, where much larger lattices can be studied.

The magnetic field introduced into the t-J Hamiltonian via the Peierls construction, affects only the hopping of the electrons. Within FTLM the results
are obtained by the numerical derivation with respect to α.

Using the standard Trotter-Suzuki decomposition and the *worldline* representation of the QMC for the fermionic models, magnetic field enters matrix
elements concerning the hole hopping. The plaquette weights along the hole
worldline obtain an additional phase factor. Taking the field derivatives explicitly the expresson for the orbital susceptibility can be written as

$$\chi_d = -\chi_0 \frac{\langle S^2 \rangle}{\beta}, \qquad (2)$$

where S is the projected area of the hole worldline. χ_d can be thus measured
without the presence of a magnetic field. This is just another consequence of

J. Bonča et al. (eds.), Open Problems in Strongly Correlated Electron Systems, 451–453.

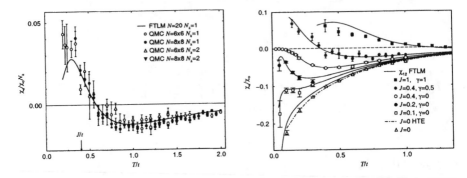

Figure 1   Orbital susceptibility χ_d vs. T (left) for one hole obtained via QMC (dots) and FTLM (line) for $J = 0.4t$; different J and γ (right). For comparison also results of high-temperature expansion (HTE) for $J = 0$ are shown (dash-dotted line).

the more general fluctuation–dissipation theorem. In doped systems we are hindered by the well known "fermionic sign problem" of the QMC, not present in the undoped case. Even though S^2 is strictly positive, the thermal average $\langle S^2 \rangle$ can become negative because of correlations between the Monte Carlo sign and the magnitude of the area S. For QMC the sizes of considered systems were 6×6 and 8×8.

With FTLM a few mobile holes on a system of tilted squares with up to 20 sites and periodic boundary conditions were considered. It is nontrivial to incorporate Landau phases corresponding to a homogeneous B, being at the same time compatible with periodic boundary conditions. This is possible only for quantized magnetic fields.

In Fig. 1, χ_d obtained via both methods is presented. For $T \gg t$, the response is diamagnetic and proportional to T^{-3} as well as essentially J-independent [1]. The most striking effect is that the orbital response below some temperature T_p turns from diamagnetic to paramagnetic, consistent with the preliminary results obtained via the FTLM [1]. In order to locate the origin of this phenomenon, results for different J and anisotropies γ are also shown. It appears that T_p scales with γJ, i.e. at $J = 0$ the response is clearly diamagnetic at all T, and for $\gamma = 0$, $J > 0$ no crossing is observed with either method.

At lower temperatures $T < T_d \ll T_p$, the diamagnetic behavior is expected to be restored. This follows from the argument that at $T \to 0$ a hole in an AFM should behave as a quasiparticle with a finite effective mass, exhibiting a cyclotron motion in $B \neq 0$, leading to $\chi_d(T \to 0) \to -\infty$ [1]. Numerically it is easiest to test this conjecture for a single hole and $\gamma = 0$. This is also true for $J = 0$.

In Fig. 2 also results for χ_d for finite doping $c_h > 0$ are presented. For nearly empty band $c_h > 0.7$ the χ_d is diamagnetic and weakly dependent on T. In this dilute regime strong correlations are unimportant, thus Landau diamagnetism

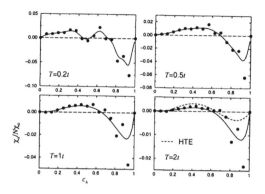

Figure 2 χ_d vs. c_h for several T and $J = 0.4t$. The last graph contains also a 4th order HTE result (dotted). Canonical (dots) and grand-canonical (line) values for all c_h are obtained with FTLM on 16 sites.

is expected. At moderate temperatures $T > J$ and for an intermediate-doping $0.2 < c_h < 0.7$ the χ_d is dominated by a paramagnetic response. There is a weak diamagnetism at $c_h < 0.2$ and $T > T_p$, while the paramagnetic regime extends to $c_h = 0$ for $T < T_p$. For low temperatures $T \ll J$ quite pronounced oscillations in $\chi_d(c_h)$ appear and can be partly attributed to finite-system effects.

The explanation can go in the direction proposed by [5], that at low doping $c_h \to 0$ we are dealing with quasiparticles (with a diamagnetic response), being a bound composite of charge (holon) and spin (spinon) elementary excitations. The binding appears to be quite weak and thus easily destroyed by finite T or c_h, enabling the independent and apparently paramagnetic response of constituents.

References

[1] D. Veberič, P. Prelovšek, and I. Sega, Phys. Rev. B **57**, 6413 (1998).

[2] D. Veberič, P. Prelovšek, and H.G. Evertz, to be published.

[3] J. Jaklič and P. Prelovšek, Adv. Phys. **49**, 1 (2000); cond-mat/9803331.

[4] H.G. Evertz, to be publ. in *Numerical Methods for Lattice Quantum Many-Body Problems*, ed. D.J. Scalapino (Perseus books, Frontiers in Physics).

[5] R.B. Laughlin, J. Low Temp. Phys. **99**, 443 (1995).

Index

Anderson impurity model, 101–102, 109, 327, 329, 413, 415
Anderson localization transition, 73
Anderson model, 101, 104–105, 108, 372, 380–381, 413, 415
anisotropy, 7, 34, 73, 97–99, 112, 155–156, 212, 266, 337, 389–390, 392, 395–396
antiferromagnet, 33–36, 39–41, 69–70, 72, 75–77, 113–116, 163–164, 171, 197, 243, 248–251, 311, 313–314, 452
antiferromagnetic correlation, 47, 125–126, 343
ARPES, 3–6, 33–34, 36, 38, 41, 44–45, 47–48, 53, 71, 120, 122–123, 125, 168–171, 303, 419
ballistic transport, 269, 273, 275–276, 278, 319, 433, 435
BaVS$_3$, 387
BCS wavefunction, 173, 176, 179–180
Bechgaard salts, 263, 278
Berry phase, 138, 194, 197–198
Bethe ansatz, 104, 109, 275, 286, 317, 348, 350, 413, 415
Bethe-Salpeter equation, 364–365, 415
bipartite lattice, 52, 135–139, 294
Bogolyubov-De Gennes, 253, 255
bond order wave, 151, 155, 158–160
Born approximation, 33, 51, 54, 113, 441
Bose condensate, 103
bosonic fields, 51
bosonization, 14, 205, 283–285, 288–291, 348
bound pairs, 362, 367, 369
bound state, 66, 142, 395–397, 434–435, 437–438
boundary exponent, 283, 286
breathing phonons, vi, 81, 83, 85, 87–89
Brinkmann-Rice transition, 447
BSCCO, 3, 59, 120–121, 125, 168–170
CaV$_4$O$_9$, 197
chaos, 276, 278, 317
charge
    density waves, 3, 146–147, 151, 153, 155, 157–160, 210

fluctuations, 107, 112–113, 146, 238, 243
ordering, 78, 152–154, 157–159, 226, 311–312
chemical potential, 3–4, 8, 15, 19, 21, 24, 27–28, 31–32, 35, 41, 103, 105, 112, 114, 123, 125–127, 163–164, 166, 176, 218, 255, 283–284, 288–289, 327, 372–373, 376, 379, 425, 448
colossal magnetoresistance, 217, 220, 226–227, 237, 247–248, 251, 325, 395
conductance
    magnetic field dependence, 438
    plateau, 433–434
    spin dependent, 433
    steps, 433
    sum rule, 413
    temperature dependence, 436
    threshold, 433
constant-energy scans, 60–61, 63
Cooper pairing, 19, 126, 403–404, 406, 425
Coulomb
    blockade, 438
    interaction, 70, 78, 141–142, 149, 153–154, 211, 219, 228, 231, 238, 259, 264, 267, 308, 312, 373, 381, 413–415, 435, 437, 443
    repulsion, 70, 72, 151, 154, 156–157, 304, 307, 312, 337, 340–341, 344, 361, 373, 381–382, 414, 435–436
coupled chains, 14, 281
crystal field, 430
cuprates
    overdoped, 119, 125
    underdoped, 3–10, 13, 19, 33–34, 41, 59, 69, 71, 73, 75–76, 119, 121–122, 125–127, 133, 141, 164, 166, 171, 293, 300, 337, 361, 391, 441–442, 449
d-wave, 5, 8, 18–21, 65, 69, 71, 74, 77, 130, 140, 173–183, 337, 343
density of states, 17–18, 20, 33, 35, 72, 124, 170–171, 232, 341–343, 354–355, 358, 372, 374, 376–377, 384, 406, 408, 410, 426, 448

density
  correlation function, 85, 149, 352
  fluctuation spectra, 81, 84
  response, 81–83, 85–87, 89–90, 307
density-matrix renormalization-group, 141–143,
    145, 174, 181–182, 192, 288–290, 348–349,
    353
diamagnetic susceptibility, 431, 451–453
dielectric response, 304, 306
disorder, 60, 62, 64–65, 82, 92, 129, 131–134, 137,
    139, 154, 194, 198, 200, 217, 220–225,
    227–228, 231–235, 237, 240–241, 299,
    312–313, 315, 334, 362, 376, 391, 409–410,
    417, 425–427, 433–434, 442, 448–449
distortion, 66, 113, 120, 154, 158, 232, 237,
    242–243, 409–411, 429, 431–432
domain-wall/stripe, 149
double exchange, 70, 218, 227, 239
Drude
  weight, 73, 168, 269–270, 273–275, 279–280,
    326, 372, 376–377
dual theory, 129, 131–133, 135, 139, 240, 293, 432
dynamic structure factor, 85
dynamical cluster approximation, 337
dynamical mean field theory, 72, 325–328,
    332–335, 353–354, 381, 383
dynamical stripes, 119–122, 125, 130–131, 133
electric susceptibility, 347–353, 358
electron momentum distribution, 33, 37
electron transport, 264, 416
electron-energy loss, 303–304, 308
electron-phonon coupling, 70, 88, 126, 153–154,
    157, 421–423
electronic structure, 43–44, 119–121, 234, 303,
    326, 334, 425, 429–432
exact diagonalization, 32, 34, 37, 39, 53, 55, 66, 81,
    90, 115, 153, 163, 192, 278, 355, 357, 443
exciton, 69–70, 303, 308
Falicov-Kimball model, 371–373, 379, 447, 449
Fano structure, 89
Fermi patches, 34
Fermi surface, 3–9, 14–16, 18–20, 23–24, 27–31,
    33–34, 37, 39–40, 43–49, 72–74, 77, 111,
    113–115, 120–122, 125, 131, 169–171, 209,
    264, 266–268, 270, 293, 298, 325, 343,
    404–405, 408, 417–418, 441–442
  electron-like, 44, 120, 125
  large, 34
  small, 37, 39
Fermi-gas, 361, 403, 405, 408
Fermi-liquid, 82–83, 85, 111, 119, 125, 131,
    163–164, 254, 263, 337, 361–364, 367,
    369–370, 387–388, 390, 403, 416, 418–419
  marginal, 163–164, 167–171, 419
  normal, 119, 125
fermion-magnon coupling, 35
ferromagnetic metallic phase, 227, 247

ferromagnetic spin-orbital model, 395
finite-size scaling, 180–181, 350
finite-temperature Lanczos method, 33–35, 163,
    451
FLST, 175–178, 181
fluctuation exchange approximation, 338, 102, 130,
    338, 367–369
free energy, 164, 166, 275, 316, 325, 363, 407, 449
frustrated phase separation, 141, 149
gap equation, 115, 426
gauge theory, 293–294, 296
Goldstone, 194, 207, 240, 293, 396
Green function, 112–113, 287, 289, 339–343, 355,
    404, 426
Gutzwiller, 154
Haldane gap, 92, 96
Haldane phase, 91–92, 95–99
half-breathing mode, 87–88
Hall constant, 33, 269–270, 273, 279–281, 372
Hartree-Fock, 285, 287, 290–291
  approximation, 287
  solutions, 141–142
heavy fermion, 110, 325, 335, 381, 384–385, 403
Heisenberg ladder, 15, 187
Heisenberg model, 34, 179–181, 190, 198, 240,
    273, 275–277, 294, 303, 305, 392, 396
high temperature superconductivity, 43–44, 46–49,
    81, 173–174, 293, 325, 370, 403, 421
hole pocket, 29, 32–34, 39–41
holon, 36, 293, 295, 298, 300–301, 303, 306–307,
    418, 453
Hubbard
  band, 24, 27, 81, 122, 326, 329
  gap, 26, 31, 241, 307
  model, 18, 23–24, 31–32, 65, 81, 136, 141, 163,
    219, 225, 253, 275, 277–278, 281, 283, 285,
    289–291, 303–305, 307, 326–327, 329,
    337–338, 340, 344, 347–348, 353–355,
    357–358, 362, 367, 381, 421, 425, 448
  operator, 112
hysteresis behavior, 247
impurity-hole bound state, 66
incoherent, 8, 10, 24, 74, 77, 81–82, 87, 113–114,
    171, 236, 330, 332, 441
incommensurate structure, 45, 48, 59–60, 120, 266,
    313
infinite dimensions, 31, 337, 347–348, 353–354,
    358, 371, 374
inverse photoemission, 33, 35
irreducible vertex functions, 365, 369
Ising model, 187–190, 217, 221–223, 311–312,
    314, 316, 326
Jahn-Teller effect, 218, 225, 229, 237, 242, 246,
    248, 409, 429
Jordan-Wigner transformation, 277
Josephson coupling, 142
Knight shift, 71

Kohn singularity, 403, 405
Kondo
  effect, 101, 362, 372–373
  lattice, 238
  model, 102, 104, 109
  temperature, 101, 106, 414–416
Kosterlitz Thouless transition, 200, 338, 342, 406
$LaCoO_3$, 429
ladder
  two-leg, 92
ladders
  even-leg, 92, 95–97, 99
  four-leg, 91–92, 97–99
  odd-leg, 94
  two-leg, 92, 97–99, 311–312
$LaMnO_3$, 429
Lanczos method, 33–35, 164, 173, 175–176, 178–183, 456
Landau theory, 13–16, 207, 332
Landauer-Büttiker formalism, 433
lattice model of spinless fermions, 283, 288–289
leading edge shift, 5, 125
loop algorithm, 51–53, 190, 193–194, 197–199, 201
LSCO, 119
Luttinger
  liquid, 14–15, 205, 263–264, 266, 268, 278, 283–284, 361, 434
  theorem, 24, 30–32, 47, 49, 113–114
Luttinger-Ward functional, 103, 106–107, 414, 416
magnetotransport, 417, 419
magnon-phonon coupling, 242
magnons, 39, 41, 70, 153, 235, 237–238, 245, 253
manganites, 217–225, 227–228, 230–231, 235–237, 245, 251, 361, 395, 409–411
Marshall signs, 138, 179
maximum entropy method, 53, 55, 191
metal-insulator transition, 23–24, 69, 73, 77, 152, 154, 217, 219, 222–225, 236, 247, 280, 362, 367–368, 370–371, 379, 381, 383, 385, 388, 411, 447
metallic phase, 171, 217–218, 223, 227, 235, 247, 269, 373, 383–384, 387, 391, 411
mode-mode coupling, 69, 75, 77
Monte Carlo method, 24, 51, 163, 174, 187, 193–195, 200–201, 218, 220–221, 223, 311, 313–316, 347, 357, 451–452
Mott charge gap, 14–16, 18, 20
Mott insulator, 14, 16, 69–76, 78, 122, 173, 228, 280, 303, 326, 330, 387–388, 392, 429
Mott transition, 280, 325–326, 328–330, 332, 334, 349, 353, 358, 383–385, 387, 390
Mott-Hubbard, 26, 31, 81, 253, 267, 273, 279, 304, 338, 340, 354, 368, 449
NMR relaxation, 71, 75, 268
nodal fermions, 129–132, 139
non-crossing approximation, 35, 106, 414–415
nonmagnetic impurities, 59–60, 62, 65

numerical renormalization group, 357, 381–382
one dimensional model, 174, 219, 221, 273–276, 278, 281–282, 293
one-band model, 174, 182, 303–305, 307, 338, 444, 448
orbital
  disorder, 232–233, 240, 410
  dynamics, 237, 241
  lattice coupling, 238, 242
  order, 227, 232, 388, 409–410
order parameter, 7, 13, 20, 62, 65, 91–92, 95–96, 99, 111, 113, 130, 132, 134–135, 137–138, 173–178, 180–182, 207, 209, 338–339, 341–342, 344, 349, 361, 369
organic metals, 263
p-wave, 426
pair resonant states, 370
pairing, 27, 47, 127, 174, 179, 181, 405–407
  interaction, 69–70, 74–75
paramagnon, 70, 253
parquet approach, 361–362, 364, 369
parquet equation, 361, 365, 367–370
Peierls transition, 18, 38, 138, 151, 153, 155–158, 312, 451
percolation, 224, 251
periodic Anderson model, 372, 380
phase separation, 141, 149, 165, 218–220, 225, 248–249, 251
phase shift, 104, 144–146
phonon renormalization, 90
photoemission, 3, 27–28, 33, 35, 43–44, 46–49, 53, 69, 71, 119–120, 124–125, 152, 165, 168, 174, 284–285, 303–304, 306–308, 337, 441–442
polaronic insulator, 409, 411
pseudogap, 3–11, 31, 33–37, 40, 69, 71, 74–78, 119–121, 124–126, 164, 171, 293, 300, 337, 342, 441–442, 448–449
  large, 10, 124–126
  small, 119, 125–126
quantum critical
  behavior, 361–363, 370
  point, 194, 197–199, 332, 352, 362, 364, 370, 417
  regime, 194, 198, 332–333
quantum dot
  open, 438
quantum
  Hall effect, 203
  magnets, 193
  Monte Carlo, 24, 51, 163, 165, 174–176, 181, 193–195, 200–201, 337–338, 347, 357, 451–452
  wire, 433–434, 438
quarter-filled band, 120, 123, 151–153, 155, 157, 264, 311

quasiparticle, 7, 10–11, 24, 27–28, 31–32, 51,
    53–55, 69–70, 73–77, 85, 113–114, 116,
    130–132, 168, 206–207, 209, 254–258,
    329–330, 332, 335, 340, 347–348, 356–358,
    361–362, 364, 369, 382–385, 404, 408,
    417–418, 452–453
random-tiling model, 447
renormalization group, 13–15, 104, 141–142,
    174–175, 284, 332, 334, 347–348, 357–358,
    362, 381–382
resistivity, 13, 65, 71, 168, 217, 227, 247–251,
    264–267, 329–330, 371–372, 375, 379,
    387–391, 419
    residual, 65
resonance
    singlet, 433
    triplet, 433
resonant valence bond, 14, 21, 81–82, 86, 91–96,
    99, 390
s+d mixing, 111
saddle points, 16–17, 19–21
scanning tunneling microscopy, 65
Schrieffer Wolff transformation, 413
Schwinger boson, 34, 38
Schwinger-Dyson equation, 365, 368–369
self energy, 84, 86, 243, 415
self-consistent Born approximation, 33–41, 51,
    54–55, 113
self-energy, 6, 23, 85, 113–114, 238, 284, 287–289,
    339–342, 364–365, 374, 441
sign problem, 16, 51–52, 165, 174, 201, 218, 338,
    347, 452
singlet, 66, 74, 87, 92–96, 106–107, 142, 146, 149,
    176, 200, 208, 258, 297, 304–306, 362, 367,
    385, 404, 406, 430, 432–433, 436–438
slave boson, 34, 83, 86, 101–102, 104, 106, 233,
    241, 243, 410, 413
sound mode, 82
spectral function, 5–6, 8, 11, 24–31, 33–36, 53–55,
    82, 88–89, 104–105, 108, 113, 163–164,
    168–171, 283, 285, 287–288, 290–291,
    303–305, 326, 329–330, 332, 375, 382–385,
    416, 418, 441
    one-particle, 284
spin
    1 chain, 92–93, 95
    charge separation, 129, 135, 239–240, 294
    dynamics, 53, 64, 66, 136, 154, 168–169, 234,
        237, 240–241, 245, 273
    fermion interaction, 240
    gap, 13, 20, 74, 141, 159, 198, 200, 348–349,
        352, 397
    ladder, 91–93, 95, 97, 99, 190, 198
    lattice relaxation, 168, 442
    orbital model, 395
    Peierls transition, 138, 151, 155, 158, 312
    polaron, 441–442

spin correlations, 144
    stiffness, 195, 239–240, 243, 245, 319
    triplet superconductivity, 253
    wave velocity, 194–195
spinless fermions, 34, 38, 238, 274, 277, 283–284,
    289, 317
spinon, 82, 84–87, 90, 131–132, 139, 293, 295,
    297–298, 300–303, 305–306, 418, 453
Sr_2CuO_3, 303
Sr_2RuO_4, 425
$SrCuO_2$, 303
staggered susceptibility, 194, 313
static stripes, 120, 123, 131, 133, 147, 149
string order, 91–92, 95–96, 99
string states, 37
stripe phase, 46, 122, 125–126, 129, 131–135, 181,
    210, 293, 300
sum rule, 31, 37–39, 86, 104, 114, 168–169, 321,
    332
superconducting
    fluctuations, 75, 418–419
    gap, 4–6, 10–11, 111, 120–121, 124–126, 255,
        257, 342
    mechanism, 69, 71
superconductivity, 13, 18, 21–22, 43, 59, 66, 70–71,
    77–78, 111, 125–126, 129–131, 133,
    140–142, 146, 164, 173–175, 180–182, 203,
    212, 253, 293, 300, 338, 341–342, 345, 387,
    403, 419, 421–422
superexchange, 70–71, 77, 227–228, 239, 241, 328,
    395, 409–410
superfluid density, 65
superfluidity, 253, 403–405
superlattice, 152, 211
symmetry breaking, 13, 15–16, 70, 135, 210, 293,
    295, 297, 299, 341, 361–362, 368–369
t-J model, 30, 32, 51–52, 66, 125, 141, 153, 163,
    183, 275, 293, 441
T-matrix approximation, 106–108, 413
    conserving, 108
thermoelectric power, 163–164, 167, 372
three-band model, 421–422
time-ordered, 104
Tomonaga-Luttinger model, 284
topological
    excitations, 129, 131–134, 205
    quantum number, 95, 99
transition-metal oxides, 247
triangular lattice, 311–312, 314, 328, 389, 392
triplet, 66, 92, 208, 253–255, 258, 305, 367, 387,
    406, 425, 430, 433, 436–438
two-dimensional electron gas, 433
Umklapp, 17, 65
van Hove, 41
    singularities, 71
    singularity, 19, 425–427
vibronic states, 409–410

Wigner-crystal, 133
Wilson ratio, 125
YBCO, 59, 64, 76, 111, 165–167
YbInCu$_4$, 371

YbInCu$_4$, 371
Zeeman energy, 211, 372, 434
Zhang-Rice singlet, 304
Zhang-Rice, 87, 304–305

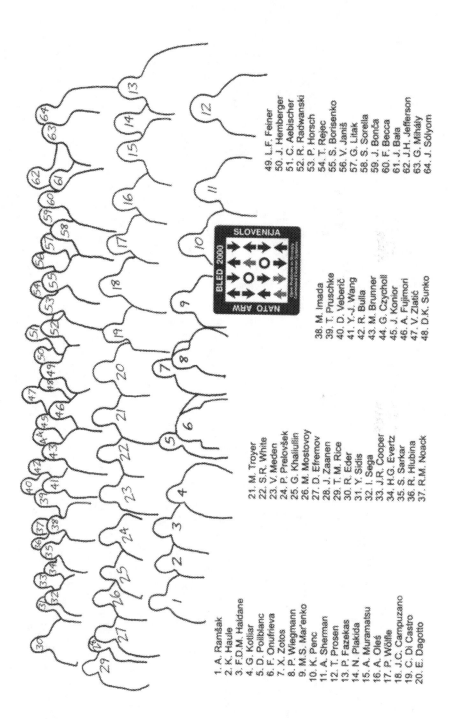

1. A. Ramšak
2. K. Haule
3. F.D.M. Haldane
4. G. Kotliar
5. D. Polßblanc
6. F. Onufrieva
7. X. Zotos
8. P. Wiegmann
9. M.S. Marenko
10. K. Penc
11. A. Sherman
12. T. Prosen
13. P. Fazekas
14. N. Plakida
15. A. Muramatsu
16. A. Oleš
17. P. Wölfle
18. J.C. Campuzano
19. C. Di Castro
20. E. Dagotto

21. M. Troyer
22. S.R. White
23. V. Meden
24. P. Pretovšek
25. G. Khaliulin
26. M. Mostovoy
27. D. Efremov
28. J. Zaanen
29. T. M. Rice
30. R. Eder
31. Y. Sidis
32. I. Sega
33. J.R. Cooper
34. H.G. Evertz
35. S. Sarkar
36. R. Hlubina
37. R.M. Noack

38. M. Imada
39. T. Pruschke
40. D. Veberič
41. Y.-J. Wang
42. R. Bulla
43. M. Brunner
44. G. Czycholl
45. J. Konior
46. A. Fujimori
47. V. Zlatić
48. D.K. Sunko

49. L.F. Feiner
50. J. Hemberger
51. C. Aebischer
52. R. Radwanski
53. P. Horsch
54. T. Rejec
55. S. Borisenko
56. V. Janiš
57. G. Lisk
58. S. Sorella
59. J. Bonča
60. F. Becca
61. J. Bala
62. J.H. Jefferson
63. G. Mihály
64. J. Sólyom